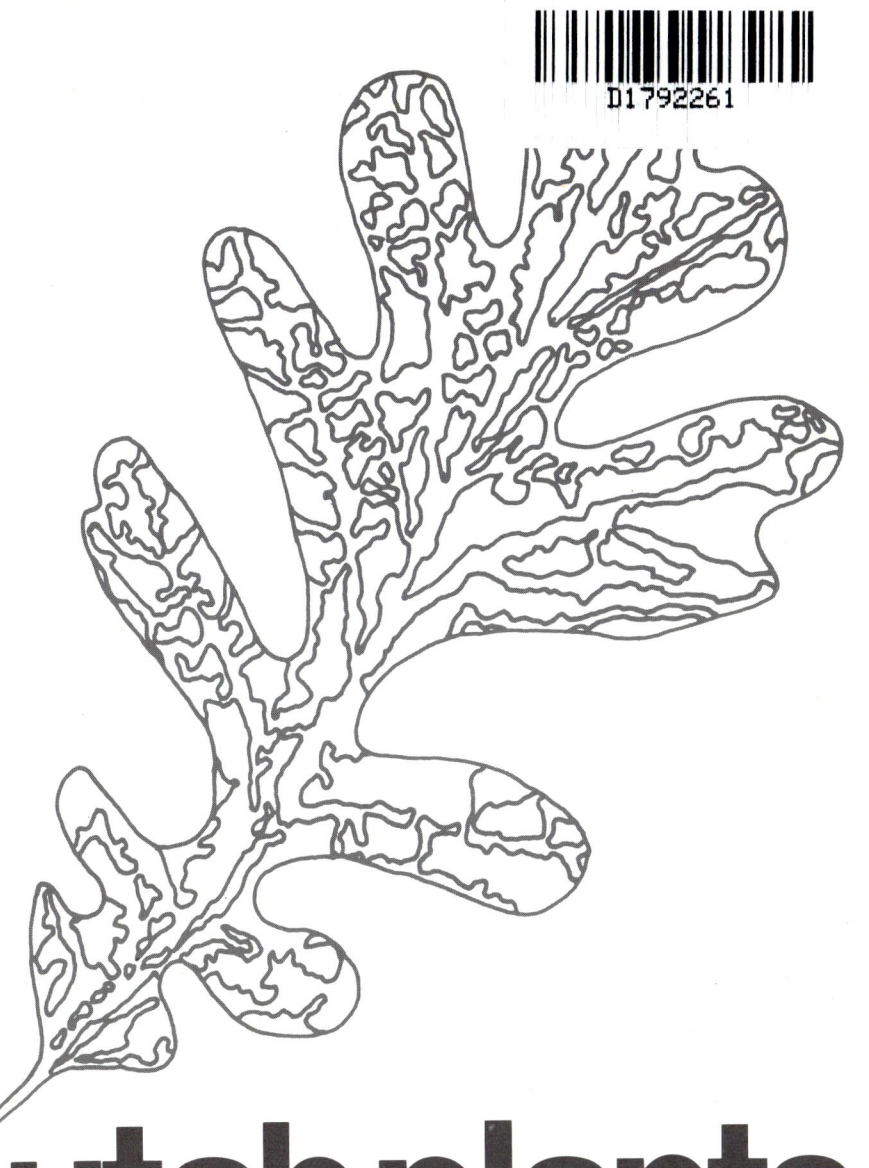

utah plants
tracheophyta

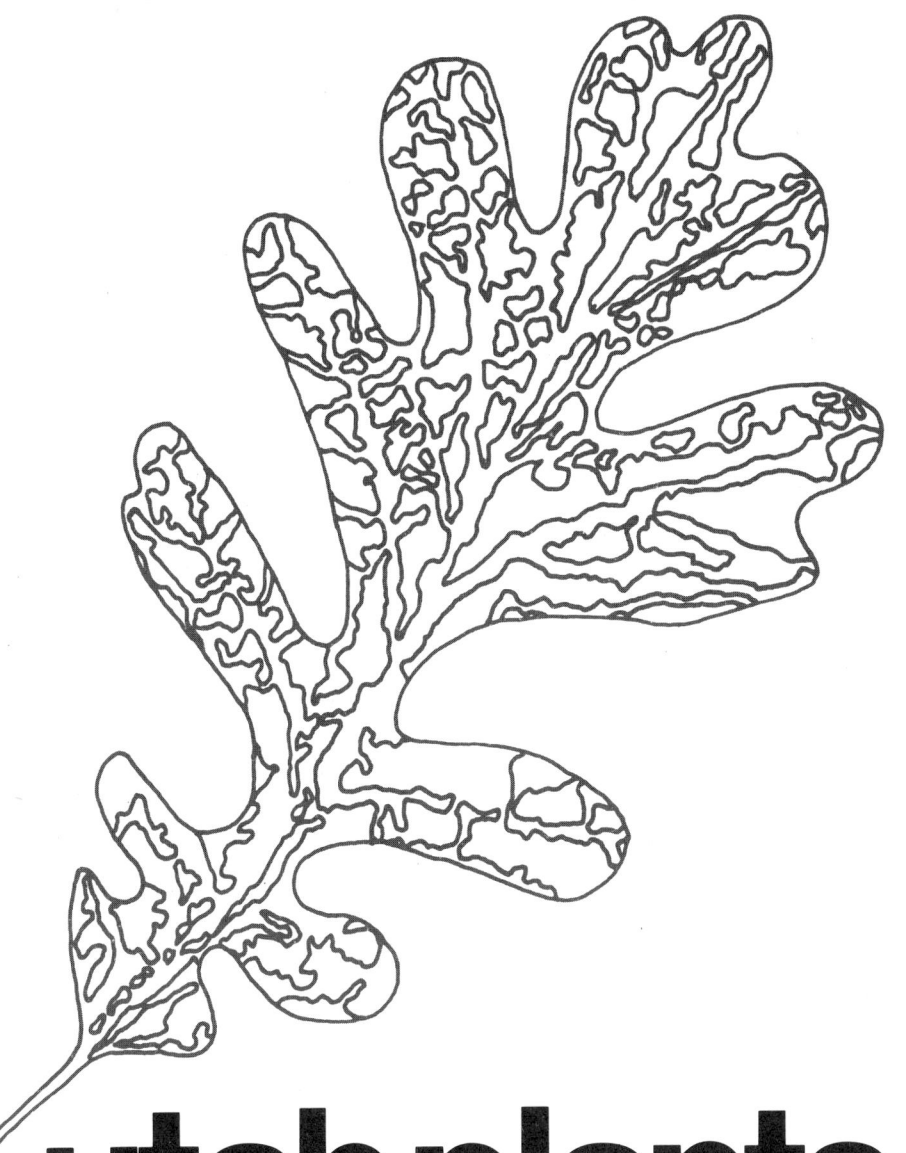

utah plants
tracheophyta
stanley l. welsh
glen moore

Third Edition
Brigham Young University Press
Provo, Utah
1973

Library of Congress Cataloging in Publication Data
Welsh, Stanley L
  Utah plants: Tracheophyta.
    Second ed. published in 1965 under title: Common Utah Plants.
Bibliography: p. 435
    1. Botany—Utah. 2. Plants—Identification.
I. Moore, Glen, 1917—   joint author. II. Title.
QK189.W4   582'.09'792   73-78085
ISBN 0-8425-0238-6

Library of Congress Catalog Card Number: 73-78085
International Standard Book Number: 0-8425-0237-8
© 1973 Brigham Young University Press. All rights reserved
Brigham Young University Press, Provo, Utah 84602
New, revised, and enlarged edition
Printed in the United States of America
73 2.5Mp .5Mh 12862

Table of Contents

Introduction	iv
Division Tracheophyta - Vascular Plants	1
Subdivision Lycopsida	3
Isoetaceae - Quillwort Family	3
Selaginellaceae - Spikemoss Family	3
Subdivision Sphenopsida	4
Equisetaceae - Horsetail Family	4
Subdivision Pteropsida	4
Class Filicinae	4
Marsileaceae - Pepperwort Family	5
Ophioglossaceae - Addertongue Family	5
Polypodiaceae - Fern Family	6
Salviniaceae - Salvinia Family	11
Class Gymnospermae	11
Araucariaceae - Araucaria Family	13
Cephalotaxaceae - Plumyew Family	14
Cupressaceae - Cypress Family	14
Cycadaceae - Cycas Family	16
Ephedraceae - Ephedra Family	16
Ginkgoaceae - Ginkgo Family	17
Pinaceae - Pine Family	17
Taxaceae - Yew Family	22
Taxodiaceae - Taxodium Family	22
Class Angiospermae, Subclass Dicotyledoneae	23
Aceraceae - Maple Family	36
Aizoaceae - Carpetweed Family	37
Amaranthaceae - Amaranth Family	38
Anacardiaceae - Cashew Family	38
Apocynaceae - Dogbane Family	39
Asclepiadaceae - Milkweed Family	40
Balsaminaceae - Balsam Family	42
Begoniaceae - Begonia Family	42
Berberidaceae - Barberry Family	43
Betulaceae - Birch Family	44
Bignoniaceae - Bignonia Family	45
Boraginaceae - Borage Family	45
Cactaceae - Cactus Family	63
Callitrichaceae - Waterstarwort Family	65
Campanulaceae - Bellflower Family	66
Capparidaceae - Caper Family	67
Caprifoliaceae - Honeysuckle Family	68
Caryophyllaceae - Pink Family	71
Celastraceae - Stafftree Family	78
Ceratophyllaceae - Hornwort Family	78

Chenopodiaceae - Goosefoot Family	78
Compositae - Composite Family	86
Convolvulaceae - Morning Glory Family	136
Cornaceae - Dogwood Family	138
Crassulaceae - Stonecrop Family	138
Cruciferae - Mustard Family	139
Cucurbitaceae - Gourd Family	156
Dipsacaceae - Teasel Family	157
Elaeagnaceae - Oleaster Family	158
Ericaceae - Heath Family	159
Euphorbiaceae - Spurge Family	160
Fagaceae - Beech Family	162
Fumariaceae - Fumitory Family	163
Gentianaceae - Gentian Family	164
Geraniaceae - Geranium Family	167
Haloragaceae - Watermilfoil Family	168
Hippocastanaceae - Horsechestnut Family	169
Hydrophyllaceae - Waterleaf Family	169
Hypericaceae - St. Johnswort Family	176
Juglandaceae - Walnut Family	176
Krameriaceae - Krameria Family	177
Labiatae - Mint Family	177
Leguminosae - Pea Family	185
Lentibulariaceae - Bladderwort Family	220
Limnanthaceae - Falsemermaid Family	220
Linaceae - Flax Family	221
Loasaceae - Loasa Family	221
Loganiaceae - Logania Family	222
Loranthaceae - Mistletoe Family	223
Magnoliaceae - Magnolia Family	224
Malvaceae - Mallow Family	224
Martyniaceae - Martynia Family	230
Meliaceae - Mahogany Family	230
Moraceae - Mulberry Family	230
Nyctaginaceae - Four O'Clock Family	231
Nymphaeaceae - Waterlily Family	233
Oleaceae - Olive Family	234
Onagraceae - Evening Primrose Family	237
Orobanchaceae - Broomrape Family	242
Oxalidaceae - Woodsorrel Family	243
Papaveraceae - Poppy Family	243
Plantaginaceae - Plantain Family	246
Polemoniaceae - Phlox Family	247
Polygalaceae - Milkwort Family	253
Polygonaceae - Buckwheat Family	253
Portulacaceae - Purslane Family	278
Primulaceae - Primrose Family	282

Pyrolaceae - Wintergreen Family ... 284
Ranunculaceae - Buttercup Family ... 286
Resedaceae - Mignonette Family ... 293
Rhamnaceae - Buckthorn Family ... 293
Rosaceae - Rose Family ... 294
Rubiaceae - Madder Family ... 305
Rutaceae - Rue Family ... 306
Salicaceae - Willow Family ... 308
Santalaceae - Sandalwood Family ... 312
Saururaceae - Lizardtail Family ... 312
Saxifragaceae - Saxifrage Family ... 313
Scrophulariaceae - Figwort Family ... 319
Solanaceae - Potato Family ... 335
Tamaricaceae - Tamarix Family ... 340
Tropaeolaceae - Tropaeolum Family ... 340
Ulmaceae - Elm Family ... 340
Umbelliferae - Carrot Family ... 342
Urticaceae - Nettle Family ... 352
Valerianaceae - Valerian Family ... 353
Verbenaceae - Verbena Family ... 354
Violaceae - Violet Family ... 356
Zygophyllaceae - Caltrop Family ... 356
Class Angiospermae, Subclass Monocotyledoneae ... 357
Agavaceae - Agave Family ... 359
Alismataceae - Water Plantain Family ... 360
Amaryllidaceae - Amaryllis Family ... 360
Araceae - Arum Family ... 364
Cannaceae - Canna Family ... 364
Commelinaceae - Spiderwort Family ... 364
Cyperaceae - Sedge Family ... 365
Gramineae - Grass Family ... 370
Hydrocharitaceae - Frogbit Family ... 393
Iridaceae - Iris Family ... 393
Juncaceae - Rush Family ... 396
Juncaginaceae - Arrowgrass Family ... 398
Lemnaceae - Duckweed Family ... 399
Liliaceae - Lily Family ... 399
Najadaceae - Waternymph Family ... 407
Orchidaceae - Orchid Family ... 407
Potamogetonaceae - Pondweed Family ... 411
Ruppiaceae - Ditchgrass Family ... 412
Sparganiaceae - Burreed Family ... 412
Typhaceae - Cattail Family ... 414
Zannichelliaceae - Horned Pondweed Family ... 414
Illustrated Glossary ... 417
Selected References ... 435
Index ... 437

Introduction

This manual contains keys to approximately 2,500 species of indigenous, adventive, and cultivated plants that occur in Utah. It is not exhaustive, nor is it intended to be. Those which have been included are generally the most representative of widespread members of the genera within the state. Future editions of this work will include all indigenous species as well as the most common cultivated plants.

Though specifically designed for students of elementary taxonomy at the university level, this manual can also be a valuable resource for the serious high school student and interested layman.

A key is simply a device whereby two alternative choices are compared and one of them is discarded. The species or plant group to which any unknown plant belongs is ultimately arrived at by reading the various alternative couplets which lead through the key to division, family, genus, and species. An unknown plant can be identified by starting at the beginning of the key and determining which of the two choices described is most appropriate, la or lb. At the end of each choice is either the name of the group or a number that refers the reader to another pair or couplet of alternatives. Once the family has been determined, keys to genera refer the reader to the correct genus. Within each genus is a key to the included species. The generic name, together with the specific name, constitutes the name of the plant. Ordinarily the name is not considered complete unless it is accompanied by the name of the author of the specific name (e.g., **Rosa woodsii** Lindl. is the name of our common indigenous rose).

Major categories—subdivisions, classes, and subclasses—appear in phylogenetic sequence. However, families and genera are arranged alphabetically. Species appear in the order in which they occur in the key.

An illustrated glossary is provided for the benefit of students unfamiliar with botanical terminology.

The authors wish to express gratitude to all who have contributed suggestions and encouragement. We are especially grateful to Kay H. Thorne, who provided the bulk of the illustrations, to the personnel of the Brigham Young University Press for their editorial assistance, and to the following individuals for their valuable contributions: James L. Reveal (**Isoëtes, Selaginella, Equisetum, Townsendia, Eriogonum**), Floyd H. Coles (**Asclepias**), Larry C. Higgins (Boraginaceae), Dorde Wright (Cactaceae), Benjamin W. Wood (**Symphoricarpos**), C. A. Hanson (**Atriplex** perennials), L. K. Shumway (**Atriplex** annuals), Loran C. Anderson (**Chrysothamnus**), Clyde Blauer (**Helianthus**), Glen T. Nebeker (**Geranium**), Duane Atwood (Hydrophyllaceae), D. B. Dunn (**Lupinus**), J. A. M. Jefferies (**Sphaeralcea**), J. B. Karren (**Rumex**), L. B. Barnett (**Scirpus**), Seville Flowers (Gramineae), and Roberta Wilson (**Iris**).

Division Tracheophyta—Vascular Plants

Plants with a well-developed vascular system (xylem and phloem), leaves (either macrophylls or microphylls), and roots; reproduction by means of spores or seeds, the latter borne in cones or flowers.

Key to the Subdivisions, Classes, and Subclasses

1a. Plants with small scalelike leaves, usually with a single vein (microphylls); reproduction by means of spores; flowers or woody cones lacking. (2)
1b. Plants with large leaves, usually with more than a single vein (macrophylls), if the leaves scalelike, then otherwise different from above; reproduction by spores or seeds, the latter borne in flowers or cones. *Subdivision PTEROPSIDA* (3)

2a. Stems not jointed; leaves green and imbricated, not whorled or forming a sheath at the node; plants either aquatic and grasslike or terrestrial and mosslike. *Subdivision LYCOPSIDA*, p. 3.
2b. Stems jointed and fluted; leaves not green, reduced to a whorl of connate scales at the nodes; plants neither grass- nor mosslike. *Subdivision SPHENOPSIDA*, p. 4.

3a. Plants fernlike with broad leaves, or free-floating aquatics with small overlapping leaves; reproduction by spores; flowers and woody cones lacking. *Class FILICINEAE*, p. 4.
3b. Plants generally not as above; reproduction by seeds; flowers or cones present. (4)

4a. Seeds not borne in ripening carpels, but naked and borne on the surface of a scale, these borne crowded together on an axis and forming a cone; true flowers absent. *Class GYMNOSPERMAE*, p. 11.
4b. Seeds borne in ripening carpels; plants producing true flowers. *Class ANGIOSPERMAE* (5)

5a. Cotyledons 2; stems mostly increasing in diameter by means of a cambium between the xylem and phloem; leaves mostly net veined; flower parts in 4s or 5s. *Subclass DICOTYLEDONEAE*, p. 23.
5b. Cotyledons 1; stems usually lacking a cambium or, if cambium present, it produces entire vascular bundles; leaves mostly parallel veined; flower parts usually in 3s. *Subclass MONOCOTYLEDONEAE*, p. 357.

Subdivision Lycopsida

Plants with scale- or grasslike leaves, these short and imbricated or long and grasslike; sporangia solitary, subtended by a sporophyll, heterosporous.

Key to the Families

1a. Plants aquatic, submerged in ponds or lakes, or occasionally growing on exposed mud, grasslike; leaves long and slender, from a broadly clasping base; sporangia at base of leaves. ISOETACEAE, p. 3.
1b. Plants terrestrial, growing in dry, rocky situations; leaves small and scalelike; sporangia borne in terminal cones. SELAGINELLACEAE, p. 3.

ISOETACEAE — QUILLWORT FAMILY

Herbaceous perennial aquatics or plants of wet places; tufted, quill-like leaves arising from a cormlike subterranean axis; spores of 2 types—microspores, which are typically borne on the inner leaves, and megaspores, which are borne on the outer leaves; lakes and ponds, usually at high elevations.

Isoëtes L. Quillwort (Contributed by James L. Reveal)

1a. Plants amphibious, growing on mud; leaves 15–30 cm long, stomata present; Dry Lake, Cache Co. **I. howellii** Engelm.
1b. Plants immersed in water or, if amphibious, not of Dry Lake. (2)
2a. Megaspores with scattered, low tubercles; leaves less than 15 cm long; ligule cordate; common. **I. bolanderi** Engelm.
2b. Megaspores more or less densely beset with spines or jagged crests or high ridges; Uinta Mountains. (3)
3a. Megaspores 0.25–0.5 mm wide, covered with spines; leaves thin. **I. echinospora** Durieu
3b. Megaspores 0.5–0.8 mm wide, covered with crests or ridges; leaves coarse. **I. lacustris** L.

SELAGINELLACEAE — SPIKEMOSS FAMILY

Plants low and creeping, mosslike in habit and appearance; leaves numerous, imbricated, spirally arranged, a small ligule at the base; heterosporous; the sporophylls green, sharply keeled, and forming a 4-angled terminal cone; micro- and megasporangia axillary, orange or yellowish.

Selaginella Beauv. (Contributed by James L. Reveal)

1a. Stems compactly branched, forming cushions. (2)
1b. Stems loosely branched, forming flat mats. (4)

Class FILICINEAE

2a. Leaves dissimilar, the lower ones of stem longer than those on upper side; terminal setae of leaves 0.5–2 mm long; Uinta and La Sal mountains. **S. densa** Rydb.
2b. Leaves similar; setae 0.5 mm long or less. (3)
3a. Terminal setae 0.2–0.4 mm long; common. **S. watsonii** Underw.
3b. Terminal setae lacking or up to 0.2 mm long; endemic to Zion National Park. **S. utahensis** Flowers
4a. Leaves shortly adnate-decurrent, 2.5–3 mm long, with setae 0.3–0.4 mm long; stems 2–3 mm broad; southwestern Utah. **S. underwoodii** Hieron.
4b. Leaves sessile, 1 mm long, with setae lacking or up to 0.1 mm long; stems 1 mm broad; eastern Utah. **S. mutica** D. C. Eaton ex Underw.

Subdivision Sphenopsida

Plants with hollow jointed stems; leaves whorled at the nodes, minute and toothlike; reproductive bodies in spikelike terminal cones with sporangia borne on underside of peltate scales; common in moist situations at middle and lower elevations.

EQUISETACEAE — HORSETAIL FAMILY

A single family treated, with the characteristics of the subdivision.

Equisetum L. (Contributed by James L. Reveal)

1a. Stems all alike and green, simple and unbranched. (2)
1b. Stems dimorphic, a sterile, highly branched green one and an unbranched, flesh-colored one appearing before the other; common. Meadow Horsetail. **E. arvense** L.
2a. Stems robust, 2–15 dm tall; common. (3)
2b. Stems small and slender, 1–3 dm high; Cedar Breaks, Iron Co. Mottled Scouring Rush. **E. variegatum** Schleich. ex Weber & Mohr
3a. Aerial stems perennial and evergreen; cones with an apiculate tip; sheaths with 2 bands; common (including **E. prealtum** Raf.). Common Scouring Rush. **E. hyemale** L.
3b. Aerial stems annual; cones blunt; sheaths with a single band; common (including **E. kansanum** Schaff.). Smooth Scouring Rush. **E. laevigatum** A. Br.

Subdivision Pteropsida

CLASS FILICINEAE

Plants from rhizomatous bases; leaves large and foliaceous, with circinate vernation; sporangia 1-loculed, borne marginally or dorsally, homosporous. Ferns.

Ophioglossaceae

Key to the Families

1a. Spores borne in sporangia on green, aerial leaves; plants terrestrial. (2)
1b. Spores borne in sporocarps (these usually below ground or water level); plants aquatic, often free floating. (3)
2a. Spore-bearing leaves strikingly different from the vegetative leaves; sporangium without an annulus, opening by a transverse, gaping slit. OPHIOGLOSSACEAE, p. 5.
2b. Spore-bearing leaves much like the vegetative leaves, at most having narrower segments; sporangium with an annulus. POLYPODIACEAE, p. 6.
3a. Leaves palmately divided into 4 leaflets, cloverlike with long petioles; plants rooting in mud, often in shallow ponds or lakes, the leaves floating. MARSILEACEAE, p. 5.
3b. Leaves entire or 2-lobed, sessile; plants small, branched, 0.5–2 cm long, floating on water. SALVINIACEAE. p. 11.

MARSILEACEAE — PEPPERWORT FAMILY

Perennial, rhizomatous aquatics; leaves long-petiolate, palmately 4-foliolate; segments triangular or cuneate; sporocarps borne on short stipes, subglobose to oblong, flattened, hard and bony, opening by 2 valves.

Marsilea L. Pepperwort

A single genus and species treated, with the characteristics of the family; rather widely distributed in lakes, ponds, and streams. **M. vestita** Hook. & Grev.

OPHIOGLOSSACEAE — ADDERTONGUE FAMILY

Plants with simple or compound leaves; sporangia either continuous in form of a band on the margin of a slender sporophyll, or distinct, ovoid to globose, without a true annulus, but opening by a transverse slit, each sporangium gaping at maturity.

Botrychium Sw. Moonwort, Grape Fern

1a. Leaves with distinct petioles, often short; plants mostly slender. (2)
1b. Leaves sessile or nearly so; plants stouter with ample foliage. (3)
2a. Petiole half the length of blade or longer; pinnae rhomboidal to narrowly fan shaped, entire to crenate; plants small, mostly 3–5 (15) cm tall; blade simple to pinnately divided; not definitely known from Utah, to be sought in high mountains. Little Grape Fern. **B. simplex** A. S. Hitchc.
2b. Petiole short, mostly 2–4 mm long; pinnae oblong, crenate to lobed,

Class FILICINEAE

        obtuse, usually distant; plants larger, mostly 7–10 (28) cm tall; blade oblong, pinnately divided. Matricary Grape Fern. **B. matricariaefolium** A. Br. ex Koch

3a. Pinnae lunate or fan shaped, mostly overlapping, entire or upper ones lobed; blade oblong; sterile leaf merely bent over the fertile segments in vernation; high mountains. Moonwort. **B. lunaria** (L.) Sw.

3b. Pinnae ovate, rhomboidal to oblong, not or only slightly overlapping, lobed to pinnately divided. (4)

4a. Blades ovate to oblong in outline; lower pinnae approximate, not conspicuously larger or longer than the upper ones, broadly ovate to ovate-oblong, obtuse; in bud, apex of sterile blade bent over but not clasping fertile portion; Uinta Mountains. Northern Grape Fern. **B. boreale** (Fries) Milde

4b. Blades deltoid or deltoid-ovate in outline; lower pinnae distant, conspicuously larger and longer than the upper. (5)

5a. Sterile blade inserted above middle of whole plant; lower pinnae mostly oblong; in bud, apex of sterile blade bent over and clasping straight or inclined fertile blade; high mountains. **B. matricariaefolium** A. Br. ex Koch

5b. Sterile blade inserted higher, near summit of whole plant; lower pinnae oblong to lanceolate; in bud, both sterile and fertile portions completely reflexed; Deep Creek Mountains. Lance-leaved Grape Fern. **B. lanceolatum** (S. G. Gmel.) Angstrom

Polypodiaceae — Fern Family

        Rhizomatous perennials; leaves with circinate vernation in bud, all alike (except in **Cryptogramma**, which has longer fertile leaves, with narrower segments); sori naked or covered by an indusium or by recurved leaf margin, with or without a membranous false indusium; sporangium with an annulus; spores all alike.

1a. Sori on veins, not marginal. (2)
1b. Sori marginal, without true indusia, exposed or protected by reflexed or recurved leaf margin. (9)
2a. Sori round in outline. (3)
2b. Sori elongated, horseshoe shaped, or continuous on veins. (7)
3a. Indusia arising from under the sori (inferior). (4)
3b. Indusia lacking, or present and attached above sori on a little hump (superior). (5)
4a. Indusia attached all around, at maturity splitting from top and center into spreading segments, starlike. **Woodsia**
4b. Indusia attached on one side only, covering the sori like hoods which bend backward at maturity. **Cystopteris**

Polypodiaceae

5a. Indusia lacking. **Polypodium**
5b. Indusia present, superior. (6)
6a. Indusia shield shaped, attached at center and spreading over the sori. **Polystichum**
6b. Indusia lunate, attached by a notch at one side. **Dryopteris**
7a. Sporangia borne in lines on the veins; indusium lacking. **Pityrogramma**
7b. Sporangia submarginal, curved across the vein, or on the margins of the veins; indusium usually present. (8)
8a. Sori curved across veins; indusium hoodlike, sometimes lacking. **Athyrium**
8b. Sori straight, on sides of veins, oblique to the margin and midvein; indusium hoodlike. **Asplenium**
9a. Leaf margins flat or only slightly recurved, the blades glabrous, hairy, scaly, or mealy beneath. **Notholaena**
9b. Leaf margins strongly reflexed or recurved. (10)
10a. Sori attached to and covered by the thin, membranous, strongly reflexed tips of leaflet lobes. **Adiantum**
10b. Sori more or less continuous, marginal or submarginal, on the leaf proper and more or less covered by the recurved margin; membranous false indusium present in some species. (11)
11a. Leaves of 2 kinds, the fertile ones longer than the sterile ones and with narrower segments. **Cryptogramma**
11b. Leaves all alike. (12)
12a. Plants not tufted; leaves single, 2.5–6 dm long, from a horizontal rhizome; growing in soil. **Pteridium**
12b. Plants densely tufted, small to medium sized; mostly growing in crevices or among rocks. (13)
13a. Leaves glabrous or nearly so, the ends of veins not thickened; membranous false indusium wanting. **Pellaea**
13b. Leaves densely hairy or scaly beneath, the ends of veins thickened; membranous false indusium present in some species. **Cheilanthes**

Adiantum L. Maidenhair Fern

1a. Leaves orbicular or reniform in outline, primary division dichotomous; Zion National Park, also uncommon in mountains. **A. pedatum** L.
1b. Leaves ovate to ovate-lanceolate in outline, primary division pinnate; widely distributed in southern Utah. **A. capillus-veneris** L.

Asplenium L. Spleenwort

1a. Leaves irregularly forking, segments slender; plants grasslike; extreme eastern Utah. **A. septentrionale** (L.) Hoffm.

Figs. 1-2. 1. **Isoëtes howellii**, x .36. 2. **Equisetum arvense**, x .21

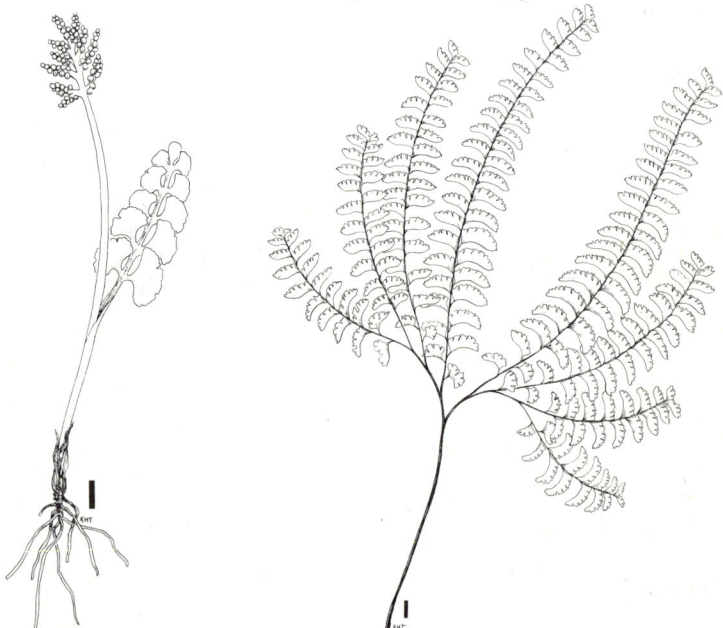

Figs. 3-4. 3. **Botrychium lunaria**, x .37. 4. **Adiantum pedatum**, x .21

Polypodiaceae

1b. Leaves pinnately compound. (2)
2a. Leaves 2- to 3-pinnately compound or pinnatifid; southwestern Utah. **A. adiantium-nigrum** L.
2b. Leaves 1-pinnately compound. (3)
3a. Petioles dark reddish brown and shiny throughout; Wasatch Mountains. **A. trichomanes** L.
3b. Petioles brown below, green or yellowish above; Wasatch Mountains. **A. viride** Huds.

Athyrium Roth   Lady Fern

1a. Leaves ample, the pinnae spreading, often at right angles; pinnules close, narrowly to broadly decurrent; sori oblong to elongate, curved or hooked; indusium present. **A. felix-foemina** (L.) Roth ex Mertens
1b. Leaves skeletonlike with prominent midribs, the pinnae strongly oblique; pinnules distant; segments narrow with wide sinuses, narrowly decurrent; sori round; indusium lacking. **A. americanum** (Butters) Maxon

Cheilanthes Sw.   Lipfern

1a. Leaves glabrous. **C. siliquosa** Maxon
1b. Leaves hairy or with scales, or both. (2)
2a. Leaves with dense white or brown scales only; southern Utah. **C. covillei** Maxon
2b. Leaves distinctly hairy, scales often present. (3)
3a. Leaves densely tomentose, scales lacking, ovate to oblong-lanceolate; ultimate segments oblong-oval to obovate; widely distributed. **C. feei** Moore
3b. Leaves with both hairs and scales, lanceolate to oblong-lanceolate. (4)
4a. Ultimate leaf segments oblong, distant; northern Utah. **C. gracillima** D. C. Eaton ex Torr.
4b. Ultimate leaf segments rounded or obovate, close together; southeastern Utah. **C. eatonii** Baker ex Hook. & Baker

Cryptogramma R. Br.   Rockbrake

1a. Petioles yellowish; leaves firm and thick, densely tufted; widespread in mountains. **C. crispa** (L.) R. Br. ex Hook.
1b. Petioles chestnut brown, at least below; leaves thin and delicate, usually tufted; Wasatch Mountains. **C. stelleri** (S. G. Gmel.) Prantl.

Cystopteris Bernh.   Bladderfern

1a. Leaves to 40 cm long, ovate-lanceolate to oblong-lanceolate, the lowest pinnae shorter; Utah's most common fern. **C. fragilis** (L.) Bernh.

Class FILICINEAE

1b. Leaves larger, to 100 cm long, triangular-lanceolate, the lowest pinnae the longest; uncommon, but widely distributed. **C. bulbifera** (L.) Bernh.

Dryopteris Adans.   Woodfern

Medium-sized ferns from thick rhizomes; leaves 2-pinnate to further divided, the pinnae mostly lanceolate; pinnules numerous; indusium attached to one side by a deep notch, lunate or horseshoe shaped, large and conspicuous; widely distributed in mountains. Male Fern. **D. felixmas** (L.) Schrott

Notholaena R. Br.   Cloakfern

1a. Leaves densely white-tomentose, not mealy; southwestern Utah. **N. parryi** D. C. Eaton
1b. Leaves glabrous or mealy with a waxy powder, hairs lacking. (2)
2a. Leaves entirely glabrous, green; segments few and large; southwestern Utah. **N. jonesii** Maxon
2b. Leaves abundantly mealy beneath; segments small and numerous. (3)
3a. Rachis sharply flexuous (zigzag); southeastern Utah. **N. fendleri** Kuntze
3b. Rachis straight or nearly so; southeastern Utah. **N. limitanea** Maxon

Pellaea Link   Cliffbrake

1a. Leaves 2-pinnate; segments narrowly oblong, spine tipped; southern Utah. **P. longimucronata** Hook.
1b. Leaves 1-pinnate; segments broader, not spine tipped. (2)
2a. Segments oblong, rhomboidal or deltoid-ovate, asymmetric, mitten shaped, cleft or divided; widespread in mountains. **P. breweri** D. C. Eaton
2b. Segments oblong, thin, entire or the lower ones cleft or parted; uncommon, but widely distributed. **P. suksdorfiana** Butters

Pityrogramma Link   Goldfern

Plants densely tufted, 10–30 cm tall; petioles chestnut brown, shiny; blades 3- to 5-angled in outline, pinnate, the lower pinnae divided again; segments white- or yellow-mealy beneath; sori continuous along the veins; indusium wanting; southwestern Utah. **P. triangularis** (Kaulf.) Maxon

Polypodium L.   Polypody

Small- to medium-sized ferns from creeping, nodulose, branched, scaly rhizomes; leaves simple to pinnatifid or nearly pinnately compound, up to 15 (2) dm long, the veins free; sori round in outline, on the veins; indusia lacking; widely distributed. **P. hesperium** Maxon

Polystichum Roth   Hollyfern

1a. Petiole very short; pinnae extending nearly to the rhizome, long

Class GYMNOSPERMAE

spinulose-dentate or -serrate, the base truncate, asymmetric, auricled on the upper side; mountains throughout Utah. **P. lonchitis** (L.) Roth ex Rom.

1b. Petiole one-fifth as long as blade, or longer; pinnae sharply serrate, the lower ones deeply cleft or parted at base, only upper ones auricled; mountains throughout Utah. **P. scopulinum** (D. C. Eaton) Maxon

Pteridium Scop. Bracken

Large, coarse ferns from stout creeping rhizomes; leaves pinnate or more divided, veins pinnate; margins of segments revolute and covering the sori, which are continuous along the margins; true indusium present, on side toward blade, inconspicuous; one of Utah's most common ferns. **P. aquilinum** (L.) Kuhn

Woodsia R. Br. Woodsia

1a. Leaves smooth, not hairy or glandular; frequent on moist or dry shaded cliffs and ledges. **W. oregana** D. C. Eaton
1b. Leaves glandular. (2)
2a. Leaf margins hyaline or ciliate, surface minutely glandular but not hairy; southern Utah. **W. mexicana** Fee
2b. Leaf margins not hyaline or ciliate, surface glandular, hairy; widespread. **W. scopulina** D. C. Eaton

SALVINIACEAE — SALVINIA FAMILY

Plants small, floating on water; stem branched; leaves 2, opposite, very small, 0.5–2 mm long; sporocarps very small, globose or ovoid, very soft and thin, 1-celled, 2 to several on a common stalk, heterosporous.

Azolla Lam. Mosquito Fern

A single genus and species treated, with the characteristics of the family; plants small and mosslike, deltoid in outline, upper surfaces very resistant to water; leaves green or green with a red border; uncommon, but locally abundant. **A. caroliniana** Willd.

CLASS GYMNOSPERMAE

Key to the Families

1a. Leaves pinnately compound, with circinate vernation; plants palmlike. CYCADACEAE, p. 16.
1b. Leaves simple or reduced to a membranous sheath and more or less connate, without circinate vernation; plants not palmlike. (2)
2a. Stems jointed; leaves scalelike, in 2s or 3s; branches green. EPHEDRACEAE, p. 16.

Figs. 5-6   5. **Asplenium trichomanes**, x .48.   6. **Cryptogramma crispa**, x .32

Figs. 7-8.   7. **Cystopteris fragilis**, x .22.   8. **Pteridium aquilinum**, x .21

Araucariaceae

2b. Stems not jointed; leaves various, needlelike or linear or broad, if scalelike then closely overlapping; branches not green. (3)
3a. Leaves broad, fan shaped, dichotomously veined; cones drupelike, covered with an aril. GINKGOACEAE, p. 17.
3b. Leaves needlelike, linear, scalelike, solitary or in fascicles; cones not usually provided with an aril. (4)
4a. Leaves alternate or subopposite. (5)
4b. Leaves opposite, whorled or spirally arranged. (6)
5a. Leaves linear, 2-ranked, not decurrent; cones berrylike or drupelike, the aril red and cup shaped. TAXACEAE, p. 22
5b. Leaves awl shaped, decurrent, rigid, radially spreading; cones large and woody, lacking an aril. ARAUCARIACEAE, p. 13.
6a. Leaves opposite or whorled. (7)
6b. Leaves spirally arranged. (9)
7a. Leaves in symmetric whorls at ends of twigs. TAXODIACEAE (**Sciadopitys**), p. 23.
7b. Leaves opposite or whorled, not as above. (8)
8a. Leaves evergreen, small, scalelike, and decurrent, if needlelike then leaves jointed and very sharp. CUPRESSACEAE, p. 14.
8b. Leaves deciduous, twisting, and appearing 2-ranked, decurrent, the needlelike leaves soft, not sharp. TAXODIACEAE (**Metasequoia**), p. 22.
9a. Leaves scalelike, not over 12 mm long, subulate and rigid. (10)
9b. Leaves needlelike, solitary or in fascicles. (11)
10a. Leaves scattered spirally; cone scales peltate. TAXODIACEAE (**Sequoiadendron**), p. 23.
10b. Leaves closely imbricated, sharply pointed; cone scales with an incurved spine. ARAUCARIACEAE, p. 13.
11a. Branches opposite; cones drupelike, covered with a green or purplish aril. CEPHALOTAXACEAE, p. 14.
11b. Branches in regular or irregular whorls; cones lacking an aril. (12)
12a. Foliage evergreen. PINACEAE, p. 17.
12b. Foliage deciduous. (13)
13a. Leaves many, crowded on a short spur; cone scales flat. PINACEAE (**Larix**), p. 18.
13b. Leaves not on a spur, 2-ranked; cone scales peltate. TAXODIACEAE (**Taxodium**), p. 23.

ARAUCARIACEAE — ARAUCARIA FAMILY

Trees with whorled branches; leaves spirally arranged, mostly awl shaped; plants usually dioecious; pistillate cones large and woody, the scales bearing 1 seed, which lacks a wing.

Class GYMNOSPERMAE

Araucaria Juss.
1a. Mature leaves with obscure midrib, about 12 mm long; greenhouse. Norfolk Island Pine. **A. excelsa** R. Br.
1b. Mature leaves with prominent midrib, about 6 mm long; greenhouse. New Caledonian Pine. **A. columnaris** Hook.

CEPHALOTAXACEAE — PLUMYEW FAMILY

Evergreen shrubs; branches opposite; leaves linear, dense, spirally arranged, appearing 2-ranked, with 2 broad, glaucous bands beneath; cones drupelike, greenish or purplish.

Cephalotaxus Sieb. & Zucc.   Plumyew
A single genus and species treated, with the characteristics of the family. Japanese Plumyew. **C. drupacea** Sieb. & Zucc.

CUPRESSACEAE — CYPRESS FAMILY

Monoecious or dioecious evergreen trees or shrubs; leaves opposite or whorled, scalelike or linear; cones woody or fleshy, the staminate ones small; ovuliferous scales paired or whorled, with 1 to several ovules near the base.
1a. Branches not arranged in flat sprays, the twigs extending more or less in all directions; native or cultivated. (2)
1b. Branches arranged in flat sprays, the twigs more or less in a single plane; all cultivated. (3)
2a. Cones dry at maturity, the scales woody and finally separating, long persistent on tree; seeds numerous under each scale; cultivated. **Cupressus**
2b. Cones berrylike, the scales fleshy, not separate at maturity; seeds few; native and cultivated. **Juniperus**
3a. Cones subglobose, the scales shield shaped; seeds few under each scale. **Chamaecyparis**
3b. Cones oblong, the scales imbricated or valvate; seeds 2 per scale. (4)
4a. Bark of trunk exfoliating in plates; cone scales 4 or 6; leaves appearing in whorls of 4. **Libocedrus**
4b. Bark of trunk not exfoliating in plates, shredded; cone scales 8 or more, rarely 6; leaves obviously paired. **Thuja**

Chamaecyparis Spach   False Cypress
Trees to 60 m tall; branchlets flattened; bark dark red-brown, smooth on younger branches; leaves bright green above, glaucous below; staminate cones oblong; pistillate cones globose, about 8 mm in diameter, reddish brown, the 8–10 scales with short conical projection at apex; cultivated. Port Orford Cedar. **C. lawsoniana** (A. Murr.) Parl.

Cupressaceae

Cupressus L. Cypress

Trees to 25 m tall; branchlets extending more or less in all directions, not flattened; bark reddish, shredded, persistent; leaves scalelike; cones nearly globular, with woody scales, these separate at maturity, persistent; seeds numerous under each scale; cultivated. Arizona Cypress. **C. arizonica** Greene

Juniperus L. Juniper

1a. Plants spreading to prostrate shrubs. (2)
1b. Plants erect or ascending trees or shrubs. (5)
2a. Leaves needle- to awllike, jointed to twig, in whorls of 3; native and cultivated. Common Juniper. **J. communis** L.
2b. Leaves scalelike or, if needlelike, then decurrent on twig, opposite or in whorls of 3. (3)
3a. Shrubs spreading, open, to 2 m high; main branches ascending; secondary branchlets spreading to slightly ascending; cultivated. Pfitzer Juniper. **J. chinensis** L. var. **pfitzeriana** Mast.
3b. Shrubs low, compact, frequently less than 6 dm high; main branches close to ground; secondary branchlets frequently strongly ascending. (4)
4a. Leaves dark green, obtuse or acutish, strong disagreeable odor when bruised; cultivated. Tam Juniper. **J. sabina** L. var. **tamariscifolia** Ait.
4b. Leaves bluish green to steel blue (often pinkish in winter), acute or cuspidate, less strong odor; cultivated. Creeping Juniper. **J. horizontalis** Moench.
5a. Leaves needlelike, jointed at base, upper leaf surface bearing a single white band. (6)
5b. Leaves mostly scalelike, decurrent at base, not jointed, upper surface usually lacking a white band. (7)
6a. Leaves concave above, rounded or slightly keeled below; trees or spreading shrubs, usually without pendulous branches; cultivated. Common Juniper. **J. communis** L.
6b. Leaves narrowly grooved above, conspicuously keeled below; trees with pendulous branches; cultivated. Needle Juniper. **J. rigida** Sieb. & Zucc.
7a. Plants shrubby, usually as broad as high; leaves glaucous, giving tree a bluish color, in whorls of 3; cultivated. Meyer Juniper. **J. squamata** Lamb. var. **meyeri** Rehd.
7b. Plants erect trees, or treelike in outline; leaves glaucous or green, opposite or in whorls of 3. (8)
8a. Leaves all needlelike. (9)
8b. Leaves scalelike, at least some of them. (10)
9a. Leaves all opposite, or nearly all; cultivated. Virginia Juniper. **J. virginiana** L.
9b. Leaves in whorls of 3; cultivated. Chinese Juniper. **J. chinensis** L.

15

Class GYMNOSPERMAE

10a. Leaf margins minutely denticulate; branchlets coarse; cones large, 6–10 mm in diameter. Utah Juniper. **J. osteosperma** (Torr.) Little
10b. Leaf margins entire; branchlets usually slender; cones usually less than 6 mm in diameter. (11)
11a. Scale leaves obtuse; needlelike leaves usually in whorls of 3; cultivated. Chinese Juniper. **J. chinensis** L.
11b. Scale leaves acute; needlelike leaves usually opposite. (12)
12a. Needlelike leaves about 2 mm long; cones usually more than 6 mm in diameter, several-seeded; cultivated. Greek Juniper. **J. excelsa** Bieb.
12b. Needlelike leaves 4–6 mm long; cones usually less than 6 mm in diameter, the seeds solitary or few. (13)
13a. Mature scale leaves overlapping those directly above; cones ripening first season; cultivated. Virginia Juniper. **J. virginiana** L.
13b. Mature scale leaves rarely overlapping those directly above, or only slightly so; cones ripening second season; native and cultivated. Rocky Mountain Juniper. **J. scopulorum** Sarg.

Libocedrus Endl.   Incense Cedar

Trees to 60 m tall; bark bright reddish, exfoliating in plates; branchlets flattened into sprays; cones with 3 pairs of scales; cultivated. **L. decurrens** Torr.

Thuja L.   Arborvitae

1a. Branchlets in vertical sprays; cone scales thick, strongly curved; cultivated in several forms. Oriental Arborvitae. **T. orientalis** L.
1b. Branchlets in horizontal sprays; cone scales thin, nearly straight; cultivated. American Arborvitae. **T. occidentalis** L.

CYCADACEAE — CYCAS FAMILY

Palmlike plants with mostly unbranched, tuberlike stems; leaves pinnately compound, with circinate vernation; dioecious; both staminate and pistillate cones modified (in **Cycas**).

Cycas L.

A single genus and species treated, with the characteristics of the family; cultivated greenhouse plant. Sago-Palm. **C. revoluta** Thunb.

EPHEDRACEAE — EPHEDRA FAMILY

Shrubs with jointed, grooved green stems; leaves scalelike, opposite or whorled, usually dioecious; staminate cones with 2–8 stamens; pistillate cones with 1–3 ovules enclosed in an envelope.

Ephedra L.   Mormon Tea
1a. Scale leaves 3 per node; bracts of fruiting cones clawed, 7–10 mm

Pinaceae

wide; plants olive green, usually low and spreading. **E. torreyana** S. Wats.
1b. Scale leaves 2 per node; fruiting bracts not clawed, 3–5 mm wide; plants various, but often yellow-green and erect or ascending. (2)
2a. Scale base dark brown, persistent; plants frequently erect. (3)
2b. Scale base grayish, deciduous; plants frequently spreading. (4)
3a. Peduncles of ovulate spikes lacking or very short; stems not viscid; widely distributed, possibly the most common species in Utah. **E. viridis** Coville
3b. Peduncles of ovulate spikes to 2 cm long; stems often viscid; southeastern Utah. **E. cutleri** Peebles
4a. Seeds mostly paired, brown, smooth; southern and western Utah. **E. nevadensis** S. Wats.
4b. Seeds mostly single, grayish or light brown, wrinkled; southwestern Utah. **E. fasciculata** A. Nels.

Ginkgoaceae — Ginkgo Family

Deciduous, broad-leaved trees; leaves alternate or on spurs, fan shaped, dichotomously veined; dioecious; cone drupelike, to 2.5 cm long, covered with an aril.

Ginkgo L.
   A single genus and species in the family; known only in cultivation.
G. biloba L.

Pinaceae — Pine Family

Resinous evergreen, or rarely deciduous, trees or shrubs; leaves spirally arranged, solitary, or in fascicles, usually linear or needlelike; plants usually monoecious; staminate cones with many stamens; pistillate cones with 2 ovules on the inner surface of each scale.

1a. Leaves deciduous, borne in clusters on short, spur branches. **Larix**
1b. Leaves persistent, variously disposed. (2)
2a. Leaves in fascicles or clusters, solitary only on long shoots. (3)
2b. Leaves solitary, not in fascicles or clusters. (4)
3a. Leaves 5 per cluster or less, with a deciduous or persistent sheath at base. **Pinus**
3b. Leaves usually many per cluster, without a sheath. **Cedrus**
4a. Leaves terete, a basal sheath present, soon lacking. **Pinus**
4b. Leaves variously flattened or shaped, but not terete. (5)
5a. Leaves flattened, not leaving a persistent base upon the twig when they fall; cones upright on branches, the scales deciduous from axis. **Abies**
5b. Leaves flattened or quadrangular, leaf base persistent or not; cones pendulous or reflexed. (6)

Class GYMNOSPERMAE

6a. Branchlets roughened by persistent leaf bases; bracts of cone scales not exserted. (7)
6b. Branchlets not roughened, the leaf bases not persistent; bracts of cone scales 3-toothed, long-exserted. **Pseudotsuga**
7a. Leaves quadrangular, or rarely flattened (in **P. omorika** and **P. sitchensis**); cones usually much more than 2.5 cm long. **Picea**
7b. Leaves conspicuously flattened; cones usually less than 2 cm long. **Tsuga**

Abies Mill. Fir

1a. Leaves glaucous, grayish or bluish green; native. (2)
1b. Leaves dark green, shining above. (3)
2a. Leaves mostly 3–5.5 cm long; branchlets nearly glabrous; usually below 8,000 ft. elevation. White Fir. **A. concolor** Hoopes
2b. Leaves mostly 2.5–4 cm long; branchlets pubescent; usually above 8,000 ft. elevation. Alpine Fir. **A. lasiocarpa** Nutt.
3a. Leaf apex sharp pointed; cultivated. Needle Fir. **A. holophylla** Maxim.
3b. Leaf apex rounded. (4)
4a. Winter buds not resinous; cones reddish brown, 12–15 cm long; cultivated. Nordman Fir. **A. nordmanniana** Spach
4b. Winter buds resinous; cones not reddish brown, 5–10 cm long. (5)
5a. Leaves 3–5.5 cm long; cones bright green, 5–10 cm long; cultivated. Giant Fir. **A. grandis** Lindl.
5b. Leaves 1–2.5 cm long; cones violet-purple, 5–7 cm long; cultivated. Balsam Fir. **A. balsamea** Mill.

Cedrus Loud. Cedar

1a. Leading shoots and branchlets pendulous; cultivated. Deodar Cedar. **C. deodara** Loud.
1b. Leading shoots and branchlets erect. (2)
2a. Leaves 2.5–3 cm long; branchlets glabrous; cultivated. Cedar of Lebanon. **C. libani** Loud.
2b. Leaves usually less than 2.5 cm long; branchlets pubescent; cultivated. Atlas Cedar. **C. atlantica** Manetti

Larix Mill. Larch

1a. Needles sharp pointed; cone bracts longer than scales; cultivated. Western Larch. **L. occidentalis** Nutt.
1b. Needles soft; cone bracts shorter than scales; cultivated. European Larch. **L. decidua** Mill.

Picea A. Dietr. Spruce

1a. Leaves flattened; cultivated. (2)
1b. Leaves quadrangular; cultivated or native. (3)

Pinaceae

2a. Leaves usually 12 mm long or less, obtuse; branchlets pubescent. Serbian Spruce. **P. omorika** Purkyne
2b. Leaves 12–25 mm long, very sharp pointed; branchlets glabrous. Sitka Spruce. **P. sitchensis** Carr.
3a. Cones usually less than 6 cm long. (4)
3b. Cones mostly more than 6 cm long. (6)
4a. Branchlets glabrous; cone scales soft, apically rounded; cultivated. White Spruce. **P. glauca** Voss
4b. Branchlets pubescent. (5)
5a. Leaves 6–20 mm long; cones 1.5–3.5 cm long, the scales slightly denticulate, the tip rounded; cultivated. Black Spruce. **P. mariana** BSP
5b. Leaves 15–25 mm long; cones 3.5–9 cm long, the scales wedge shaped, apically notched; native, rare in cultivation. Engelmann Spruce. **P. engelmannii** (Parry) Engelm.
6a. Leaves somewhat flattened, broader than high in cross section; cultivated. Korean Spruce. **P. koyamai** Shiras.
6b. Leaves usually quite quadrangular and nearly square in cross section. (7)
7a. Leaves 1–1.5 cm long, the tip obtuse; cones 6–9 cm long, the scales rounded; cultivated. Oriental Spruce. **P. orientalis** (L.) Link
7b. Leaves 1–3 cm long, the tip acute; cones to 18 cm long, the scales notched. (8)
8a. Leaves 1–2 cm long; cones 10–18 cm long; cultivated. Norway Spruce. **P. abies** (L.) Karst.
8b. Leaves 2.5–3 cm long; cones 6–10 cm long; cultivated and native. Blue Spruce (state tree). **P. pungens** Engelm.

Pinus L. Pine

1a. Leaves in bundles of 5. (2)
1b. Leaves not in bundles of 5. (5)
2a. Leaves serrulate. (3)
2b. Leaves entire. (4)
3a. Leaves 12–20 cm long, slender, drooping; cultivated. Himalayan Pine. **P. griffithii** McClelland
3b. Leaves less than 12 cm long; cultivated. White Pine. **P. strobus** L.
4a. Leaves 3–7 cm long, the sheaths early deciduous; cones 7–20 cm long, unarmed; native, widely distributed. Limber Pine. **P. flexilis** James
4b. Leaves 2–4 cm long, the sheaths persisting for 2–3 years; cones 5–9 cm long, armed with spines; native, mostly in southern Utah. Bristle-cone Pine. **P. aristata** Engelm.
5a. Leaves borne singly, rarely 2 per bundle; native, mostly in the Great Basin. Single-leaf Pinyon. **P. monophylla** Torr. & Frem.

Figs. 9-10. 9. **Juniperus communis,** x .3. 10. **Thuja plicata,** x .28

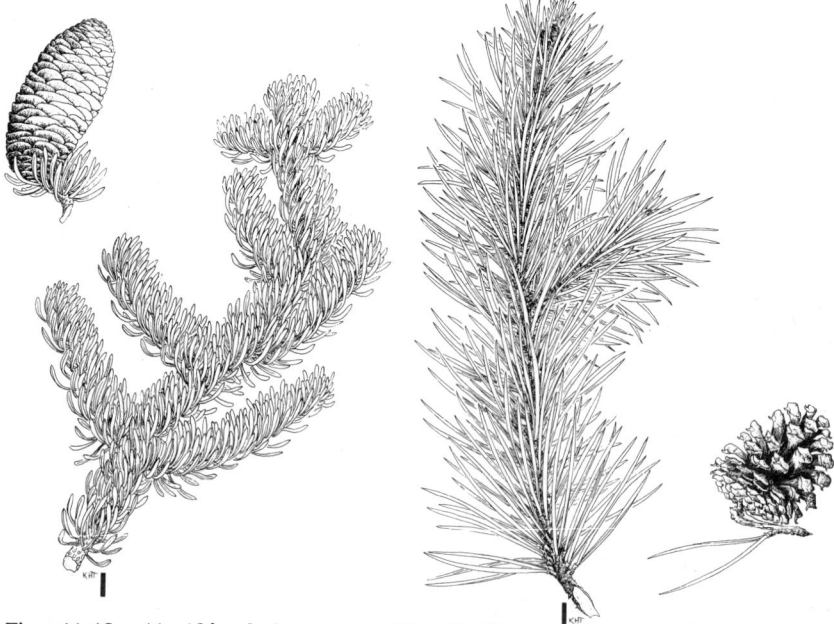

Figs. 11-12. 11. **Abies lasiocarpa,** x .35. 12. **Pinus contorta,** x .28

Pinaceae

5b. Leaves borne in bundles of 2–3. (6)
6a. Leaves borne in bundles of 3, at least some. (7)
6b. Leaves borne in bundles of 2, rarely some with 3. (8)
7a. Cones 15–30 cm long; bark smells of vanilla; buds not covered with resin droplets; cultivated. Jeffrey Pine. **P. jeffreyi** A. Murr
7b. Cones 7–15 cm long; bark smells of turpentine; buds often with resin droplets; cultivated and native. Ponderosa Pine. **P. ponderosa** Laws
8a. Sheath at base of leaf deciduous; native, mostly in the Colorado Basin. Pinyon Pine. **P. edulis** Engelm.
8b. Sheath at base of leaf persistent. (9)
9a. Leaves 8 cm long or more. (10)
9b. Leaves 8 cm long or less. (13)
10a. Plants shrublike, more or less round in shape, to 6 m tall; cones usually less than 5 cm long; cultivated. Japanese Red Pine. **P. densiflora** Sieb. & Zucc.
10b. Plants erect, treelike; cones usually over 5 cm long. (11)
11a. Leaves 12–18 cm long, brittle, breaking when bent; cultivated. Red Pine. **P. resinosa** Ait.
11b. Leaves 9–18 cm long, stiff but not breaking when bent. (12)
12a. Leaves bluish green, sharply serrulate; cones 3.5–5.5 cm long; cultivated. Chinese Pine. **P. tabulaeformis** Carr.
12b. Leaves dark green, minutely serrulate; cones 5–9 cm long; cultivated. Austrian Pine. **P. nigra** Arnold
13a. Plants shrubs; cultivated. Mugo Pine. **P. mugo** Turra
13b. Plants treelike. (14)
14a. Leaves not exceeding 2.5 cm long; cultivated. Jack Pine. **P. banksiana** Lamb.
14b. Leaves exceeding 2.5 cm long. (15)
15a. Upper trunk cinnamon colored; cones short stalked; cultivated. Scots Pine. **P. sylvestris** L.
15b. Upper trunk not cinnamon colored; cones sessile or subsessile. (16)
16a. Cones persistent and abundant on trees; leaves mostly 2.5–5 cm long; native, rare in cultivation. Lodgepole Pine. **P. contorta** Loud.
16b. Cones not persistent, soon deciduous and few or none present; leaves mostly 5–9 cm long; cultivated. Mugo Pine. **P. mugo** Turra

Pseudotsuga Carr. Douglas Fir

Trees to 100 m tall; trunk to 10 m in diameter; bark deeply fissured; leaves straight, with a short petiole; cone 5–10 cm long, bracts 3-toothed, exserted; native and cultivated, an important lumber tree. **P. menziesii** (Mirb.) Franc

Tsuga Carr. Hemlock

Evergreen trees to 25 m tall; leaves 1–2 cm long, flat, tapering

Class GYMNOSPERMAE

from base to apex; cones 1–2 cm long; cultivated. Canadian Hemlock. **T. canadensis** (L.) Carr.

TAXACEAE — YEW FAMILY

Evergreen shrubs; leaves needlelike, flattened, alternate or rarely opposite, often 2-ranked; plants dioecious; cone berry- or drupelike, surrounded by a fleshy, cup-shaped scarlet aril; cultivated.

Taxus L. Yew
1a. Young foliage yellow striped. English Yew. **T. baccata** L.
1b. Young foliage green. (2)
2a. Shrubs low, spreading, usually as wide as tall or wider. (3)
2b. Shrubs erect or ascending, taller than wide. (5)
3a. Leaves gradually acuminate. English Yew. **T. baccata** L.
3b. Leaves abruptly pointed. (4)
4a. Mature branchlets reddish brown. Japanese Yew. **T. cuspidata** Sieb. & Zucc.
4b. Mature branchlets olive green. Hybrid Yew. **T. x media** Rehd.
5a. Leaves gradually acuminate. English Yew. **T. baccata** L.
5b. Leaves abruptly pointed. (6)
6a. Plants treelike, usually with a central stem, loose, open, broadly pyramidal. Japanese Yew. **T. cuspidata** Sieb. & Zucc.
6b. Plants upright or ascending shrubs, narrowly pyramidal or columnar. Hybrid Yew. **T. x media** Rehd.

TAXODIACEAE — TAXODIUM FAMILY

Evergreen or deciduous trees; leaves spirally arranged, though often 2-ranked, or in symmetric whorls at ends of twigs; cones woody, with thickened, wide-spreading scales, the pistillate with scales bearing 2–9 ovules; all species cultivated.

1a. Conspicuous leaves arranged in symmetric whorls at ends of twigs. **Sciadopitys**
1b. Conspicuous leaves alternate or opposite. (2)
2a. Leaves opposite. **Metasequoia**
2b. Leaves alternate, mostly spirally arranged. (3)
3a. Leaves scalelike, subulate and rigid. **Sequoiadendron**
3b. Leaves needlelike, flat, not subulate and rigid. **Taxodium**

Metasequoia Miki Dawn Redwood
Deciduous trees to 35 m tall and 3 m in diameter; the branchlets deciduous as in **Taxodium** but opposite; bark fibrous, reddish, becoming dark gray; leaves flat, about 6 mm long, appearing opposite, actually decussate; cones less than 25 mm long. **M. glyptostroboides** Hu & Cheng

Class ANGIOSPERMAE

Sciadopitys Sieb. & Zucc. Umbrella Pine
Trees to 40 m tall; bark shredded, gray; leaves of 2 kinds, 1 small, scalelike and scattered, but crowded at ends of branches and bearing in their axils linear, flat leaves 7–15 cm long; cones oblong-ovate, woody, 7–12 cm long, the pistillate with scales bearing 7–9 ovules; cultivated, usually indoors. **S. verticillata** (Thunb.) Sieb. & Zucc.

Sequoiadendron Buchholz. Big-Tree
Trees to 100 m tall and 3–10 m in diameter; bark fibrous, red; leaves scalelike, 4–12 mm long, subulate; cones ovoid, 5–9 cm long, dark reddish brown; cultivated. **S. giganteum** (Lindl.) Buchholz.

Taxodium Rich. Bald Cypress
Trees to 50 m tall, with a strongly buttressed base; branchlets deciduous; leaves 2-ranked, linear-lanceolate, alternate, 1–1.5 cm long; cones globose, about 2.5 cm broad; cultivated. **T. distichum** (L.) Rich.

CLASS ANGIOSPERMAE
SUBCLASS DICOTYLEDONEAE

Key to the Families

1a. Perianth consisting of a single whorl, arbitrarily called sepals, or none. KEY I, p. 23.
1b. Perianth consisting of 2 whorls (sepals and petals). (2)
2a. Corolla of separate petals. (3)
2b. Corolla of united petals, at least near the base. KEY IV, p. 33.
3a. Stamens numerous, more than twice as many as the petals. KEY II, p. 27.
3b. Stamens few, not more than twice as many as the petals. KEY III, p. 29.

KEY I. Perianth consisting of a single whorl.

1a. Plants parasitic on the branches of trees or shrubs, rooting in the host, usually yellow-green. LORANTHACEAE, p. 223.
1b. Plants not parasitic on branches of trees, rooting in soil. (2)
2a. Plants trees, shrubs, or vines. (3)
2b. Plants herbaceous. (26)
3a. Leaves opposite. (4)
3b. Leaves alternate. (11)
4a. Plants trailing vines. (5)
4b. Plants trees or shrubs. (6)
5a. Leaves compound; stamens and pistils many; flowers showy. RANUNCULACEAE (**Clematis**), p. 288.

23

Subclass DICOTYLEDONEAE

5b. Leaves simple, deeply palmately lobed; stamens and pistils 5 or fewer; flowers not showy. MORACEAE (**Humulus**), p. 231.

6a. Leaves compound. (7)
6b. Leaves simple. (8)

7a. Leaflets 3 (5); fruit a double samara; flowers in dense fascicles. ACERACEAE, p. 36.
7b. Leaflets 3–7 or more; fruit a simple samara; flowers in racemes or panicles. OLEACEAE (**Fraxinus**), p. 235.

8a. Leaves palmately lobed or parted; fruit a double samara. ACERACEAE, p. 36.
8b. Leaves various but usually pinnately veined and entire; fruit a simple samara, capsule, or drupe. (9)

9a. Ovary superior; fruit a simple samara, capsule, or drupe. OLEACEAE, p. 234.
9b. Ovary inferior (appearing so in Elaeagnaceae); fruit a drupe. (10)

10a. Flowers in corymbose cymes, perfect; ovary usually 2-celled; stamens usually 5. CORNACEAE, p. 138.
10b. Flowers solitary or in axillary clusters, often imperfect; ovary 1-celled; stamens 4–8. ELAEAGNACEAE, p. 158.

11a. Leaves compound. (12)
11b. Leaves simple. (14)

12a. Leaves spinulose toothed, evergreen. BERBERIDACEAE, p. 43.
12b. Leaves entire, or with serrate margins, not spinulose, deciduous. (13)

13a. Leaflets entire; fruit a legume; flowers conspicuous. LEGUMINOSAE, p. 185.
13b. Leaflets toothed; fruit drupaceous; flowers inconspicuous. JUGLANDACEAE, p. 176.

14a. Plants trailing vines. POLYGONACEAE, p. 253.
14b. Plants trees or shrubs. (15)

15a. Flowers of one or both sexes in catkins; plants monoecious or dioecious. (16)
15b. Flowers not in catkins, perfect or imperfect. (21)

16a. Perianth lacking. (17)
16b. Perianth present. (19)

17a. Plants desert shrubs in saline soils; leaves entire, terete or nearly so; fruit a utricle. CHENOPODIACEAE (**Sarcobatus**), p. 84.
17b. Plants of moist situations in a variety of soils; leaves mostly not entire, the blades flat; fruit not a utricle. (18)

18a. Plants monoecious; staminate flowers attached to bract of catkin. BETULACEAE, p. 44.

Key to Families

18b. Plants dioecious; staminate flowers attached to axis of catkin. SALICACEAE, p. 308.
19a. Fruit multiple, fleshy and elongate or globose and woody. MORACEAE, p. 230.
19b. Fruit an acorn or a nut. (20)
20a. Fruit a nut enclosed in a leafy involucre, or in conelike catkins; plants usually of moist situations if native. BETULACEAE, p. 44.
20b. Fruit an acorn, chestnut, or beechnut, with a basal cup, spiny bur, or subtended by bracts; plants seldom of moist situations unless cultivated. FAGACEAE, p. 162.
21a. Ovary inferior, or appearing so. ELAEAGNACEAE, p. 158.
21b. Ovary superior. (22)
22a. Ovary of 1 carpel; fruit an achene or berry. (23)
22b. Ovary of 2 or more carpels; fruit an achene or otherwise, but not a berry. (24)
23a. Plants spinose, introduced (or native) shrubs with a small but showy perianth; fruit a berry. BERBERIDACEAE, p. 43.
23b. Plants unarmed native shrubs, the perianth showy or not; fruit a plumose achene. ROSACEAE (**Cercocarpus, Coleogyne**), p. 297, 299.
24a. Plants normally trees; fruit a samara. ULMACEAE, p. 340.
24b. Plants normally shrubs; fruit an achene or utricle. (25)
25a. Flowers perfect, subtended by a cuplike involucre; stamens 6–9; fruit an achene. POLYGONACEAE (**Eriogonum**), p. 254.
25b. Flowers usually imperfect, not subtended by a cuplike involucre; stamens 1–5; fruit a utricle. CHENOPODIACEAE, p. 78.
26a. Ovary inferior. (27)
26b. Ovary superior. (34)
27a. Plants aquatic; leaves entire and in whorls, or the immersed ones dissected. HALORAGACEAE, p. 168.
27b. Plants terrestrial; leaves various but not as above. (28)
28a. Ovary 3-loculed, 3-winged; plants monoecious; staminate flowers with 2 sepals and 2 petals; pistillate with 2 to many imbricated petaloid parts. BEGONIACEAE, p. 42.
28b. Ovary 1- or 2-loculed; plants usually perfect or dioecious; flowers otherwise different from above. (29)
29a. Ovary 2-loculed, 1 ovule in each cell; fruit 2-seeded. (30)
29b. Ovary 1-loculed, 1- to 2-ovuled, sometimes with several carpels but with a single ovule developing; fruit 1- or 2-seeded. (31)
30a. Perianth united; leaves opposite or whorled; flowers usually in cymes. RUBIACEAE, p. 305.

Subclass DICOTYLEDONEAE

30b. Perianth of separate segments; leaves alternate or basal; flowers in umbels. UMBELLIFERAE, p. 342.

31a. Flowers sessile, in involucrate heads; anthers united into a tube around style. COMPOSITAE, p. 86.
31b. Flowers not sessile or in involucrate heads; anthers distinct. (32)

32a. Leaves alternate. SANTALACEAE, p. 312.
32b. Leaves opposite. (33)

33a. Leaves pinnately parted or compound, at least the cauline ones sheathing at the base; dioecious. VALERIANACEAE, p. 353.
33b. Leaves simple, not sheathing at the base; usually perfect. NYCTAGINACEAE, p. 231.

34a. Pistils several to many per flower; stamens usually 10 to many. RANUNCULACEAE, p. 286.
34b. Pistils 1 per flower; stamens 1 to many, usually 10 or less. (35)

35a. Plants aquatic, usually more or less submerged. (36)
35b. Plants terrestrial, sometimes growing in moist or wet soil. (37)

36a. Leaves entire, opposite, often crowded into terminal rosettes. CALLITRICHACEAE, p. 65.
36b. Leaves largely dissected, whorled. CERATOPHYLLACEAE, p. 78.

37a. Perianth lacking entirely; in some Euphorbiaceae, involucre looks like a hypanthium with petals. (38)
37b. Perianth present. (39)

38a. Inflorescence spicate, subtended by a conspicuous petaloid involucre, the whole resembling a single flower. SAURURACEAE, p. 312.
38b. Inflorescence of a cyathium (cup-shaped involucre subtending clusters of staminate flowers consisting of single pedicellate stamens and pistillate flowers consisting of single, 3-lobed pistils), which often bears petaloid appendages and glands. EUPHORBIACEAE, p. 160.

39a. Flowers perigynous, the ovary enclosed in or seated in a floral tube. (40)
39b. Flowers hypogynous, the ovary not enclosed in a floral tube. (42)

40a. Stipules present; leaves alternate. ROSACEAE, p. 294.
40b. Stipules none; leaves opposite. (41)

41a. Stamens many; fruit opening by a circumscissile lid. AIZOACEAE, p. 37.
41b. Stamens 3–5; fruit indehiscent. NYCTAGINACEAE, p. 231.

42a. Perianth showy; sepals and petals both present in bud, but the sepals caducous and represented only by scars during anthesis; plants with milky juice. PAPAVERACEAE, p. 243.
42b. Perianth showy or not showy; plants various but not as above. (43)

Key to Families

43a. Styles and stigmas single. (44)
43b. Styles and stigmas more than 1. (45)
44a. Perianth neither tubular nor corollalike. URTICACEAE, p. 352.
44b. Perianth tubular, corollalike. NYCTAGINACEAE, p. 231.
45a. Leaves mostly deeply palmately 5- to 7-lobed, or with 5–7 leaflets; flowers imperfect. MORACEAE (**Cannabis, Humulus**), pp. 230, 231.
45b. Leaves mostly entire, or shallowly lobed; flowers perfect or imperfect. (46)
46a. Ovary with more than 1 locule. (47)
46b. Ovary with a single locule. (48)
47a. Leaves opposite. AIZOACEAE, p. 37.
47b. Leaves alternate. EUPHORBIACEAE, p. 160.
48a. Leaves opposite; ovules and seeds more than 1; fruit a capsule. (49)
48b. Leaves alternate or opposite; ovules and seeds solitary; fruit an achene or utricle. (50)
49a. Capsule opening by means of a circumscissile lid. AIZOACEAE, p. 37.
49b. Capsule opening by valves or teeth. CARYOPHYLLACEAE, p. 71.
50a. Flowers borne in a cup-shaped involucre. POLYGONACEAE, p. 253.
50b. Flowers not borne in an involucre. (51)
51a. Plants with a stipular sheath (ocrea) above each node. POLYGONACEAE, p. 253.
51b. Plants lacking an ocrea. (52)
52a. Perianth mostly with 6 segments; stamens 3, 6, or 9. POLYGONACEAE, p. 253.
52b. Perianth with 1, 4, or 5 segments. (53)
53a. Bracts subtending flowers scarious, usually awn tipped; plants not scurfy. AMARANTHACEAE, p. 38.
53b. Bracts subtending flowers not scarious, not awn tipped; plants usually scurfy. CHENOPODIACEAE, p. 78.

KEY II. Corolla of separate petals; stamens more than twice as many as the petals.

1a. Ovary inferior or partly so. (2)
1b. Ovary superior. (7)
2a. Petals numerous; stems thick and succulent, spiny; leaves lacking or caducous. CACTACEAE, p. 63.
2b. Petals few; stems not thick and succulent, though occasionally spiny; leaves present and conspicuous during growing season. (3)
3a. Ovary only partly inferior. (4)

Subclass DICOTYLEDONEAE

- 3b. Ovary wholly inferior. (5)
- 4a. Leaves opposite; fruit a capsule. SAXIFRAGACEAE, p. 313.
- 4b. Leaves alternate; fruit a pome. ROSACEAE, p. 294.
- 5a. Plants woody; fruit fleshy. ROSACEAE, p. 294.
- 5b. Plants herbaceous; fruit a capsule. (6)
- 6a. Flowers perfect; plants usually rough-hairy; capsules not winged. LOASACEAE, p. 221.
- 6b. Flowers imperfect; plants smooth; capsules 3-winged. BEGONIACEAE, p. 42.
- 7a. Plants aquatic, with floating leaves. NYMPHAEACEAE, p. 233.
- 7b. Plants terrestrial, often growing in moist or wet soils, but leaf blades not floating. (8)
- 8a. Plants trees or shrubs. (9)
- 8b. Plants herbaceous, or woody only at base. (12)
- 9a. Shrubs or small trees with recurved thorns and pinnate leaves; Fruit a legume. LEGUMINOSAE (**Acacia**), p. 187.
- 9b. Shrubs or trees without thorns or pinnate leaves; fruit various but not a legume. (10)
- 10a. Stipular scars encircling the stem; perianth parts, stamens, and pistils spirally arranged. MAGNOLIACEAE, p. 224.
- 10b. Stipular scars not encircling the stem; perianth parts and stamens in whorls; pistils usually 1. (11)
- 11a. Filaments united into a tube around the pistil. MALVACEAE, p. 224.
- 11b. Filaments not united into a tube, distinct. ROSACEAE, p. 294.
- 12a. Sepals 2. (13)
- 12b. Sepals more than 2. (14)
- 13a. Sepals persistent; plants somewhat succulent. PORTULACACEAE, p. 278.
- 13b. Sepals caducous; plants not succulent. PAPAVERACEAE, p. 243.
- 14a. Filaments united into a tube around the pistil. MALVACEAE, p. 224.
- 14b. Filaments not united into a tube, distinct. (15)
- 15a. Maturing ovary open apically, the seeds exposed; flowers irregular. RESEDACEAE, p. 293.
- 15b. Maturing ovary closed apically, the seeds not exposed; flowers usually regular. (16)
- 16a. Stamens attached to the margin of a hypanthium. ROSACEAE, p. 294.
- 16b. Stamens attached at base of ovary. (17)

Key to Families

17a. Leaves opposite. HYPERICACEAE, p. 176.
17b. Leaves alternate or basal, but still alternate. (18)

18a. Leaves compound with 3–5 leaflets; ovary short-stipitate. CAPPARIDACEAE (**Polanisia**), p. 68.
18b. Leaves various but not as above; ovary sessile. RANUNCULACEAE, p. 286.

KEY III. Corolla of separate petals; stamens few, not more than twice as many as the petals.

1a. Flowers with more than 1 pistil. (2)
1b. Flowers with a single pistil. (4)

2a. Plants succulent, fleshy. CRASSULACEAE, p. 138.
2b. Plants not succulent. (3)

3a. Stamens inserted on a hypanthium, perigynous. ROSACEAE, p. 294.
3b. Stamens inserted at base of ovary, hypogynous. RANUNCULACEAE, p. 286.

4a. Plants usually with tendrils, trailing vines; monoecious. CUCURBITACEAE, p. 156.
4b. Plants lacking tendrils or, if tendrils present, then flowers perfect, usually not trailing. (5)

5a. Styles 2–5, distinct to near the base. (6)
5b. Styles 1, sometimes lobed or divided at apex. (18)

6a. Plants trees or shrubs. (7)
6b. Plants herbaceous. (11)

7a. Leaves scalelike. TAMARICACEAE, p. 340.
7b. Leaves not scalelike, well developed. (8)

8a. Ovary inferior. (9)
8b. Ovary superior. (10)

9a. Leaves alternate; stipules present. SAXIFRAGACEAE (**Ribes**), p. 316.
9b. Leaves opposite; stipules lacking. SAXIFRAGACEAE (**Jamesia**), p. 315.

10a. Leaves alternate; fruit a drupe. ANACARDIACEAE, p. 38.
10b. Leaves opposite; fruit a double samara. ACERACEAE, p. 36.

11a. Plants aquatic. HALORAGACEAE, p. 168.
11b. Plants terrestrial. (12)

12a. Ovary inferior (partly so in some Saxifragaceae). (13)
12b. Ovary superior. (14)

13a. Flowers in umbels; fruit a schizocarp. UMBELLIFERAE, p. 342.
13b. Flowers variously disposed but not in umbels; fruit a many-seeded capsule or berry. SAXIFRAGACEAE, p. 313.

29

Subclass DICOTYLEDONEAE

14a. Leaves compound; leaflets 3. OXALIDACEAE, p. 243.
14b. Leaves simple. (15)

15a. Leaves opposite. CARYOPHYLLACEAE, p. 71.
15b. Leaves alternate. (16)

16a. Sepals 2; plants succulent. PORTULACACEAE, p. 278.
16b. Sepals more than 2; plants not succulent. (17)

17a. Flowers irregular; capsule open at the top before maturity. RESEDACEAE, p. 293.
17b. Flowers regular; capsules opening at maturity. CARYOPHYLLACEAE, p. 71.

18a. Ovary inferior. (19)
18b. Ovary superior (sometimes enclosed by, but not adnate to, the floral tube). (22)

19a. Plants herbaceous. (20)
19b. Plants woody, shrubs. (21)

20a. Flowers 4-merous. ONAGRACEAE, p. 237.
20b. Flowers 5-merous. SAXIFRAGACEAE, p. 313.

21a. Leaves opposite; flowers cymose; stamens 4. CORNACEAE, p. 138.
21b. Leaves alternate; flowers racemose or solitary; stamens 5. SAXIFRAGACEAE, p. 313.

22a. Plants trees or shrubs. (23)
22b. Plants herbaceous. (37)

23a. Flowers irregular. (24)
23b. Flowers regular. (27)

24a. Petals 3, the lower 2 forming a keel. POLYGALACEAE, p. 253.
24b. Petals 4 or 5, the lower 2 sometimes forming a keel. (25)

25a. Flowers papilionaceous, the lower 2 petals often united and forming a keel; fruit a legume. LEGUMINOSAE, p. 185.
25b. Flowers not papilionaceous, the lower 2 petals not forming a keel; fruit a 3-valved, frequently 1-seeded capsule, or a spiny, 1-seeded pod. (26)

26a. Plants low-growing shrubs; leaves simple; stamens usually 4. KRAMERIACEAE, p. 177.
26b. Plants large shrubs or trees; leaves palmately compound; stamens 5-8. HIPPOCASTANACEAE, p. 169.

27a. Leaves compound, with 2 or more leaflets. (28)
27b. Leaves simple, sometimes deeply divided or parted. (33)

28a. Leaves 2-pinnate; flowers purplish. MELIACEAE, p. 230.

Key to Families

28b. Leaves 1-pinnate; flowers variously colored. (29)

29a. Leaves trifoliolate, alternate; fruit a samara with the wing continuous around it. RUTACEAE, p. 306.
29b. Leaves usually not trifoliolate; fruit various or, if samaroid, then the wing not continuous around it and the leaves opposite. (30)

30a. Leaves opposite. (31)
30b. Leaves alternate. (32)

31a. Leaflets 3–9 or more; fruit a samara; plants usually trees; leaves deciduous. OLEACEAE (**Fraxinus**), p. 235.
31b. Leaflets 2; fruit not a samara; plants shrubs; leaves persistent. ZYGOPHYLLACEAE (**Larrea**), p. 357.

32a. Leaves spiny toothed; sepals usually 6 in 2 whorls; petals usually 6 in 2 series; stamens 6 or 12. BERBERIDACEAE (**Mahonia**), p. 43.
32b. Leaves not spiny toothed; sepals usually 5; petals 5 or less; stamens 5 or 10. LEGUMINOSAE, p. 185.

33a. Plants subshrubs at most; fruit stipitate. CRUCIFERAE (**Stanleya**), p. 154.
33b. Plants well-developed trees or shrubs; fruit not stipitate (except in Rutaceae, which has entire caducous leaves and blue flowers). (34)

34a. Flowers blue, 8–14 mm long; herbage glandular-punctate; fruit stipitate. RUTACEAE (**Thamnosma**), p. 308.
34b. Flowers variously colored, usually not blue, usually less than 8 mm long; herbage not glandular-punctate; fruit not stipitate. (35)

35a. Plants with leaves reduced to spines (foliage leaves in their axils), these sometimes 3-pronged; perianth conspicuous, usually 9- to 12-parted; fruit a berry; ovary 1-loculed. BERBERIDACEAE (**Berberis**), p. 43.
35b. Plants not spiny or, if so, not as above; perianth usually inconspicuous, usually 8- to 10-parted; fruit a capsule or a drupe; ovary 2- to 5-loculed. (36)

36a. Stamens opposite the petals; ovary 2- to 3-loculed. RHAMNACEAE, p. 293.
36b. Stamens alternate with petals or more numerous; ovary 2- to 5-loculed. CELASTRACEAE, p. 78.

37a. Sepals 2 or 3. (38)
37b. Sepals 4, 5, or more. (42)

38a. Plants succulent. (39)
38b. Plants not succulent. (40)

39a. Sepals 2, equal or nearly so. PORTULACACEAE, p. 278.

Subclass DICOTYLEDONEAE

39b. Sepals 3, 2 small and green, the third petaloid and produced backward into a spur. BALSAMINACEAE, p. 42.

40a. Sepals caducous; stamens 6–12; leaves entire. PAPAVERACEAE, p. 243.

40b. Sepals persistent; stamens 6; leaves dissected. (41)

41a. Petals 4, in 2 unlike pairs; stamens diadelphous, in 2 sets of 3. FUMARIACEAE, p. 163.

41b. Petals 3, alike; stamens in one set. LIMNANTHACEAE, p. 220.

42a. Flowers irregular. (43)
42b. Flowers regular. (45)

43a. Flowers papilionaceous; fruit a legume. LEGUMINOSAE, p. 185.
43b. Flowers not papilionaceous; fruit a capsule or separating into 3 indehiscent, 1-seeded carpels at maturity. (44)

44a. Leaves peltate; fruit separating into 3 indehiscent, 1-seeded segments at maturity. TROPAEOLACEAE, p. 340.
44b. Leaves not peltate; fruit a 1-loculed, 3-valved capsule. VIOLACEAE, p. 356.

45a. Leaves compound. (46)
45b. Leaves simple. (49)

46a. Leaves mostly basal, deeply pinnatifid. GERANIACEAE, p. 167.
46b. Leaves alternate or opposite. (47)

47a. Leaves opposite; leaflets 10–16. ZYGOPHYLLACEAE (**Tribulus**), p. 357.
47b. Leaves alternate; leaflets usually 3. (48)

48a. Stamens 10; herbage with a sour taste. OXALIDACEAE, p. 243.
48b. Stamens 2–6; herbage with a sharp biting taste. CAPPARIDACEAE, p. 67.

49a. Sepals and petals 4; stamens 6 (4 plus 2), 4, or rarely 2. CRUCIFERAE, p. 139.
49b. Sepals and petals mostly 5; stamens 5–10. (50)

50a. Leaves with stipules; carpels tailed in mature fruit, separating from each other as 1-seeded, indehiscent segments. GERANIACEAE, p. 167.
50b. Leaves lacking stipules; carpels various but, if as above, then not long tailed. (51)

51a. Ovary 10-celled, with 1 ovule per cell. LINACEAE, p. 221.
51b. Ovary 2- to 5-celled. (52)

52a. Plants annual; leaves dissected; fruit 5-seeded, breaking into indehiscent nutlets. LIMNANTHACEAE, p. 220.
52b. Plants perennial; leaves simple or scalelike; fruit a many-seeded capsule. PYROLACEAE, p. 284.

Key to Families

KEY IV. Corolla of united petals.

1a. Ovary inferior, or partly so. (2)
1b. Ovary superior. (10)
2a. Stamens more than 5; anthers opening by terminal pores. ERICACEAE, p. 159.
2b. Stamens 5 or less; anthers not opening by terminal pores. (3)
3a. Stamens united by anthers. (4)
3b. Stamens separate. (6)
4a. Plants usually tendril bearing; fruit a pepo. CUCURBITACEAE, p. 156.
4b. Plants not tendril bearing; fruit capsular or an achene. (5)
5a. Flowers in involucrate heads; stamens adnate to corolla. COMPOSITAE, p. 86.
5b. Flowers not in involucrate heads; stamens not adnate to corolla. CAMPANULACEAE, p. 66.
6a. Leaves alternate. CAMPANULACEAE, p. 66.
6b. Leaves opposite or whorled. (7)
7a. Stamens 1–3; flowers irregular; plants frequently ill scented. VALERIANACEAE, p. 353.
7b. Stamens 4–5; flowers regular or irregular. (8)
8a. Herbs with ovary 1-loculed; flowers in involucrate heads; fruit an achene. DISPACACEAE, p. 157.
8b. Herbs or shrubs with ovary 2- to 5-loculed; flowers not in involucrate heads; fruit not an achene. (9)
9a. Shrubs with broad leaves; ovary 3- to 5-loculed; leaves opposite or perfoliate, not whorled or with stipules. CAPRIFOLIACEAE, p. 68.
9b. Shrubs, or more usually herbs, if shrubs, the flowers in globose heads; ovary 2-loculed; leaves opposite and having stipules, or whorled and lacking them. RUBIACEAE, p. 305.
10a. Stamens more than 5. (11)
10b. Stamens 5 or less. (19)
11a. Corolla segments distinctly united, cup or tube shaped. (12)
11b. Corolla segments not distinctly united, usually united only near base or only part of segments united. (14)
12a. Pistils more than 1; stamens twice as many as the petals. CRASSULACEAE, p. 138.
12b. Pistils 1 per flower; stamens the same number as the petals. (13)
13a. Anther opening by a terminal pore. ERICACEAE, p. 159.
13b. Anther opening by a longitudinal slit. PYROLACEAE, p. 284.

Subclass DICOTYLEDONEAE

14a. Corolla irregular. (15)
14b. Corolla regular. (17)

15a. Petals 5; fruit a legume. LEGUMINOSAE, p. 185.
15b. Petals 3 or 4; fruit not a legume. (16)

16a. Leaves entire; petals 3. POLYGALACEAE, p. 253.
16b. Leaves dissected; petals 4, in 2 unlike pairs. FUMARIACEAE, p. 163.

17a. Flowers minute, in dense heads or spikes; stamens 10 to many; fruit a legume. LEGUMINOSAE, p. 185.
17b. Flowers not as above; fruit various but not a legume. (18)

18a. Filaments united well above the base; stamens many; leaves simple. MALVACEAE, p. 224.
18b. Filaments united at the base only; stamens 10; leaves compound. OXALIDACEAE, p. 243.

19a. Plants parasitic, devoid of chlorophyll. (20)
19b. Plants usually not parasitic, chlorophyllous. (21)

20a. Flowers regular, minute; slender trailing, twining vines. CONVOLVULACEAE (**Cuscuta**), p. 137.
20b. Flowers irregular; stems erect, not trailing or twining. OROBANCHACEAE, p. 242.

21a. Corolla irregular. (22)
21b. Corolla regular. (27)

22a. Ovary with 1 ovule per locule, appearing 4-loculed, often 4-lobed; fruit consisting of 4 indehiscent, 1-seeded nutlets. (23)
22b. Ovary with more than 1 ovule per locule, usually neither 4-loculed nor 4-lobed; fruit a capsule. (24)

23a. Ovary 4-lobed; style arising from between lobes, cleft apically. LABIATAE, p. 177.
23b. Ovary not, or only slightly 4-lobed; style arising from apex of ovary, not cleft apically. VERBENACEAE, p. 354.

24a. Plants aquatic; leaves often dissected; corolla spurred; stamens 2. LENTIBULARIACEAE, p. 220.
24b. Plants usually terrestrial or, if aquatic, then other than above. (25)

25a. Plants trees or shrubs; seeds winged; capsule linear to cigar shaped. BIGNONIACEAE, p. 45.
25b. Plants herbaceous, or woody only at base; seeds not winged; capsules various but not as above. (26)

26a. Ovary 1-loculed; plants strongly viscid-pubescent; fruit a woody capsule with 2 recurved, spinelike appendages. MARTYNIACEAE, p. 230.

Key to Families

26b. Ovary 2-loculed; plants not usually viscid-pubescent; fruit not as above. SCROPHULARIACEAE, p. 319.

27a. Plants with milky juice; pistils 2, separate at base, united by stigmas and/or styles. (28)

27b. Plants usually without milky juice; pistils 1. (29)

28a. Styles united; stamens appressed around stigma. APOCYNACEAE, p. 39.

28b. Styles distinct below; stamens adnate to stylar column. ASCLEPIADACEAE, p. 40.

29a. Ovary 1-loculed, 1-ovuled; styles and stigmas 1; fruit hard, dry; the apparent corolla really a calyx subtended by an involucre. NYCTAGINACEAE, p. 231.

29b. Ovary and fruit various, but not as above; both calyx and corolla present. (30)

30a. Stamens as many as the corolla lobes and opposite them. PRIMULACEAE, p. 282.

30b. Stamens as many as, or fewer than, the corolla lobes and alternate with them. (31)

31a. Corolla small (2 mm or less), scarious, veinless; capsule opening by a lid; leaves parallel veined, or nearly so. PLANTAGINACEAE, p. 246.

31b. Corolla various, but not as above; capsule usually not opening by a lid; leaves usually distinctly net veined. (32)

32a. Ovary 4-lobed, 4-loculed; fruit consisting of 4 nutlets at maturity. BORAGINACEAE, p. 45.

32b. Ovary not 4-lobed, 1- to 3-loculed; fruit not as above. (33)

33a. Style 3-cleft; ovary 3-loculed; fruit a 3-valved capsule. POLEMONIACEAE, p. 247.

33b. Style not 3-cleft; ovary 1- to 2-loculed; fruit not as above. (34)

34a. Ovary 1-loculed. (35)
34b. Ovary with 2 or more locules. (36)

35a. Leaves opposite or whorled, entire; styles 1 or none; plants mostly glabrous; inflorescence not scorpioid. GENTIANACEAE, p. 164.

35b. Leaves usually alternate, if opposite not entire; styles 2 or single and 2-cleft apically; plants mostly hairy; inflorescence usually scorpiod. HYDROPHYLLACEAE, p. 169.

36a. Stems trailing or twining. CONVOLVULACEAE, p. 136.
36b. Stems not trailing or twining. (37)

37a. Styles 2 or 2-branched apically; fruit a capsule; plants herbaceous (woody in **Eriodictyon**). HYDROPHYLLACEAE, p. 169.

Subclass DICOTYLEDONEAE

37b. Styles 1; stigma entire or merely lobed apically; fruit a capsule, drupe, or berry; plants herbs, shrubs, or small trees. (38)
38a. Stamens 2. (39)
38b. Stamens more than 2 or others represented by staminodes. (40)
39a. Plants woody. OLEACEAE, p. 234.
39b. Plants herbaceous. SCROPHULARIACEAE (Veronica), p. 334.
40a. Shrubs, herbs, or rarely trailing vines; leaves alternate. (41)
40b. Shrubs; leaves opposite or whorled. LOGANIACEAE, p. 222.
41a. Plants tall-stemmed, hairy, biennial herbs; inflorescence a dense spike of yellow flowers. SCROPHULARIACEAE (Verbascum), p. 334.
41b. Plants shrubs, herbs, or trailing vines; inflorescence various but not a dense spike. SOLANACEAE, p. 335.

ACERACEAE — MAPLE FAMILY

Important ornamental and native trees and shrubs; leaves opposite, simple or compound; flowers perfect, imperfect, or polygamous, regular, small in terminal or lateral racemes or panicles; petals 4–5 or lacking; stamens 4–10, usually 8; ovary superior; fruit a flat, winged, double samara.

Acer L. Maple
1a. Leaves compound. (2)
1b. Leaves simple. (3)
2a. Leaves all compound, 7–15 cm long or more, coarsely serrate; native and cultivated trees to 20 m; river bottom forests throughout Utah. Box Elder. **A. negundo** L.
2b. Leaves occasionally 3-foliolate, 5–8 cm long, serrate; native shrubs to small trees; stream sides and mountain slopes. Rocky Mountain Maple. **A. glabrum** Torr.
3a. Leaves 3- to 5-lobed, tips of lobes rounded; small cultivated trees. Hedge Maple. **A. campestris** L.
3b. Leaves various, tips of leaf lobes acute to acuminate. (4)
4a. Leaf lobes divided at least three-fourths the distance to midrib. (5)
4b. Leaf lobes divided less than one-half the distance to midrib. (6)
5a. Leaves mostly 5-lobed, some 3-lobed, coarsely serrate; lower leaf surface pale green to silvery; petiole 3.5–7 cm long; trees to 30 m. Silver Maple. **A. saccharinum** L.
5b. Leaves 5- to 9-lobed or -parted, serrate; lower leaf surface green; petioles usually less than 4.5 cm long; small, delicate, ornamental trees. Japanese Maple. **A. palmatum** Thunb.
6a. Leaves mostly 5-lobed or more, few appearing 3-lobed, 7–12 cm wide. (7)

6b. Leaves prominently 3-lobed, although they may be subtended by 2 much smaller lobes, 2.5–12 cm wide. (9)
7a. Margin of leaf lobes serrate; wings of fruit spreading; shrubs or trees to 10 m. Vine Maple. **A. circinatum** Pursh
7b. Margin of leaf lobes entire to dentate; trees to 30 m tall. (8)
8a. Lower leaf surface much lighter than above; wings of fruit hanging down. Sugar Maple. **A. saccharum** Marsh
8b. Lower leaf surface close to same green as above; wings of fruit spreading at nearly 180°. (Note: There are several purple forms of this species; a common one in our area is var. schwedleri Nichols.) Norway Maple. **A. platanoides** L.
9a. Central lobe about twice as long as those on either side; lower leaf surface light green; shrubs or small trees to 6 m tall. **A. ginnala** Maxim.
9b. Central lobe about equal to other lobes in size. (10)
10a. Lobe margins entire; leaves mostly 5–8 cm long; fruit wings diverging at a narrow angle; native trees to 10 m tall; foothills and mountains. Big Tooth Maple. **A. grandidentatum** Nutt.
10b. Lobe margin serrate to dentate. (11)
11a. Leaves mostly less than 8 cm across. (12)
11b. Leaves mostly over 8 cm across. (13)
12a. Sinus between lobes narrow; deeply 3-lobed to 3-parted; native shrubs or trees to 6 m tall; widespread in mountains. Rocky Mountain Maple. **A. glabrum** Torr.
12b. Sinus between lobes wide; shallowly lobed; occasionally cultivated trees to 30 m tall. Red Maple. **A. rubrum** L.
13a. Undersurface of leaves lighter green than above; fruit glabrous, in pendulous racemes; branchlets glabrous; lobes crenate-serrate; trees to 35 m tall. Sycamore Maple. **A. pseudoplatanus** L.
13b. Undersurface of leaves about same green as above; fruit with bristles, in short racemes; branchlets pubescent early; lobes remotely dentate; trees to 10 m tall. Devil's Maple. **A. diabolicum** Koch

AIZOACEAE — CARPETWEED FAMILY

Glabrous, succulent, much-branched annual herbs; stems mostly prostrate, 10–70 cm long; leaves fleshy, opposite, simple, entire; flowers axillary, borne singly; calyx 5-lobed, usually horned at back of each lobe; corolla none; stamens 1 to many, showy; ovary half-inferior, 3- to 5-loculed; fruit opening by a circumscissile lid.

Sesuvium L. Sea Purslane

A single genus and species treated, with the characteristics of the family; locally abundant, usually in saline soils. **S. verrucosum** Raf.

Subclass DICOTYLEDONEAE

AMARANTHACEAE — AMARANTH FAMILY

Herbs; leaves entire, alternate or opposite, simple; flowers perfect or imperfect, inconspicuous, with 3 dry, scarious, persistent bracts; calyx commonly of 5 sepals, persistent, usually scarious; petals none; stamens as many as the sepals; ovary superior, 1-loculed, usually with 2–3 stigmas; fruit a utricle, indehiscent or circumscissile.

1a. Leaves largely opposite; plants white stellate-woolly or villous. **Tidestromia**
1b. Leaves alternate; plants quite or nearly glabrous. (2)
2a. Ovary with 1 ovule; filaments of stamens distinct; weedy species of waste places. **Amaranthus**
2b. Ovary with 2 or more ovules; filaments united at base; occasional in cultivation. **Celosia**

Amaranthus L. Amaranth

1a. Inflorescence entirely of axillary clusters. (2)
1b. Inflorescence of terminal or axillary clusters. (3)
2a. Sepals 4–5; seeds 1.5 mm broad; stems prostrate. **A. graecizans** L.
2b. Sepals 1–3; seeds less than 1 mm broad. **A. albus** L.
3a. Sepals exceeding the utricle, all spreading at maturity, oblong-linear, obtuse, truncate or emarginate, often aristate. **A. retroflexus** L.
3b. Sepals shorter to slightly longer than the utricle, erect, lanceolate, tapering gradually to the apical aristate tip. (4)
4a. Stamens 5; seeds circular in outline, 1 mm in diameter or less, black. **A. hybridus** L.
4b. Stamens usually less than 5; seeds elliptic, usually over 1 mm in diameter, brown-black. **A. powellii** S. Wats.

Celosia L. Cockscomb

Annual herbs; branches terminated by dense, chaffy, highly colored spikes; flowers perfect; several cultivated garden forms. **C. argentea** L.

Tidestromia Standl.

Annual herbs; stems branched, stellate-pubescent; leaves opposite, petiolate, entire; flowers perfect, clustered in the axils of the leaves; perianth 5-parted; stamens 5. **T. lanuginosa** (Nutt.) Standl.

ANACARDIACEAE — CASHEW FAMILY

Shrubs or small trees; leaves simple or more commonly pinnately compound; alternate; flowers borne in racemes or panicles and regular, monoecious, dioecious, or polygamous; sepals and petals 3–6, the petals inserted on a hypogynous disk; stamens 3–6, alternate with the petals; ovary 1-loculed; fruit a drupe.

Apocynaceae

1a. Plants coarse, branched shrubs or small trees or, if slender and with 3 leaflets, then many branched; leaflets 1–15, variously shaped and toothed; fruit pubescent, red. **Rhus**
1b. Plants slender, unbranched, or seldom-branched shrubs; leaflets 3, rhombic-ovate, coarsely toothed, 3–12 cm long; fruit glabrous, white. **Toxicodendron**

Rhus L. Sumac

1a. Leaves ternate (rarely 5-foliate); flowers from scaly-bracted aments of the previous season, opening before the leaves; widely distributed at lower elevations. Squawbush. **R. trilobata** Nutt.
1b. Leaves with 5–15 leaflets; flowers in terminal panicles from the season's growth; locally abundant in the foothills, also in cultivation. Smooth Sumac. **R. glabra** L.

Toxicodendron Mill. Poison Ivy

Slender, unbranched or seldom-branched shrubs; leaflets 3, rhombic-ovate, coarsely toothed, 3–12 cm long; fruit white, glabrous; widely distributed in moist situations at lower elevations throughout Utah. (Note: Beware of this plant; it produces severe dermatitis in most humans.) **T. radicans** L.

APOCYNACEAE — DOGBANE FAMILY

Caulescent perennials with milky juice; leaves simple, entire, opposite or alternate; flowers perfect, regular; petals 5, partly united; stamens 5; ovary superior, of 2 1-loculed carpels, joined by the styles; fruit of 2 pods and a large number of seeds, often with long hair.

1a. Leaves alternate. **Amsonia**
1b. Leaves opposite. (2)
2a. Plants erect; flowers in cymes. **Apocynum**
2b. Plants trailing; flowers solitary. **Vinca**

Amsonia Walt.

1a. Leaves broadly ovate, acuminate at tips; corolla lobes white or nearly so; pods not constricted between the seeds; northeastern and southwestern Utah. **A. jonesii** Woodson
1b. Leaves linear to lanceolate; pods constricted between the seeds. (2)
2a. Stem and leaves glabrous. (3)
2b. Stem and leaves tomentose; southeastern Utah. **A. tomentosa** Torr. & Frem. ex Frem.
3a. Corolla tube 7–10 mm long; leaves narrowly ovate, or the upper oblong-lanceolate to ovate-lanceolate; southeastern Utah. **A. brevifolia** Gray
3b. Corolla tube (9) 10–17 mm long; leaves oblong-lanceolate, or the upper linear-lanceolate; southeastern Utah. **A. eastwoodiana** Rydb.

Subclass DICOTYLEDONEAE

Apocynum L.   Dogbane
1a. Corolla 2–3 times as long as calyx, the lobes spreading or recurved; leaves spreading or drooping. (2)
1b. Corolla usually less than 2 times as long as calyx, the lobes erect or slightly spreading; leaves ascending or slightly spreading. (3)
2a. Corolla usually over 5 mm long, the lobes often recurved; leaves drooping. **A. androsaemifolium** L.
2b. Corolla not over 5 mm long, the lobes merely spreading; leaves spreading. **A. medium** Greene
3a. Leaves, at least the upper, petiolate, cuneate to rounded at base; bracts of inflorescence usually inconspicuous. **A. cannabinum** L.
3b. Leaves quite or nearly sessile, cordate to rounded at base; bracts of inflorescence often subfoliaceous. **A. sibiricum** Jacq.

Vinca L.   Myrtle
Trailing evergreen ornamentals; flowers blue; corolla lobes twisted; stamens adnate to middle of the tube; seeds not hairy. **V. minor** L.

Asclepiadaceae — Milkweed Family

Perennial herbs with milky juice, erect or decumbent, with deep, thick, hard roots; leaves opposite or whorled, entire, rarely alternate; flowers perfect; calyx and corolla with 5 parts, these generally strongly reflexed; stamens 5, attached to the stigma, a corona of 5 hornlike appendages between corolla and stamens; pollen cohering in waxy masses (pollinia); pistils superior, 2, attached by the stigma; fruit a follicle; seeds with a tuft of hairs.

1a. Stems not twining, erect or decumbent; flowers in terminal and/or lateral clusters. **Asclepias**
1b. Stems twining; flowers lateral, solitary or clustered. (2)
2a. Corona lacking; corolla lobes hoodlike. **Astephanus**
2b. Corona present, double—the outer a mere ring, the inner of 5 fleshy, hoodlike scales. **Funastrum**

Asclepias L.   Milkweed   (Contributed by Floyd H. Coles)
1a. Corolla lobes erect or spreading, not reflexed. (2)
1b. Corolla lobes distinctly reflexed in mature flowers. (3)
2a. Leaves lanceolate; stems more than 20 cm tall; flowers large, sepals 3.1–5.2 mm long; growing in flats, desert swales, and on sandy or rocky hillsides with pinyon, juniper, or oak, widely distributed. **A. asperula** (Decne.) Woodson
2b. Leaves narrowly linear; stems less than 20 cm tall; flowers small, sepals 3–4 mm long; San Juan and Washington cos. **A. rusbyi** (Vail) Woodson
3a. Corolla yellow, green, or white. (4)

Asclepiadaceae

- 3b. Corolla red, pink, or purple. (11)
- 4a. Hoods yellow, green, or white; stem erect or suberect; leaves linear or lanceolate. (5)
- 4b. Hoods purple; stem decumbent; leaves ovate; eastern and southern Utah, frequently on clay soils. **A. cryptoceras** S. Wats.
- 5a. Leaves narrowly linear, whorled. (6)
- 5b. Leaves lanceolate, opposite to alternate. (7)
- 6a. Leaves less than 4 mm wide, margins more or less revolute; mostly in sandy soils, widely distributed; very poisonous to stock. **A. subverticillata** (Gray) Vail
- 6b. Leaves more than 4 mm wide, margins not revolute; roadsides and waste places, northern Utah. **A. fascicularis** Decne. ex DC.
- 7a. Stems less than 25 cm tall; leaves conspicuously white-ciliate with curved hairs. (8)
- 7b. Stems more than 25 cm tall; leaves not conspicuously white-ciliate. (9)
- 8a. Leaves inconspicuously pilosulose beneath, sessile or subsessile, narrowly lanceolate; hood twice as long as anthers; dry, gravelly hills, San Juan Co. **A. involucrata** Engelm.
- 8b. Leaves conspicuously tomentulose beneath, short-petiolate, broadly ovate to ovate-lanceolate; hoods about same length as anthers; sandy soil, southern and southeastern Utah. **A. macrosperma** Eastw.
- 9a. Leaves lanceolate; hoods truncate. (10)
- 9b. Leaves broadly ovate; hoods not truncate; sandy soils, southeastern Utah. **A. latifolia** Raf.
- 10a. Leaves more than 3 cm wide, broadly ovate to ovate-lanceolate; mostly in washes, Washington Co. **A. erosa** Torr.
- 10b. Leaves less than 3 cm wide, lanceolate to linear-lanceolate; many habitats, eastern Utah; Utah's most poisonous milkweed. **A. labriformis** M. E. Jones
- 11a. Leaves narrowly linear or oblong; inflorescences terminal or subterminal (axillary in **A. curassavica**). (12)
- 11b. Leaves ovate to lanceolate; inflorescences terminal or lateral. (14)
- 12a. Corolla red to orange; stems more than 30 cm tall. (13)
- 12b. Corolla purple; stems less than 20 cm tall; sandy and gravelly soil, San Juan Co. **A. cutleri** Woodson
- 13a. Stems villous or hirsute; leaves more or less alternate, sessile or subsessile; pods pubescent; mostly in sandy soils, southern Utah. **A. tuberosa** L.
- 13b. Stems glabrous; leaves opposite, short-petiolate, thin; pods glabrous; greenhouse ornamental. **A. curassavica** L.
- 14a. Leaves ovate; stems less than 20 cm tall; clay hills, rocky slopes, and sandy soils, Emery and Wayne cos. **A. ruthiae** Maguire
- 14b. Leaves lanceolate; stems more than 20 cm tall. (15)

Subclass DICOTYLEDONEAE

15a. Plants glabrous; leaves narrowly lanceolate. (16)
15b. Plants distinctly pubescent; leaves broadly lanceolate. (17)
16a. Flowers orange to scarlet; greenhouse ornamental. **A. curassavica** L.
16b. Flowers purplish; marshes, northern Utah. **A. incarnata** L.
17a. Follicles bearing soft subulate processes; hoods attenuate-acuminate; widespread, weed of waste places and cultivated land. **A. speciosa** Torr.
17b. Follicles without subulate processes; hoods not attenuate-acuminate; widely distributed, mostly in pinyon-juniper association. **A. hallii** Gray

Astephanus R. Br.

Plants suffrutescent, the stems twining or spreading; leaves opposite, pale gray-green; umbels peduncled, few- to many-flowered; calyx lobes narrowly lanceolate, 1.5 mm long; corolla yellowish, 2–2.5 mm long; follicles terete, long-acuminate; drifting sand hills near Saint George, Washington Co. **A. utahensis** Engelm.

Funastrum Fourn.

Plants twining, woody; leaves opposite, 1.5–5 cm long; umbels 5- to many-flowered; corolla purplish, 5–6 mm long; follicles long-attenuate, 7–10 cm long; southern Utah. **F. heterophyllum** (Engelm.) Standl.

BALSAMINACEAE — BALSAM FAMILY

Succulent herbs; leaves simple, alternate, opposite or whorled; flowers perfect, very irregular; sepals 3—the lateral 2 small and green, the third large and petaloid, prolonged backward as a spur; petals 5 or 3; stamens 5; ovary superior, 5-loculed; fruit an explosive capsule.

Impatiens L. Snapweed

A single genus and species treated, with the characteristics of the family; common greenhouse and house plants. **I. balsamina** L.

BEGONIACEAE — BEGONIA FAMILY

Perennial, succulent herbs (annual if planted out); leaves alternate; flowers monoecious, staminate with 2 petaloid sepals and 2 petals; stamens numerous; perianth of pistillate flowers with 2–5 or more segments; ovary inferior, 2- to 3-loculed; fruit a capsule.

Begonia L.
1a. Roots fibrous; potted indoor or bedding plant. **B. semperflorens** Link & Otto
1b. Roots tuberous; garden or rarely indoor plant grown for large double flowers. Tuberous Begonia. **B. tuberhybrida** Voss

BERBERIDACEAE — BARBERRY FAMILY

Shrubs or undershrubs; leaves alternate, simple or compound; flowers perfect, hypogynous, regular, yellow or whitish, solitary or racemose; sepals and petals imbricate, in several series (often in 3s); stamens opposite the petals; anthers opening by pores; pistil a single carpel; ovary 1-loculed; fruit a few-seeded berry.

1a. Leaves simple, at least some (occasionally with 3–5 leaflets in Mahoberberis). (2)
1b. Leaves compound. (3)
2a. Branches armed. **Berberis**
2b. Branches unarmed, leaf margins with spines. **Mahoberberis**
3a. Leaves 1-pinnately compound; leaflets spiny toothed. **Mahonia**
3b. Leaves more than 1-pinnately compound; at least ultimate leaflets entire. **Nandina**

Berberis L. Barberry

1a. Plants evergreen or half-evergreen; spines 3-parted; fruit blue to black. (2)
1b. Plants deciduous; spines simple to 3-parted; fruit red to purple. (3)
2a. Leaves slightly reticulate beneath; flowers 2–6; fruit black, slightly bloomy; stigma sessile; cultivated. Sargent Barberry. **B. sargentiana** Schneid.
2b. Leaves with veinlets not apparent; flowers about 15; fruit black; style distinct; cultivated. Juliana Barberry. **B. julianae** Schneid.
3a. Flowers 1–3, borne in clusters along the stem; spines simple; cultivated. Japanese Barberry. **B. thunbergii** DC.
3b. Flowers in many-flowered racemes to 5 cm long; spines usually 3-parted; cultivated. Common Barberry. **B. vulgaris** L.

Mahoberberis Schneid.

Unarmed, half-evergreen shrubs to 2 m tall; leaves simple to pinnately compound, serrate to spiny; not known to bloom, occasionally cultivated; a hybrid between **Mahonia aquifolium** and **Berberis vulgaris**. **M. neubertii** Schneid.

Mahonia Nutt.

1a. Leaves glaucous to blue glaucous, stiff-spiny; berries yellowish to red-purple at maturity; southern Utah. Fremont Mahonia. **M. fremontii** Torr.
1b. Leaves green to glaucous; berries purple to bluish at maturity. (2)
2a. Plants less than 30 cm tall; leaflets 3–7, minutely papillose beneath; common in canyons. Oregon Grape. **M. repens** G. Don
2b. Plants usually much more than 30 cm tall. (3)
3a. Plants to 2 m tall; leaflets 5–9, glossy green above, light green be-

Subclass DICOTYLEDONEAE

neath; racemes clustered, 5–7.5 cm long; fruit blue, glaucous; cultivated. Holly Mahonia. **M. aquifolium** Nutt.

3b. Plants to 4 m tall; leaflets 9–15, to 12.5 cm long, yellow-green above, glaucescent beneath, rather rigid; racemes 7.5–15 cm long, upright, densely flowered, with acute bracts 1.5 mm long; fruit 10 mm long, blue-black; cultivated. **M. bealei** Carr.

Nandina Thunb.

Small evergreen shrubs to 2.5 m tall; leaves 2- to 3-pinnately compound; ultimate leaflets entire; flowers small, in terminal racemes; sepals and petals many, white; stamens 6; berries red; cultivated. **N. domestica** Thunb.

BETULACEAE — BIRCH FAMILY

Trees or shrubs; leaves alternate, simple; flowers in catkins, each scale subtending 1 or 2 flowers; perianth none or minute; staminate flowers drooping at anthesis, usually much longer than the pistillate; stamens 2–10, fastened directly to the subtending bract; pistillate catkins ovoid to short-cylindric; ovary inferior, 2-celled and 4-ovuled, with 2 styles, maturing into a 1-seeded nut or nutlet, or small samara.

1a. Leaves 1-serrate, rarely obscurely 2-serrate; pistillate catkin of several to many deciduous bracts. **Betula**
1b. Leaves 2-serrate; pistillate catkin of several persistent bracts, or a single nut subtended by a prominent involucre. (2)
2a. Plants shrubs, seldom over 2 m tall; fruit a large nut, enclosed in an involucre; cultivated. **Corylus**
2b. Plants trees or large shrubs, 3–8 m tall; fruit of several nutlets that fall from a persistent, conelike, pistillate catkin; native trees of stream banks. **Alnus**

Alnus Gaertn. Alder

Trees or shrubs, 3–8 m tall; fruit of several nutlets that fall from persistent pistillate catkins; common along streams at middle and higher elevations. Thin-leaved Alder. **A. tenuifolia** Nutt.

Betula L. Birch

1a. Plants shrubs, seldom over 2 m tall; leaves seldom over 2 cm long, rounded at apex; boggy meadows and stream sides at higher elevations. **B. glandulosa** Michx.
1b. Plants treelike shrubs, over 2 m tall; leaves 2–4 cm long, acute or acuminate at apex; stream sides and seeps at middle and lower elevations, common. River Birch. **B. occidentalis** Hook.

Corylus L. Hazelnut

Shrubs seldom over 3 m tall; fruit a large nut, enclosed in a foliose involucre; cultivated. Hazelnut. **C. avellana** L.

BIGNONIACEAE — BIGNONIA FAMILY

Trees, shrubs, or woody vines; leaves alternate or opposite, simple or compound; flowers perfect, irregular, mostly showy, in terminal or axillary thyrse, panicles or racemes; calyx 5-toothed; corolla 5-lobed; fertile stamens usually 4 (5); ovary superior, 1- or 2-loculed; styles 1; stigma bilobed; fruit a capsule.

1a. Leaves compound; woody vines. **Campsis**
1b. Leaves simple; trees or shrubs. (2)
2a. Fertile stamens 2; leaves ovate; normally trees. **Catalpa**
2b. Fertile stamens 4; leaves linear; shrubs. **Chilopsis**

Campsis Lour.   Trumpet-Vine

Woody vines climbing by aerial roots; leaves opposite; leaflets 9–11, serrate; flowers orange and scarlet; corolla tubular-funnelform, 5–10 cm long, 2.5–5 cm broad. **C. radicans** (L.) Seem.

Catalpa Scop.   Catalpa
1a. Leaves glabrous beneath. **C. bungei** C. A. Meyer
1b. Leaves pubescent beneath. (2)
2a. Apex of leaves long-acuminate; panicles few-flowered. Western Catalpa. **C. speciosa** Warder
2b. Apex of leaves abruptly acuminate; panicles many-flowered. **C. bignonioides** Walt.

Chilopsis G. Don

Deciduous shrubs or small trees; flowers in terminal racemes, white tinged with purple, to 4 cm long; fruit slender, 1.5–3 dm long; native to southwestern deserts. Desert Willow. **C. linearis** (Cav.) Sweet

BORAGINACEAE — BORAGE FAMILY
(Contributed by Larry C. Higgins)

Annual, biennial, or perennial herbs, usually bristly-hairy; leaves simple, alternate or sometimes opposite or whorled, entire, and pubescent, hispid, or setose; flowers perfect, regular, solitary, or cymose; cymes glomerate-racemose, frequently 1-sided and coiled (scorpioid), usually with bracts between, beside, or opposite to flowers; calyx usually 5-lobed or 5-parted, usually persistent, the lobes valvate; corolla united, 5-lobed, commonly crested or with appendages in the throat; stamens 5, alternate with corolla lobes, borne upon the corolla tube; ovary superior, 2-carpelled, usually 4-ovuled, often deeply 4-lobed, appearing 4-loculed, becoming tough or bony at maturity; fruit commonly breaking up into 4 1-seeded nutlets; style simple or 2-cleft, arising at apex of fruit, or from between nutlets, or from a prolongation of the receptacle (gynobase); endosperm absent or scarce; embryo straight or curved; fruit often necessary for accurate determination.

Subclass DICOTYLEDONEAE

1a. Style 2-cleft; stigmas 2, distinct; flowers solitary or clustered, in the stem forks; plants low-growing perennials; leaves hispid, pungent-pubescent. **Coldenia**
1b. Style simple; stigmas united. (2)
2a. Style arising from pericarp at apex of fruit, falling away with the nutlets; stigma annular-peltate, bearing on top a conic or cylindric, simple or lobed appendage. **Heliotropium**
2b. Style arising from between lobes of fruit (nutlets) and attached to the receptacle or gynobase; stigma capitate, with no appendages. (3)
3a. Nutlets with uncinate or barbed prickles on back, margins, or apex. (4)
3b. Nutlets without hooked or barbed prickles. (7)
4a. Nutlets without definite margins, subglobose, dorsal surface quite uniformly covered with barbed prickles; widely distributed, foul-smelling weed. **Cynoglossum**
4b. Nutlets with a definite margin, the prickles confined to this (back may be muricate or tuberculate). (5)
5a. Nutlets stellate-spreading, attached at the apical (radicle) end, armed with hooked appendages; small, slender annuals. **Pectocarya**
5b. Nutlets erect, incurved, or weakly divergent, attached at or below the middle (i.e., toward the cotyledon end). (6)
6a. Plants annual; pedicels erect or nearly so; styles surpassing the nutlets; subulate gynobase about as long as nutlets. **Lappula**
6b. Plants biennial or perennial; pedicels reflexed in fruit; pyramidal gynobase about one-half as long as nutlets. **Hackelia**
7a. Corolla irregular, the upper lobes usually longer than lower ones; stamens not all equal in length. **Echium**
7b. Corolla regular or nearly so. (8)
8a. Calyx much enlarged in fruit, becoming conspicuously veiny, folded, or flattened; stems procumbent, angled, with stiff retrorse bristles on angles. **Asperugo**
8b. Calyx only slightly if at all enlarged in fruit, not becoming veiny, folded, or flattened; stems various, but not as above. (9)
9a. Nutlet attachment surrounded by a swollen ring, leaving a distinct pit on the gynobase; fields and waste places. (10)
9b. Nutlet attachment neither surrounded by a rim nor leaving a pit. (11)
10a. Stamens with a dorsal appendage, densely crowded around style; corolla rotate. **Borago**
10b. Stamens without a dorsal appendage, included within the tubular corolla; fields and waste places. **Anchusa**
11a. Corolla normally blue (aberrant white-flowered plants are found occasionally), or reddish in the bud state. (12)

Boraginaceae

11b. Corolla white, greenish white, yellow, or orange. (14)
12a. Dorsal face of nutlets oblique, encircled by an upturned flange or rim, this often irregularly toothed; depressed-pulvinate plants, seldom over 7 cm tall; alpine areas of Utah. **Eritrichium**
12b. Dorsal face of nutlets, if present, not encircled by an upturned flange or rim; plants not depressed-pulvinate, usually over 7 cm tall; most species below alpine areas in Utah. (13)
13a. Corolla lobes convolute in the bud; nutlets basally attached to a flat gynobase; corolla salverform. **Myosotis**
13b. Corolla lobes imbricate in the bud; nutlets obliquely attached to a convex gynobase; corolla with a tube and usually a campanulate throat, not salverform. **Mertensia**
14a. Nutlets attached above the base along a usually open, generally basally forked ventral groove or slit, or by a triangular opening in the pericarp. **Cryptantha**
14b. Nutlets lacking a distinct ventral groove or opening in the pericarp, this usually replaced by an elevated ventral keel. (15)
15a. Plants perennial; nutlets attached by a broad, smooth, quite basal, noncaruncular attachment; nutlets ovoid, smooth, and shiny; corolla usually yellow or orange. **Lithospermum**
15b. Plants annual; nutlets attached by a caruncular scar borne upon or at basal end of ventral keel, the attachment usually lateral or suprabasal; nutlets usually rough. (16)
16a. Corolla white; cotyledons entire. **Plagiobothrys**
16b. Corolla orange or yellow, the tube definitely longer than the calyx; cotyledons 2-lobed. **Amsinckia**

Amsinckia Lehm.   Fiddleneck

1a. Corolla tube 20-nerved below insertion of stamens; calyx lobes unequal in width and reduced in numbers (2, 3, or 4) by fusion; nutlets tesselate; one collection in Davis Co., all others in Washington Co. **A. tesselata** Gray
1b. Corolla tube 10-nerved below insertion of stamens; calyx lobes 5, distinct. (2)
2a. Corolla orange-yellow, 7–20 mm long, well exserted beyond calyx; plants usually green; stems hirsute-bristly, but with few or no fine appressed hairs; mostly Washington Co., but perhaps in western tier of counties. **A. intermedia** Fisch. & Mey.
2b. Corolla pale yellow, 4–7 mm long, little or not at all exserted beyond calyx lobes; leaves pubescent, with appressed or ascending hairs; northern Utah. **A. retrorsa** Suksd.

Anchusa L.   Bugloss

Perennial herbs, from taproot, up to 100 cm high; stems erect, but branching from near base; plants coarsely hirsute, hairs often pustulate

47

Subclass DICOTYLEDONEAE

at base; basal leaves 8–20 cm long, oblanceolate; stem leaves lanceolate; inflorescence a panicled, scorpioid raceme; calyx 5–8 mm long, with lanceolate to narrowly triangular lobes about as long as the tube; corolla about 10 mm long, dark blue; nutlets 2–3 mm long, rugose or granulate, inserted by their bases on a flat gynobase; roadsides and waste places, native to Eurasia but naturalized in the eastern United States and as far west as Utah. **A. officinalis** L.

Asperugo L.   Catchweed

Procumbent annuals; stems slender, 2–6 dm long, diffusely branched, with short-hispid retrorse hairs; leaves 1–4 cm long, obovate to oblanceolate, scabrous, obtuse to acutish at apex; fruiting calyx 8–15 mm wide; corolla small, 2–3 mm long, blue, purple, or purplish red; nutlets obliquely ovoid, about 4 mm long, granulate-tuberculate; introduced European plant of waste places, found presently in 5 counties near Great Salt Lake. **A. procumbens** L.

Borago (Tourn.) L.   Borage

Annuals (?) with erect stem 5–8 dm tall, with ascending or spreading branches; leaves entire, oblong to obovate, 5–11 cm long, rounded to acute at apex, upper clasping, lower narrowed to winged petiole; pedicels spreading or recurving, 2–5 cm long; calyx lobes linear-lanceolate, 7–10 mm long; corolla 15–20 mm broad, bright blue; anther beak dark purple, about 6–7 mm long; nutlets 4 mm long, ovoid, erect, attached by their bases to the flat receptacle; scar or attachment large, concave; European plant, escaping from gardens, thus (sparingly) naturalized in Utah (Cache Co.). **B. officinalis** L.

Coldenia L.

1a. Fruit nearly globose, unlobed, breaking apart at maturity into quarter sections, each quarter forming a nutlet; leaves ovate to elliptic, white-tomentose, obscurely veined; limited and rare in Washington Co. **C. canescens** DC.
1b. Fruit deeply 4-lobed, the lobes joined only by their inner angle, each lobe forming a nutlet; leaves not tomentose. (2)
2a. Plants perennial; leaves not evidently nerved, lanceolate to linear, usually very pungently setose; base of petiole expanded, indurate, usually villous; flowers solitary in leaf axils; nutlets finely warty, ovate; usually on sand dunes and dry slopes, southern Utah, mostly within the Colorado drainage, but including Washington Co. **C. hispidissima** (T. & G.) Gray
2b. Plants annual; leaves with evidently impressed nerves, ovate or obovate to nearly orbicular; base of petiole not expanded or indurate or villous; flowers in dense clusters at forks of stem; nutlets smooth or granulate; in sandy or alkaline soil, northern Utah and the western tier of counties. **C. nuttallii** Hook.

Boraginaceae

Cryptantha Lehm. Catseye
- 1a. Plants annual; stems slender; distribution various. KEY I, p. 49.
- 1b. Plants perennial or at least biennial; stems coarse; distribution various. KEY II, p. 52.

KEY I: Plants annual; stems slender.

- 1a. Nutlets with margins decidedly winged or knifelike. (2)
- 1b. Nutlets with margins rounded or angled, but never winged or knifelike. (6)
- 2a. Pedicels usually evident, slender, 1–4 mm long; nutlets heteromorphous; sandy flats and rocky ridges of the Lower Sonoran Zone, southwestern Utah. C. racemosa (S. Wats.) Greene
- 2b. Pedicels obscure or none, less than 1 mm long. (3)
- 3a. Nutlets heteromorphous, the odd nutlet abaxial; among rocks and shrubs, southwestern Utah. **C. inaequata** Johnst.
- 3b. Nutlets homomorphous or, if slightly heteromorphous, the odd nutlet axial. (4)
- 4a. Nutlets 1 or rarely 2; calyx obliquely conic at base; corolla conspicuous; flowers scented; desert washes and ridges, in sandy or rocky soil, southwestern Utah. C. utahensis (Gray) Greene
- 4b. Nutlets 4; calyx symmetric; corolla inconspicuous. (5)
- 5a. Nutlets heteromorphous, the axial one wingless; sandy or gravelly places in Upper and Lower Sonoran zones, southern Utah. **C. pterocarya** (Torr.) Greene var. **pterocarya**
- 5b. Nutlets homomorphous, all winged; sandy or gravelly deserts, Washington and Kane cos. C. pterocarya (Torr.) Green var. **cycloptera** (Greene) Macbr.
- 6a. Nutlets all smooth. (7)
- 6b. Nutlets all rough, or at least some of them so. (14)
- 7a. Hairs on calyx uncinate or decidedly arcuate; southwestern Utah. **C. flaccida** (Dougl.) Greene
- 7b. Hairs on calyx straight. (8)
- 8a. Nutlets with eccentric groove; flowers in 2-ranked naked spikes; sandy to rocky soils, northern Utah. C. affinis (Gray) Greene
- 8b. Nutlets with centrally located groove. (9)
- 9a. Nutlets broadly ovate. (10)
- 9b. Nutlets oblong-ovate to lanceolate. (12)
- 10a. Spikes usually geminate; inflorescence projected above leafy mass of plant and well defined; mainly in transition zone, on open slopes or partial shade, western half of Utah, as far south as Washington Co. **C. torreyana** (Gray) Greene
- 10b. Spikes usually solitary, not sharply defined from leafy peduncular stems. (11)

Subclass DICOTYLEDONEAE

11a. Nutlets homomorphous; on dry slopes and ridges of open pine forests and sagebrush flats, northwestern quarter of Utah. **C. ambigua** (Gray) Greene

11b. Nutlets slightly heteromorphous, odd (axial) nutlet somewhat larger; dry, sandy or gravelly soil, mostly in mountains, northeastern Utah; closely related to **C. kelseyana** and **C. ambigua,** and intergrading with both. **C. pattersonii** (Gray) Greene

12a. Style reaching one-fourth to three-fourths the height of nutlets; calyx densely appressed hispid-villous, commonly lacking conspicuous spreading bristles; dry, usually brushy, slopes and ridges, in Utah south and east of a line drawn from northwest corner of Juab Co. diagonally to southwest corner of Wyoming. **C. gracilis** Osterh.

12b. Style almost reaching nutlet tips or surpassing them. (13)

13a. Margins of nutlets acute, at least above the middle; sandy or rocky slopes and plains of arid transition zone, central Utah from east to west border. **C. watsoni** (Gray) Greene

13b. Margins of nutlets rounded or obtuse; sagebrush plains or pinyon-juniper community, northeastern and southwestern Utah. **C. fendleri** (Gray) Greene

14a. Nutlets decidedly heteromorphous. (15)
14b. Nutlets decidedly homomorphous. (22)

15a. Mature calyces strongly appressed to the flattened rachis, decidedly gibbous on axial side, persistent; sandy to gravelly deserts, southwestern Utah. **C. dumetorum** Greene

15b. Mature calyces somewhat spreading, not at all gibbous. (16)

16a. Odd nutlet abaxial, surpassed by style. (17)
16b. Odd nutlet axial, the style surpassed by, or occasionally reaching to, nutlet tips. (20)

17a. Spikes bracteate throughout; calyx persistent; dry, sandy soils of the Lower Sonoran Zone, southern Utah counties, west of Colorado River. **C. micrantha** (Torr.) Johnst.

17b. Spikes naked or nearly so; calyx deciduous. (18)

18a. Pedicels slender, 1–4 mm long; southwestern Utah. **C. racemosa** (S. Wats.) Greene

18b. Pedicels stout and obscure, less than 1 mm long. (19)

19a. Nutlets 1.3–1.7 mm long; calyx 2–3 mm long; southwestern Utah. **C. inaequata** Johnst.

19b. Nutlets about 1 mm long; sandy and gravelly washes, Lower Sonoran Zone, just over border in both Arizona and Nevada, to be sought in Beaver Dam Wash and valleys of the Virgin River. **C. augustifolia** (Torr.) Greene

20a. Odd nutlet spinulose-muricate; calyx lobes conspicuously thickened;

Boraginaceae

seedy and gravelly places, Colorado River drainage of southeastern Utah. **C. crassisepala** (T. & G.) Greene var. **elachantha** Johnst.
20b. Odd nutlet more or less granulate; calyx lobes moderately thickened. (21)
21a. Nutlets ovate, smoothish or sparsely tuberculate, odd one about 1.9 mm long; northeastern Utah. **C. pattersonii** (Gray) Greene
21b. Nutlets lanceolate or narrowly ovate, coarsely tuberculate, odd one 2–2.6 mm long; on sandy plains or rocky hillsides, Tooele Co. and the eastern tier of counties. **C. kelseyana** Greene
22a. Calyx circumscissile; sandy to gravelly soils, from Sonoran to arid transition zone, western and southern tiers of counties. **C. circumscissa** (H. & A.) Johnst.
22b. Calyx not circumscissile. (23)
23a. Style surpassing the nutlets; southwestern Utah. **C. micrantha** (Torr.) Johnst.
23b. Style not surpassing the nutlets, but nearly reaching tips of nutlets. (24)
24a. Corolla conspicuous, 2–5 mm broad. (25)
24b. Corolla inconspicuous, 0.5–2 mm broad. (26)
25a. Abaxial nutlet developing; gynobase one-third to one-half the height of nutlet; sandy to gravelly slopes and ridges of Lower Sonoran Zone, southwesternmost Utah. **C. decipiens** (M. E. Jones) Heller
25b. Axial nutlet developing; gynobase about two-thirds the height of nutlet; flowers scented; southwestern Utah. **C. utahensis** (Gray) Greene
26a. Ovules 2; nutlet and calyx bent; sandy or gravelly slopes or wash bottoms, southern half of Utah. **C. recurvata** Coville
26b. Ovules 4; nutlet and calyx straight. (27)
27a. Nutlets usually solitary, abaxial; gynobase one-third to one-half the height of nutlet; southwestern Utah. **C. decipiens** (M. E. Jones) Heller
27b. Nutlets usually 4. (28)
28a. Nutlets decidedly ovate, with low, rounded tuberculations; northwestern Utah. **C. ambigua** (Gray) Greene
28b. Nutlets more or less lanceolate. (29)
29a. Stems spreading-hirsute; sandy desert washes or slopes of the Lower Sonoran Zone, southern Utah. **C. barbigera** (Gray) Greene
29b. Stems strigose. (30)
30a. Nutlets verrucose or verrucose-muricate; sandy or gravelly slopes and washes, Lower Sonoran Zone, deserts of southern Utah. **C. nevadensis** Nels. & Kenn.
30b. Nutlets spinulose-muricate; dry sagebrush plains, usually in sandy soils, northern Utah. **C. scoparia** A. Nels.

Subclass DICOTYLEDONEAE

    Key ii. Plants perennial or biennial; stems coarse.

1a. Corolla tube elongate, obviously surpassing the calyx; flowers usually heterostylous. (2)
1b. Corolla tube short, scarcely if at all surpassing the calyx; flowers not heterostylous. (17)
2a. Nutlets smooth and shining. (3)
2b. Nutlets more or less roughened. (8)
3a. Corolla yellow. (4)
3b. Corolla white. (5)
4a. Inflorescence an elongate cylindric thyrse; nutlets lanceolate with acute margins, usually only 1 developing; sandy soils, eastern Utah. C. flava (A. Nels.) Payson
4b. Inflorescence a large terminal cluster with 1 or more remote, much smaller lateral clusters; nutlets broadly ovate, with winged margins, usually all 4 maturing; tolerant of soil types, central to western Utah. C. confertiflora (Greene) Payson
5a. Inflorescence capitate, 0.1–0.4 dm long; corolla limb 6–8 mm broad, tube only slightly surpassing the calyx; nutlets lanceolate; sandy or sandy-loam soil of transition zone, south central Utah. C. capitata (Eastw.) Johnst.
5b. Inflorescence elongate, 0.4–4.4 dm long; corolla limb 8–17 mm broad, the tube distinctly longer than the calyx (except in C. barnebyi); nutlets ovate. (6)
6a. Ventral surface of leaves glabrous; clay soils, Upper Sonoran Zone, southwestern Utah. C. semiglabra Barneby
6b. Ventral surface of leaves strigose or setose-hispid. (7)
7a. Corolla limb 13–17 mm broad, crests at base of tube absent; nutlets 3–3.5 mm long; in clay or clay-loam soil, known only from near head of Cottonwood Wash on the San Raphael Swell, Emery Co. C. johnstonii Higgins
7b. Corolla limb 8–11 mm broad, crests at base of tube conspicuous; nutlets 3.5–4.5 mm long; endemic on white, barren shale knolls, lower Uintah Co. C. barnebyi Johnst.
8a. Nutlets uniformly muricate or papillose (occasionally in C. jonesiana also with some inconspicuous ridges). (9)
8b. Nutlets more or less rugose or tuberculate, or sometimes with a few inconspicuous murications. (11)
9a. Leaves spatulate, hispid with pustulate bristles; corolla 10–15 mm long, the fornices low and broad; endemic on barren clay hills of the San Rafael Swell, Emery Co. C. jonesiana (Payson) Payson
9b. Leaves oblanceolate, strigose with pustulate hairs small or lacking; corolla 7–13 mm long, the fornices elongate. (10)
10a. Murications on nutlet rounded; corolla 9–13 mm long; inflorescence narrow, white-setose at maturity; open slopes and ridges in sandy

Boraginaceae

or sandy clay-loam soils, southeastern quarter of Utah. **C. fulvocanescens** (S. Wats.) Payson var. **fulvocanescens**

10b. Murications on nutlet with 1 or 2 setose projections; corolla 7–9 mm long; inflorescence broader and usually yellowish-setose at maturity; usually on alkali- or salt-charged heavy clay soils, south central to southwestern Utah. **C. fulvocanescens** (S. Wats.) Payson var. **echinoides** (M. E. Jones) Higgins

11a. Ventral or inner surface of nutlets smooth or nearly so; open hills and ridges, on white or red shale in Emery, Uintah, and Duchesne cos. **C. rollinsii** Johnst.

11b. Ventral surface of nutlets distinctly roughened. (12)

12a. Leaves conspicuously pustulate ventrally; corolla tube 12–16 mm long; calyx segments 7–10 mm long in anthesis; open ridges and flats, in sandy to clay soil, east central Utah (Grand Co. only?). **C. longiflora** (A. Nels.) Payson

12b. Leaves sparsely if at all pustulate ventrally; corolla tube 5.5–12 mm long; calyx segments 3.5–7 mm long in anthesis. (13)

13a. Inflorescence 0.1–0.4 dm tall; corolla tube 10–12 mm long; margins of nutlets not in contact; plants less than 1.5 dm tall; sandy to even heavy clay soils on flats and open ridges, east central Utah. **C. paradoxa** (A. Nels.) Payson

13b. Inflorescence 0.5–3 dm tall; corolla 5–10 mm long; margins of nutlets in contact or nearly so; plants usually over 1.5 dm tall. (14)

14a. Scar of nutlets surrounded by an elevated margin but tightly closed; style 1–2 mm long; calyx 3.5–4 mm long in anthesis; pinyon-juniper community on sandy or clay soil, southern one-third of Utah. **C. bakeri** Payson

14b. Scar of nutlets open; style 3–8 mm long; calyx 4.5–7 mm long in anthesis. (15)

15a. Scar of nutlet conspicuously open and surrounded by a definite elevated margin; tolerant of a wide variety of soils, on open slopes and ridges, mostly in the pinyon-juniper community, almost throughout Utah. **C. flavoculata** (A. Nels.) Payson

15b. Scar of nutlet slightly open, without elevated or conspicuous margin. (16)

16a. Leaves linear-spatulate; nutlets sharply and deeply rugose; corolla tube 5.5–7 mm long, the fornices low and broad; sandy or sandy-loam soils of Upper Sonoran Zone, Emery, Grand, Wayne, and San Juan cos. **C. tenuis** (Eastw.) Payson

16b. Leaves obovate or broadly oblanceolate; nutlets with rounded ridges and tubercles; corolla tube 7–10 mm long, the fornices long-papillose; usually on heavy clay soils with **Atriplex** in east central Utah, as far south as Garfield Co., west side of the Colorado River. **C. wetherillii** (Eastw.) Payson

17a. Nutlets more or less roughened, muricate, rugose, or tuberculate, at least on dorsal surface. (22)

53

Subclass DICOTYLEDONEAE

17b. Nutlets not roughened, rugose, muricate, or tuberculate, but smooth on dorsal surface. (18)
18a. Fruit conical, ovoid, or lanceolate; nutlets in contact by their margins, or nearly so; style exceeding mature fruit 5–6 mm; corolla tube 5–7 mm long; Uintah Co. **C. barnebyi** Johnst.
18b. Fruit depressed-globular; nutlets not in contact by their margins; style exceeding mature fruit 1–3 mm; corolla tube 2.5–3 mm long. (19)
19a. Ventral surface of leaves strigose or setose; petioles ciliate margined; leaves tufted at base. (20)
19b. Ventral surface of leaves glabrous; petioles not ciliate margined; leaves not tufted at base; sandy or gravelly soil, in pinyon-juniper community of southeastern Utah. **C. jamesii** (Torr.) Payson var. **pustulosa** (Rydb.) Harringt.
20a. Stems branched from base as well as from above; sandy soils, south central half of Utah. **C. jamesii** (Torr.) Payson var. **disticha** (Eastw.) Payson
20b. Stems simple, not branched above base. (21)
21a. Stems 1–4.4 dm long, usually twice as long as basal tufted leaves; wide variety of soils, but mainly sands in southern Utah, west of the Colorado River. **C. jamesii** (Torr.) Payson var. **multicaulis** (Torr.) Payson
21b. Stems 0.2–0.9 dm long, usually not exceeding basal tuft of leaves; wide variety of soils, commonly in pinyon-juniper community, southern Utah, west of the Colorado River. **C. jamesii** (Torr.) Payson var. **setosa** (M. E. Jones) Johnst.
22a. Ventral surface of nutlets smooth, or nearly so. (23)
22b. Ventral surface of nutlets rugose or variously wrinkled. (26)
23a. Nutlets bordered by a conspicuous wing; robust plants, 5–10 dm tall, with long ebracteate spikes; sandy or gravelly soils of transition zone, central to southwestern Utah. **C. setosissima** (Gray) Payson
23b. Nutlets not conspicuously winged, but sometimes with an acute margin; plants usually shorter and caespitose. (24)
24a. Corolla tube 7–9 mm long; calyx 6–9 mm long in anthesis; central and northeastern Utah. **C. rollinsii** Johnst.
24b. Corolla tube 2–6 mm long; calyx 2.5–6 mm long in anthesis. (25)
25a. Nutlets distinctly muricate or tuberculate between the rugae and near margins; in gravelly-loam or clay soils, west central to northwestern Utah. **C. rugulosa** (Payson) Payson
25b. Nutlets scarcely or not at all muricate between the rugae; strictly erect, conspicuously hispid perennials; clay or shale of the transition zone, northeastern Utah. **C. stricta** (Osterh.) Payson
26a. Nutlets conspicuously muricate or, in **C. humilis**, also with a few irregular ridges. (27)

Boraginaceae

26b. Nutlets not exclusively muricate, but rugose or tuberculate. (32)
27a. Pubescence of leaves silky-strigose or strigillose, but not subtomentose or tomentose; heavy clay soils, Duchesne and Uintah cos. C. breviflora (Osterh.) Payson
27b. Pubescence of leaves distinctly subtomentose or tomentose, and also setose in C. humilis. (28)
28a. Leaves 2.5 cm long or longer; calyx 3–5 mm long in anthesis; corolla tube 3–5 mm long; plants 0.4–2.5 dm tall; widely distributed. (29)
28b. Leaves 0.5–2.5 cm long; calyx 2–2.5 mm long in anthesis; corolla tube 1.8–2.2 mm long; plants 0.3–1 dm tall; open slopes and ridges in gravelly-loam soil, known only from southwestern Millard Co. C. compacta Higgins
29a. Leaves densely strigose as well as tomentose; calyx setose and subtomentose. (30)
29b. Leaves strigose and setose but not conspicuously tomentose; calyx evidently setose. (31)
30a. Style scarcely exceeding mature nutlets; inflorescence somewhat open at maturity; open sandy or gravelly slopes and ridges, northwestern quarter of Utah. C. humilis (Gray) Payson var. shantzii (Tidestr.) Higgins
30b. Style exceeding mature nutlets by 0.5–1.5 mm; inflorescence cylindric and congested in fruit; gravelly-loam or clay soils, pinyon-juniper community, southwestern quarter of Utah, in Great Basin drainage. C. humilis (Gray) Payson var. ovina (Payson) Higgins
31a. Style exceeding mature nutlets by 1–1.5 mm; inflorescence open and broad; plants loosely tufted; gravelly soil or talus slopes, pinyon-juniper community, west central Utah. C. humilis (Gray) Payson var. commixta (Macbr.) Higgins
31b. Style not or only slightly exceeding mature nutlets; inflorescence congested even in fruit; plants densely caespitose; mostly in heavy clay, occasionally in sandy loam, eastern Utah. C. humilis (Gray) Payson var. nana (Eastw.) Higgins
32a. Nutlet scar open some distance above base. (33)
32b. Nutlet scar closed or nearly so, without a conspicuous triangular opening toward base. (40)
33a. Scar of nutlet rather constricted some distance below middle of open portion. (34)
33b. Scar of nutlet triangular, not constricted below middle. (35)
34a. Elevated margin of nutlet scar definitely limited; pustules present on both leaf surfaces; clay or clay-loam soils, east central Utah. C. mensana (M. E. Jones) Payson
34b. Elevated margin of nutlet scar indefinitely limited; pustules present on only dorsal surface of leaf; sandy soil or rocky ledges and

Subclass DICOTYLEDONEAE

slopes, pinyon-juniper community, Wayne and San Juan cos. **C. osterhoutii** (Payson) Payson

35a. Scar of nutlet surrounded by a somewhat elevated margin. (36)
35b. Scar of nutlet with no evidence of an elevated surrounding margin. (37)
36a. Cymules elongated and so inflorescence broad; biennials or short-lived perennials; nutlets usually with an evident dorsal ridge; on clay or clay-loam soils, Lower Sonoran Zone of southwestern Utah. **C. virginensis** (M. E. Jones) Payson
36b. Cymules not elongated and thus the inflorescence narrow; long-lived, caespitose perennials; nutlets with slight or no dorsal ridge; sandy or gravelly soil, usually in transition zone, south central to southwestern Utah. **C. abata** Johnst.
37a. Style exceeding mature nutlets by 1.6 mm or more; plants usually taller than 1.3 dm (see couplets 29–31 of this key for separation of varieties). **C. humilis** (Gray) Payson
37b. Style exceeding mature nutlets by no more than 0.5 mm; plants usually less than 1.3 dm tall. (38)
38a. Corolla tube 2–2.6 mm long; nutlets 2.3–3 mm long; endemic on red Wasatch formation near Red Canyon campground, southwestern Garfield Co. **C. ochroleuca** Higgins
38b. Corolla tube 3–4 mm long; nutlets 3–3.5 mm long. (39)
39a. Ventral surface of nutlets deeply rugose and tuberculate; south-central to southwestern Utah. **C. abata** Johnst.
39b. Ventral surface of nutlets indefinitely muricate; heavy clay soils, to be expected in Summit and/or Daggett cos. **C. caespitosa** (A. Nels.) Payson
40a. Upper leaf surface with 2 distinct kinds of hairs; hairs pustulate at base. (43)
40b. Upper leaf surface uniformly appressed-strigose and lacking pustulate hairs. (41)
41a. Nutlets sharply rugose and tuberculate, the scar closed and surrounded by an elevated margin; southeastern Utah. **C. bakeri** Payson
41b. Nutlets not sharply rugose or tuberculate, the scar not surrounded by an elevated margin. (42)
42a. Corolla tube 2–2.5 mm long; style exceeding nutlets by up to 1 mm; endemic to Garfield Co. **C. ochroleuca** Higgins
42b. Corolla tube 3.5 mm long or longer; style exceeding nutlets by more than 1 mm; nutlet scar closed, straight; heavy clay soils, pinyon-juniper community, northeastern Utah. **C. sericea** (Gray) Payson
43a. Mature calyx exceeding nutlets by 2–4 mm; inflorescence broad topped; heavy clay soils, Grand Co. **C. elata** (Eastw.) Payson
43b. Mature calyx exceeding nutlets by 4–8 mm. (44)

Boraginaceae

44a. Nutlets tuberculate, scarcely if at all rugose; nutlet scar straight, open, narrowly linear; endemic to Green River shale, Uintah Co. C. grahamii Johnst.

44b. Nutlets more or less rugose; nutlet scar triangular, open; widely distributed in Utah, but mostly in western tier of counties. C. humilis (Gray) Payson

Cynoglossum (Tourn.) L. Houndstongue

Biennial plants, villous-tomentose throughout, with stout, erect stems, leafy to the top, 4–5 dm tall; lower leaves oblong to oblong-lanceolate, slender-petiolate, 15–30 cm long, 2–7 cm wide; upper leaves lanceolate, acute or acuminate, sessile, or upper mostly clasping; racemes several to many, simple or branched; bracts few or none; calyx segments ovate-lanceolate, obtuse to acutish, 5–7 mm long in fruit; corolla reddish purple, the broad tube 3–5 mm long, limb 6–8 mm broad; nutlets 4, ascending on a pyramidal gynobase, flattish on the upper surface and margined, splitting away from the gynobase at maturity but hanging attached to the subulate style, glochidiate; native to Europe and Asia, widely naturalized in North America in neglected or disturbed places. C. officinale L.

Echium L. Vipers Bugloss

Biennial, possibly perennial, hispid herbaceous plants; leaves alternate, entire; flowers blue to violet-purple, rarely rose or white, in leafy-bracted, scorpioid, spikelike racemes; calyx 5-parted, about equaling corolla tube; corolla tubular-funnelform, irregular, usually 5-lobed, throat not appendaged; stamens unequal, odd stamen included, both other pairs surpassing the lower corolla lobe and slightly unequal; ovary 4-lobed, these separating in fruit; style 2-cleft at apex; nutlets erect, rugose, about 2 mm long and attached by their bases to a flat gynobase, the scar flat or somewhat concave, not leaving a pit; native to Europe, introduced and naturalized in eastern United States, spreading thence to the Rocky Mountains, Summit Co. E. vulgare L.

Eretrichium Schrad.

Low, depressed, cushionlike perennials; plants villous, often silvery looking, about 2–4 cm tall, exclusive of flowering branches; stems short, densely clothed with small, often imbricate, leaves; flowers few in a racemelike cluster terminating the slender flowering stem, to 7 cm tall, or inflorescence compact when sessile; calyx lobes linear, ascending, 1.5–3 mm long; corolla tube equaling calyx lobes, limb variable in size (1) 4–5 (7) mm broad, bright blue or rarely white, with puberulent crests in the throat; nutlets obliquely attached to the conical gynobase, smooth, the apex obliquely truncate with an entire or obscurely toothed margin; rocky ridges in high alpine areas, 10,000 to 13,000 ft., northern Utah. E. nanum (Vill.) Schrad. var. elongatum (Rydb.) Cronq.

Figs. 13-14. 13. **Amaranthus retroflexus**, x .21. 14. **Apocynum androsaemifolium**, x .21

Figs. 15-16. 15. **Cryptantha flava**, x .25. 16. **Eretrichium nanum** var. **elongatum**, x 123.

Boraginaceae

Hackelia Opiz in Bercht.   Stickseed

1a. Surface of nutlets more or less ridged, but without prickles; branches many-flowered; brushy slopes and edges of woods, the length of Utah through the central region. **H. floribunda** (Lehm.) Johnst.
1b. Surface of nutlets more or less prickly (may be ridged); branches fewer with fewer flowers. (2)
2a. Nutlets broadly ovate; basal leaves few; stems leafy above, these leaves not conspicuously reduced in size; corolla blue; usually on moist banks and slopes, north central Utah. **H. jessicae** (McGregor) Brand
2b. Nutlets narrowly ovate; basal leaves many; stem leaves few and reduced in size; corolla white; sandy or gravelly slopes and foothills, western half of Utah. **H. patens** (Nutt.)   Johnst.

Heliotropium L.

1a. Plants not succulent, hairy, never glaucous; fruit 2-lobed, each lobe splitting into 2 nutlets; stigma capped by a tuft of bristles; flowers fragrant; annual; sandy soils, southern two-thirds of Utah. **H. convolvulaceum** (Nutt.)   Gray
1b. Plants very succulent, glabrous, usually glaucous; fruit not lobed; stigma discoid, naked; perennial. (2)
2a. Fruit 2.5 mm wide; corolla 5–16 mm broad, at most only purplish tinged at the throat; alkaline or saline areas, northern three-fourths of Utah. **H. curassavicum** L. var. **obovatum** A. DC.
2b. Fruit 1.5–2 mm wide; corolla 3–5 (7) mm broad, usually becoming distinctly purple or purplish at the throat; southwesternmost Utah. **H. curassavicum** L. var. **oculatum** (Heller)   Johnst.

Lappula Moench.   Stickseed

1a. Nutlets with marginal prickles in at least 2 rows; native to Eurasia and naturalized widely in the United States and Canada, dry plains, hillsides, and waste places, as well as cultivated ground, Summit Co. **L. echinata** Gilib.
1b. Nutlets with marginal prickles definitely in a single row. (2)
2a. Marginal prickles distinct to their bases or nearly so, not confluent to form a cupulate structure; dry hillsides and valleys throughout Utah. **L. occidentalis** (S. Wats.) Greene var. **occidentalis**
2b. Marginal prickles confluent, forming a conspicuous, smooth, cupulate structure on the back of some or all of nutlets; wide occurrence diagonally across Utah from southwest to northeast. **L. occidentalis** (S. Wats.) Greene var. **cupulata** (Gray)   Higgins

Lithospermum L.   Puccoon, Stoneseed

1a. Annual; flowers white; nutlets densely tuberculate and dull; natu-

Subclass DICOTYLEDONEAE

       ralized from Europe, grassy hillsides and grain fields, north central Utah **L. arvense** L.
- 1b. Perennial; flowers greenish to yellow; nutlets white, smooth. (2)
- 2a. Corolla 10 mm or less in length, the tube about equaling calyx, greenish or pale yellow; nutlets 4–6 mm long; dry plains and hillsides, common in northern half of Utah, less common in southern half. **L. ruderale** Dougl. ex Lehm.
- 2b. Corolla 10 mm long or more, the tube typically exceeding calyx, yellow (except **L. incisum**, which has small, cleistogamous flowers late in the season). (3)
- 3a. Styles of all flowers about same length (homostylous); stamens all borne near top of corolla tube; corolla usually over 20 (30) mm long, its lobes toothed or with a fringe of hairs; later flowers cleistogamous, much smaller; with recurved pedicels in fruit; dry plains and slopes, widely distributed throughout Utah. **L. incisum** Lehm.
- 3b. Styles of flowers of 2 lengths (heterostylous); stamens borne either at about middle or near top of corolla tube; corolla not 20 mm long, its lobes entire or nearly so; smaller cleistogamous flowers absent; root thick, containing a purplish dye; hills, canyons, and mountain slopes, southern one-third of Utah. **L. multiflorum** Torr. ex. Gray

Mertensia Roth. Bluebell
- 1a. Plants usually with prominent lateral veins in the cauline leaves; stems usually 4 dm or more (1–17) tall; normally flowering in late spring and in the summer; mostly occurring in moist, shaded situations. (2)
- 1b. Plants usually lacking prominent lateral veins in the cauline leaves (with the exception of some specimens of **M. oblongifolia** var. **nevadensis**); stems usually less than 4 dm tall; normally flowering in early spring, later when growing in the mountains, but commonly as soon as the snow and temperatures permit; mostly in fairly open situations. (7)
- 2a. Limb of corolla longer than tube; leaves usually acuminate. (3)
- 2b. Limb of corolla shorter than, or subequal to, tube; leaves usually not acuminate. (6)
- 3a. Leaves pubescent, at least on one surface. (4)
- 3b. Leaves glabrous on both surfaces. (5)
- 4a. Calyx not accrescent, margins densely ciliate, backs pubescent or glabrous; southeastern Utah. **M. franciscana** Heller
- 4b. Calyx accrescent, margins not densely ciliate, backs glabrous; Sevier Co., Utah, thence west into White Pine Co., Nev. **M. arizonica** Greene var. **subnuda** (Macbr.) Williams
- 5a. Calyx campanulate, lobes shorter than the tube; on moist stream banks and shaded places, central to southwestern Utah. **M. arizonica** Greene var. **arizonica**

Boraginaceae

5b. Calyx not campanulate, lobes longer than the tube; moist slopes and bottomland and shaded areas, central and north central Utah. **M. arizonica** Greene var. **leonardii** (Rydb.) Johnst.

6a. Leaves pubescent, at least on one surface; southeastern Utah. **M. franciscana** Heller

6b. Leaves glabrous on both surfaces, though sometimes the upper surface papillate; foothills and mountains up to 12,000 ft., in northern and southwestern Utah. **M. ciliata** (James) G. Don

7a. Filaments attached in corolla tube, with anthers not projecting beyond throat but contained within tube; north central Utah. **M. brevistyla** S. Wats.

7b. Filaments attached near throat of corolla tube, with anthers projecting beyond throat and not contained within tube. (8)

8a. Limb of corolla subequal to, or longer than, tube. (9)

8b. Limb of corolla shorter than tube. (11)

9a. Leaves pubescent on both surfaces; Bald Mountain, Summit Co. **M. viridis** A. Nels. var. **cana** (Rydb.) Williams

9b. Leaves strigose above only, or glabrous on both surfaces. (10)

10a. Filaments shorter than anthers; calyx divided nearly to the base; style usually not reaching anthers; anthers straight; alpine areas of southeastern Utah. **M. viridis** A. Nels. var. **viridis**

10b. Filaments longer than anthers; calyx not divided to near the base; style usually reaching or surpassing anthers; anthers usually curved; typically in the mountains other than in alpine or subalpine areas, in northern, southeastern, and southwestern Utah. **M. fusiformis** Greene

11a. Tube of mature corolla only slightly longer than the limb; high mountains. (12)

11b. Tube of mature corolla typically markedly longer than the limb; plains and low hills. (15)

12a. Leaves pubescent above only, or glabrous on both surfaces. (13)

12b. Leaves pubescent on both surfaces. (14)

13a. Leaves glabrous on both surfaces; high Uinta Mountains, northeastern Utah. **M. viridis** A. Nels. var. **dilatata** (A. Nels.) Williams

13b. Leaves strigose to strigillose above; mountains, southeastern Utah. **M. viridis** A. Nels. var. **viridis**

14a. Leaves usually 1.5–3 cm long; stems ascending; leaves unilateral; Bald Mountain, Summit Co. **M. viridis** A. Nels. var. **cana** (Rydb.) L. O. Williams

14b. Leaves usually longer than 3 cm (basal 2–11, though usually 4–6 cm long); stems more erect; leaves not unilateral; Uinta Mountains. **M. bakeri** Greene

15a. Leaves glabrous on both surfaces, on upper surface sometimes pustulate; northwestern quarter of Utah. **M. oblongifolia** (Nutt.) G. Don var. **nevadensis** (A. Nels.) Williams

Subclass DICOTYLEDONEAE

15b. Leaves pubescent, at least on 1 surface. (16)

16a. Leaves pubescent above, glabrous below; northeastern Utah west to Salt Lake and Utah cos. **M. oblongifolia** (Nutt.) G. Don var. **oblongifolia**

16b. Leaves pubescent on both surfaces; Rich Co. **M. oblongifolia** (Nutt.) G. Don var. **amoena** (A. Nels.) Williams

Myosotis (Dill.) L.   Forget-me-not

Perennial herbs with slender rootstocks or stolons, herbage appressed-pubescent with straight pointed hairs; stems slender, decumbent or ascending, to 4 dm long, rooting at the lower nodes; leaves oblanceolate to oblong-lanceolate, with upper stem leaves sessile, the lower narrowed to a winged petiole, to 8 cm long, to 12 mm wide; racemes loosely many-flowered; fruiting pedicels longer than the calyx; calyx with straight appressed hairs, equal lobes shorter than the tube and ovate-triangular, more or less spreading in fruit; corolla blue with a yellow eye; native to Europe and Asia, well established in the United States, but sparingly naturalized in Utah, meadows and stream margins, vicinity of Logan, Cache Co. **M. scorpioides** L.

Pectocarya DC. ex Meisn.   Combseed

1a. Nutlets orbicular or nearly so, both body and very thin conspicuous wing beset with slender uncinate bristles; fine sandy areas, Washington Co. **P. setosa** Gray

1b. Nutlets oblong or linear, the body without uncinate bristles. (2)

2a. Nutlets heteromorphic, 1 of each divergent pair with a broad, somewhat incurved, uncinate-toothed wing, the other wingless or merely margined; sandy or gravelly slopes and plains, Washington Co. **P. heterocarpa** Johnst.

2b. Nutlets homomorphic, all 4-winged, -margined, or -toothed. (3)

3a. Margin of nutlet conspicuous, the teeth confluent at base; dry, gravelly slopes and benches, Washington Co. **P. platycarpa** Munz and Johnst.

3b. Margin of nutlet very narrow or wanting, the teeth nearly or quite distinct, subulate; nutlets strongly recurved; sandy or gravelly slopes or ridges, western Utah (?). **P. recurvata** Johnst.

Plagiobothrys Fisch. & Mey.   Popcornflower

1a. Plants glabrous or nearly so; nutlets attached basally or nearly so; in heavy, usually alkaline, soils, northern Utah. **P. leptocladus** (Greene) Johnst.

1b. Plants strigose or bristly-hairy; nutlets attached laterally or obliquely basal. (2)

2a. Nutlets checkered with broad, flattened, contiguous (mosaic?), pavementlike raised areas; plants erect, hispid, with terminal, bractless, scorpioid cymes; washes and rocky desert slopes, southwestern Utah. **P. jonesii** Gray

Cactaceae

2b. Nutlets not checkered, the back rugose, raised areas scattered or none. (3)
3a. Leaves charged with conspicuous purple dye, particularly at midribs and margins; calyx with a weakened ring which allows it to break loose (circumscissile); lobes short, strongly pressed together at maturity; sandy to gravelly slopes and plains, southwestern Utah. Bloodweed. **P. arizonicus** (Gray) Greene
3b. Leaves green, lacking conspicuous purple dye; calyx without weakened ring. (4)
4a. Basal leaves crowded into a rosette, none opposite; plants slender, erect, loosely branched, not producing flowers near the base; nutlets incurved, contracted at both ends, somewhat cruciform, transverse ridges very broad; grassy slopes and meadows, northern Utah. **P. tenellus** (Nutt.) Gray
4b. Basal leaves distinct, at least not in well-developed rosettes, lower leaves opposite; nutlets not incurved, but highly variable; moist soils in sandy or clay areas, western Utah, apparently mostly in mountainous areas. **P. scouleri** (H. & A.) Johnst.

CACTACEAE — CACTUS FAMILY

(Contributed by Dorde Wright)

Plants perennial; stems thick and succulent, flat or cylindric; spines or glochids, or both, borne in areoles; leaves none or vestigial, fleshy, caducous; flowers perfect, regular; perianth segments and stamens indefinitely numerous; ovary inferior, 1-loculed; fruit a dry or fleshy berry.

1a. Areoles with glochids and usually spines; young stems with small, fleshy leaves; stems composed of a series of flat or cylindric joints (opuntiae). **Opuntia**
1b. Areoles lacking glochids; stems without leaves or joints (cereae). (2)
2a. Flowers and spines borne in same areoles; stems ribbed (except in **Pediocactus**), tuberculate. (4)
2b. Flowers not from spiniferous areoles; stems tuberculate, not ribbed. (3)
3a. Central spine hooked, dark; tubercles not grooved; fruit bright red, elongated. **Mamillaria**
3b. Central spine not hooked; tubercles grooved on upper side in mature plants; flowers pink, rarely yellow; fruit dull red, green, or yellow, not elongated. **Coryphantha**
4a. Flowers from side of plant or below apex, with a tube; spines not hooked. **Echinocereus**
4b. Flowers from top of plant, lacking a tube. (5)
5a. Stem ribbed, may be somewhat tuberculate on ribs; spines hooked or curved (except in **E. johnsonii**). **Echinocactus**

Subclass DICOTYLEDONEAE

5b. Stem not ribbed; tubercles separate, not grooved; spines not curved or hooked. **Pediocactus**

Coryphantha Lem.

Spines not hooked; tubercles grooved on upper side in mature plants; flowers pink, rarely yellow; fruit dull red, green or yellow, not elongated. **C. vivipara** (Nutt.) Britt. & Rose

Echinocactus Link & Otto   Fishhook Cactus

1a. Spines, at least some, hooked or curved. (2)
1b. Spines not hooked or curved, red or yellow; plants to 25 cm high and 7.5 cm wide. **E. johnsonii** Parry ex Engelm.
2a. Plants large, to 1 m tall or more; wide central spine curved, cross-ribbed, red, stout. **E. acanthodes** Lem.
2b. Plants medium in size, to 45 cm tall and 10 cm wide; dwarf varieties smaller (two varieties are recognized: var. **spinosior**, with 1 hooked central spine and pink or purple flowers, or less often white flowers; var. **parviflorus**, with several hooked central spines and yellow or pink to purple flowers, less often white). **E. whipplei** Engelm. & Bigel.

Echinocereus Engelm.   Hedgehog Cactus

1a. Flowers red; plants forming large clumps; spines short, light colored. **E. triglochidiatus** Engelm.
1b. Flowers pink to purple; plants solitary or forming small clumps; spines 1 or some of them long. (2)
2a. Central spine 1, erect, straight, rigid. **E. fendleri** (Engelm.) Rumpler
2b. Central spines 2–6, may be curved, flattened, or deflexed. **E. engelmannii** (Parry) Rumpler

Mamillaria Haw.   Ball Cactus

Central spine hooked, dark, tubercles not grooved; fruit bright red, elongated. **M. tetrancistra** Engelm.

Opuntia Mill.   Prickly Pear

1a. Joints cylindric, tuberculate; spines sheathed; glochids minute (cylindropuntia, cholla). (2)
1b. Joints flattened (except in **O. fragilis**), not tuberculate; glochids numerous; spines sheathless (platyopuntia, prickly pear). (4)
2a. Fruit fleshy at maturity, persistent, yellow, tuberculate, with cup-like cavity at apex. **O. whipplei** Engelm. & Bigel.
2b. Fruit dry at maturity, deciduous. (3)
3a. Tubercles short, 1–2 times as long as broad; longer joints 10–15 cm long; distinct trunk in mature plant. **O. echinocarpa** Engelm. & Bigel.
3b. Tubercles elongated, 3 to several times as long as broad; longer

Callitrichaceae

joints 15–45 cm long; trunk none, or short. **O. acanthocarpa** Engelm. & Bigel.
4a. Fruit fleshy at maturity, persistent. (5)
4b. Fruit dry at maturity, deciduous. (7)
5a. Plants low, sprawling; joints obovate with elongated base, yellow-green, to 15 cm long, very prostrate and shriveled in winter; fruit very persistent. **O. compressa** Macbr.
5b. Plants more erect; joints large, obovate, areoles farther apart; fruit not as persistent; spines reddish brown, at least partly, the largest long and stout. (6)
6a. Joints bluish, flat. **O. covillei** Britt. & Rose
6b. Joints green, somewhat thick; central spines 1–4. **O. phaeacantha** Engelm. ex Gray
7a. Plants small, forming masses; joints small, to 5 cm long, turgid, very easily detached; flower small, relatively few petals. **O. fragilis** (Nutt.) Haw.
7b. Plants medium to large; joints 5–20 cm long, 4–10 cm broad. (8)
8a. Joints spineless, or almost so, but bearing glochids. (9)
8b. Joints bearing spines as well as glochids. (10)
9a. Joints bluish, heart shaped or spatulate; flowers pink with red filaments, white anthers and stigmas or, rarely, flowers white. **O. basilaris** Engelm. & Bigel.
9b. Joints green to bluish, obovate; flowers yellow or, less often, pink. **O. aurea** Baxter
10a. Plants medium in size, usually less than 15 cm long. (11)
10b. Plants larger; joints obovate, 12–20 cm long, stout, white at maturity, flattened and twisted. **O. nicholii** L. Benson
11a. Spines many; joints green. (12)
11b. Spines few, confined to upper portion of joint; joints lead colored. **O. erinacea** Engelm.
12a. Spines flattened in cross section, numerous, flexible, white at maturity; plants erect; joints elliptic-oblong; flowers pink to purple, or yellow; fruit very spiny. **O. erinacea** Engelm.
12b. Spines circular in cross section, gray, yellow, or red, deflexed, pungent, longer at margin of joint; plants sprawling; joints circular to obovate; flowers yellow or pink. **O. polyacantha** Haw.

Pediocactus Britt. & Rose

Stem not ribbed; tubercles separate, not grooved; spines not curved or hooked. **P. simpsonii** (Engelm.) Britt. & Rose

CALLITRICHACEAE — WATERSTARWORT FAMILY

Small, slender aquatics; leaves simple, opposite, entire, floating or submerged, less than 12 mm long; flowers inconspicuous, solitary,

Subclass DICOTYLEDONEAE

axillary; calyx and corolla lacking; bracts often 2; stamens 1; ovary superior, 4-loculed; styles 2.

Callitriche L.   Waterstarwort
1a. Fruit about as wide as high. **C. heterophylla** Pursh
1b. Fruit about two-thirds as wide as high (closely related to the above, which may not be distinct); shallow water, montane through alpine regions. **C. palustris** L.

CAMPANULACEAE — BELLFLOWER FAMILY

Annual or perennial herbs; leaves alternate; flowers regular or irregular, perfect, the petals united, epigynous; petals, sepals, and stamens 5, the latter alternate with corolla lobes; styles 1; ovary 1- to 5-loculed, with axile or parietal placentation; fruit a capsule with numerous seeds.

1a. Corolla irregular. (2)
1b. Corolla regular. (4)
2a. Anthers distinct, all alike; filaments united; fruit a capsule. **Nemacladus**
2b. Anthers united into a tube, with 2 shorter than the others, the orifice of the tube thus oblique or appearing lateral; filaments united; fruit various. (3)
3a. Flowers sessile in the axils of foliaceous bracts, but appearing long stalked because of the much elongated, linear, or subulate hypanthium; capsule dehiscent by long slits on the sides. **Downingia**
3b. Flowers pedicellate; hypanthium and fruit fusiform to ellipsoid or globose; fruit indehiscent, or dehiscent by apical valves. **Lobelia**
4a. Plants perennial or biennial. (5)
4b. Plants annual. (6)
5a. Capsules opening on the sides by valves or pores; native or cultivated. **Campanula**
5b. Capsules opening at the top by valves; cultivated. **Platycodon**
6a. Corollas shallowly lobed; bracts opposite the flowers; capsule opening near the base. **Heterocodon**
6b. Corollas deeply lobed; bracts subtending several flowers; capsule opening near the middle or near the summit. **Triodanis**

Campanula L.   Bellflower
1a. Stems slender, 1–5 dm tall; native or cultivated. (2)
1b. Stems stout, 5–10 dm tall; cultivated ornamentals. (4)
2a. Flowers nodding; native and in cultivation. **C. rotundifolia** L.
2b. Flowers erect; native or cultivated. (3)
3a. Basal and cauline leaves similar in shape, lower ones long-petiolate; cultivated. Tussock Bellflower. **C. carpatica** Jacq.
3b. Basal and cauline leaves dissimilar in shape, the lower ones subpetiolate and elliptic to oblanceolate. **C. parryi** Gray

Capparidaceae

4a. Plants biennial. Canterbury Bells. **C. medium** L.
4b. Plants perennial. Peach-leaf Bellflower. **C. persicifolia** L.

Downingia Torr.

Annuals to 20 cm tall or more; leaves alternate, cauline, linear; inflorescence 1- to 10-flowered; corolla 4–7 mm long, light blue or purplish, the lower lip variegated; mature capsule 2–4 cm long, linear. **D. laeta** (Greene) Greene

Heterocodon Nutt.

Simple or sparingly branched annuals, 5–30 cm tall, glabrous or occasionally hispid on leaf margins and stem angles; leaves somewhat clasping, sharply toothed, to 1 cm long; calyx divided to the hypanthium; flowers regular, blue, 3–6 mm long; hypanthium commonly spreading hispid; not definitely known from Utah. **H. rariflorum** Nutt.

Lobelia L.

Diffuse and half-trailing annuals, to 3 dm tall; leaves alternate, reduced upward; flowers 12–20 mm broad, on slender pedicels; corolla light blue or violet with white or yellowish throat, irregular; anthers united into a tube or ring around the style; ovary inferior, 2-loculed; fruit a 2-valved capsule; cultivated border plant. **L. erinus** L.

Nemacladus Nutt.

Diffusely branched annuals, 7–20 cm tall; leaves mostly basal, the cauline leaves minute, sessile; inflorescence racemose; corolla irregular; calyx 2 mm long, the lobes linear to narrowly triangular; capsule about half-inferior; southwestern Utah. **N. glanduliferus** Jepson

Platycodon A. DC. Balloon Flower

Plants perennial; stems erect, branched above, glabrous, 6–10 dm tall; leaves ovate to ovate-lanceolate, 2–8 cm long, sharply dentate; flowers regular, mostly solitary, 5–8 cm broad, blue, lilac, or white; cultivated. **P. grandiflorum** A. DC.

Triodanis Raf. Venus Lookingglass

Plants annual; stems erect, 1–6 dm tall, simple or occasionally sparingly branched; leaves sessile, cordate-clasping; flowers regular; calyx divided to the hypanthium, lobes 5–8 mm long (smaller on cleistogamous flowers); open flowers purple to lavender, 8–13 mm long; capsules to 1 cm long, 2- to 3-loculed. **T. perfoliata** (L.) Nieuwl.

Capparidaceae — Caper Family

Annual herbs; leaves alternate, palmately compound, rarely simple above; flowers in racemes, regular, perfect; sepals 4; petals 4, distinct; stamens 6 to many; ovary superior, sessile, or more usually stipitate, 1-loculed, with 2 parietal placentae; carpels 2; fruit a 2-valved capsule.

Subclass DICOTYLEDONEAE

- 1a. Stamens 8–16, of unequal length; herbage glandular-villous, very clammy. **Polanisia**
- 1b. Stamens 6, of equal length. (2)
- 2a. Fruits 10–80 mm long, longer than broad, many-seeded; plants usually 2–20 dm tall. **Cleome**
- 2b. Fruits 2–8 mm long, broader than long, few-seeded; plants usually 0.4–3 dm tall. **Cleomella**

Cleome L. Beeplant

- 1a. Petals yellow; leaflets 5–7; a large- and a small-flowered form exist; especially prevalent in sandy areas and in southern Utah but probably found in every county at 2,800–6,000 ft. Yellow Beeplant, Stinking Mustard. **C. lutea** Hook.
- 1b. Petals dark to pale purple, rarely white; leaflets 3. (2)
- 2a. Petals purple to pale purple, in some populations occasionally white; fruit 2–6 cm long, 2.5–6 mm wide; seeds 3–4 mm long; probably in most counties from 2,700–6,500 ft. Rocky Mountain Beeplant. **C. serrulata** Pursh var. **serrulata**
- 2b. Petals dark purple; fruit 6–7 cm long, 3.5–5.5 mm wide; seeds 4–4.5 mm long; Salt Lake, Wasatch, and Uintah cos., at 6,000–8,000 ft. **C. serrulata** Pursh var. **angusta** (M. E. Jones) Tidestr.

Cleomella DC.

- 1a. Leaflets 2–7 mm wide (average 4 mm), 3.5 times as long as wide; petioles of lower leaves 6–27 mm long (average 11 mm); sandy washes and dry clay hillsides, 4,000–6,000 ft., known from Kane, Wayne, Grand, and Garfield cos. **C. palmerana** M. E. Jones
- 1b. Leaflets 1–3 mm wide (average 2 mm), 5.5 times as long as wide; petioles of lower leaves 2–12 mm long (average 6.4 mm); with dry desert shrubs at 4,800 ft. on Beryl Desert in Iron Co. **C. plocosperma** S. Wats.

Polanisia Raf. Clammyweed

Plants annual, with stems 20–80 cm tall; herbage glandular-pubescent and strongly scented; leaves trifoliolate; leaflets 2–5 cm long, elliptic or lanceolate; petals yellowish white to white, 8–12 mm long; stamens numerous (8–16), long-exserted; filaments purple; pods sessile or nearly so, 3.5–7 cm long, somewhat flattened, tipped by persistent style; seeds numerous, the cleft of seed open; usually in sandy stream beds, 1,200–6,500 ft., southern Utah. Clammyweed. **P. trachysperma** T. & G.

CAPRIFOLIACEAE — HONEYSUCKLE FAMILY

Shrubs or small trees; leaves opposite, simple or compound; flowers regular or irregular; calyx 4- to 5-toothed; corolla 4- to 5-lobed; ovary inferior, 1- to 6-celled; fruit berrylike, capsular, or drupaceous.

Caprifoliaceae

1a. Corolla flat, regular; style short or absent. (2)
1b. Corolla tubular, often irregular; style elongated. (3)
2a. Leaves pinnately compound. **Sambucus**
2b. Leaves simple, often lobed. **Viburnum**
3a. Stamens 4–5; fertile locules of ovary 1-ovuled. (4)
3b. Stamens 5; all locules fertile, 2- to many-ovuled. (6)
4a. Plants trailing subshrubs; flower pairs on long terminal peduncles. **Linnaea**
4b. Plants upright shrubs; flower pairs axillary or forming terminal corymbs. (5)
5a. Stamens 4–5, almost equal; ovary 4-loculed, 2 locules sterile. **Symphoricarpos**
5b. Stamens didynamous, 4; ovary 4-loculed. **Kolkwitzia**
6a. Flowers large, 1 to several, in axillary cymes. **Weigela**
6b. Flowers small, in axillary pairs or whorls. **Lonicera**

Kolkwitzia Graebn. Beautybush
　　Plants shrubs to 6 m tall; leaves broadly ovate, 2.5–7 cm long, acuminate, nearly entire; flowers very showy, spring; cultivated ornamental. **K. amabilis** Graebn.

Linnaea Gronov. Twinflower
　　Evergreen, trailing subshrubs; leaves roundish, 6–12 mm long, with a few crenate teeth; flowers white, tinged with rose-purple; northern Utah, at middle and higher elevations. **L. borealis** L.

Lonicera L. Honeysuckle
1a. Plants trailing; cultivated ground cover. Japanese Honeysuckle. **L. japonica** Thunb.
1b. Plants upright shrubs. (2)
2a. Branches with solid, white pith. (3)
2b. Branches hollow. (5)
3a. Leaves 6–12 cm long; fruit black; flowers with large bracts; along streams at middle and higher elevations. Bush Honeysuckle. **L. involucrata** Banks & Spreng.
3b. Leaves 2.5–8 cm long; fruit red. (4)
4a. Leaves fringed with hairs; petiole less than 6 mm long; cultivated shrubs to 2.5 m tall; flowers before the leaves, very fragrant. Winter Honeysuckle. **L. fragrantissima** Lindl. & Paxt.
4b. Leaves generally lacking marginal hairs; petiole 6–12 mm long; low, native shrubs to 1 m or more; canyons and mountains at middle and higher elevations. Utah Honeysuckle. **L. utahensis** S. Wats.
5a. Leaves mostly under 2.5 cm long, ovate-elliptic; corolla rose colored, about 12 mm long; cultivated shrubs to 4 m. Blue Leaf Honeysuckle. **L. korolkowii** Stapf.
5b. Leaves mostly over 2.5 cm long. (6)

Subclass DICOTYLEDONEAE

6a. Plants essentially glabrous; flowers not yellowish; upright shrubs to 3 m tall; corolla pink to white, 12–20 mm long; common in cultivation. Tatarian Honeysuckle. **L. tatarica** L.
6b. Plants with lower leaf surface densely pubescent; flowers yellowish, or tinged pinkish. (7)
7a. Leaves broadly ovate; corolla about 6 mm long, whitish or yellowish white, often reddish tinged; fruit dark red; cultivated shrubs to 3 m tall. European Fly Honeysuckle. **L. xylostema** L.
7b. Leaves elliptic to ovate-oblong; corolla white, changing to yellow, about 12 mm long; fruit dark red, rarely yellow; cultivated shrubs to 2 m tall. Morrow Honeysuckle. **L. morrowii** Gray

Sambucus L. Elder

1a. Cyme flat, umbellike; pith white. (2)
1b. Cyme paniclelike; pith usually brown. (4)
2a. Flowers white in cymes to 25 cm wide; leaflets mostly 7; fruit purple-black; occasionally cultivated shrubs to 4 m tall. American Elder. **S. canadensis** L.
2b. Flowers yellowish white in cymes to 20 cm wide; leaflets 3–7. (3)
3a. Leaflets usually 5, elliptic to elliptic-ovate, sparingly hairy on veins beneath; fruit lustrous black; flowers highly odoriferous; cultivated shrubs or trees to 10 m tall. European Elder. **S. nigra** L.
3b. Leaflets 5–7, oblong to oblong-lanceolate, glabrous; fruit glaucous, blue-black; canyons and mountains. Elderberry. **S. coerulea** Raf.
4a. Inflorescence corymbose, about as broad as high; fruit black; flowers yellowish white; canyons and mountains. Blackbead Elder. **S. melanocarpa** Gray
4b. Inflorescence paniclelike; fruit red. Red Elder. **S. racemosa** L.

Symphoricarpos Juss. Snowberry (Contributed by Benjamin W. Wood)

1a. Corolla short, open-campanulate, slightly ventricose on lower side, less than 6 mm long; lobes equal or nearly equal to length of tube. (2)
1b. Corolla oblong-campanulate, tubular-funnelform or salverform, symmetric; lobes about one-third as long as tube. (4)
2a. Fruit white. (3)
2b. Fruit red; introduced and occurring as a fugitive from cultivation. Coralberry. **S. orbiculatis** Moench.
3a. Stamens inserted; cultivated. Common Snowberry. **S. albus** (L.) Blake
3b. Stamens exserted; no specimen has been seen from the state but it should be looked for. Wolfberry. **S. occidentalis** Hook.
4a. Corolla 6–10 mm long, oblong-campanulate or funnelform-campanulate. (5)
4b. Corolla 9–13 mm long, tubular-funnelform or salverform. (7)

Caryophyllaceae

5a. Plants trailing; western Utah. Trailing Snowberry. **S. parishii** Rydb.
5b. Plants erect; common in mountains. (6)
6a. Young twigs glabrous or pubescent, hairs short and curved; corolla 6–8 mm long. Snowberry. **S. vaccinioides** Rydb.
6b. Young twigs pubescent, hairs short and straight; corolla 7–10 mm long. **S. rotundifolius** Gray
7a. Corolla 10–13 mm long, salverform; semiarid regions of southern Utah. **S. longiflorus** Gray
7b. Corolla 9–13 mm long, tubular-funnelform; mountain areas. (8)
8a. Young twigs and foliage glabrous; corolla lobes spreading. Mountain Snowberry. **S. oreophilus** Gray
8b. Young twigs and foliage pubescent; corolla lobes not spreading. **S. utahensis** Rydb.

Viburnum L.

1a. Leaves lobed, reddish in fall; cultivated shrubs to 4 m tall. Snowball Bush. **V. opulus** L.
1b. Leaves not lobed. (2)
2a. Leaves evergreen, entire or obscurely denticulate; petiole 12–25 mm long; shrubs to 3 m tall. **V. rytidophyllum** Hemsl.
2b. Leaves not as above; petiole over 2.5 cm long. (3)
3a. Shrubs erect; leaves 2.5–10 cm long, shiny green, irregularly toothed. **V. burkwoodii** Burkwood
3b. Shrubs spreading, to 3 m tall; cymes flat, 6–12 cm across; sterile flowers marginal; fruit ellipsoid, red, turning blue-black. (4)
4a. Leaves 2.5–10 cm long; winter buds scaly. **V. tomentosum** Thunb.
4b. Leaves 10–20 cm long; winter buds naked. **V. alnifolium** Marsh

Weigela Thunb.

Deciduous ornamental shrubs to 3 m tall; branches with 2 rows of hairs; leaves short-petiolate, elliptic-obovate, 5–10 cm long; corolla funnelform, to 4 cm long; cultivated in many varieties. **W. japonica** A. DC.

CARYOPHYLLACEAE — PINK FAMILY

Annual or perennial herbs with opposite, entire, simple leaves; flowers usually perfect, regular; sepals 4–5; petals 4–5; stamens usually as many as, or twice as many as, the petals; ovary superior, 1-loculed or incompletely 3- to 5-loculed; placenta free, central; styles 2–5, rarely 1; fruit a capsule.

1a. Calyx of united sepals. (2)
1b. Calyx of more or less separate sepals. (6)
2a. Ribs of calyx twice as many as calyx teeth, ending both in teeth apices and in sinuses. (3)

Subclass DICOTYLEDONEAE

- 2b. Ribs of calyx as many as calyx teeth, the calyx 5-ribbed, 5-nerved, nerveless, or striate-nerved. (4)
- 3a. Styles usually 3. **Silene**
- 3b. Styles usually 5. **Lychnis**
- 4a. Calyx scarious between green nerves. **Gypsophila**
- 4b. Calyx not at all scarious. (5)
- 5a. Calyx subtended by bractlets. **Dianthus**
- 5b. Calyx not subtended by bractlets. **Saponaria**
- 6a. Stipules present, conspicuous. **Spergularia**
- 6b. Stipules none. (7)
- 7a. Petals 2-cleft or parted. (8)
- 7b. Petals entire or shallowly notched (2-cleft in **Arenaria kingii**). (9)
- 8a. Styles usually 3; capsule short, ovoid or oblong. **Stellaria**
- 8b. Styles usually 5; capsule elongate, cylindric. **Cerastium**
- 9a. Styles as many as sepals and alternate with them. **Sagina**
- 9b. Styles 3, opposite sepals. **Arenaria**

Arenaria L.   Sandwort

- 1a. Capsule dehiscent by 6 teeth. (2)
- 1b. Capsule dehiscent by 3 teeth. (11)
- 2a. Leaves broad, ovate or obovate to lanceolate or oblanceolate. (3)
- 2b. Leaves narrowly linear and more or less pungent, setaceous, or subulate. (4)
- 3a. Plants strongly perennial, without rhizomes; aerial stems matted, caespitose, or pulvinate, from a definite taproot which is frequently subligneous; 10 cm or much more in height; seeds essentially smooth, shining black; southern Utah. **A. lanuginosa** (Michx.) Rohrb.
- 3b. Plants with slender, extensively produced, horizontal rhizomes; aerial stems simply branched, not matted, caespitose, or pulvinate, and not arising from a definite taproot; leaves ovate to elliptic-oblong, usually obtuse; seeds more or less strophiolate; capsule 5 mm long or less, not inflated. **A. lateriflora** L.
- 4a. Sepals obtuse or merely acutish or apiculate. (5)
- 4b. Sepals acute to acuminate. (6)
- 5a. Inflorescence open, at least at full maturity, more or less dichotomous; leaves aculeate or ascending, stiffly pungent; plants typically glaucous and matted; northwestern Utah. **A. aculeata** S. Wats.
- 5b. Inflorescence congested, subcongested, or umbellate. **A. congesta** Nutt. ex T. & G.
- 6a. Inflorescence open, at least at full maturity, more or less dichotomous. (7)
- 6b. Inflorescence congested, subcongested, or proliferate. (10)
- 7a. Sepals narrowly acute to acuminate. (8)
- 7b. Sepals broadly acute. (9)

Caryophyllaceae

8a. Stems glandular-pubescent with 5 or more pairs of leaves; glands oval-oblong, about 0.5 mm long; southeastern Utah. **A. fendleri** Gray
8b. Stems glabrous or glandular; leaves essentially basal; glands oblong-truncate, 1–2 mm long; desert regions, northeastern Utah. **A. eastwoodiae** Rydb.
9a. Stems mostly 20–40 cm tall, woody at base; sepals 4.5–6.6 mm long; petals entire, erose, or retuse; glands 1–2 mm long; lower Great Basin. **A. macrodenia** S. Wats.
9b. Stems mostly 20 cm tall or less, not woody at base; sepals 3.6–4.5 (6.0) mm long; petals entire, erose, retuse, or cleft to the base; glands about 0.5 mm long; Great Basin. **A. kingii** (S. Wats.) M. E. Jones
10a. Sepals 3–5 (6.0) mm long, acute. **A. congesta** Nutt. ex T. & G.
10b. Sepals 5.5–8 mm long, acuminate; stems not leafy, scabrous-puberulent; northeastern Utah. **A. hookeri** Nutt. ex T. & G.
11a. Plants annual, puberulent, caespitose or spreading, 2–10 cm tall; sepals 3-nerved, acuminate or strongly acute; inflorescence several-flowered; seeds reniform, tuberculate, 0.5–0.7 mm broad; leaves 3-nerved; high mountains. **A. rubella** (Wahl.) Smith
11b. Plants perennial. (12)
12a. Sepals broadly obtuse; leaves rigid, linear-subulate, triquetrous, abruptly acute or obtuse; petals obovate, 4–8 mm long, well surpassing sepals; flowering stems 10 cm tall or less; seed 0.6–0.8 mm long, indistinctly papillate-tuberculate or nearly smooth; mountains. **A. obtusiloba** (Rydb.) Fern.
12b. Sepals acuminate, acute, or acutish; leaves imbricate or internodes short, narrowly linear or subulate, less than 2 cm long, only primary leaves present. (13)
13a. Plants weakly caespitose or spreading perennials, 2–5 (10) cm tall; sepals 3-nerved; inflorescence several-flowered; seed reniform, tuberculate. (14)
13b. Plants caespitose or mat forming, from a more or less heavy taproot and ligneous caudex, entire plant glandular-puberulent; leaves 5–10 mm long, 3-nerved; seed 1.0–1.3 mm broad, reniform, checkered-tuberculate, black-brown; mountains. **A. nuttallii** Pax
14a. Leaves 3-nerved; plants strongly puberulent, seldom glabrous; seed 0.4–0.7 mm broad. **A. rubella** (Wahl.) Smith
14b. Leaves 1-nerved; plants wholly glabrous; seed 0.7–1.0 mm broad; south central Utah. **A. filiorum** Maguire

Cerastium L.   Chickweed

1a. Petals equaling or shorter than sepals; plants perennial; flowers cymose; pedicels at length longer than calyx. (2)
1b. Petals decidedly longer than sepals; plants annual, viscid-pubescent; sepals 3–4 mm long. (3)

Subclass DICOTYLEDONEAE

- 2a. Plants biennial or perennial, viscid-pubescent; upper leaves oblong, lower oblong-spatulate, 10–25 mm long, acute or obtuse; petals 2-cleft; plants simple or usually tufted, decumbent or ascending, 10–40 cm long; lawn and pasture weed. **C. vulgatum** L.
- 2b. Plants perennial with creeping stems, grayish-tomentose, with clearly twisted hairs; leaves linear-lanceolate, 12.5–18 mm long, about 3 mm wide; flowers showy; rock garden plant. **C. tomentosum** L.
- 3a. Pedicels not over twice the length of calyx; sepals about 4 mm long; petals 5–7 mm long; moist places. **C. brachypodum** (Engelm.) Robins.
- 3b. Pedicels much longer than calyx; sepals 4–5 mm long; petals nearly twice as long as calyx; moist, shaded places. **C. nutans** Raf.

Dianthus L.   Pink

- 1a. Flowers in dense terminal clusters, odorless, highly variable in color and form; cultivated. Sweet William. **D. barbatus** L.
- 1b. Flowers not in dense clusters, usually solitary or in 2s or 3s, fragrant. (2)
- 2a. Petals fringed, rose or pink, with striate or darker center; calyx usually purplish; cultivated. Garden Pink. **D. plumarias** L.
- 2b. Petals dentate, rose, purple, or white, 2.5–10 cm across; both single and double forms exist; fragrant, with a clove odor; cultivated. Carnation. **D. caryophyllus** L.

Gypsophila L.   Babysbreath

- 1a. Plants 6–10 dm tall or more, perennial. **G. paniculata** L.
- 1b. Plants usually less than 4 dm tall, annual. **G. elegans** Bieb.

Lychnis L.   Campion

- 1a. Plants usually over 20 cm tall; stems with more than 1 flower (usually 3 or more); seldom in alpine areas. (2)
- 1b. Plants usually less than 20 cm tall; stems with 1 flower (rarely 2 or 3); usually in alpine areas. (4)
- 2a. Flowers perfect and in a dense terminal head, scarlet or white; stem simple or slightly branched, usually loose-hairy; leaves 5–10 cm long, rounded or cordate, and usually clasping at base; to 90 cm tall; cultivated. **L. chalcedonica** L.
- 2b. Flowers dioecious or polygamous, or, if perfect, few in number; to 60 cm tall. (3)
- 3a. Calyx 12–18 mm long; plants dioecious or polygamous; flowers white to pinkish; leaves ovate-lanceolate or broader, upper sessile, lower short-petiolate; waste places above 7,500 ft., eastern Utah. **L. alba** Mill.
- 3b. Calyx 10–13 mm long; flowers few, perfect, white or purplish;

Caryophyllaceae

leaves oblanceolate to a margined petiole at base, linear and sessile above; dry slopes in mountains. **L. drummondii** (Hook.) S. Wats.
4a. Flowers erect in anthesis; calyx slightly inflated in fruit; alpine regions. **L. kingii** S. Wats.
4b. Flowers nodding in anthesis; calyx much inflated in fruit; alpine regions. **L. apetala** L.

Sagina L. Pearlwort

Matted perennials, 2–10 cm tall; leaves 4–15 mm long, linear-filiform; flowers axillary; sepals 5, 1.3–3 mm long; petals little shorter than sepals; stamens usually 4 or 5; styles 4 or 5, alternate with the sepals; capsule dehiscent, 4- to 5-valved; moist places. **S. saginoides** (L.) Britt.

Saponaria L. Soapwort
1a. Plants annual; flowers in a broad, open, corymblike cyme; petals rose or pale red, without appendages; calyx strongly 5-angled; dry, sandy or gravelly soil. Cow Soapwort. **S. vaccaria** L.
1b. Plants perennial; flowers in dense terminal cymes; petals pink or white, appendaged at claw; calyx cylindric or nearly so; stamens often petaloid; fields and waste places, cultivated and escaping. Bouncing Bet. **S. officinalis** L.

Silene L. Catchfly
1a. Plants pulvinate-caespitose, rarely over 6 cm tall; flowers pink or purplish, borne 1 per stem; found at high altitudes, mostly above 9,000 ft. **S. acaulis** L.
1b. Plants taller; flowers usually more than 1 per stem; habitat variable. (2)
2a. Calyx tube less than 10 mm long (usually about 7 mm). (3)
2b. Calyx tube at least 10 mm long (if shorter, then both pubescent and with conspicuous veins). (4)
3a. Plants annual, erect, glabrous or nearly so, with glutinous band near middle of upper internodes; cauline leaves reduced; calyx glabrous, with conspicuous veins; fields and waste places throughout Utah. **S. antirrhina** L.
3b. Plants perennial, usually decumbent, pubescent below, glandular-pubescent above; cauline leaves not reduced; calyx pubescent; stream banks and moist places in wooded areas throughout Great Basin. **S. menziesii** Hook.
4a. Plants over 15 cm tall (if less than 15 cm, then not rhizomatous and always more than 10 cm tall). (5)
4b. Plants less than 15 cm and usually less than 10 cm tall, arising from a branched system of rhizomes; south central Utah. **S. petersonii** Maguire
5a. Blades of petals deeply dissected into 4–6 lobes; petal appendages usually 4, acute; extreme northern Utah. **S. oregana** S. Wats.

Figs. 17-18. 17. **Linnaea borealis**, x .41. 18. **Lonicera involucrata**, x .31

Figs. 19-20. 19. **Sambucus racemosa**, x .26. 20. **Silene antirrhina**, x .21

Carophyllaceae

5b. Blades of petals not deeply dissected; petal appendages 2. (6)
6a. Branches ascending from root and basal stem; mountains, southern Utah. **S. scouleri** Hook.
6b. Branches decumbent at base, prostrate portion often buried. (7)
7a. Calyx veins conspicuous; calyx tubular, papery, considerably inflated in fruit, with each lobe 2 mm long and nearly always ciliate; open forests in rocky places, throughout (?) Utah. **S. douglasii** Hook.
7b. Calyx veins inconspicuous; calyx tubular-campanulate, constricted below the ovary, often purplish, glandular-pubescent; southwestern Utah. **S. verecunda** S. Wats.

Spergularia J. & C. Presl. Sandspurry

Annual herbs with linear, nearly terete, fleshy leaves; inflorescence a terminal cyme; sepals 5; petals 5, entire, sometimes none; stamens 2–10; styles 3; ovary 1-celled; capsule 3-valved; wet places in saline areas. **S. marina** (L.) Griseb.

Stellaria L. Chickweed
1a. Plants annual. (2)
1b. Plants perennial. (3)
2a. Stems procumbent, 10–40 cm long; leaves ovate, 0.5–3.5 cm long; sepals 4–6 mm long; waste places and shaded lawns. **S. media** (L.) Cyrill.
2b. Stems erect or ascending; leaves mostly basal, lower ovate, 4 mm long, upper sessile, linear-lanceolate, 6–10 mm long; flowers few; sepals 3 mm long, scarious margined; dry, grassy places. **S. nitens** Nutt.
3a. Plants glandular-pubescent, at least above; petals 6–8 mm long; leaves 5–10 cm long; mountains. **S. jamesiana** Torr.
3b. Plants not glandular; petals lacking or less than 6 mm long; leaves rarely if ever over 5 cm long. (4)
4a. Petals equaling or surpassing the sepals in length. (5)
4b. Petals lacking or minute, not over one-half as long as sepals. (6)
5a. Cyme diffuse, many-flowered, pedicels finally spreading or deflexed; leaves linear or lanceolate-linear, spreading or ascending, 2–6 cm long, widest at or above middle; seeds smooth; low meadows. **S. longifolia** Muhl.
5b. Cyme usually few-flowered, pedicels erect or ascending; leaves ascending; bracts of cyme small and scarious, especially above, usually not leaflike; moist places in mountains. **S. longipes** Goldie
6a. Bracts of inflorescence small and scarious, divaricate or reflexed at maturity; sepals scarious margined, about half as long as the capsule. **S. umbellata** Turcz.
6b. Bracts of inflorescence foliose, resembling upper leaves, not scarious; branches of inflorescence erect or ascending; leaves ovate,

77

Subclass DICOTYLEDONEAE

not over 12 mm long; plants caespitose, no running rootstock; stems diffuse. **S. obtusa** Engelm.

Celastraceae — Stafftree Family

Plants shrubs; leaves alternate or opposite, simple; flowers regular, perfect; calyx of 4–5 sepals; petals 4–5, distinct; stamens 4–5 or 8–10; ovary 1, superior, 2- to 5-loculed, the stigma 2- to 5-lobed; fruit a capsule or follicle.

1a. Leaves opposite. **Pachystima**
1b. Leaves alternate. (2)
2a. Herbage scabrous, yellowish; leaves very thick, persistent, elliptic to nearly orbicular; flowers in panicles, these mostly terminal; stamens commonly 5. **Mortonia**
2b. Herbage not scabrous, pale green; leaves deciduous; flowers axillary; stamens often more than 5. **Forsellesia**

Forsellesia Greene   Greasebush

1a. Stipules more than 0.5 mm long, frequently adnate to a persistent, often glandular, thickened base; western and southern Utah. **F. nevadensis** (Gray) Greene
1b. Stipules less than 0.5 mm long, without a glandular base; eastern and central Utah. **F. meionandra** (Kochne) Heller

Mortonia Gray

Low evergreen shrubs, 9–12 dm tall; inflorescence 3–6 cm long; calyx lobes 2 mm long; petals obovate, 3 mm long; capsule 4 mm long; southwestern Utah. **M. utahensis** (Cov.) A. Nels.

Pachystima Raf.   Mountain Lover

Prostrate or decumbent shrubs; leaves opposite, serrulate; sepals and petals 4; carpels 2; middle elevations. **P. myrsinites** (Pursh) Raf.

Ceratophyllaceae — Hornwort Family

Submerged or emergent aquatics, the stem length equaling the water depth; leaves whorled and finely dissected; flowers monoecious, regular, inconspicuous, solitary and sessile in the axils, subtended by a perianthlike involucre, the perianth wanting; stamens many; ovary superior, 1-carpeled, 1-loculed, 1-ovuled; fruit an achene.

Ceratophyllum L.   Hornwort

A single genus and species treated, with the characteristics of the family; lakes, ponds, and slow streams. **C. demersum** L.

Chenopodiaceae — Goosefoot Family

Plants herbs, subshrubs, or shrubs; leaves simple, alternate or opposite; flowers inconspicuous, monoecious, dioecious, polygamous, or

Chenopodiaceae

perfect; calyx persistent, 1- to 5-lobed; corolla none; stamens opposite the calyx lobes and the same number, or fewer; pistil with 1–3 stigmas, superior, 1-celled and 1-ovuled; fruit a utricle.

1a. Leaves scalelike; stems fleshy, jointed, with short internodes. (2)
1b. Leaves not scalelike; stems not fleshy or jointed. (3)
2a. Branches opposite; plants strictly herbaceous, usually less than 3 dm tall. **Salicornia**
2b. Branches alternate; plants woody at base, to 10 dm tall. **Allenrolfea**
3a. Plants densely white-pubescent, becoming golden brown with age; leaves linear, revolute; widely distributed native forage plant. **Eurotia**
3b. Plants various but not as above. (4)
4a. Leaves subulate, spine tipped, at least at maturity; an abundant tumbleweed of disturbed soils. **Salsola**
4b. Leaves not spine tipped (may be tipped with a hairlike bristle). (5)
5a. Leaves abruptly narrowed into a weak bristle; weed of wide distribution in the sheep ranges at lower elevations. **Halogeton**
5b. Leaves not produced into a bristle. (6)
6a. Flowers imperfect, monoecious or dioecious; perianth lacking (except as indicated in 7a and 7b). (7)
6b. Flowers perfect or polygamous; perianth present. (9)
7a. Shrubs with spiny branchlets; leaves fleshy, narrowly linear; staminate flowers in catkinlike spikes, lacking a perianth; pistillate flowers with a perianth; fruit with a horizontal wing; tall, bright green shrubs of saline soils. **Sarcobatus**
7b. Shrubs, subshrubs, or herbs without spiny branches or, if spiny, then otherwise different from above; staminate flowers with a perianth; pistillate flowers lacking a perianth but enclosed by 2 bracts. (8)
8a. Bracts dorsally compressed, variously tuberculate or winged (flat in **A. confertifolia**); pubescence of inflated hairs or none; plants perennial or annual; rounded axillary buds lacking. **Atriplex**
8b. Bracts laterally compressed, lacking appendages; pubescence of simple or branched hairs; plants perennial; rounded axillary buds present. **Grayia**
9a. Fruit largely concealed in the perianth lobes; stamens more than 3. (10)
9b. Fruit largely exposed; perianth lobes 1 or rarely 3; stamens 1–3. (13)
10a. Perianth developing conspicuous, horizontal, scarious wings in fruit. **Kochia**
10b. Perianth not developing conspicuous horizontal scarious wings (may be horned, hooked, or rarely winged). (11)
11a. Plants pilose; perianth lobes armed with a slender, curved spine. **Echinopsilon**

Subclass DICOTYLEDONEAE

11b. Plants glabrous, scurfy, glandular or puberulent but not pilose. (12)
12a. Leaves fleshy, often nearly terete, narrowly linear, glabrous. **Suaeda**
12b. Leaves not fleshy, flat, various in shape, usually other than linear, usually scurfy or pubescent. **Chenopodium**
13a. Perianth lobes 1–3; stamens 1–3; leaves entire; fruit winged. **Corispermum**
13b. Perianth lobes 1; stamens 1; leaves with hastate bases; fruit not winged. **Monolepis**

Allenrolfea Kuntze   Pickleweed

A widely distributed species in moist, saline soils. **A. occidentalis** (S. Wats.) Kuntze

Atriplex L.   Saltbush   (Contributed by C. A. Hanson, perennials, and L. K. Shumway, annuals)

1a. Plants perennial. (2)
1b. Plants annual. (17)
2a. Leaves dentate; Washington Co. (3)
2b. Leaves entire, rarely dentate but, if so, then dioecious herbs not as above. (4)
3a. Plants monoecious; leaves green, the teeth to 3 mm long. **A. semibaccata** R. Br.
3b. Plants dioecious; leaves white, the teeth to 10 mm long. **A. hymenelytra** (Torr.) S. Wats.
4a. Leaf base subhastate; shrubs to 4 m tall; Washington Co. (5)
4b. Leaf base attenuate to rounded; herbs or shrubs usually less than 2 m tall. (6)
5a. Branchlets terete. **A. lentiformis** (Torr.) S. Wats.
5b. Branchlets angled. **A. torreyi** S. Wats.
6a. Bracts with 4 lateral wings or 4 rows of teeth; plants at least one-third woody and usually not spinose. (7)
6b. Bracts without lateral wings; herbs, subshrubs, or spinose shrubs. (10)
7a. Leaves less than 8 mm wide; bract tip without lateral teeth. (8)
7b. Leaves more than 8 mm wide; bract tip with or without lateral teeth. (9)
8a. Bracts over 9 mm wide, with tips not exceeding the wings; staminate flowers yellow; shrubs to 2 m; widely distributed. Fourwing Saltbush. **A. canescens** (Pursh) Nutt.
8b. Bracts under 9 mm wide, with tips exceeding the wings; staminate flowers mostly brown; from one-third woody to completely so; northern Great Basin. **A bonnevillensis** Hanson
9a. Leaves 1–9 to first flowering branch of current growth; bract wings denticulate; staminate flowers brown, in panicles of 25–250 glomer-

Chenopodiaceae

ules; mostly on talus along the canyons of the Colorado River and its tributaries. **A. garrettii** Rydb.

9b. Leaves 7–28 to first flowering branch of current growth; bract wings deeply dentate; staminate flowers yellow, in panicles of 500–5,000 glomerules; Navajo Bridge, Arizona, to be expected along the Colorado River Canyon rim in southern Utah. **A. navajoensis** Hanson

10a. Plants shrubs; spines present; bracts foliose, united only at base, the surfaces lacking appendages; staminate flowers yellow; widely distributed. Shadscale. **A. confertifolia** (Torr. & Frem.) S. Wats.

10b. Plants herbs or subshrubs; spines lacking or, if present, very weak; bracts not foliose, these at least one-third united and the surfaces appendaged; staminate flowers yellow or brown. (11)

11a. Lower leaves opposite or subopposite; plants more or less prostrate; widely distributed on the Mancos Shale formation of eastern Utah. (12)

11b. Lower leaves alternate; plants ascending to erect; distribution mostly other than eastern Utah. (14)

12a. Leaves less than 4 mm wide; bracts appendaged on lower one-third; staminate flowers in spikes; plants prostrate, usually less than 1.5 dm tall. Mat-Atriplex. **A. corrugata** S. Wats.

12b. Leaves more than 4 mm wide; bracts various; staminate flowers paniculate; plants over 1.5 dm tall. (13)

13a. Leaves light gray-green; bracts 5–9 mm wide, heavily tuberculate; eastern Utah, south of the Uinta Mountains. Castle Valley Clover. **A. cuneata** A. Nels.

13b. Leaves green; bracts 2–5 mm wide, without tubercles or these less than 1 mm long; Daggett Co. **A. gardneri** (Moq.) Dietr.

14a. Terminal teeth of bracts half-united, not subtended by lateral teeth; staminate flowers mostly brown; saline soils, northern Great Basin. **A. falcata** (M. E. Jones) Standl.

14b. Terminal teeth of bracts free, subtended by lateral teeth; staminate flowers yellow (rarely brown in **A. tridentata**); usually on highly saline soils, mostly distributed in the Colorado drainage (except for some **A. tridentata**). (15)

15a. Leaves 5–15 times longer than wide; bracts in spikes; local endemic, near Cisco, Grand Co. **A. welshii** Hanson

15b. Leaves wider (except in **A. tridentata**); bracts usually in panicles; not near Cisco, Grand Co. (16)

16a. Leaves oblong-ovate to orbicular, over 10 mm wide; staminate glomerules 500–5,000 to a panicle; extreme southeastern Utah. **A. obovata** Moq.

16b. Leaves linear to oblong, under 10 mm wide; staminate glomerules less than 500 to a panicle; Great Basin and northern Colorado River drainage. **A. tridentata** Kuntze

Subclass DICOTYLEDONEAE

17a. Leaves bright green or greenish, triangular-hastate, 3–12 cm long or more. (18)
17b. Leaves gray or whitish with a fine scurf, at least on the lower surface, variously shaped but not triangular-hastate, 5 cm long or less. (19)
18a. Pistillate flowers of two kinds, one kind ebracteolate; fruiting bracts 8 mm broad or more. **A. hortensis** L.
18b. Pistillate flowers all alike and bracteolate; fruiting bracts 5 mm broad or less. **A. patula** L.
19a. Leaves linear or linear-oblong, entire; plants 10–30 cm tall, much branched, with slender branches. **A. wolfii** S. Wats.
19b. Leaves variously shaped, but not linear or linear-oblong and entire. (20)
20a. Fruiting bracts with orbicular, deeply laciniate-dentate margins, 2–4 mm in diameter. **A. elegans** (Moq.) D. Dietr.
20b. Fruiting bracts variously shaped, but not as above. (21)
21a. Staminate flowers in terminal panicles, the pistillate axillary; bracts orbicular, margins entire, 10–16 mm long. **A. graciliflora** M. E. Jones
21b. Staminate flowers among the pistillate ones (at least in part); bracts not orbicular, less than 10 mm long. (22)
22a. Fruiting bracts broadest at or below the middle. **A. rosea** L.
22b. Fruiting bracts broadest above the middle. (23)
23a. Leaves deltoid to elliptic. **A. truncata** (Torr.) Gray
23b. Leaves ovate or rounded-ovate. (24)
24a. Bracts not toothed at summit; leaves strongly 3-nerved. **A. powellii** S. Wats.
24b. Bracts irregularly toothed at summit; leaves triangular-ovate, rhombic or rounded-ovate. **A. argentea** Nutt.

Chenopodium L. Goosefoot

1a. Leaves and inflorescence bearing numerous glands, not farinose. (2)
1b. Leaves and inflorescence not glandular, 1 or both farinose or occasionally puberulent. (3)
2a. Inflorescence spicate or paniculate; flowers in irregular glomerules; fruit glandular-dotted. **C. ambrosioides** L.
2b. Inflorescence loosely dichotomous; flowers solitary or in loose clusters, some pedicellate; fruit not glandular-dotted. **C. botrys** L.
3a. Perianth becoming bright red and fleshy at maturity. **C. capitatum** (L.) Asch.
3b. Perianth not bright red or very fleshy at maturity. (4)
4a. Leaf blades all narrowly linear to linear-lanceolate, 1-nerved, entire. **C. leptophyllum** Nutt.
4b. Leaf blades broader, if narrowly lanceolate, then 3-nerved, often toothed or lobed. (5)

Chenopodiaceae

5a. Seeds all, or nearly all, vertical in fruit. (6)
5b. Seeds all, or nearly all, horizontal in fruit. (8)
6a. Leaf blades densely white-farinose beneath, at least when young. **C. glaucum** L.
6b. Leaf blades glabrous or very sparsely farinose. (7)
7a. Stems erect, 3–10 dm tall; leaf blades dentate; glomerules of flowers in dense elongate spikes. **C. rubrum** L.
7b. Stems prostrate or ascending, 2 dm tall or less; leaf blades hastately lobed, otherwise entire; glomerules of flowers solitary or subspicate. **C. humile** Hook.
8a. Leaf blades narrowly lanceolate or narrowly oblong, entire or somewhat basally lobed. **C. pratericola** Rydb.
8b. Leaf blades lance-ovate, broadly oblong, oval or broader, frequently as broad as long, margins mostly toothed or lobed. (9)
9a. Leaf blades cordate or subcordate at base, nearly glabrous, 5–15 cm long, with 1–3 (4) large teeth on each side. **C. gigantospermum** Aellen
9b. Leaf blades rounded or truncate to attenuate at base, glabrous or farinose, seldom to 5 cm long or, if so, then not cordate or subcordate at base, margin entire or variously toothed or lobed. (10)
10a. Leaf blades as broad as long, or nearly so. (11)
10b. Leaf blades much longer than broad, usually 2–4 times as long. (12)
11a. Leaf blades densely farinose, at least below, thick; plants diffusely branched from base, seldom over 2.5 dm tall. **C. incanum** (S. Wats.) Heller
11b. Leaf blades usually glabrate, thin; plants slender, often over 2.5 dm tall. **C. fremontii** S. Wats.
12a. Fruit free from the seed. **C. atrovirens** Rydb.
12b. Fruit adherent to the seed. (13)
13a. Plants bright green, not at all farinose. **C. paganum** Reichenb.
13b. Plants pale blue-green; leaves sparingly to densely farinose, at least on lower surface. (14)
14a. Seeds 1.3–2 mm broad, smooth and shiny; plants not ill scented. **C. album** L.
14b. Seeds 0.8–1 mm broad, roughened, not very shiny; plants ill scented. **C. berlandieri** Moq.

Corispermum L.   Bugseed

Caulescent annuals with branching stems and alternate leaves; locally abundant in sandy, usually saline soils at lower elevations. **C. hyssopifolium** L.

Echinopsilon Moq.

Herbs with narrow to terete leaves; naturalized from the Old World. **E. hyssopifolium** (Pall.) Moq.

Subclass DICOTYLEDONEAE

Eurotia Adans. Winterfat

A single species, distinctive because of its white villous pubescence; widely distributed, especially at lower elevations. **E. lanata** (Pursh) Moq.

Grayia H. & A. Hopsage, Applebush

1a. Shrubs with spinescent branches; fruiting bracts glabrous, not keeled, over 6 mm wide; widely distributed in desert regions. Spiny Hopsage. **G. spinosa** (Hook.) Moq.
1b. Shrubs without spinescent branches; fruiting bracts scurfy-pubescent, keeled at maturity, less than 6 mm wide; widely distributed, but only locally common, especially on clay soils derived from shale formations. (Note: Frequently confused with **Atriplex** species.) **G. brandegei** Gray

Halogeton C. A. Meyer

A single, introduced species; widely distributed in the winter rangelands of sheep; poisonous to livestock. **H. glomeratus** (Bieb.) C. A. Meyer

Kochia Roth. Summer Cypress

1a. Plants annual; leaves thin, petiolate, blades lance-linear; a tall, weedy plant of roadsides and waste places. **K. scoparia** (L.) Schrad.
1b. Plants perennial; leaves subterete, blades linear; widely distributed, low elevations, usually clay soils. **K. americana** S. Wats.

Monolepis Schrad. Poverty Weed

A single species reported; prostrate to ascending annual with hastate leaf bases; waste places throughout Utah. **M. nuttallianus** (Schult.) Greene

Salicornia L. Samphire

1a. Plants annual, erect or spreading; common around highly saline seeps or springs. **S. rubra** A. Nels.
1b. Plants perennial, erect or decumbent, frequently rooting at the nodes; salt marshes and seeps. **S. pacifica** Standl.

Salsola L. Russian Thistle

A single introduced species, frequently growing into large, rounded tumbleweeds; waste places throughout Utah. **S. kali** L.

Sarcobatus Nees Greasewood

Shrubs locally abundant in saline soils at low and lower middle elevations. **S. vermiculatus** (Hook.) Torr.

Suaeda Forsk. Seepweed

1a. Calyx lobes with hornlike appendages. (2)
1b. Calyx lobes with no appendages, the sepals flat, round, or somewhat keeled. (3)
2a. Leaves, at least in the inflorescence, broadest at the base; flowers and leaves crowded; saline soils. **S. depressa** (Pursh) S. Wats.

Figs. 21-22. 21. **Stellaria jamesiana**, x .2. 22. **Atriplex tridentata**, x .2

Figs 23-24. 23. **Chenopodium album**, x .23. 24. **Monolepis nuttallianus**, x .25

Subclass DICOTYLEDONEAE

2b. Leaves narrowed at the base, those of the inflorescence linear to linear-lanceolate; flowers and leaves not crowded; saline soils. **S. occidentalis** S. Wats.
3a. Plants annual or, if perennial, not suffrutescent at base. **S. nigra** (Raf.) Macbr.
3b. Plants perennial, suffrutescent at base. (4)
4a. Plants green; leaves strongly flattened. **S. torreyana** S. Wats.
4b. Plants glaucous; leaves subterete. **S. fruticosa** (L.) Forsk.

COMPOSITAE — COMPOSITE FAMILY

Broad-leaved, usually deciduous herbs or shrubs with alternate, opposite, whorled, simple, compound, or pinnatifid leaves; flowers in involucrate heads, these solitary or several in corymbose or cymose clusters; flowers few to many on a common receptacle, surrounded by green bracts forming a cup-shaped, cylindric, or urn-shaped involucre; heads entirely of tubular corollas, entirely of ligulate (ray) corollas, or with both tubular corollas forming a central disk (disk flowers) and an outer radiating row of ligulate corollas (rays or ray flowers); calyx, when present, crowning the summit of the ovary and modified as a pappus of capillary bristles, scales, or a crown; stamens alternate with corolla lobes; filaments free, the anthers united, forming a tube or almost separate; ovary inferior, of 2 carpels, 1-loculed, with a single ovule; styles 1, 2-cleft, passing through the anther tube; fruit an achene. (Note: The individual flowers in some genera have no bracts at the base of the ovaries and, after the achenes have fallen, the receptacle is naked; in other genera each flower has a scalelike or bristly bract at the base of the ovary and, after the achenes fall, the receptacle is chaffy.)

1a. Corollas all raylike; plants usually with milky juice. KEY I, p. 86.
1b. Corollas not all raylike, some or all of them tubular; juice seldom if ever milky. (2)
2a. Corollas all tubular; no ray flowers present, or the rays vestigial and minute. KEY II, p. 88.
2b. Corollas not all tubular; ray flowers present. (3)
3a. Pappus of capillary bristles, at least in part. KEY III, p. 93.
3b. Pappus of awns or scales, or lacking. (4)
4a. Pappus lacking. KEY IV, p. 94.
4b. Pappus present, of awns or scales: KEY V, p. 96.

KEY I. Corollas all raylike; plants usually with milky juice.

1a. Pappus lacking. **Atrichoseris**
1b. Pappus present. (2)
2a. Pappus, at least in part, of plumose bristles. (3)
2b. Pappus of simple bristles, of awns, or of scales. (5)

Compositae

3a. Achenes not beaked, truncate at apex; involucres usually less than 15 mm long. **Stephanomeria**
3b. Achenes tapering or beaked at apex; involucres usually more than 15 mm long. (4)
4a. Leaves pinnatifid; corollas white or pinkish; involucre with an outer series of short bractlets; southern Utah. **Rafinesquia**
4b. Leaves not pinnatifid, entire; corollas yellow or purplish; involucre lacking short outer bractlets; widespread. **Tragopogon**
5a. Pappus of 1–3 series of unawned or awned scales. (6)
5b. Pappus of capillary bristles. (7)
6a. Pappus of 2 or 3 series of unawned scales; corollas blue, closing by midmorning. **Cichorium**
6b. Pappus scales in a single series, awned; corollas yellow, not closing by midmorning. **Microseris**
7a. Achenes more or less flattened; stems leafy; heads in panicles or in umbellate clusters. (8)
7b. Achenes not flattened; stems leafy or scapose; heads solitary or variously disposed. (9)
8a. Involucres cylindric or ovoid-cylindric; achenes beaked; flowers yellow or blue. **Lactuca**
8b. Involucres broadly campanulate to hemispheric; achenes not beaked; flowers yellow. **Sonchus**
9a. Corollas pink or purplish. **Lygodesmia**
9b. Corollas yellow or yellowish, white or cream colored. (10)
10a. Leaves all basal; heads solitary on scapose peduncles. (11)
10b. Leaves not all basal, the stems leafy; heads not on scapose peduncles. (13)
11a. Achenes not beaked, truncate; pappus bristles barbellate. **Microseris**
11b. Achenes beaked or tapering to apex; pappus not of barbellate bristles. (12)
12a. Achenes 10-ribbed or 10-nerved, not spinulose; involucral bracts usually imbricated in several series. **Agoseris**
12b. Achenes 4- to 5-ribbed, spinulose, especially near apex; principal bracts in a single series, the outer much shorter. **Taraxacum**
13a. Achenes ridged or tuberculate between the angles; southwestern Utah. (14)
13b. Achenes striate between the angles; widely distributed. (15)
14a. Plants depressed annuals with crustaceous-margined leaves, not stipitate-glandular; achenes abruptly beaked, transversely ridged between the ribs. **Glyptopleura**
14b. Plants erect, lacking crustaceous-margined leaves, conspicuously stipitate-glandular above; achenes tapering to a beak, not transversely ridged. **Calycoseris**
15a. Pappus bristles early deciduous, more or less united below and

Subclass DICOTYLEDONEAE

        falling together, only a few of the stout outer ones may be persistent. **Malacothrix**
- 15b. Pappus bristles persistent or tardily deciduous, and then falling separately. (16)
- 16a. Pappus tan to brown; involucral bracts not thickened. **Hieracium**
- 16b. Pappus white or whitish; involucral bracts somewhat thickened at base or on midrib. **Crepis**

        KEY II. Corollas all tubular; no ray flowers present.

- 1a. Heads unisexual, the pistillate heads with 1–4 flowers enclosed in involucre; involucre burlike or nutlike, only style tips exserted. (2)
- 1b. Heads not unisexual (staminate and pistillate flowers in same head or at least some perfect); involucre not burlike or nutlike. (4)
- 2a. Involucral bracts of the staminate heads perfect; fruiting involucres burlike, covered with hooked appendages. **Xanthium**
- 2b. Involucral bracts of the staminate heads united; fruiting involucres various but, if burlike, then lacking hooked appendages. (3)
- 3a. Fruiting involucre with several transverse, scarious wings; leaves or their lobes linear-filiform. **Hymenoclea**
- 3b. Fruiting involucre lacking transverse wings; leaves and their lobes not linear-filiform. **Ambrosia**
- 4a. Stamens not united by their anthers; flowers always unisexual, the pistillate corollas none or much reduced. (5)
- 4b. Stamens with united anthers or rarely not united in some species with perfect flowers, at least some flowers usually perfect. (7)
- 5a. Achenes long-villous; leaves or their lobes linear-filiform. **Oxytenia**
- 5b. Achenes not long-villous; leaves or their lobes not linear-filiform. (6)
- 6a. Pistillate flowers subtended by large, chaffy scales simulating inner involucral bracts; achenes with pectinate or winged margins. **Dicoria**
- 6b. Pistillate flowers subtended by chaffy scales or these lacking; achenes without pectinate or toothed wings. **Iva**
- 7a. Involucral bracts with translucent, usually yellow or orange dots; pappus of 8–15 scales, each soon dissected into several bristles. **Dyssodia**
- 7b. Involucral bracts without distinct dots; pappus various, but not as above. (8)
- 8a. Pappus of capillary bristles, at least in part, these smooth, scabrous, barbellate, or plumose. (9)
- 8b. Pappus lacking or, if present, not of capillary bristles. (37)
- 9a. Leaves opposite or whorled, some or all cauline. (10)
- 9b. Leaves alternate, at least basally, or basal and actually alternate. (14)

Compositae

10a. Corollas yellow; involucral bracts in 1 series or in 2 series, but all equal in length. **Arnica**
10b. Corollas white, ochroleucous, flesh colored, blue, or purple; involucral bracts in 2 to several series. (11)
11a. Pappus double—the outer series of short scales, the inner series of capillary bristles; shrubs with white bark. **Hofmeistera**
11b. Pappus single, or else plants herbaceous; shrubs or herbs. (12)
12a. Corollas of some or all flowers bilabiate; leaves spinulose-dentate; flowers pink; arid sites in Kane and Washington cos. **Perezia**
12b. Corollas not bilabiate; leaves not or seldom spinulose-dentate; flowers pink-white or cream; various distribution. (13)
13a. Achenes 5-angled or 5-ribbed; involucral bracts subequal or in 2 series. **Eupatorium**
13b. Achenes 10-angled or 10-ribbed; involucral bracts imbricated in several series of different lengths. **Brickellia**
14a. Leaves spinescent, usually with spiny teeth or lobes, rarely entire but then with spine-tipped apex. (15)
14b. Leaves entire, denticulate or lobed, lacking spines. (18)
15a. Pappus of narrow scales; receptacle with bristles or scales; flowers orange or orange-yellow; cultivated and escaping. **Carthamnus**
15b. Pappus bristles plumose or merely barbellate; flowers not orange or orange-yellow. (16)
16a. Pappus bristles plumose (rarely some otherwise); receptacle densely bristly. **Cirsium**
16b. Pappus bristles merely barbellate. (17)
17a. Receptacle densely bristly, not fleshy or honeycombed; heads nodding. **Carduus**
17b. Receptacle not bristly or scarcely so, fleshy and honeycombed; heads not nodding. **Onopardum**
18a. Receptacle with dense bristles or narrow, chaffy scales between disk flowers. (19)
18b. Receptacle naked or at most short-hairy, never with dense bristles or scales. (21)
19a. Involucral bracts with hooked spines; lower leaves large (resembling rhubarb), cordate at base. **Arctium**
19b. Involucral bracts without spines, or spines not hooked; lower leaves not large and cordate at base. (20)
20a. Receptacle chaffy except in center; plants small, woolly. **Filago**
20b. Receptacle chaffy throughout; plants not small and woolly. **Centaurea**
21a. Heads unisexual; plants dioecious (staminate flowers may have styles but ovary does not develop). (22)
21b. Heads with at least central flowers perfect. (24)
22a. Plants shrubs or else woody at base, not tomentose; leaves usually

89

Subclass DICOTYLEDONEAE

toothed or lobed; involucral bracts not strongly scarious margined. **Baccharis**
22b. Plants herbaceous, more or less tomentose; leaves entire; involucral bracts strongly scarious, at least along margins. (23)
23a. Pappus bristles of pistillate flowers united at base and falling together; pappus bristles of staminate flowers usually club shaped at apex; plants usually less than 30 cm tall; basal leaves commonly in a rosette; cauline leaves reduced and different in shape; leaves usually tomentose on both sides. **Antennaria**
23b. Pappus of pistillate flowers separate at base and falling separately; pappus bristles of staminate flowers not club shaped at apex; plants mostly over 30 cm tall; leaves all alike, usually green and glabrate above. **Anaphalis**
24a. Involucral bracts scarious or hyaline (only partly so in **Pluchea**). (25)
24b. Involucral bracts herbaceous, at least in the center. (27)
25a. Involucral bracts subscarious; corollas purplish; plants not tomentose. **Pluchea**
25b. Involucral bracts scarious; corollas rarely purplish; plants tomentose. (26)
26a. Plants perennial, subdioecious; pistillate heads usually with a few central, perfect flowers. **Anaphalis**
26b. Plants annual or biennial, not dioecious; heads all alike, the marginal flowers pistillate and central ones perfect. **Gnaphalium**
27a. Involucral bracts in a single series, a few very short ones may be present at the very base. (28)
27b. Involucral bracts of 2 or more series, these often of different lengths. (31)
28a. Plants woody, shrubs; involucral bracts 4–6 per head. **Tetradymia**
28b. Plants herbaceous; bracts more than 6 per head. (29)
29a. Plants annual; heads with inner flowers perfect, the outer pistillate. **Conyza**
29b. Plants perennial; heads with all flowers perfect. (30)
30a. Style branches with a tuft of hairs near the truncate apex; involucral bracts in 1 series only (a few short bracts may be present). **Senecio**
30b. Style branches without a tuft of hairs near the truncate apex; involucral bracts actually in 2 or more series. **Erigeron**
31a. Pappus double, the outer series of short scales, the inner ones of capillary bristles; shrubs with white bark. **Hofmeistera**
31b. Pappus simple or else the plants herbaceous. (32)
32a. Plants annual. (33)
32b. Plants perennial, often woody. (35)
33a. Plants low, depressed, scurfy pubescent herbs; leaves broadly ovate or roundish, entire or toothed. **Psathyrotes**
33b. Plants not as above. (34)

Compositae

34a. Leaves all entire. **Aster**
34b. Leaves toothed or lobed, at least the lower. **Conyza**
35a. Involucral bracts in more or less vertical rows. **Chrysothamnus**
35b. Involucral bracts not in vertical rows. (36)
36a. Plants woody or else the leaves spinulose toothed; involucral bracts in 2 or more imbricate series. **Haplopappus**
36b. Plants herbaceous, the leaves not spinulose toothed; involucral bracts usually subequal. **Erigeron**
37a. Receptacles with bristles or chaffy scales among the flowers. (38)
37b. Receptacles naked or short-hairy. (47)
38a. Receptacles densely bristly. **Centaurea**
38b. Receptacles with chaffy scales. (39)
39a. Leaves with spiny margins. **Carthamnus**
39b. Leaves without spiny margins. (40)
40a. Plants low woolly annuals; outer bracts boat shaped and enclosing the achenes. **Stylocline**
40b. Plants various, but not low and woolly; outer bracts various, but not usually enclosing the achenes. (41)
41a. Involucral bracts in 2 distinct sets—the outer herbaceous, the inner differing in shape and texture; leaves opposite, at least below, or alternate. (42)
41b. Involucral bracts not in 2 unlike sets; leaves alternate or basal. (44)
42a. Leaves alternate throughout; outer involucral bracts about 5, spreading, herbaceous, the inner (1–3 subtending pistillate flowers) larger and broader, becoming strongly accrescent and hooded in fruit. **Dicoria**
42b. Leaves opposite, at least below; outer involucral bracts various, but not as above, not accrescent and hooded in fruit. (43)
43a. Inner involucral bracts united to the middle or above, forming a cup. **Thelesperma**
43b. Inner involucral bracts not united, or united only at base. **Bidens**
44a. Involucral bracts in 1 series, boat shaped, each bract enclosing a marginal flower; rays short, yellow. **Madia**
44b. Involucral bracts in 1 or more series, not boat shaped and enclosing marginal flowers; rays lacking. (45)
45a. Plants woody shrubs; mostly along the canyons of the Colorado and Green rivers. **Encelia**
45b. Plants herbaceous; widely distributed. (46)
46a. Receptacles high-conical, mostly over 3 cm long; stems leafy. **Rudbeckia**
46b. Receptacles merely convex, much less than 3 cm long; leaves all basal. **Enceliopsis**
47a. Pappus none. (48)
47b. Pappus present. (52)

Subclass DICOTYLEDONEAE

48a. Leaves opposite, some cauline, somewhat connate at base; east central Utah. **Flaveria**
48b. Leaves alternate or basal. (49)
49a. Heads numerous, in spikes, racemes, or panicles; anthers with acute tips; receptacles flat; plants woody or herbaceous. **Artemisia**
49b. Heads solitary on ends of stems, or sometimes corymbose or capitate; anthers with rounded tips; receptacles convex or conic; plants herbaceous, or woody only at base. (50)
50a. Plants annual; heads solitary; leaves green and glabrous. **Matricaria**
50b. Plants perennial; heads corymbose or capitate; leaves usually silvery-canescent. (51)
51a. Involucral bracts in 2-3 series; heads capitate or, if corymbose, the leaves 2-pinnatifid or lobed at apex only. **Tanacetum**
51b. Involucral bracts in 4-5 series; heads corymbose, the leaves not lobed or only somewhat 1-pinnate at base. **Chrysanthemum**
52a. Receptacle chaffy. (53)
52b. Receptacle not chaffy, sometimes merely hairy. (57)
53a. Pappus of numerous flattened bristles. **Baccharis**
53b. Pappus of 1-4 teeth, scales, or awns. (54)
54a. Achenes with pectinate or toothed wings. **Dicoria**
54b. Achenes without pectinate or toothed wings. (55)
55a. Plants shrubby; pappus awns or teeth not retrorsely barbed; mostly of the canyons of the Colorado and Green rivers. **Encelia**
55b. Plants herbaceous; pappus awns or teeth retrorsely barbed. (56)
56a. Inner involucral bracts united to about the middle, forming a cup. **Thelesperma**
56b. Inner involucral bracts not united. **Bidens**
57a. Plants dioecious. **Baccharis**
57b. Plants not dioecious. (58)
58a. Pappus of 2-8 caducous awns; plants usually strongly glutinous. **Grindelia**
58b. Pappus various, but not of 2-8 caducous awns. (59)
59a. Leaves and involucre conspicuously punctate with translucent oil glands. **Dyssodia**
59b. Leaves and involucre sometimes impressed-punctate, but without translucent oil glands. (60)
60a. Pappus of 12 or more scalelike segments, these nearly or quite as long as achene. **Vanclevea**
60b. Pappus of fewer than 12 scalelike segments or else much shorter than achene. (61)
61a. Achenes strongly compressed; pappus of 1 or 2 slender awns. **Laphamia**
61b. Achenes not compressed or, if so, then pappus not of 1 or 2 slender awns. (62)

Compositae

62a. Pappus a crown with margins entire or of short scales united into a crown. (63)
62b. Pappus not as above. (64)
63a. Plants annual; heads solitary; all flowers in head perfect; leaves green and glabrous. **Matricaria**
63b. Plants perennial; heads corymbose or capitate, rarely solitary; some marginal flowers pistillate only; leaves mostly silvery-canescent. **Tanacetum**
64a. Involucral bracts with a thin, scarious, white, yellow, or purplish margin and tip. (65)
64b. Involucral bracts without a scarious, colored margin and tip. (66)
65a. Plants tomentose; pappus of 10 or more scales. **Hymenopappus**
65b. Plants not tomentose; pappus of 8 scales. **Bahia**
66a. Corollas yellow. **Eriophyllum**
66b. Corollas white, flesh colored or purplish. (67)
67a. Plants scapose; leaves roundish, entire, or crenate. **Chamaechaenactis**
67b. Plants leafy stemmed; leaves not roundish and entire or subentire. (68)
68a. Pappus scales with a strong midrib; leaves lanceolate or linear, entire; southern Utah. **Palafoxia**
68b. Pappus scales nerveless or essentially so; leaves, at least in part, toothed to pinnatifid; widely distributed. **Chaenactis**

KEY III. Corollas not all raylike; pappus of capillary bristles.

1a. Rays white, pink, violet, or purple, not yellow. (2)
1b. Rays yellow or orange-yellow. (7)
2a. Pappus of a single subplumose bristle and a short, scarious cup, or of numerous unequal bristles; low winter annuals; involucral bracts subequal. **Monoptilon**
2b. Pappus of numerous bristles; plants various, but seldom low winter annuals; involucral bracts imbricate or subequal. (3)
3a. Pappus, at least of disk flowers, of several to many rigid bristles; achenes pubescent with 2-forked hairs or the hairs barbed at apex. **Townsendia**
3b. Pappus, at least of disk flowers, of many capillary bristles, at least in part; achenes glabrous or pubescent with simple hairs. (4)
4a. Rays very inconspicuous, shorter than the tube and scarcely if at all exceeding their pappus; central perfect flowers few; plants annual. **Conyza**
4b. Rays usually conspicuous, longer than the tube and pappus; central perfect flowers several to many; plants annual, biennial, or perennial. (5)
5a. Involucres usually strongly graduated; rays comparatively broad; style tips ovate and acute to subulate, usually lanceolate. (6)

93

Subclass DICOTYLEDONEAE

5b. Involucres subequal, rarely somewhat graduated; rays usually narrow; style tips very short, triangular, rounded, or obtuse. **Erigeron**
6a. Plants perennial, rhizomatous, or annual or, if from a caudex, then ordinarily less than 10 cm tall. **Aster**
6b. Plants from a caudex or taproot. **Machaeranthera**
7a. Leaves opposite, at least below. (8)
7b. Leaves alternate throughout. (10)
8a. Plants subshrubs. **Laphamia**
8b. Plants herbaceous. (9)
9a. Leaves with stiff marginal bristles; involucre and leaves with conspicuous oil glands. **Pectis**
9b. Leaves without stiff marginal bristles; involucre and leaves without oil glands. **Arnica**
10a. Pappus of 2–8 stiff, caducous bristles; plants usually glutinous. **Grindelia**
10b. Pappus of numerous, usually soft, persistent bristles. (11)
11a. Pappus of about 20 twisted, flattish bristles. **Amphipappus**
11b. Pappus of numerous, straight, capillary bristles. (12)
12a. Pappus double, the inner of numerous bristles, the outer sometimes scalelike. (13)
12b. Pappus not double, of subequal capillary bristles only. (14)
13a. Leaves essentially filiform. **Conyza**
13b. Leaves not filiform, linear-oblong or broader. **Heterotheca**
14a. Involucral bracts in distinct vertical ranks. (15)
14b. Involucral bracts not in distinct vertical ranks. (16)
15a. Outer involucral bracts with loose herbaceous tips; erect stems perennial; leaves deciduous. **Chrysothamnus**
15b. Outer involucral bracts without loose herbaceous tips; erect stems annual; leaves persistent. **Petradoria**
16a. Involucral bracts in 1 series, frequently with some smaller bracts at base; style branches truncate apically. **Senecio**
16b. Involucral bracts neither in 1 series nor with smaller bracts at base; style branches without truncate tips. (17)
17a. Heads small, the involucres usually less than 6 mm high, usually very numerous and densely paniculate, rarely racemose or corymbose; plants rhizomatous, fibrous rooted. **Solidago**
17b. Heads medium to large, the involucres usually more than 6 mm high, neither very numerous nor densely paniculate; plants with taproots, occasionally also rhizomatous. **Haplopappus**

KEY IV. Corollas not all tubular; ray flowers present; pappus lacking.

1a. Rays white, sometimes yellow at base. (2)

Compositae

1b. Rays yellow, sometimes partly purplish or maroon. (5)
2a. Receptacle naked. (3)
2b. Receptacle with chaffy scales. (4)
3a. Receptacle broad and flattish; involucral bracts with a dark brown submarginal line. **Chrysanthemum**
3b. Receptacle convex, conic, or hemispheric; involucral bracts without a dark brown submarginal line. **Matricaria**
4a. Heads small, numerous, in dense, flattish or rounded cymose panicles; plants perennial. **Achillea**
4b. Heads comparatively large, solitary or few; plants annual. **Anthemis**
5a. Receptacles not chaffy. (6)
5b. Receptacles chaffy, at least toward the margin. (11)
6a. Heads 1- or 2-flowered, in dense glomerate clusters, sessile in the forks of the stem, or terminal and leafy involucrate; eastern Utah. **Flaveria**
6b. Heads several- to many-flowered, solitary on terminal peduncles. (7)
7a. Plants woolly. (8)
7b. Plants not woolly. (9)
8a. Rays persistent, becoming papery. **Baileya**
8b. Rays not persistent. **Eriophyllum**
9a. Involucre and leaves with translucent oil glands. **Pectis**
9b. Involucre and leaves without translucent oil glands. (10)
10a. Rays conspicuous; involucral bracts acuminate, without scarious margins. **Bahia**
10b. Rays minute; involucral bracts obtuse, with scarious margins. **Tanacetum**
11a. Ray achenes partly or wholly enfolded by their involucral bracts; plants annual, glandular-viscid above. **Madia**
11b. Ray achenes not conspicuously enfolded by their involucral bracts; plants perennial or, if annual, not glandular above. (12)
12a. Involucre distinctly double, the outer bracts herbaceous, the inner ones broader and united to about the middle. **Thelesperma**
12b. Involucre not double. (13)
13a. Plants scapose perennials; leaves broad, silvery-pubescent, entire; heads very broad. **Enceliopsis**
13b. Plants leafy stemmed or subscapose; leaves various but not broad and silvery-pubescent; heads broad or narrow. (14)
14a. Plants subscapose; leaves variously dissected or sagittate; heads broad. **Balsamorrhiza**
14b. Plants with stems definitely leafy; leaves usually not dissected or sagittate. (15)
15a. Plants shrubby; achenes conspicuously ciliate on the margins, notched at the apex, very flat. **Encelia**

Subclass DICOTYLEDONEAE

15b. Plants herbaceous; achenes not conspicuously ciliate on the margins. (16)
16a. Achenes 2-winged. **Verbesina**
16b. Achenes not 2-winged. **Viguiera**

  Key v. Corollas not all tubular; ray flowers present;
    pappus of awns or scales.

1a. Receptacle chaffy. (2)
1b. Receptacle not chaffy, either naked or bristly. (14)
2a. Receptacle bearing a row of chaffy scales between the ray flowers and the outer disk flowers, otherwise naked; pappus of 10–20 slender scales. **Layia**
2b. Receptacle chaffy throughout; pappus not of 10–20 slender scales. (3)
3a. Pappus of awns only, without scales. (4)
3b. Pappus, at least in part, of scales. (8)
4a. Achenes flat and obcompressed; awns retrorsely hispid. **Bidens**
4b. Achenes not obcompressed; awns not retrorsely hispid. (5)
5a. Achenes plump; pappus of 2 caducous awns. **Helianthus**
5b. Achenes flat, very strongly compressed; pappus various. (6)
6a. Plants scapose; heads large, solitary. **Enceliopsis**
6b. Plants leafy stemmed; heads medium sized, usually several. (7)
7a. Plants shrubby; achenes narrowly white margined, the margin not continuous between weak awns. **Encelia**
7b. Plants herbaceous annuals; achenes strongly white margined, the margin continuous between stout awns. **Geraea**
8a. Achenes very flat, strongly compressed. (9)
8b. Achenes not very flat, usually much thickened. (10)
9a. Plants scapose. **Enceliopsis**
9b. Plants leafy stemmed. **Helianthella**
10a. Pappus caducous (of 2 awns and rarely some scales). **Helianthus**
10b. Pappus persistent. (11)
11a. Plants shrubs. **Viguiera**
11b. Plants herbaceous. (12)
12a. Inner involucral bracts united to middle into a cup. **Thelesperma**
12b. Inner involucral bracts not united into a cup. (13)
13a. Receptacle conic; rays neuter. **Rudbeckia**
13b. Receptacle merely convex; rays pistillate. **Wyethia**
14a. Rays white or purple. (15)
14b. Rays yellow, sometimes marked with purple. (20)
15a. Pappus a short crown. **Chrysanthemum**
15b. Pappus of awns or scales. (16)
16a. Pappus of 1 awn and a denticulate crown. **Monoptilon**

Compositae

16b. Pappus of 2 or several awns or scales. (17)
17a. Plants dwarf woolly annuals. **Eriophyllum**
17b. Plants annual or perennial, not woolly. (18)
18a. Leaves glandular-punctate; plants perennial, from a thick caudex, the caudex branches woolly-villous; known from Duchesne and Emery cos. **Parthenium**
18b. Leaves not glandular-punctate; plants slender annuals or, if perennial and with a caudex, the caudex branches not woolly-villous. (19)
19a. Pappus of numerous awns or scales; involucral bracts conspicuously scarious margined. **Townsendia**
19b. Pappus of 4 or 5 stiff awns; involucral bracts obscurely scarious margined. **Rigiopappus**
20a. Receptacle densely bristly or hairy. (21)
20b. Receptacle naked. (23)
21a. Heads very small, with 12 flowers or less. **Gutierrezia**
21b. Heads medium sized, with more than 12 flowers. (22)
22a. Pappus of 5–10 often aristate scales. **Gaillardia**
22b. Pappus of 15–18 awns and as many shorter bristles or awns. **Acamptopappus**
23a. Pappus a mere crown or of caducous awns. (24)
23b. Pappus persistent, of awns or scales. (26)
24a. Pappus of 2–8 caducous awns. **Grindelia**
24b. Pappus a short crown. (25)
25a. Leaves entire, bristly margined. **Pectis**
25b. Leaves 2-pinnate or 3-pinnate. **Tanacetum**
26a. Pappus of 1 or 2 awns or scales, with or without a crown. (27)
26b. Pappus of 4 to many awns or scales. (28)
27a. Pappus of a single awn, without a crown. **Laphamia**
27b. Pappus of 1–2 awns and a low crown. **Pectis**
28a. Pappus of about 20 slender, twisted awns; rays 1 or 2, small. **Amphipappus**
28b. Pappus of 4–16 twisted awns or scales; rays usually several. (29)
29a. Pappus of 4 or 5 stiff, narrowly lanceolate awns; achenes linear, transversely rugulose. **Rigiopappus**
29b. Pappus of scales or setose-dissected awns. (30)
30a. Pappus of several scales dissected nearly to base; dwarf woolly annuals. **Syntrichopappus**
30b. Pappus awns or scales not dissected or else plants perennial or woody. (31)
31a. Pappus of several more or less united scales; rays broad, papery, and persistent. **Psilostrophe**
31b. Pappus not of united scales; rays not papery and persistent (occasionally so in **Hymenoxys**). (32)
32a. Leaves and involucre with conspicuous oil glands. **Dyssodia**

Subclass DICOTYLEDONEAE

32b. Leaves and involucre without conspicuous oil glands. (33)
33a. Achenes slender, elongate-clavate. (34)
33b. Achenes stouter, oblong or obovoid. (35)
34a. Plants woolly. **Eriophyllum**
34b. Plants not woolly. **Bahia**
35a. Involucral bracts spreading or reflexed; receptacle convex; leaves decurrent. **Helenium**
35b. Involucral bracts appressed; receptacle not convex; leaves not decurrent. (36)
36a. Pappus of numerous scales; stems leafy; leaves linear or linear-spatulate, entire, 2.5 mm wide or less. **Gutierrezia**
36b. Pappus of about 5 scales; leaves lobed or, if entire, broader and mostly or entirely basal. **Hymenoxys**

Acamptopappus Gray   Goldenhead

Desert shrubs with white bark, 2–9 dm tall; leaves alternate, linear-spatulate, 5–20 mm long; heads subglobose, 7–10 mm high, discoid, the involucre 5–6 mm high; pappus persistent, of 30–40 flattened scales and bristles. **A. sphaerocephalus** (Harv. & Gray) Gray

Achillea L.   Yarrow

Strongly scented perennial herbs from rhizomes; stems 10–60 cm tall; leaves alternate, finely 2-pinnatifid; involucral bracts with a greenish keel; rays 2–6 mm long, white or rose to red-purple. **A. millefolium** L.

Agoseris Raf.   Mountain Dandelion

1a. Plants annual. **A. heterophylla** (Nutt.) Greene
1b. Plants perennial. (2)
2a. Beak of achenes short and stout, scarcely half as long as body. **A. glauca** (Pursh) D. Dietr.
2b. Beak of achenes slender, as long as, or longer than, the body, rarely slightly shorter. (3)
3a. Beak of achenes twice as long as body or more; inner involucral bracts, even in young fruit, about twice as long as the outer; involucre in fruit 3–3.5 cm high. **A. grandiflora** (Nutt.) Greene
3b. Beak of achenes not twice as long as body; inner involucral bracts more evenly graduated; involucre in fruit 2.5–3 cm high or less. (4)
4a. Flowers light yellow, often turning pinkish. **A. arizonica** Greene
4b. Flowers deep orange or brownish red changing to purple. **A. aurantiaca** (Hook.) Greene

Ambrosia L.   Ragweed, Bursage

1a. Pistillate involucre nutlike, with 4–6 teeth or tubercles in a single series below the tip. (2)
1b. Pistillate involucre bearing numerous long prickles in several series. (3)

Compositae

2a. Plants annual; leaves thin, at least the lower usually 2-pinnatifid. **A. artemisiifolia** L.
2b. Plants perennial; leaves thickish, mostly 1-pinnatifid. **A. coronopifolia** T. & G.
3a. Plants annual, herbs; leaves 5-cleft to 3-pinnatifid; fruit 5–8 mm long, armed with flattish lance-subulate spines; common, especially in eastern and southern Utah. **A. acanthicarpa** Hook.
3b. Plants shrubs; known from southwestern Utah. (4)
4a. Leaves densely canescent-strigillose, 2- to 3-pinnately divided into small, mostly ovate to obovate divisions; fruit glandular to sparsely pilose. **A. dumosa** (Gray) Payne
4b. Leaves greenish above, densely whitish- or canescent-tomentulose beneath, crenate-serrate to pinnatifid; fruit villous to tomentose. **A. eriocentra** (Gray) Payne

Amphipappus T. & G. Chaffbush
Shrubs with spinescent branches, 3–6 dm tall; leaves alternate, 5–12 mm long; heads 4–5 mm high; pappus of about 15–20 short white scales. **A. fremontii** T. & G.

Anaphalis DC. Pearly Everlasting
Perennial herbs from rhizomes, 20–80 cm tall, gray-tomentose; leaves alternate, 5–15 cm long, glabrate and green above, tomentose below; involucre 5–7 mm high, woolly at base; involucral bracts papery, with milk-white tips. **A. margaritacea** (L.) Gray

Antennaria Gaertn. Pussytoes
1a. Plants rarely over 5 cm tall; heads usually solitary, sessile, or subsessile among the leaves or on short peduncles rarely over 3 dm long. (2)
1b. Plants over 5 cm tall; heads usually several on stems rising well above basal leaves. (4)
2a. Leaves less than 10 cm long; heads sessile or on stems 1 cm long or less. **A. rosulata** Rydb.
2b. Leaves mostly over 10 mm long; heads often on stems over 1 cm long. (3)
3a. Plants with spreading leafy stolons; pappus bristles of staminate flowers clavate; inner involucral bracts of pistillate heads oblong, obtuse or acute; heads usually more than 1. **A. parvifolia** Nutt.
3b. Plants lacking stolons; pappus bristles of staminate flowers not at all clavate; inner involucral bracts of pistillate heads linear-lanceolate, acuminate to attenuate at apex; heads 1. **A. dimorpha** (Nutt.) T. & G.
4a. Leaves, at least on flowering stems, soon glabrous and green above. (5)
4b. Leaves permanently tomentose on both sides. (6)

Figs. 25-26. 25. **Salicornia pacifica,** x .44.   26. **Achillea millefolium,** x .22

Figs. 27-28.   27. **Agoseris aurantiaca,** x .21.   28. **Anaphalis margaritacea,** x .2

Compositae

5a. Involucral bracts whitish near apex. **A. neglecta** Greene
5b. Involucral bracts greenish brown or brown near apex. **A. umbrinella** Rydb.
6a. Basal leaves, at least some, over 2.5 cm long; plants lacking horizontal stolons. (7)
6b. Basal leaves mostly less than 20 mm long; plants with spreading leafy stolons. (8)
7a. Involucres 4–5 mm high; leaves mostly 3–6 cm long; stems seldom over 25 cm tall. **A. luzuloides** T. & G.
7b. Involucres 6 mm long or more; leaves mostly over 6 cm long; stems often over 24 cm tall. **A. anaphaloides** Rydb.
8a. Pistillate heads 8–10 mm high; involucres 6–9 mm high; stems seldom over 15 cm tall. **A. parvifolia** Nutt.
8b. Pistillate heads 5–8 mm high; involucres 5–6 mm high, rarely longer; stems over 15 cm tall. (9)
9a. Involucral bracts, at least the inner, rose colored or pink. (10)
9b. Involucral bracts white to brown at apex, not at all rose or pink. (11)
10a. Leaves cuneate-spatulate, the blade not distinct from petiole; outer bracts of pistillate heads brown, inner pink. **A. concinna** E. Nels.
10b. Leaves oblanceolate, spatulate, or obovate, the petioles distinct; all bracts of pistillate heads pink. **A. rosea** (D. C. Eaton) Greene
11a. Involucral bracts green or brown at base, white above; leaves usually not over 1 cm long. **A. microphylla** Rydb.
11b. Involucral bracts dark brownish green at base, green or brown or only slightly whitish above; leaves mostly over 1 cm long. **A. media** Greene

Anthemis L. Camomile

1a. Rays yellow; pappus a short crown; disk commonly more than 12 mm broad; cultivated ornamental, escaping and persisting. Yellow Camomile. **A. tinctoria** L.
1b. Rays white; pappus lacking disk commonly less than 10 mm broad; weedy plant of disturbed soils. May Weed **A. cotula** L.

Arctium L. Burdock

Coarse biennial herbs, 1–1.5 m tall; lower leaves 2–3 dm long, mostly cordate, with hollow petioles; heads usually many, racemose; flowers all tubular and perfect; corollas purplish or pink; involucre globose, bracts with inwardly hooked tips; pappus of short, narrow, deciduous scales. **A. minus** Schk.

Arnica L. Arnica

1a. Cauline leaves mostly 5–12 pairs. (2)
1b. Cauline leaves mostly 2–4 pairs. (3)
2a. Involucral bracts obtuse to acutish, bearing an apical tuft of hairs at or within the tip. **A. chamissonis** Less.

Subclass DICOTYLEDONEAE

- 2b. Involucral bracts sharply acute, lacking a tuft of hairs at the tip. **A. longifolia** D. C. Eaton
- 3a. Heads discoid, lacking rays. **A. parryi** Gray
- 3b. Heads with ray flowers. (4)
- 4a. Pappus subplumose, brownish; basal leaves much reduced. (5)
- 4b. Pappus merely barbellate, white or nearly so; basal leaves enlarged or reduced. (6)
- 5a. Cauline leaves ovate or broader, or sometimes merely lance-elliptic; heads narrow. **A. diversifolia** Greene
- 5b. Cauline leaves more variable; heads subhemispheric. **A. mollis** Hook.
- 6a. Leaf blades truncate to cordate basally; basal leaves seldom tufted. **A. cordifolia** Hook.
- 6b. Leaf blades ordinarily attenuate basally; basal leaves ordinarily tufted. (7)
- 7a. Heads mostly with 7–10 rays; lower cauline leaves sessile or nearly so. **A. rydbergii** Greene
- 7b. Heads with mostly 10–20 rays; lower cauline leaves commonly petiolate. (8)
- 8a. Bases of old leaves bearing dense tufts of long brown hairs in the axils; disk corollas with both stipitate glands and nonglandular hairs. **A. fulgens** Pursh
- 8b. Bases of old leaves lacking hairs or with few white hairs; disk corollas with stipitate glands only. **A. sororia** Greene

Artemisia L. Sagebrush, Wormwood

- 1a. Plants herbaceous throughout, or woody only at base. (2)
- 1b. Plants shrubs. (6)
- 2a. Receptacles densely hairy between flowers. **A. absinthium** L.
- 2b. Receptacles not evidently hairy between flowers. (3)
- 3a. Leaves glabrous, dark green; stems erect, 3–13 dm tall. (4)
- 3b. Leaves thinly or densely canescent or tomentose. (5)
- 4a. Stems single or few, stout and coarse, leafy, without odor; leaves 2-pinnately divided, the lobes laciniately toothed; annual or biennial. **A. biennis** Willd.
- 4b. Stems numerous, slender; leaves linear, aromatic, entire or slightly toothed; perennial. **A. dracunculus** L.
- 5a. Plants shrubby at base; leaf blades small, 3–12 mm long, 2-ternately to quinately divided into small, crowded lobes, silvery silky-canescent; heads small and numerous in racemes or open panicles. **A. frigida** Willd.
- 5b. Plants herbaceous throughout; leaf blades larger, 3–11 cm long, entire, toothed or pinnatifid with few widely spaced lobes. **A. ludoviciana** Nutt.

6a. Shrubs spinescent; leaves crowded, mostly 1 cm long or less, closely 3- to 5-palmately parted, the segments 3-lobed, white-tomentose; heads in glomerate racemes on short, leafy branchlets. Bud Sagebrush. **A. spinescens** D. C. Eaton
6b. Shrubs not spinescent; leaves linear, spatulate, cuneate or narrowly lanceolate, entire, 3-toothed or 3-parted. (7)
7a. Leaves linear-filiform, entire or 3-parted, ample, fine and lax, silvery white-canescent; heads very small; mostly sandy soil. Old Man Sagebrush. **A. filifolia** Torr.
7b. Leaves linear to broader, firm; heads small. (8)
8a. Leaves 3-parted near apex, the lobes linear to narrowly spatulate. **A. tripartita** Rydb.
8b. Leaves entire or 3-toothed at apex. (9)
9a. Leaves linear-oblong to linear-lanceolate, 3–5 cm long, entire or only a few 3-toothed; inflorescence leafy. **A. cana** Pursh
9b. Leaves 3 cm long or less, 3-toothed; inflorescence mostly tall and naked. (10)
10a. Leaves oblong to linear-cuneate, 1–1.5 cm long, mostly finely tridentate with some entire; pale whitish, with tall narrow panicles. **A. bigelovii** Gray
10b. Leaves cuneate, 1–3 cm long. (11)
11a. Plants medium to tall shrubs, 0.5–4 m tall, pale grayish; panicles not naked. Big Sagebrush. **A. tridentata** Nutt.
11b. Plants low shrubs, 1–2 dm tall, rarely taller, dull grayish-tomentose; naked, tall, narrow spikelike panicles extending above the herbage. Black Sagebrush. **A. nova** A. Nels.

Aster L. Aster

1a. Plants annual; stem leaves linear, linear-lanceolate, or oblong, entire; rays short and inconspicuous or lacking; wet or damp saline soil. (2)
1b. Plants perennial; leaves various; rays conspicuous. (3)
2a. Involucral bracts linear; rays usually lacking. **A. brachyactis** Blake
2b. Involucral bracts oblong, obtuse; rays about 2 mm long. **A. frondosus** (Nutt.) T. & G.
3a. Stems low, usually 10 cm high or less, numerous from a branching caudex; leaves usually less than 1 cm long. (4)
3b. Stems taller, arising from rhizomes; leaves mostly more than 1 cm long. (5)
4a. Leaves hispid-ciliate, at least the upper surface densely glandular. **A. hirtifolius** Blake
4b. Leaves, at least some, not or only inconspicuously ciliate. **A. arenosus** (Heller) Blake
5a. Involucral bracts dry and chartaceous, with thin or scarious tips, strongly keeled dorsally. (6)

Subclass DICOTYLEDONEAE

5b. Involucral bracts with distinct, thickened, herbaceous tips or herbaceous throughout, seldom keeled dorsally. (7)

6a. Involucral bracts all acute; plants often more than 5 dm tall; leaves green. **A. engelmannii** (D. C. Eaton) Gray

6b. Involucral bracts obtuse, at least the outer ones; plants mostly 3–5 dm tall; leaves glaucous. **A. glaucodes** Blake

7a. Rays 5–13 (21); involucral bracts tending to be keeled; lower stem leaves reduced. (8)

7b. Rays commonly more than 21, or involucral bracts not keeled, or the lower stem leaves well developed, or all of these. (9)

8a. Rays white; plants glandular or glandular-hairy. **A. paucicapitatus** Robins.

8b. Rays pink-purple to violet; plants not glandular (except in inflorescence). **A. perelegans** Nels. & Macbr.

9a. Involucre and peduncles glandular. **A. integrifolius** Nutt.

9b. Involucre and peduncles not glandular. (10)

10a. Plants of bogs, slender; leaves commonly 2–5 mm wide. **A. junciformis** Rydb.

10b. Plants seldom or not of bogs, slender or otherwise; leaves mostly more than 10 mm wide, at least some. (11)

11a. Pubescence of stems occurring in decurrent lines below leaf bases; inflorescence leafy, generally large. **A. hesperius** Gray

11b. Pubescence of stems uniform, seldom in lines; inflorescence various. (12)

12a. Involucral bracts with loose or recurved, spinulose tips; rays white. (13)

12b. Involucral bracts appressed or spreading, not spinulose; rays pink to lavender, violet, or white. (14)

13a. Plants with creeping rhizomes. **A. falcatus** Lindl. in Hook.

13b. Plants with very short rhizomes and clustered stems. **A. pansus** (Blake) Cronq.

14a. Involucral bracts strongly graduated, at least the outer ones obtuse, much shorter than the inner, not foliaceous. **A. chilensis** Nees

14b. Involucral bracts not strongly graduated or, if strongly graduated, then strongly acute; bracts acute or, if obtuse, then foliaceous. (15)

15a. Achenes and plants glabrous, or nearly so, except for pubescence in the inflorescence in some plants; leaves often glaucous. **A. laevis** L.

15b. Achenes usually more or less hairy; leaves and stems variously hairy or subglabrous, scarcely glaucous. (16)

16a. Inflorescence a narrow, long, leafy panicle, usually with numerous heads; leaves mostly more than 7 times as long as wide; rays pink or white. **A. eatonii** (Gray) Howell

16b. Inflorescence with a few heads or, if many-headed, then shorter and more open; leaves various; rays usually blue or violet. (17)

Compositae

17a. Leaves and involucral bracts relatively small and narrow; main cauline leaves less than 1 cm wide, more than 7 times as long as wide; involucral bracts not foliaceous. **A. occidentalis** (Nutt.) T. & G.
17b. Leaves and involucral bracts relatively large; main cauline leaves mostly over 1 cm wide, less than 7 times as long as wide; involucral bracts often foliaceous (includes **A. foliaceus** Lindl. in DC.). **A. subspicatus** Nees

Atrichoseris Gray

Glabrous, winter annuals; leaves obovate, spinulose toothed, often spotted; heads white; involucre of 12–15 equal, scarious-margined bracts in 2 series; pappus lacking; southwestern Utah. **A. platyphylla** Gray

Baccharis L. Baccharis

1a. Branches slender, dense, strongly angled, broomlike; leaves few, very small, 5–15 mm long, soon falling. **B. sergiloides** Gray
1b. Branches stouter and fewer, striate or angled, not broomlike; leaves numerous, 20–50 mm long or more, persistent. (2)
2a. Pappus of pistillate flowers 8–12 mm long, far exceeding the styles. **B. emoryi** Gray
2b. Pappus of pistillate flowers 3–4 mm long, equaling the styles. (3)
3a. Heads in small clusters, terminating numerous, short, lateral branchlets. **B. viminea** DC.
3b. Heads in terminal corymbs or panicles. **B. glutinosa** (R. & P.) Pers.

Bahia Lag. Bahia

1a. Leaves 1- to 3-ternately dissected, the ultimate lobes oblong to oblanceolate; pappus lacking; middle elevations in mountains. **B. dissecta** (Gray) Britt.
1b. Leaves entire, the blades lanceolate to elliptic, or ovate; pappus present. (2)
2a. Involucral bracts attenuate, sparsely pubescent; plants merely puberulent, not stipitate-glandular; eastern Utah. **B. ourolepis** Blake
2b. Involucral bracts obtuse or acute, stipitate-glandular; plants stipitate-glandular. (3)
3a. Stems scapelike; leaves mostly basal; leaf blades ovate to lanceolate. **B. nudicaulis** Gray
3b. Stems not scapelike, leafy throughout, though reduced upward; leaf blades elliptic to oblong. **B. oblongifolia** Gray

Baileya Harv. & Gray   Wild Marigold

1a. Plants annual; stems leafy throughout. **B. pleniradiata** Harv. & Gray
1b. Plants biennial; stems leafy below, nearly scapose above. **B. multiradiata** Harv. & Gray

Subclass DICOTYLEDONEAE

Balsamorrhiza Nutt. Balsamroot

1a. Leaf blades sagittate, entire or merely crenate; herbage permanently soft-hairy. Arrow-leaf Balsamroot. **B. sagittata** (Pursh) Nutt.
1b. Leaf blades pinnately cleft to divided; herbage green, not or only rarely soft-hairy. (2)
2a. Plants large; leaf segments mostly 4.5–11 cm long, dark green, glandular-viscid. Large-leaf Balsamroot. **B. macrophylla** Nutt.
2b. Plants small to moderate in size; leaf segments commonly 1.5–3 cm long. (3)
3a. Involucral bracts, at least some, abruptly narrowed to a slender, elongate apex; cauline leaves conspicuous, 1- to 2-pinnate. Smaller Balsamroot. **B. hirsuta** Nutt.
3b. Involucral bracts gradually tapering to the apex; cauline leaves lacking or inconspicuous, sometimes toothed or pinnatifid. Hooker Balsamroot. **B. hookeri** Nutt.

Bidens L. Beggar's Ticks

1a. Leaves simple, sessile, entire or serrate; heads hemispheric, nodding; rays present, short or lacking. **B. cernua** L.
1b. Leaves pinnately parted or divided with 3–5 leaflets, petiolate; heads flat or convex, not nodding; rays lacking. **B. frondosus** L.

Brickellia Ell. Brickellbush

1a. Leaves very narrowly lanceolate to ovate-lanceolate; heads 3- to 5-flowered. **B. longifolia** S. Wats.
1b. Leaves deltoid-ovate to elliptic or spatulate; heads 10- to 60-flowered. (2)
2a. Leaves elliptic to spatulate, entire or subentire. **B. oblongifolia** S. Wats.
2b. Leaves deltoid-ovate or ovate. (3)
3a. Petioles mostly 1–7 cm long, half as long as blades. **B. grandiflora** (Hook.) Nutt.
3b. Petioles short, mostly less than 1 cm long. (4)
4a. Involucral bracts glabrous, green with pale stripes. **B. californica** (T. & G.) Gray
4b. Involucral bracts glandular-pubescent or hispidulous. (5)
5a. Stems glandular-villous; achenes 4–4.5 mm long. **B. microphylla** (Nutt.) Gray
5b. Stems lanate or glandular-puberulent; achenes 3.5 mm long. (6)
6a. Stems finely lanate. **B. watsonii** Robins.
6b. Stems glandular-puberulent. **B. scabra** (Gray) A. Nels. ex Robins.

Carduus L. Musk Thistle

Biennial herbs, 0.5–2 dm tall; leaves to 4 dm long and 1.5 dm wide; heads solitary, nodding, the disk 4–8 cm wide (when pressed); involucral bracts with spreading or reflexed spiny tips; flowers pink-purple; widespread in Utah. **C. nutans** L.

Figs 29-30.   29. **Antennaria rosea**, x .26.   30. **Arnica mollis**, x .2

Figs. 31-32.   31. **Balsamorrhiza sagittata**, x .2.   32. **Bidens cernua**, x .2

Subclass DICOTYLEDONEAE

Carthamnus L. Safflower

Annual glabrous herbs, 2–10 dm tall; leaves ovate, clasping; heads all discoid, with chaffy receptacles; corollas yellow-orange; pappus of scales, 3–5 mm long; occasional weed. **C. tinctorius** L.

Centaurea L. Knapweed

1a. Plants annual; corollas blue, the marginal ones enlarged, deeply lobed, and spreading. Bachelor's Buttons. **C. cyanus** L.
1b. Plants perennial, from rhizomes; corollas purplish or pinkish, short, the marginal ones neither enlarged nor spreading. Russian Knapweed. **C. repens** L.

Chaenactis DC.

1a. Plants alpine perennials, with 1 to several rosettes, usually less than 1 dm tall; mountains of northern Utah. **C. alpina** (Gray) M. E. Jones
1b. Plants annual, biennial, or perennial, 0.5–3 dm tall; seldom alpine. (2)
2a. Involucral bracts alternate; herbage scurfy-puberulent, not tomentose, glandless or nearly so; receptacle with some bristles; southwestern Utah. **C. carphoclinia** Gray
2b. Involucral bracts obtuse to acuminate; herbage tomentose or glandular, at least when young; receptacle naked. (3)
3a. Leaves entire and linear, or pinnatifid with few unequal lobes; herbage soon glabrate; southern Utah. **C. fremontii** Gray
3b. Leaves 2-pinnate, at least some, with numerous lobes; herbage more or less tomentose. (4)
4a. Involucres 10–15 mm long; anthers included; outer involucral bracts with loose tips; mostly of western and southwestern Utah. **C. macrantha** D. C. Eaton
4b. Involucres 6–10 mm high; anthers exserted; outer involucral bracts scarcely loose tipped. (5)
5a. Plants annual; pappus of 4–5 scales; outermost corollas larger than inner. **C. stevioides** H. & A.
5b. Plants biennial or perennial; pappus with more than 5 scales; outer corollas not enlarged. **C. douglasii** (Hook.) H. & A.

Chamaechaenactis Rydb.

Perennial herbs from a branching caudex; leaves all basal, simple, 8–12 mm long; heads solitary, scapose, the scapes to 8 cm tall; involucres 10–16 mm long; rays lacking; disk flowers flesh colored; pappus of about 8 scales; eastern Utah. **C. scaposa** (Eastw.) Rydb.

Chrysanthemum L. Chrysanthemum

1a. Heads solitary or few, large, the disk commonly 1–2 cm broad; rays

Compositae

1–2 cm long; weedy plant of roadsides and disturbed soils. Oxeye Daisy. **C. leucanthemum** L.
1b. Heads numerous, small, the disk commonly less than 1 cm broad; rays usually lacking; weedy plant of roadsides. Costmary. **C. balsamita** L.

Chrysothamnus Nutt. Rabbitbrush (Contributed by Loran C. Anderson)

1a. Flowers white; alkaline areas, Millard to Box Elder Co. and westward. **C. albidus** (M. E. Jones) Greene
1b. Flowers pale to dark yellow. (2)
2a. Leaves resinous-punctate, terete; sandy washes, Spring Wash (Castle Cliff), Washington Co. and west. **C. paniculatus** (Gray) Greene
2b. Leaves neither resinous-punctate nor truly terete. (3)
3a. Twigs glabrous or finely puberulent. (4)
3b. Twigs tomentose, often densely compacted. (14)
4a. Achenes glabrous or glandular only at distal end. (5)
4b. Achenes densely pubescent. (7)
5a. Flowers less than 9 mm long, noticeably surpassing the pappus. (6)
5b. Flowers over 10 mm long, but hidden by abundant pappus; San Raphael Swell, Emery Co. **C. pulchellus** (Gray) Greene
6a. Involucre 10–13 mm high, the bracts strongly aligned; Uintah Co. south to San Juan Co., then southwestward in mountain areas to Iron Co. **C. depressus** Nutt.
6b. Involucre 5–7 mm high, the bracts not strongly aligned; south side of Uinta Mountains of Duchesne and Wasatch cos., then south in mountains and foothills of central portions of Utah into Arizona. **C. vaseyi** (Gray) Greene
7a. Involucral bracts narrowly acuminate; dry flats over much of Utah, often seen in almost pure strands with **Eurotia lanata** in western tier of counties. **C. greenei** (Gray) Greene
7b. Involucral bracts obtuse to acute. (8)
8a. Heads somewhat turbinate, 5- to 6-flowered; shrubs large with flat leaves 4–8 mm wide; drainage areas of the Colorado-Green River system from Kane Co. northeastward. **C. linifolius** Greene
8b. Heads narrowly cylindric, 3- to 5-flowered; shrubs of various sizes with usually twisted leaves 1–6 mm wide (**C. viscidiflorus** complex). (9)
9a. Leaves and upper stem glabrous. (10)
9b. Leaves and upper stem pubescent. (12)
10a. Shrubs over 4 dm tall; leaves 1–5 mm wide; scattered, but mostly over southern part of Utah. **C. viscidiflorus** (Hook.) Nutt.
10b. Shrubs less than 4 dm tall; leaves 1–2 mm wide. (11)
11a. Leaves linear, 1–2 mm wide; occurrence questionable but, if it occurs, only in northern part of Utah. **C. viscidiflorus** (Hook.) Nutt.

Subclass DICOTYLEDONEAE

11b. Leaves filiform, 1 mm wide; dry flats, mostly in western and extreme southern part of Utah. **C. viscidiflorus** (Hook.) Nutt. var. **stenophyllus** (Gray) Hall

12a. Leaves 2.5–6 mm wide; common in mountains and foothills throughout Utah. **C. viscidiflorus** (Hook.) Nutt. var. **lanceolatus** (Nutt.) Greene

12b. Leaves 2 mm wide or less. (13)

13a. Bracts with thickened greenish spot near tip; dry flats, mostly east of Wasatch Mountains and central high plateaus. **C. viscidiflorus** (Hook.) Nutt. var. **elegans** (Greene) Blake

13b. Bracts lacking thickened spot near tip; dry flats, mostly west of Wasatch Mountains and central high plateaus. **C. viscidiflorus** (Hook.) Nutt. var. **puberulus** (D. C. Eaton) Jepson

14a. Bracts very attenuate; inflorescence mostly racemose (**C. parryi** complex). (15)

14b. Bracts obtuse to acute; inflorescence mostly cymose (**C. nauseosus** complex). (18)

15a. Flowers 10 or more per head; most collections are from western Kane Co., but to be expected in the Henry, La Sal, and Abajo mountains. **C. parryi** (Gray) Greene var. **parryi**

15b. Flowers 9 or fewer per head. (16)

16a. Upper leaves longer than inflorescence; flowers 8–10 mm long, pale yellow; north of the Uinta Mountains in Daggett Co. and eastward. **C. parryi** (Gray) Greene var. **howardii** (Parry) Kittell in Tidestr. & Kittell

16b. Upper leaves shorter than inflorescence; flowers 9–12 mm long, clear yellow. (17)

17a. Tips of bracts straight; heads 11–15 mm long; common south of Uinta Mountains through the central high plateaus. **C. parryi** (Gray) Greene var. **attenuatus** (M. E. Jones) Kittell in Tidestr. & Kittell

17b. Tips of bracts recurved; heads 14–19 mm long; occasional in high mountains of southwestern Utah. **C. parryi** (Gray) Greene var. **nevadensis** (Gray) Kittell in Tidestr. & Kittell

18a. Involucres puberulent or tomentose; leaves usually grayish white. (19)

18b. Involucres glabrous; leaves usually greenish yellow. (22)

19a. Leaves 3–10 mm wide; bracts mostly obtuse; Wasatch and western Uinta mountains, Cache Co. to Sanpete Co. **C. nauseosus** (Pall.) Britt. var. **salicifolius** (Rydb.) Hall

19b. Leaves narrower than 3 mm; bracts various, usually acute. (20)

20a. Involucre over 10 mm long; corolla lobes villous; dry sand in Kane and Iron cos. north to Tooele Co. **C. nauseosus** (Pall.) Britt. var. **turbinatus** (M. E. Jones) Blake

20b. Involucre less than 10 mm long; corolla lobes glabrous. (21)

21a. Corolla lobes 1–2 mm long; style appendage longer than stigmatic

Compositae

portion; scattered throughout state, mostly in mountains. **C. nauseosus** (Pall.) Britt. var. **albicaulis** (Nutt.) Rydb.

21b. Corolla lobes 0.5–1 mm long; style appendage shorter than stigmatic portion; occasional in western tier of counties. **C. nauseosus** (Pall.) Britt. var. **gnaphaloides** (Greene) Hall

22a. Achenes glabrous; dry areas in southern and western tiers of counties. **C. nauseosus** (Pall.) Britt. var. **leiospermus** (Gray) Hall

22b. Achenes pubescent. (23)

23a. Leaves 1–3 mm wide, 3- to 5-nerved; widespread, but sporadic, more frequent in southern counties. **C. nauseosus** (Pall.) Britt. var. **graveolens** (Nutt.) Hall

23b. Leaves less than 1 mm wide, 1-nerved. (24)

24a. Involucre 7–8 mm long; corolla lobes glabrous; widespread and abundant. **C. nauseosus** (Pall.) Britt. var. **consimilis** (Greene) Hall

24b. Involucre 9–10 mm long; corolla lobes loosely villous; dry sand, Kane Co. **C. nauseosus** (Pall.) Britt. var. **junceus** (Greene) Hall

Cichorium L. Chickory

Plants perennial from a fleshy taproot; stems 3–9 dm tall; lower leaves spatulate, the upper ones smaller, clasping and often auriculate at the base; heads 2.5–4 cm across, sessile, single or 2–4 in cluster; corollas spreading, blue or sometimes pink or white, showy; escaped cultivation. Blue Sailors. **C. intybus** L.

Cirsium Mill.

1a. Plants partly dioecious; heads unisexual; perennial with creeping rootstocks; introduced weed. Canada Thistle. **C. arvense** (L.) Scop.

1b. Plants with perfect flowers; biennial or perennial, seldom if ever with creeping rootstocks. (2)

2a. Leaves roughened-hairy above; stems conspicuously winged by decurrent, spiny leaf bases; biennial, introduced weed. Bull Thistle. **C. vulgare** (Savi) Airy-Shaw

2b. Leaves villous to floccose, arachnoid, tomentose, or glabrous above, not at all roughened-hairy; leaf bases only shortly if at all decurrent; native. (3)

3a. Basal rosettes to 10 dm across, the mature leaves often more than 10 cm wide, green and glabrate or glabrous; heads small, with tapering, recurved spines; alcoves and grottos, rarely along stream courses, in Grand, San Juan, Wayne, Garfield, and Kane cos. Rydberg Thistle. **C. rydbergii** Petrak

3b. Basal rosettes often less than 5 dm across, the mature leaves mostly less than 8 cm wide, more or less tomentose, floccose, or arachnoid on one or both surfaces; heads various; not or seldom of alcoves or grottos, of various habitats. (4)

4a. Involucral bracts densely arachnoid to arachnoid-tomentose. (5)

Subclass DICOTYLEDONEAE

4b. Involucral bracts glabrous or sparingly arachnoid or tomentose, usually along margins. (8)
5a. Middle involucral bracts spreading, the outer ones reflexed. (6)
5b. Middle involucral bracts commonly appressed, the outer appressed or merely spreading. (7)
6a. Involucres 2.5–3 cm high, with stout spines commonly 6–15 mm long; leaves commonly white-tomentose on both sides. New Mexico Thistle. **C. neomexicanum** Gray
6b. Involucres commonly 1.5–2.5 cm high, with slender spines mostly 3–9 mm long; leaves commonly greenish above; closely related and possibly not distinct from the above. Utah Thistle. **C. utahense** Petrak
7a. Inner involucral bracts with conspicuously dilated, scarious tips; high mountains. Parry Thistle. **C. parryi** (Gray) Petrak
7b. Inner involucral bracts merely attenuate, not or scarcely dilated apically; reported from high elevations in mountains. **C. scopulorum** (Greene) Cockerell in Daniels
8a. Outer involucral bracts spinulose-ciliate; subalpine regions. (9)
8b. Outer involucral bracts entire; various distribution but seldom subalpine. (10)
9a. Involucral bracts very spiny, about equal; leaves undulate, very spiny. **C. catonii** (Gray) Robins.
9b. Involucral bracts sparingly spiny, markedly graduated; leaves flattish, weakly spiny. **C. clavatum** (M. E. Jones) Petrak
10a. Outer or middle involucral bracts with a conspicuous, glutinous, dorsal ridge. (11)
10b. Outer or middle involucral bracts lacking a conspicuous, glutinous, dorsal ridge. (14)
11a. Middle stem leaves strongly decurrent, forming broad wings along the stem; plants only slightly arachnoid. **C. calcareum** (M. E. Jones) Woot. & Standl.
11b. Middle stem leaves not or seldom decurrent, not forming wings on the stem; plants more or less persistently tomentose. (12)
12a. Heads large, 3–5 cm high, solitary or few. **C. undulatum** (Nutt.) Spreng.
12b. Heads smaller, to 3 cm high, several to many. (13)
13a. Flowers pale, cream colored; leaves greenish, more or less floccose, or glabrate above. **C. canovirens** (Rydb.) Petrak
13b. Flowers bright, ordinarily rose to purplish; leaves persistently tomentose. **C. canescens** Nutt.
14a. Leaves glabrous or glabrate on both sides. (15)
14b. Leaves persistently tomentose, at least beneath. (17)
15a. Spines of middle involucral bracts about 1 cm long, longer than body of bracts; low elevations in southern Utah. **C. rothrockii** (Gray) Petrak

Compositae

15b. Spines of middle involucral bracts commonly 3–7 mm long, shorter than body of bracts; mostly of middle elevations and of broader distribution. (16)
16a. Middle stem leaves strongly decurrent, the stem winged. **C. calcareum** (M. E. Jones) Woot. & Standl.
16b. Middle stem leaves only slightly or not at all decurrent, the stem not winged. **C. bipinnatum** (Eastw.) Petrak
17a. Middle involucral bracts with stout, yellowish spines 10–20 mm long. **C. nidulum** (M. E. Jones) Petrak
17b. Middle involucral bracts with shorter, slender spines. (18)
18a. Flowers bright red or crimson. **C. arizonicum** (Gray) Petrak
18b. Flowers pink-purple. **C. pulchellum** (Greene) Woot. & Standl.

Crepis L. Hawksbeard

1a. Plants glabrous, dwarf; leaves mostly entire; alpine areas at high elevations, possibly throughout the state. **C. nana** Richards.
1b. Plants not completely glabrous or dwarf; leaves generally toothed to pinnatifid; various elevations. (2)
2a. Stems and leaves glabrous or sometimes hispidulous but never tomentose; leaves essentially basal; moist places. (3)
2b. Stems and leaves tomentose or pubescent; at least a few well-developed cauline leaves; drier places. (5)
3a. Involucres glandular-pubescent. (4)
3b. Involucres not at all glandular; Uinta Mountains and Grand Co. in eastern Utah, Kane Co. in the south, and Wayne Co. north through Box Elder and Cache cos. **C. runcinata** T. & G. var. **glauca** (Nutt.) Babc. & Stebbins
4a. Basal leaves narrowly obovate, oblanceolate, or spatulate, 0.5–3.5 cm wide and 4–8 times as long; reported from Kane, Sanpete, and Salt Lake cos. **C. runcinata** T. & G. var. **runcinata**
4b. Basal leaves obovate, 3–8 cm wide and 2–4 times as long; through center of Utah from north to south. **C. runcinata** T. & G. var. **hispidulosa** (Howell) Babc. & Stebbins
5a. Involucres with long blackish setae; stems with long yellowish setae, at least at base, nonglandular; Box Elder Co. south to eastern Beaver Co., with an arm extending into Duchesne Co. (Note: Some forms of **C. atrabarba** may key here, but may be distinguished by their deeply pinnatifid leaves bearing narrow entire segments.) **C. modocensis** Greene
5b. Involucres and stems with setae sparse or lacking or, if setose, then glandular. (6)
6a. Inner involucral bracts glabrous; north central Utah, Piute, Grand, and San Juan cos. **C. acuminata** Nutt.
6b. Inner involucral bracts tomentose, may be also setose, glandular. (7)
7a. Involucre thick-cylindric, 5–10 mm wide at anthesis; plants 1–3

Subclass DICOTYLEDONEAE

dm tall; setae glandular if present; north central Utah south to Garfield Co., then to southwestern Utah, with collections also in La Sal (Grand Co.) and Abajo (San Juan Co.) mountains. **C. occidentalis** Nutt.

7b. Involucre narrow-cylindric, 3–5 mm wide at anthesis; plants up to 6 dm tall; setae not glandular if present. (8)

8a. Leaf segments linear or narrowly lanceolate, mostly entire; achenes generally greenish; Uintah Co. to Salt Lake Co., Garfield Co. **C. atrabarba** Heller

8b. Leaf segments mostly broader, lanceolate or deltoid, usually toothed; achenes yellowish to brown; south end of Great Salt Lake drainage and southern Utah. **C. intermedia** Gray

Conyza Less.

Weedy annuals, 2–10 dm tall; leaves numerous, the lower to 10 cm long and 1 cm wide, becoming narrower upward; heads numerous, with involucres 3–4 mm high, the bracts imbricate; rays short and inconspicuous, white; widespread. Horseweed. **C. canadensis** (L.) Cronq.

Dicoria T. & G.

1a. Upper leaves broadly ovate to oval; Washington Co. **D. canescens** Gray

1b. Upper leaves lanceolate to linear; southern Utah. **D. brandegei** Gray

Dyssodia Cav.

Perennial herbs, 0.5–2 dm tall; leaves alternate, pinnately divided into linear, spinulose-tipped segments, these bearing translucent glands; heads solitary on elongate bracteate peduncles; involucres 4–7 mm tall, the bracts in one series, these conspicuously glandular apically; rays yellow; pappus of few to several scales, these awn tipped; low elevations along the Colorado River and its tributaries. Dogweed. **D. thurberi** (Gray) A. Nels.

Encelia Adans.

1a. Heads numerous, in panicles; leaves mostly basal, closely and densely white-tomentose. **E. farinosa** Gray

1b. Heads solitary, on long pubescent peduncles; stems leafy, hispidulous-canescent. **E. frutescens** Gray

Enceliopsis (Gray) A. Nels.

1a. Plants hispid-scabrous; rays lacking; heads nodding in fruit; leaves oval to obovate; dry plains and mesas, southern Utah. **E. nutans** (Eastw.) A. Nels.

1b. Plants white-tomentose; rays present. (2)

2a. Stems scapose; leaf blades orbicular to spatulate; rays 1–2.5 cm long; Colorado River Basin and southern Utah. **E. nudicaulis** (Gray) A. Nels.

Compositae

2b. Stems leafy; leaf blades rhomboidal-obovate; rays 3.5–4 cm long; dry saline plains and foothills, southern Utah. **E. argophylla** (S. Wats.) A. Nels.

Erigeron L. Fleabane

1a. Plants annual or biennial. (2)
1b. Plants perennial from a woody caudex or rhizome. (7)
2a. Rays very short and inconspicuous, white or pinkish. (3)
2b. Rays conspicuous, purple, violet, or white. (4)
3a. Heads very small and numerous, about 3 mm high, in corymbose or paniculate clusters, the bracts graduated; stems single, strict with erect branches above, 2–12 dm tall; waste places. **Conyza canadensis** (L.) Cronq.
3b. Heads medium sized, fewer, 6–8 mm high, in leafy racemose clusters, the bracts not graduated; stems usually rather weak, 2 to several, 1–3 dm tall; damp or wet soil. **E. lonchophyllus** Hook.
4a. Heads large, 2–5 cm across, solitary or few; rays very slender, 100–150; stems simple or branched above, 1.5–6 dm tall; open places, waysides and meadows. **E. glabellus** Nutt.
4b. Heads small, 1–3 cm across; rays narrow, 30–100; stems branched from the base, 1–4 dm tall. (5)
5a. Stems several to many, caespitose, ascending, densely leafy throughout. (6)
5b. Stems (at least the central) erect, lateral ones widely ascending or prostrate, often rooting; stem leaves few, small, and widely spaced; dry soil. **E. flagellaris** Gray
6a. Pappus without outer scales; rays about 100; dry soil. **E. divergens** T. & G.
6b. Pappus bristles with minute outer scales; rays 30–65; dry, sandy or gravelly soil. **E. bellidiastrum** Nutt.
7a. Stems 3 dm tall or more, simple or with few branches; stem leaves ovate to lanceolate; heads solitary or few, rather large; disk 10–15 mm across. (8)
7b. Stems mostly less than 3 dm tall, either branched near base or caespitose; stem leaves narrow, mostly linear, often reduced or lacking. (12)
8a. Rays broad, more than 1 mm wide; bracts loose above. (9)
8b. Rays narrow, less than 1 mm wide; bracts appressed, except at the tips; leaves ciliate on margins. (10)
9a. Bracts glandular-puberulent; leaves ovate to oblong or ovate-lanceolate; moist soil, mountain slopes and valleys. **E. peregrinus** (Pursh) Greene
9b. Bracts villous, hairs black at base; leaves ovate to oblong; moist soil, high mountain valleys. **E. coulteri** T. C. Porter

Subclass DICOTYLEDONEAE

10a. Bracts hirsute, not glandular; open places, waysides and meadows. **E. glabellus** Nutt.
10b. Bracts densely glandular-puberulent. (11)
11a. Upper leaves narrowly lanceolate; open places in woods, often in shade. **E. speciosus** (Lindl.) DC.
11b. Upper leaves ovate or elliptic-ovate; open places in woods, often in shade. **E. macranthus** Nutt.
12a. Leaves 1–3 times ternately dissected; alpine areas. **E. compositus** Pursh
12b. Leaves entire or slightly dentate or serrate. (13)
13a. Plants densely silvery-strigillose; stems leafy; heads solitary, rather large; involucre 3–10 mm high; disk 15–20 mm across; rays few and rather broad. **E. argentatus** Gray
13b. Plants not densely silvery-strigillose; heads smaller; disk 1 cm across or less; rays numerous, narrow. (14)
14a. Involucres densely villous with soft, spreading hairs; low plants less than 10 cm tall; basal leaves spatulate to oblanceolate; rocky soil and crevices, higher slopes to subalpine regions. **E. simplex** Greene
14b. Involucres hispid, strigose or glandular, not villous. (15)
15a. Stems slender, weak and decumbent; heads mostly solitary; rays white, or occasionally purplish. (16)
15b. Stems mostly short and stout or, if slender, erect or ascending, not decumbent; rays mostly blue, violet, or purple, but sometimes white. (17)
16a. Basal leaves narrowly lanceolate or oblanceolate, 3-nerved, exceeding the stem leaves; heads 1 to several on each stem; rays about 20; dry soil, foothills and lower mountain slopes. **E. eatonii** Gray
16b. Basal leaves linear-oblanceolate, 1-nerved; heads 1–2 on each stem; rays about 50; dry plains and hillsides. **E. engelmannii** A. Nels.
17a. Stems glabrous or with appressed hairs; plants low, less than 2 dm tall, often very short. (18)
17b. Stems pubescent with spreading hairs. (21)
18a. Stems very slender, filiform, glabrous; basal leaves linear-oblanceolate to filiform, becoming subulate above; disk 10 mm broad or less; involucre glandular-puberulent; rocky places, crevices and ledges, middle elevations and above. **E. arenarioides** (D. C. Eaton) Rydb.
18b. Stems short and stout; basal leaves obovate, spatulate, or oblanceolate; alpine areas. (19)
19a. Heads small; disk about 10 mm broad; involucre 4–5 mm high, very finely glandular; leaves glabrous; rocky soil, subalpine to alpine areas. **E. leiomeris** Gray
19b. Heads larger; disk 12–15 mm broad; involucre 6–8 mm high. (20)
20a. Stems strigose above; cauline leaves very reduced, 0–2; basal leaves

Compositae

oblanceolate, obtuse-rounded; bracts glandular-puberulent and slightly strigose as well. **E. controversus** Greene

20b. Stems with appressed hairs; cauline leaves reduced, several; basal leaves oblanceolate to linear-oblanceolate, acute; bracts glandular-puberulent and hirsute; subalpine to alpine areas. **E. ursinus** D. C. Eaton

21a. Plants scapose or nearly so, 10 cm tall or less; heads solitary. (22)

21b. Plants with leafy stems, several-headed, 10–30 cm tall. (23)

22a. Pappus without short outer scales; subalpine to alpine areas. **E. vetensis** Rydb.

22b. Pappus with conspicuous paleaceous outer scales; dry soil, plains, foothills, and mesas. **E. concinnus** (H. & A.) T. & G.

23a. Basal leaves narrowly linear-oblanceolate, linear-spatulate, or linear, acute or acuminate; plants canescent throughout; dry, rocky hillsides and open slopes. **E. caespitosus** Nutt.

23b. Basal leaves oblanceolate to obovate, obtuse or slightly acute; plants strongly hirsute. (24)

24a. Outer pappus of inconspicuous lanceolate or bristlelike scales; dry, rocky soil, open hillsides and mesas. **E. pumilus** Nutt.

24b. Outer pappus of conspicuous broader scales; dry soil, plains, foothills, and mesas. **E. concinnus** (H. & A.) T. & G.

Eriophyllum Log. Woolly Daisy

1a. Plants perennial, from a branching caudex; rays yellow; reported from western Utah. **E. lanatum** (Pursh) Forbes

1b. Plants annual, from a slender taproot; rays white to pinkish or yellow; southwestern Utah. (2)

2a. Rays yellow; pappus scales not awn tipped. **E. wallacei** Gray

2b. Rays white; pappus scales awn tipped. **E. lanosum** Gray

Eupatorium L. Joe Pye Weed

Perennial herbs; stems stout, 6–15 dm tall; leaves mostly whorled in 3s or 4s, 5–10 cm long, coarsely serrate; heads in flat-topped corymbs; flowers all tubular, perfect, purple or reddish purple; pappus of numerous capillary bristles. **E. maculatum** L.

Filago L.

Annual white-woolly herbs, 0.5–3 dm tall; leaves linear-oblong to spatulate, sessile; heads 3–4 mm high, discoid, the bracts woolly; outer flowers lacking a pappus, subtended by bracts; inner flowers with pappus of capillary bristles, lacking bracts; reported from southern Utah. **F. californica** Nutt.

Flaveria Juss.

Annual herbs, glabrous or villous at nodes; stems 2–6 dm tall; leaves opposite, linear to lanceolate, serrulate to entire; heads clustered;

Subclass DICOTYLEDONEAE

involucres 5-6 mm high, with 2-5 bracts; rays 1-2 mm long, usually 1 per head; pappus lacking; known from central eastern Utah. **F. campestris** J. R. Johnston

Gaillardia Foug. Blanketflower
1a. Stems scapose; leaves all basal, elliptic, entire. **G. parryi** Greene
1b. Stems not scapose; leaves, at least some, cauline or, if scapose, usually toothed or lobed. (2)
2a. Disk flowers yellow. (3)
2b. Disk flowers purple. (4)
3a. Leaf blades entire, obovate to oblanceolate. **G. spathulata** Gray
3b. Leaf blades, at least some, usually toothed or lobed or, if entire, then narrower than above. (5)
4a. Stem leafy throughout; leaves pinnatifid; plants perennial. **G. flava** Rydb.
4b. Stem leaves reduced upward, subscapose; leaves toothed to pinnatifid; plants annual. **G. arizonica** Gray
5a. Teeth of disk corollas lance-acuminate; usually montane or cultivated. **G. aristata** Pursh
5b. Teeth of disk corollas triangular, merely acute; warm deserts. (6)
6a. Upper leaves, at least, deeply pinnatifid, with narrow divisions. **G. pinnatifida** Torr.
6b. Upper leaves entire, or sometimes shallowly lobed. **G. gracilis** A. Nels.

Geraea T. &. G. Desert-Sunflower

Annual herbs with alternate entire to dentate leaves; leaves lanceolate to ovate or oblanceolate, acute, 1-7 cm long; heads solitary or few, showy; involucre 7-12 mm high, the bracts white villous-ciliate; rays golden; reported from southwestern Utah. **G. canescens** T. & G.

Glyptopleura D. C. Eaton
1a. Leaves with conspicuous, whitish, scarious-thickened margin, this toothed; rays short, barely exceeding the bracts; western and southern Utah. **G. marginata**
1b. Leaves with narrow white margin, the teeth acuminate; rays elongate, much longer than the bracts; southwestern Utah. **G. setulosa** Gray

Gnaphalium L. Cudweed

1a. Heads very small, leafy bracted; involucral bracts subequal, greenish to brown; plants to 1.5 dm tall or less. (2)
1b. Heads medium, not leafy bracted; involucral bracts strongly unequal, white or yellowish; plants commonly 30 cm tall or more. (3)
2a. Leaves loosely floccose-tomentose, oblong, spatulate, or oblanceolate. **G. palustre** Nutt.

Compositae

2b. Leaves appressed-tomentose, narrowly lanceolate to linear. **G. uliginosum** L.
3a. Leaves green and glandular above, tomentose beneath, strongly decurrent; involucral bracts all acute. **G. macounii** Greene
3b. Leaves grayish, tomentose on both sides, moderately if at all decurrent; involucral bracts obtuse. (4)
4a. Leaves not at all decurrent; involucral bracts pearly white. **G. wrightii** Gray
4b. Leaves moderately decurrent; involucral bracts straw colored or yellowish. **G. chilense** Spreng.

Grindelia Willd. Gumweed

1a. Heads discoid, lacking rays. (2)
1b. Heads with conspicuous rays. (3)
2a. Plants perennial; tips of involucral bracts much thickened and leathery. **G. fastigiata** Greene
2b. Plants annual or biennial; tips of involucral bracts only slightly thickened and leathery. **G. aphanactis** Rydb.
3a. Involucral bracts with strongly spreading or recurved tips; leaves merely toothed to entire. **G. squarrosa** (Pursh) Dunal
3b. Involucral bracts with appressed or merely erect tips; leaves all or mostly laciniate-dentate or pinnatifid. **G. laciniata** Rydb.

Gutierrezia Lag. Snakeweed

1a. Leaves linear-filiform, less than 1 mm wide; rays 1–4 (5); disk flowers 1–3; southern and southeastern Utah. **G. microcephala** (DC.) Gray
1b. Leaves linear, 1–2 mm wide; rays 3–9; disk flowers 3–8. (2)
2a. Heads commonly solitary; involucres 5–7.5 mm long, 2–5 mm broad, turbinate to cylindric; Uintah Co. **G. sarothrae** (Pursh) Britt. & Rusby var. **pomariensis** Welsh
2b. Heads commonly clustered; involucres commonly less than 5 mm long and 2 mm broad, turbinate; throughout Utah. **G. sarothrae** (Pursh) Britt. & Rusby var. **sarothrae**

Haplopappus Cass. Goldenweed

1a. Plants perennial herbs. (2)
1b. Plants low, much-branched, round-topped shrubs. (7)
2a. Plants low, caespitose, with numerous leaves crowded on the crowns of short stems; flowering stems scapose with a few reduced leaves; dry, rocky slopes and crevices. (3)
2b. Plants rather tall, more or less leafy; leaves spinulose toothed or 1- to 2-pinnatifid. (4)
3a. Leaves spatulate to narrowly oblanceolate, hispidulous; bracts acute to acuminate. **H. acaulis** (Nutt.) Gray

Subclass DICOTYLEDONEAE

3b. Leaves linear to oblanceolate, glabrous or nearly so; bracts obtuse or rounded. **H. armerioides** (Nutt.) Gray
4a. Leaves 2–3.5 cm long, not leathery; dry, rocky slopes. (5)
4b. Leaves 10–30 cm long, thick and leathery, simple, coarsely spiny toothed or nearly entire; moist saline meadows and plains. (6)
5a. Leaves simple, spinulose toothed; heads discoid. **H. nuttallii** T. & G.
5b. Leaves 1- to 2-pinnatifid; heads radiate. **H. spinulosus** (Pursh) DC.
6a. Inflorescence corymbose; heads on short or long peduncles. **H. lanceolatus** (Hook.) T. & G.
6b. Inflorescence narrow, spicate or racemose. **H. racemosus** (Nutt.) Torr.
7a. Branches erect, densely white-tomentose; leaves wavy on margins; heads discoid. **H. macronema** Gray
7b. Branches irregular, not tomentose but densely glandular-pubescent; leaves often undulate; heads radiate. (8)
8a. Leaves spatulate to oblanceolate, 2–3 mm wide. **H. watsonii** Gray
8b. Leaves broadly obovate, 7–10 mm wide. **H. rydbergii** Blake

Helianthella T. & G.

1a. Heads numerous, in a corymbose inflorescence; involucral bracts triangular-ovate, strongly unequal, acute tipped; ray flowers very short, inconspicuous, little exceeding the disk; disk flowers purple; eastern and southern Utah. **H. microcephala** (Gray) Gray
1b. Heads solitary or few; involucral bracts lanceolate to lance-linear, about equal, or the outer larger; ray flowers large, showy, much exceeding the disk; disk flowers yellow. (2)
2a. Heads erect at flowering and afterward; scales of receptacle firm; southwestern to northern Utah. **H. uniflora** (Nutt.) T. & G.
2b. Heads nodding or spreading at flowering and afterward; scales of receptacle soft; mostly of eastern Utah. **H. quinquenervis** (Hook.) Gray

Helianthus L. Sunflower (Contributed by Clyde Blauer)

1a. Plants perennial; disk flowers yellow; leaves linear-lanceolate to lanceolate; valleys, often along water courses and in moist regions, at elevations of 3,500–6,500 ft., north central Utah. **H. nuttallii** T. & G.
1b. Plants annual; disk flowers reddish brown to purplish, rarely yellow; leaves lanceolate to broadly ovate. (2)
2a. Involucral bracts linear to lanceolate; pappus of numerous unequal scales. (3)
2b. Involucral bracts broader than lanceolate; pappus normally of 2 scales. (4)
3a. Involucral bracts linear to linear-lanceolate, usually much surpassing the disk; plants to 6.5 dm tall; pappus of linear scales; sandy

Compositae

soils of desert plains and hillsides, including dunes, both stable and moving, southern two-thirds of western Utah and southern half of eastern Utah. **H. anomalus** Blake

3b. Involucral bracts linear-lanceolate, much shorter, about same height as disk; plants 1–4 dm tall; pappus of ovate scales; at elevations from 2,000–4,500 ft., southwestern Utah. **H. deserticola** Heiser

4a. Leaves ovate-lanceolate to broadly ovate, 5.5–40 cm long, 2.5–40 cm wide, usually cordate with serrate margins; involucral bracts usually ovate or broadly ovate. (5)

4b. Leaves lanceolate to ovate-lanceolate, 1.5–8 cm long, 0.4–4 cm wide, rarely cordate, usually with entire margins; involucral bracts linear-lanceolate to ovate-lanceolate. (7)

5a. Plants to 4 m tall or more, usually unbranched, with a large single head; disk 5.5 cm in diameter; leaves very large, cordate; cultivated. **H. annuus** L.

5b. Plants smaller in all parts than above, usually branched; often occurring as "weeds." (6)

6a. Leaves ovate-lanceolate to ovate, truncate at base; disk 2–3.7 cm in diameter, reddish purple to dark purple; rays 17–26; plants 0.5–2.5 m tall; occurring as the "wild" sunflower of western North America; throughout Utah at elevations up to 8,000 ft. **H. annuus** L. ssp. **lenticularis** (Dougl.) Cockerell

6b. Leaves ovate, cordate at base; disk 3–5 cm in diameter; rays 21–35; plants 1.7–3.4 m tall; occurring as the "weed" sunflower of central United States; though not found in the herbaria of Utah, it is to be expected since the majority of the cultivated ornamental sunflowers are of this subspecies. Ruderal Sunflower. **H. annuus** L. ssp. **annuus**

7a. Leaves hispidulous to strigose; southern and eastern Utah, in dry, often sandy, soils at elevations from 3,000–6,000 ft. Prairie Sunflower. **H. petiolaris** Nutt. ssp. **fallax** Heiser

7b. Leaves canescent with an appressed hoary pubescence on both sides; rare in Utah, known only from San Juan Co. **H. canus** (Britt.) Woot. & Standl.

Heterotheca Cass.   Golden Aster

1a. Plants perennial, mostly 1–5 dm tall; widely distributed and common. **H. villosa** (Pursh) Shinners

1b. Plants annual or biennial, mostly 5–10 dm tall or more; restricted in Utah. (2)

2a. Heads small, the disk 8–10 mm high in fruit; upper leaves usually with cordate-clasping bases; known from Grand Co. **H. subaxillaris** (Lam.) Britt. & Rusby

2b. Heads larger, the disk 10–12 mm high in fruit; upper leaves narrowed to the base; known from Washington Co. **H. grandiflora** Nutt.

121

Subclass DICOTYLEDONEAE

Hieracium L. Hawkweed
1a. Stems slender, 1–2 dm tall; leaves mostly basal, glabrous or very short-hairy, 4–8 cm long; heads small; involucres 5–7 mm high, black-hairy; mostly subalpine regions. **H. gracile** Hook.
1b. Stems stouter, 3–6 dm tall or taller, leafy; leaves densely long-pilose, 5–15 cm long; heads larger; involucres 7–12 mm high, whitish-hairy, somewhat glandular; foothills to mountains. **H. scouleri** Hook.

Hofmeistera Walp. Arrowleaf
Low shrubs, 3–8 dm tall; bark white, shreddy; leaves opposite below, alternate above; leaf blades triangular-lanceolate, 4–10 mm long, entire or toothed; heads few; involucres 7–9 mm high, the bracts acuminate, often with recurved tips; corollas whitish; pappus of 10–12 bristles; reported from Washington Co. **H. pluriseta** Gray

Hymenoclea T. & G.
Erect shrubs, mostly 5–10 dm tall, yellowish green; leaves alternate, filiform, 2–5 cm long; heads unisexual, the pistillate 1-flowered, the staminate several flowered; known from Washington Co. **H. salsola** T. & G.

Hymenopappus L'Her
Perennial herbs, 2–4 dm tall, leafy or subscapose; leaves clustered at base, ovate or oblong in outline, 3–8 cm long, reduced upward or lacking; involucres 7–9 mm high; rays lacking; disk flowers numerous, yellow; pappus of 1–20 thin, hyaline, obtuse scales, often short or lacking; dry, rocky soil, deserts, hillsides, and upland plains and valleys. **H. filifolius** Hook.

Hymenoxys Cass. Hymenoxys
1a. Leaves entire, all basal; heads few, rather large; dry, rocky slopes, in open places. **H. acaulis** (Pursh) Parker
1b. Leaves pinnatifid or divided in 3s, not all basal; heads small and numerous; foothills and lower mountain slopes. **H. richardsonii** (Hook.) Cockerell

Iva L. Marsh Elder
1a. Plants low, perennial, with numerous, linear-oblong to obovate leaves; heads solitary, nodding, in the axils of the upper leaves; waste places, often along river courses, in saline soil. **I. axillaris** Pursh
1b. Plants tall, annual, with fewer large ovate leaves; heads small, in dense terminal clusters; waste places. **I. xanthifolia** Nutt.

Lactuca L. Lettuce
1a. Plants perennial with spreading rootstocks; achenes lanceolate or oblong-lanceolate, the beak less than one-half as long as body; flowers blue or violet. Blue Lettuce. **L. pulchella** (Pursh) DC.

Compositae

1b. Plants annual or biennial, rarely perennial but then not with rootstocks; achenes oval or oval-oblong, the beak one-half as long as body or longer; flowers yellow, but often turning blue-purple with age. Prickly Lettuce **L. scariola** L.

Lamphamia Gray

1a. Plants pubescent, gray-green; heads discoid; known from Washington Co. **L. palmeri** Gray
1b. Plants glabrous, yellow-green; heads with rays; known from western and central Utah. **L. stansburii** Gray

Layia H. & A.   Tidy Tips
 Plants annual; leaves alternate, the lower ones laciniate or pinnately lobed or 2-pinnate; stems 1–4 dm tall, much branched, with dark, stipitate glands, especially above; rays yellow or white, 6–10 mm long; pappus of disk flowers, of 15–20 capillary bristles; receptacle chaffy; dry plains and hillsides. **L. glandulosa** H. & A.

Lygodesmia D. Don   Rush Pink
1a. Branches spine tipped; heads 7–9 mm high, 3- to 5-flowered; dry soil. Spiny Rush Pink. **L. spinosa** Nutt.
1b. Branches not spine tipped; heads 4- to 10-flowered. (2)
2a. Heads 9–16 mm high; achenes 1 cm long or less; dry, sandy or gravelly soil, on open hillsides. Slender Rush Pink. **L. juncea** (Pursh) D. Don
2b. Heads 18–22 mm high; achenes about 1.5 cm long; rather common on dry plains and hillsides. Rush Pink. **L. grandiflora** (Nutt.) T. & G.

Machaeranthera Nees   Aster

1a. Plants annual, biennial, or less often perennial herbs; heads several to many or seldom solitary, medium sized to small, the disk not over about 2.5 cm wide; involucral bracts narrow, seldom over 1.5 mm wide, very often with squarrose tips; species flowering mostly from midsummer or late summer to autumn. (2)
1b. Plants perennial from a stout taproot, either shrubby or with a stout branched woody caudex; heads mostly solitary at ends of the stems or branches, medium sized to large, the disk often over 2.5 cm wide; involucral bracts (except for **M. kingii**) usually appressed or merely loose, without squarrose tips, very often over 1.5 mm wide; species (except **M. kingii**) flowering mostly in the spring and early summer. (10)
2a. Plants mostly biennial to perennial; leaves toothed or laciniate or sometimes entire. (3)
2b. Plants mostly annual or winter annual; leaves, at least the lower ones, pinnatifid to pinnately dissected; achenes rather short and broad, mostly 1.5–3.5 mm long, thinly sericeous-strigose to densely woolly-villous. (8)

Subclass DICOTYLEDONEAE

3a. Involucral bracts narrow and elongate, mostly linear-subulate, the narrow tip green and equal to, or longer than, the chartaceous base. (4)
3b. Involucral bracts mostly broader, more linear-oblong, the green tip generally shorter than the chartaceous base. (6)
4a. Involucre and herbage cinereous, not glandular; leaves seldom over 1 cm wide, mostly 5–10 times as long as wide; mostly in alluvial soil, 1,500–7,000 ft., southern Utah. **M. tephrodes** (Gray) Greene
4b. Involucre and often also the herbage glandular or glandular-hispid, not cinereous (or only the herbage somewhat so). (5)
5a. Herbage glandular or glandular-hispid, not cinereous; leaves generally toothed; cauline leaves mostly well over 3 mm wide; plants grazed freely, 3,000–7,000 ft., southern Utah, Millard to Grand Co. and south. **M. bigelovii** (Gray) Greene
5b. Herbage cinereous to partly glabrous, but not glandular; leaves mostly entire; cauline leaves up to about 3 mm wide; reported from southern Utah. **M. mucronata** Greene
6a. Leaf surfaces mostly glabrous or glandular, not cinereous-puberulent; stem generally more or less rough-puberulent; eastern and southern Utah, Summit Co. south to Garfield and San Juan cos., then west to Washington Co. **M. linearis** Greene
6b. Leaf surfaces, at least the lower, cinereous-puberulent, sometimes also glandular; cauline leaves either well developed, or more than 4 times as long as wide, or both; involucral bracts mostly 4–8 seriate, usually strongly squarrose; heads nearly always radiate; rays with normal pappus. (7)
7a. Involucral bracts broad, mostly 1–2 mm wide; plants often perennial; mountains throughout Utah, north to Idaho and western Wyoming. **M. commixta** Greene
7b. Involucral bracts narrower, up to about 1 mm wide; plants seldom perennial; probably in most or all Utah counties. **M. canescens** (Pursh) Gray
8a. Heads middle sized; involucre 6–11 mm high, the disk 1–2.5 cm wide; achenes 2–3.5 mm long; upper leaves generally well developed and more or less pinnatifid. (9)
8b. Heads small; involucre 4–6 mm high, the disk 6–10 mm wide; achenes 1.5–2 mm long; involucral bracts mostly appressed, with short, broad, green tips; upper leaves numerous, reduced, often entire or toothed; mesas and plains, 1,000–5,000 ft., southern Utah. **M. parviflora** Gray
9a. Green tips of involucral bracts mostly long and narrow, loose or spreading; heads mostly hemispheric; involucres often distinctly hairy as well as glandular; roadsides and waste areas, to 6,000 (8,000) ft., southeastern Utah. **M. tanacetifolia** (H. B. K.) Nees
9b. Green tips of involucral bracts mostly short and broad, more or less appressed; heads turbinate; involucres strongly glandular, but

Compositae

generally not otherwise hairy; mesas and roadsides, to 4,000 ft., probably widespread from Utah Co. southward, but seldom collected. **M. tagetina** Greene

10a. Leaves all or mostly entire, or nearly so; plants herbaceous from a woody caudex; style appendage mostly about equaling the stigmatic lines. (11)

10b. Leaves evidently toothed, lanceolate or lance-linear, pubescent; plants robust, often shrubby; style appendages acute to obtuse, shorter than the stigmatic lines; head large, the disk mostly 2–4 cm wide; involucre 13–19 mm high; stem glandular and hispid, often also tomentose; involucral bracts narrow, glandular-hirtellous or pilose; Grand Co. to southern Utah. **M. tortifolia** (T. & G.) Cronq. & Keck

11a. Plants dwarf, to about 1 dm tall; the tufted and persistent basal leaves larger than cauline leaves; disk mostly 10–15 mm wide; involucral bracts, or some of them, with more or less squarrose tips; stem and peduncle glandular-hairy; involucre finely glandular; mountains of central Utah. **M. kingii** (D. C. Eaton) Cronq. & Keck

11b. Plants larger, mostly 1–3 dm tall, with leaves chiefly or wholly cauline, the basal leaves not much developed; disk 12–35 mm wide; involucral bracts appressed or merely loose; herbage and involucre glabrous to villous-puberulent, not glandular; desert regions. (12)

12a. Stems leafy to near the top, the peduncles to about 6 cm long; involucres 8–14 mm high; disk mostly 12–22 mm wide; Carbon and Garfield cos., eastern Utah (?). **M. glabriuscula** (Nutt.) Cronq. & Keck

12b. Stems leafy to about the middle or a little beyond, the peduncles mostly over 6 cm long; heads large; involucres 12–30 mm high; disk mostly 20–35 mm wide. **M. venusta** (M. E. Jones) Cronq. & Keck

Madia Molina Tarweed

Strongly scented annuals; stems simple, erect, leafy, 1–5 dm tall, hirsute or glandular above; leaves erect, lanceolate-linear to linear, 2–6 cm long; heads more or less densely clustered; involucres 6–9 mm high; flowers greenish yellow; rays 1–5, lacking in a few heads; pappus lacking. **M. glomerata** Hook.

Malacothrix DC.

1a. Involucre 12–15 mm high, strongly graduated, the bracts 3–4 mm broad; stem leaves cordate-clasping, the upper subentire to dentate or laciniate; known from Washington Co. **M. coulteri** Gray

1b. Involucre 5–12 mm high, scarcely graduated, the bracts to 1.5 mm wide; stem leaves not cordate-clasping, mostly much reduced and pinnatifid. (2)

2a. Leaf segments linear-filiform, elongate; mostly southern Utah. **M. glabrata** Gray

Subclass DICOTYLEDONEAE

2b. Leaf segments oblong to triangular, short and commonly toothed. (3)
3a. Involucral bracts long-acuminate; achenes 3 mm long or more; persistent bristles of pappus 2-8. **M. torreyi** Gray
3b. Involucral bracts merely acute; achenes less than 3 mm long; persistent bristles of pappus none. **M. sonchioides** (Nutt.) T. & G.

Matricaria L. Matricary

Annual herbs, 3-20 cm tall; leaves alternate, finely 1- to 3-pinnately dissected; heads numerous; disk 5-8 mm broad; corollas tubular, greenish yellow; pappus reduced to a short border; bracts in 2-3 imbricated series, margins scarious; aromatic, common weed of vacant lots and waste places. Pineapple Weed. **M. matricarioides** (Less.) T. C. Porter

Microseris D. Don Microseris

Acaulescent perennial herbs, 1-3 dm tall; leaves linear to linear-lanceolate; peduncles slender; heads 8- to 20-flowered; involucres 1-2 cm high; corollas all ligulate; pappus of 15-20 slender white scales, each terminating in a plumose bristle; saline plains and gravelly foothills. **M. nutans** (Geyer) Schultz

Onopardum L. Onopardum

Coarse, spiny, thistlelike biennials to 15-20 dm tall, with a broadly winged stem, the wings spiny; herbage sparingly to densely tomentose; leaves sessile and decurrent, or the lower with petioles; heads 2.5-5 cm wide, the bracts spine tipped, the receptacle fleshy, honeycombed, and often shortly bristly; central and southwestern Utah. **O. acanthium** L.

Oxytenia Nutt. Copperweed

Perennial poisonous herbs, commonly 8-15 dm tall; leaves alternate, pinnately parted, with 3-5 linear-filiform segments, or the upper entire; heads numerous, about 4-5 mm high; corollas whitish; widely distributed in southeastern Utah. **O. acerosa** Nutt.

Palafoxia Lag.

Erect annual herbs, 1.5-5 dm tall; leaves alternate, linear or linear-lanceolate, 2-6 cm long, entire; heads corymbose, discoid, 12-18 mm high; flowers white; pappus of 4 scales; Washington Co. **P. linearis** (Cav.) Lag.

Parthenium L.

Caespitose perennial herbs; leaves basal, spatulate, arising from the summit of a caudex, long-hairy in the axils; heads sessile or nearly so, solitary; rays white, very short; disk flowers white; pappus lacking; receptacle chaffy; Uintah, Duchesne, and Emery cos. **P. alpinum** (Nutt.) T. & G. var. **ligulatum** M. E. Jones

Pectis L. Chinchweed

Low, aromatic annual herbs, 0.5-2 dm long; leaves opposite, linear, 1-6 cm long and 1-2 mm wide, glandular-dotted, with marginal bristles near the base; heads solitary or cymose, the involucres about 5 mm high, the bracts in 1 series, these glandular-dotted; pappus of brownish, plumose bristles; Kane and Washington cos. **P. papposa** Harv. & Gray ex Gray

Perezia Lag.

Perennial herbs, commonly 3-8 dm tall; stem bases and often lower axils bearing copious brown woolly hair; leaves alternate, ovate-lanceolate, sessile and clasping, spinulose toothed; heads several to many, corymbose; involucres 5-10 mm high, the bracts strongly graduated; flowers lavender-pink, bilabiate; pappus of capillary bristles; Kane and Washington cos. **P. wrightii** Gray

Petradoria Greene

Perennial herbs, from a woody caudex; stems several, 0.8-3 dm tall; leaves linear to oblanceolate, 2-12 cm long, to 1 cm wide, the basal ones in rosettes; heads several to numerous, in a corymbose cluster; involucres 5-10 mm high, the bracts graduated in vertical rows; ray flowers 1-3, yellow; pappus of capillary bristles; widespread. **P. pumila** (Nutt.) Greene

Pluchea Cass.

Willowlike, silvery-silky shrubs, commonly 1-2 m tall; leaves alternate, lanceolate, 1-4 cm long, entire, acute, sessile; heads several to many, in corymbs; involucres 5-6 mm high, the bracts graduated; flowers all discoid, purplish; pappus of capillary bristles; Garfield, Kane, San Juan, and Washington cos. **P. sericea** (Nutt.) Cov.

Psathyrotes Gray

1a. Leaves entire, with long, many-celled hairs on margin and petiole; achenes subcylindric, hispidulous; southern Utah. **P. pilifera** Gray
1b. Leaves toothed or crenate, the many-celled hairs, when present, short and inconspicuous; achenes obconic, silky-pilose. (2)
2a. Plants lanate-tomentose as well as scurfy; outer involucral bracts much broader than inner ones, obovate; southwestern Utah. **P. ramosissima** (Torr.) Gray
2b. Plants scurfy-tomentose; outer involucral bracts not broader than inner ones, merely lanceolate; western Utah. **P. annua** (Nutt.) Gray

Psilostrophe DC.   Paperflower

1a. Stems densely tomentose; leaves linear to narrowly spatulate; stems woody below; Washington Co. **P. cooperi** (Gray) Greene
1b. Stems pilose to villous, not tomentose; leaves, at least some, spatulate to obovate; plants strictly herbaceous. (2)
2a. Stems densely villous; lowermost leaves obovate; ray flowers mostly over 7 mm long; east central Utah. **P. bakeri** Greene
2b. Stems loosely pilose, glabrate; lowermost leaves spatulate to nar-

Subclass DICOTYLEDONEAE

rowly oblanceolate; ray flowers often less than 7 mm long; widespread, southern Utah. **P. sparsiflora** (Gray) A. Nels.

Rafinesquia Nutt.

1a. Corollas surpassing the involucre by about 5 mm; achenes with a slender beak as long as the body; pappus brownish or dull whitish; southwestern Utah. **R. californica** Nutt.
1b. Corollas surpassing the involucre by 10–15 mm or more; achenes with stout beak shorter than the body; pappus bright white. **R. neomexicana** Gray

Rigiopappus Gray

Slender annual herbs, commonly 1–3 dm tall; leaves alternate, linear, 1–3 cm long, entire; heads solitary, terminating the branches; involucres 4–7 mm high; rays few, 1.5–2 mm long, yellowish; pappus of 3–5 awnlike scales; western Utah. **R. leptocladus** Gray

Rudbeckia L. Cone Flower

1a. Rays lacking; disk ovoid to oblong-conic; flowers dark purplish brown or nearly black; mostly in open woods in mountains. Western Cone Flower. **R. occidentalis** Nutt.
1b. Rays present, yellow. (2)
2a. Leaves entire or slightly serrate; disk black, becoming dark brown; damp soil, stream sides, ditch banks, and meadows. Blackeyed Susan. **R. hirta** L.
2b. Leaves mostly deeply 3- to 7-cleft or -divided; disk greenish yellow or dull yellow; damp soil, stream sides and meadows. Tall Cone Flower. **R. laciniata** L.

Senecio L. Groundsel

1a. Heads large, 15–25 mm high, 1–5, more or less nodding; high mountains. **S. holmii** Greene
1b. Heads smaller, less than 15 mm high, not nodding. (2)
2a. Stems more or less equally leafy throughout; leaves about the same shape. (3)
2b. Stems with leaves becoming conspicuously smaller and usually a different shape upward on the stem. (8)
3a. Plants tomentose when young, becoming arachnoid and finally glabrate with age; leaves deeply lyrate-pinnatifid with numerous lobes, the lobes irregularly toothed; dry plains and upward in mountains. **S. uintahensis** (A. Nels.) Greene
3b. Plants not tomentose; leaves entire, serrate or lobed, but not pinnatifid throughout. (4)
4a. Heads solitary or few at the tips of the stems or branches; leaves coarsely and doubly dentate. (5)
4b. Heads numerous, cymose or paniculate. (6)
5a. Plants low, procumbent, 20 cm tall or less; leaves obovate, 2.5 dm

Figs. 33-34. 33. **Cirsium arvense**, x .21.   34. **Lactuca scariola**, x .2

Figs. 35-36.   35. **Microseris nutans**, x .21.   36. **Senecio ambrosiodes**, x .22

Subclass DICOTYLEDONEAE

long or less, all except the upper ones narrowed to a petiolelike base; rocky soil, high mountains. **S. fremontii** T. & G.

5b. Plants taller, erect or ascending at base, 20–50 cm tall; leaves oblong-obovate or oval, 2–5 cm long, all but basal ones sessile and clasping by a slightly hastate base; rocky soil, high mountains. **S. blitoides** Greene

6a. Leaves laciniately lobed or pinnatifid; bracts black tipped; rocky soil, high mountains. **S. ambrosioides** Rydb.

6b. Leaves entire to coarsely dentate; stems strict and leafy, 5–15 dm tall. (7)

7a. Leaves lanceolate or elliptic, tapering to the base, 5–15 cm long, closely serrate or serrulate, or entire; petioles scarcely evident; damp or dry soil, canyons and mountain valleys. **S. serra** Hook.

7b. Leaves elongate-triangular, abruptly contracted to a petiole, 6–20 cm long, coarsely repand dentate or denticulate; along brooks, in seepage areas and other wet places, canyons and mountain meadows. **S. triangularis** Hook.

8a. Stems slender, rarely more than 30 cm tall. (9)

8b. Stems stout, usually exceeding 30 cm tall. (13)

9a. Plants permanently white-tomentose; lower and basal leaves spatulate to oval, obtuse, entire; dry, rocky soil, lower slopes to high alpine summits. **S. canus** Hook.

9b. Plants glabrous or slightly tomentose at bases of stems and petioles. (10)

10a. Basal leaves ovate to nearly round, cordate at base, 3–9 cm long, closely crenate-serrate; petioles long; stem leaves laciniate or lyrate-pinnatifid; moist soil, stream banks and meadows, canyons and high mountains. **S. pseudaureus** Rydb.

10b. Basal leaves rounded or obovate to lanceolate, tapering to the base or truncate, but not cordate. (11)

11a. Basal leaf blades broadly ovate, truncate, or subcordate, 4–7 cm long, crenate-serrate; petioles long; stem leaves oblanceolate, laciniate at base; moist or rather dry shady places, canyons and mountainsides. **S. platylobus** Rydb.

11b. Basal leaf blades ovate to obovate or cuneate-spatulate, 2–3 cm long, toothed in upper part. (12)

12a. Ray flowers orange-red; petioles of basal leaves winged; upper leaves ovate or triangular with the bases auriculate, clasping and coarsely toothed or lobed; dry or damp rocky soil and crevices, higher mountains to alpine regions. **S. crocatus** Rydb.

12b. Ray flowers yellow; petioles of basal leaves not winged; upper leaves reduced, sessile, toothed to lobed; lower stem leaves toothed to lyrate-pinnatifid; open wooded areas in mountains. **S. cymbalarioides** Nutt.

13a. Plants densely white-tomentose when young, leafy to the summit;

Compositae

heads numerous, corymbose; dry, often rocky soil, mountain slopes to alpine regions. **S. atratus** Greene

13b. Plants glabrous or only loose woolly-pubescent, sometimes becoming glabrate with age. (14)

14a. Plants very stout; stem simple, becoming 15 dm tall, glabrous; leaves lanceolate, the basal ones oblanceolate, 10–30 cm long, thick and fleshy; heads very numerous; wet meadows. **S. hydrophyllus** Nutt.

14b. Plants moderately stout; heads fewer, 3–8, rather large, in umbellate-cymose or corymbose clusters. (15)

15a. Stems solitary; plants loose-hairy or arachnoid when young; terminal head sessile or on a very short peduncle; woods or open places, canyons and foothills, lowlands and high mountains. **S. integerrimus** Nutt.

15b. Stems 1 to several; plants glabrous; terminal head on a long peduncle; woods and open places, lowlands to high mountains. **S. crassulus** Gray

Solidago L. Goldenrod

1a. Heads large, 10–12 mm high, rather few, crowded in a corymbose cluster; bracts large and leaflike; stems procumbent or decumbent; rocky places in high mountains. **S. parryi** (Gray) Greene

1b. Heads smaller, 8 mm high or less; bracts small, dry, green only at the tips; stems erect to decumbent. (2)

2a. Involucres cylindric, 6–9 mm high; heads erect, few in flat-topped or glomerulate-corymbose clusters; stems low, strict, 1–2 dm tall; dry, often rocky slopes and ridges in open places, at moderate elevations. **Petradoria pumila** (Nutt.) Greene

2b. Involucres ovoid, less than 6 mm high; heads several to many in racemes. (3)

3a. Stems densely puberulent, 2–6 dm tall, slender; leaves distant, linear-lanceolate to oblanceolate, 6–10 cm long, basal ones smaller and broader, early deciduous; branches of the racemes recurved, 1-sided; plains, canyons, and hillsides. **S. sparsiflora** Gray

3b. Stems glabrous, or nearly so. (4)

4a. Basal and lower leaves spatulate-obovate to oblanceolate, tapering to winged petioles, obtuse or acute, 4–10 cm long; stem leaves smaller; heads large; involucres 4–5 mm high; inflorescence compact; alpine areas. **S. decumbens** Greene

4b. Leaves uniformly lanceolate, petioles not winged, sharply serrate, 8–14 cm long; heads small; involucres 2–3.5 mm high; branches of the raceme widely recurved, 1-sided; lowlands and canyons. **S. canadensis** L.

Sonchus L. Sow Thistle

Coarse annual herbs with erect leafy stems, 3–15 dm tall; lower

Subclass DICOTYLEDONEAE

leaves more or less petioled, the upper ones shorter and narrower, auricled and clasping the stem; heads rather small; involucres glabrous, 9–14 mm high; flowers numerous, yellow; achenes flattened, slightly narrowed at apex; pappus of numerous soft white bristles. **S. asper** (L.) Hill

Stephanomeria Nutt.
1a. Involucres 9–13 mm high, 10- to 20-flowered; leaves deeply runcinate-pinnatifid; desert plains, southern Utah. **S. parryi** (Gray) Coville
1b. Involucres 6–10 mm high, 3- to 9-flowered. (2)
2a. Plants annual; pappus plumose in upper half; lower leaves 1- to 2-pinnatifid; dry open plains and hillsides. **S. exigua** Nutt.
2b. Plants perennial. (3)
3a. Pappus plumose nearly to the base, the lower one-fourth hirsute; lower leaves lanceolate, runcinate-pinnatifid; dry soils, deserts and foothills. **S. pauciflora** (Torr.) A. Nels.
3b. Pappus bristles plumose to the base; lower leaves runcinate; dry, often rocky soils. **S. tenuifolia** (Torr.) Hall

Stylocline Nutt.
Low, woolly annual herbs, branching from near the base, 2–8 cm tall; leaves alternate, linear, 5–8 mm long; heads small, clustered; bracts enclosing the pistillate flowers with a broad, long-woolly body produced at apex into an ovate, hyaline appendage; Kane and Washington cos. **S. micropoides** Gray

Syntrichopappus Gray
Low, floccose-woolly, winter annuals, 2–10 cm tall; leaves alternate, 0.5–2 cm long, narrowly spatulate, apically 3-toothed or entire; heads solitary at tips of branches; involucres 5–7 mm high; rays 5, yellow, 3–5 mm long; pappus of capillary bristles; southwestern Utah. **S. fremontii** Gray

Tanacetum L. Tansy
1a. Cauline leaves well developed, pinnately dissected; robust plants to 1 m tall or more; cultivated and escaping. **T. vulgare** L.
1b. Cauline leaves reduced upward, usually entire; plants usually less than 3 dm tall; indigenous in northern Utah. **T. diversifolium** D. C. Eaton

Taraxacum Haller Dandelion
Perennial herbs from a thick taproot; leaves all basal in a rosette, variable, 6–40 cm long; heads solitary; involucral bracts slender, in 2 rows; flowers yellow; pappus of soft, white capillary bristles; achenes long beaked; widely distributed. **T. officinale** Weber

Tetradymia DC. Horsebrush
1a. Primary leaves not modified as spines but often sharply pointed;

Compositae

tomentum disappearing early; heads 4-flowered; dry saline plains and sand dunes, foothills. Horsebrush. **T. glabrata** Gray
1b. Primary leaves not modified as spines; tomentum persistent, or tardily glabrate. (2)
2a. Heads 4-flowered, with 4–5 bracts; dry plains and foothills. **T. nuttallii** T. & G.
2b. Heads 5- to 9-flowered, with 5–6 bracts; dry saline plains. Spiny Horsebrush. **T. spinosa** T. & G.

Thelesperma Less.

1a. Plants 3–8 dm tall, leafy, usually branched; pappus of 2 awns; rays lacking; plants evidently uncommon, San Juan Co. **T. megapotamicum** (Spreng.) Kuntze
1b. Plants commonly less than 3 dm tall, subscapose or scapose; pappus lacking; widespread, eastern Utah. **T. subnudum** Gray

Townsendia Hook. (Contributed by James L. Reveal)

1a. Plants long-lived perennials. (2)
1b. Plants annuals or biennials. (12)
2a. Involucral bracts broadly lanceolate to ovate or elliptic, in 2–5 series. (3)
2b. Involucral bracts linear to narrowly lanceolate, in 5–7 series. (9)
3a. Plants acaulescent, rosulate-pulvinate. (4)
3b. Plants caulescent; western and south central Utah. **T. florifer** (Hook.) Gray
4a. Corolla rays yellowish to golden yellow, densely glandular abaxially; leaves spatulate to oblanceolate, 7–16 mm long, 1–3.5 mm wide, strigose; involucral bracts lanceolate, fimbriate at the tip, strigose without, 5–7 mm long, 0.9–1.8 mm wide; rays 4–7 mm long; achenes pubescent with glochidiate hairs; rare endemic known only from Sevier Co. **T. aprica** Welsh & Reveal
4b. Corolla rays white, pink, or blue, or, if drying yellowish, then not from Sevier Co. (5)
5a. Leaves linear, 4–14 mm long, 0.8–1.3 mm wide, strigose; involucral bracts lanceolate, margins ciliate, glabrous to sparsely glandular or slightly pilose without, 3.5–8 mm long, 1.2–1.8 mm wide; rays 5–7.5 mm long, 0.9–1.4 mm wide; achenes pubescent with glochidiate hairs; infrequent endemic of the Uinta Basin. **T. mensana** M. E. Jones
5b. Leaves oblanceolate to spatulate or wider. (6)
6a. Achenes glabrous; leaves spatulate to broadly elliptic, 4–15 mm long, 2.5–3.5 (5) mm wide, strigose; involucral bracts oblong to obovate, margins ciliated, glabrous or nearly so; rays 6–9 mm long, 1.3–5 mm wide, often blue; high mountains of central and northeastern Utah. **T. montana** M. E. Jones

Subclass DICOTYLEDONEAE

6b. Achenes pubescent. (7)
7a. Achenes with simple or unevenly bifurcated hairs, rarely truly glochidiate; leaves narrowly spatulate or oblanceolate, 3–8 mm long, 1.2–3 mm wide, glabrous or sparsely strigose; involucral bracts oblanceolate, ciliated margins, strigose, 3–7 mm long, 0.9–2.1 mm wide; rays 4–8 mm long, 1.5–2.1 mm wide; endemic to the Red Canyon and Pink Cliffs area, near and in Bryce Canyon and near Orderville (Garfield and Kane cos.). **T. minima** Eastw.
7b. Achenes with glochidiate hairs; leaves moderately to densely strigose. (8)
8a. Involucral bracts glabrous or the outer ones only slightly pubescent, lanceolate, 4–10 mm long, 1–2.5 mm wide, margins ciliated; leaves narrowly oblanceolate to narrowly spatulate, 1–4 cm long, 1–2.5 (4) mm wide; rays 4–7 mm long, 1–2 mm wide, rarely drying yellowish in some; infrequent to rare in west central Utah, mainly in desert ranges. **T. jonesii** (Beaman) Reveal
8b. Involucral bracts conspicuously strigose; eastern and southern Utah. **T. incana** Nutt.
9a. Involucral bracts with a tuft of tangled cilia at apex, linear, 5–10 mm long, 0.6–1.2 mm wide, pilose-strigose; leaves linear to narrowly oblanceolate, 2.5–4 cm long, 1–2 mm wide, densely strigose; rays glabrous, 6–9 mm long, 1–1.9 mm wide; infrequent and widely scattered in low foothills and mountains of northeastern Utah. **T. hookeri** Beaman
9b. Involucral bracts without a tuft of cilia at apex, narrowly lanceolate. (10)
10a. Rays densely glandular on the abaxial surface; leaves linear; Uinta Basin. **T. mensana** M. E. Jones
10b. Rays glabrous or only sparsely pubescent. (11)
11a. Disk pappus 3–6 mm long; leaves linear to oblanceolate or narrowly spatulate, 1–4 cm long, 1.3–2.6 mm wide, strigose-sericeous; involucral bracts lanceolate to nearly linear, margins ciliated, glabrous or nearly so, 3–9 mm long, 0.8–1.4 mm wide; rays 6–10 mm long, 1.2–2 mm wide; northeastern and central Utah. **T. leptotes** (Gray) Osterh.
11b. Disk pappus 6–12 mm long; leaves oblanceolate, 2.5–5 cm long, 2–3.5 mm wide, strigose; involucral bracts linear to narrowly lanceolate, margins ciliated, glabrous, 4–12 mm long, 1–2.3 mm wide; rays 8–15 mm long, 1.2–3 mm wide; south central Utah. **T. exscapa** (Richards.) T. C. Porter
12a. Disk pappus as long as, or longer than, disk corolla. (13)
12b. Disk pappus shorter than disk corolla; leaves basal and cauline, oblanceolate to spatulate, 0.5–1.5 cm long, 1–3 (5) mm wide, strigose; involucral bracts elliptic to obovate or ovate, 2–6 mm long, 1–2 mm wide, strigose-pilose; rays 4–8 mm long, 1–2.3 mm wide; achenes with glochidiate hairs; southeastern Utah. **T. annua** Beaman

Compositae

13a. Achenes with unevenly branched hairs; leaves basal and cauline, spatulate, 2–5.5 cm long, 3–8 mm wide, strigose; involucral bracts lanceolate, 4–12 mm long, 1–2.5 mm wide, strigose; rays 7–12 mm long, 1.5–3 mm wide; annuals or biennials to short-lived perennials; western and central Utah. **T. florifer** (Hook.) Gray

13b. Achenes with glochidiate hairs; eastern and southern Utah. (14)

14a. Stems among the leaves grayish white with a dense canescent pubescence; involucral bracts strigose, 3–10 mm long, 1–3.5 mm wide; rays 6–10 mm long, 1.5–3 mm wide; eastern and southern Utah (including T. arizonica Gray). **T. incana** Nutt.

14b. Stems merely lightly to moderately strigose with the red stems among the leaves obvious; involucral bracts moderately strigose to nearly glabrous, 3–7 mm long, 1.2–1.9 mm wide; rays 6–11 mm long, 1.3–3 mm wide; extreme northeastern Utah southward (rarely) to Carbon Co. **T. strigosa** Nutt.

Tragopogon L. Goatsbeard

1a. Flowers purple; roots edible; fields and waysides, escaped cultivation. Salsify. **T. porrifolius** L.

1b. Flowers yellow; roots edible; fields and waste places, escaped cultivation. Goatsbeard. **T. dubius** Scop.

Vanclevea Greene

Glutinous shrubs, with white bark, commonly 4–12 dm tall; leaves alternate, linear-lanceolate, entire or nearly so; heads solitary or several to many, more or less corymbose; involucres glutinous, the bracts graduated; flowers discoid, yellow, the styles long-exserted; sand dunes and other sandy sites, southeastern and south central Utah. **V. stylosa** (Eastw.) Greene

Verbesina L. Crownbeard

Annual herbs, 2.5–6 dm tall; leaves alternate or sometimes opposite, ovate to lanceolate, serrate to lobed; heads solitary, terminating stems and branches; involucres mostly 7–10 mm high; rays 12–20 mm long, yellow; pappus of disk flowers of 2 slender awns; widespread, eastern Utah. **V. encelioides** (Cav.) Benth. & Hook. ex Gray

Viguiera H. B. K.

1a. Plants perennial; with shrubs and trees in mountains, widespread. **V. multiflora** (Nutt.) Blake

1b. Plants annual; in saline or argillaceous soils at low elevations. (2)

2a. Plants subscapose; leaves mostly near base of plant; heads large, few; known only from Kane Co. **V. soliceps** Barneby

2b. Plants leafy, not subscapose; heads moderate, commonly several to many; more widely distributed. (3)

3a. Involucral bracts green, hispid-ciliate; leaves hispid-ciliate; known from central Utah. **V. ciliata** (Robins. & Greenm.) Blake

Subclass DICOTYLEDONEAE

3b. Involucral bracts silvery-strigose; leaves not conspicuously ciliate; southwestern Utah. V. annua (M. E. Jones) Blake

Wyethia Nutt. Mules-ears

1a. Stems white; leaves hispid-roughened; often in sand, low elevations, southern and southeastern Utah. W. scabra Hook.
1b. Stems not especially white; leaves not roughened; middle elevations in mountains. (2)
2a. Plants glabrous; upper cauline leaves clasping; northern Utah. W. amplexicaulis (Nutt.) Nutt.
2b. Plants hairy; upper cauline leaves petioled; southeastern Utah. W. arizonica Gray

Xanthium L. Cocklebur

Annual herbs; stems stout, erect, widely branched, 2-20 dm tall, brown spotted; leaves long-petiolate, broadly ovate-cordate; burs 20-25 mm long, the spines hooked, beaks incurved or hooked; waste places and cultivated lands. X. strumarium L.

CONVOLVULACEAE — MORNING GLORY FAMILY

Twining or trailing herbs, rarely suffrutescent; leaves alternate, simple, sometimes deeply lobed or parted; flowers perfect, regular, mostly 5-merous, often showy, axillary, solitary, or in cymes; sepals imbricate, distinct or partly united; pistils 1; ovary superior, typically 2-carpelled; styles 1 or 2; fruit a capsule or a pair of utricles.

1a. Plants without chlorophyll, parasitic on stems of various host plants; leaves reduced to small scales; stems twining; flowers small, the corolla white or whitish, usually with fimbriate or dentate appendages within. Cuscuta
1b. Plants with chlorophyll, not parasitic; leaves with well-developed blades. (2)
2a. Corolla imbricate in bud, white; styles 2, entire; stigmas capitate. Cressa
2b. Corolla plicate-convolute in bud; styles 1, or 2-cleft apically. (3)
3a. Stigmas 1, globose or nearly so, entire or lobed. Ipomoea
3b. Stigmas 2, linear-filiform to ovate. Convolvulus

Convolvulus L. Bindweed

1a. Calyx subtended by a pair of ovate bracts longer than the sepals (10-20 mm long); corollas over 3 cm long, white to pink; fields, roadsides, and waste places. C. sepium L.
1b. Calyx not closely subtended by bracts, these well below the calyx, the bracts linear to narrower, shorter than the sepals (not over 5 mm long); corollas less than 3 cm long, the lobes hardly if at all

Convolvulaceae

noticeable, not pointed; basal lobes of leaves entire or sparingly dentate; noxious weeds of wide distribution. **C. arvensis** L.

Cressa L.
Low, somewhat suffrutescent perennials, with branches 10–20 cm long; leaves alternate, 4–10 mm long; flowers solitary; sepals about 4 mm long; corolla white, 5–6 mm long; ovary 2-celled, pubescent; saline soils, southern Utah. **C. truxillensis** H. B. K.

Cuscuta L.   Dodder

1a. Stigmas linear-elongate, often reddish; styles equal; capsule circumscissile in a definite line; parasitic on legumes and other plants. **C. approximata** Babc.

1b. Stigmas capitate or peltate, not elongate; styles equal or unequal; capsule usually not circumscissile (except in **C. umbellata**). (2)

2a. Sepals nearly or quite distinct, closely invested with bracts; capsule globose, carrying withered corolla at apex; on species of Compositae and Leguminosae. **C. cuspidata** Engelm.

2b. Sepals definitely united, not immediately invested with bracts. (3)

3a. Corolla appendages lacking entirely; on various hosts. **C. occidentalis** Mill. ex Mill. & Nutt.

3b. Corolla appendages present, opposite stamens, more or less fringed. (4)

4a. Most flowers with 3 or 4 sepals, petals, and stamens; corolla membranous, when withered remaining on top of capsule, the lobes erect or slightly spreading, obtuse, not inflexed at tip; sepals obtuse, about one-half as long as corolla tube; on various hosts. **C. cephalanthii** Engelm.

4b. Most flowers with 5 sepals, petals and stamens; corolla lobes erect, spreading, or reflexed. (5)

5a. Corolla fleshy-papillate, its lobes commonly erect with inflexed tips; flowers 3–5 mm long; scales prominently fringed and mostly free from corolla tube, at least above; mostly on Compositae and Leguminosae. **C. indecora** Choisy

5b. Corolla not fleshy-papillate, its lobes not both erect and inflexed at tips, but may be either; flowers less than 3 mm long. (6)

6a. Corolla lobes orbicular, obtuse; flowers about 2 mm long, subsessile, in few-flowered glomerules; calyx lobes orbicular, broadly overlapping; capsule conic, mostly 1-seeded; margins of perianth lobes denticulate; on various hosts. **C. denticulata** Engelm.

6b. Corolla lobes triangular or lanceolate, acute or acuminate; flowers 2–3 mm long. (7)

7a. Scales prominent, commonly exserted; corolla lobes triangular to sublanceolate, acute, reflexed, with inflexed tips; capsules mostly depressed-globose, 2- to 4-seeded; mostly on herbaceous hosts, including grasses. **C. campestris** Yuncker

137

Subclass DICOTYLEDONEAE

7b. Scales not prominent, commonly included; corolla lobes ovate-lanceolate, scales attached to corolla tube most of their length; anthers oval, the filaments well developed; capsule globose-conic, mostly 1-seeded; host plants include **Atriplex, Suaeda, Allenrolfea,** and **Salsola. C. salina** Engelm.

Ipomoea L. Morning Glory

1a. Plants with villous or pubescent stems, annual; calyx villous; leaves entire, villous or pubescent beneath, broadly cordate-ovate, 7–12.5 cm long; flowers 1–5, axillary, funnelform, 5–7.5 cm long, purplish with lighter tube; cultivated, sometimes escaping. **I. purpurea** Lam.
1b. Plants with glabrous stems, annual or perennial. (2)
2a. Stems long trailing and rooting, the juice milky; leaves various on the same plant, mostly ovate to orbicular-ovate, 5–15 cm long, cordate or truncate at base, entire or angled and notched, or digitately lobed; flowers few to several on axillary stalks, funnelform, 4–5 cm long, rose-violet or bluish, the center darker; grown for edible thickened roots. Sweet Potato. **I. batatas** Lam.
2b. Stems twining, not trailing and rooting, tinged purple; flowers few to several, about 5 cm long, 7.5–10 cm across, color various. Heavenly Blue, perhaps Scarlet O'Hara, and some white forms belong here. **I. tricolor** Cav.

CORNACEAE — DOGWOOD FAMILY

Plants shrubs with simple, entire, opposite leaves; inflorescence cymose, often flat topped; flowers perfect, regular; sepals 4–5; petals 4–5; stamens the same number as petals and alternate; ovary inferior, 1- to 2-loculed; fruit a 1- or 2-seeded drupe.

Cornus L. Dogwood

1a. Shrubs to 3 m tall, the bark dark red; flowers white, in flat-topped cymes; fruit globose, 7–9 mm in diameter, white to bluish; widely distributed in moist situations, also cultivated. Red Osier Dogwood. **C. stolonifera** Michx.
1b. Shrubs or small trees to 6 m tall, the bark not red; flowers yellow, in umbels from the preceding season's buds; fruit oblong, about 20 mm long, scarlet, edible; cultivated. Cornelian-Cherry **C. mas** L.

CRASSULACEAE — STONECROP FAMILY

Succulent, fleshy herbs; flowers usually cymose; sepals, petals, and pistils 4 or 5, distinct or united at the base; stamens as many as petals or twice as many; ovary superior; fruit a follicle; variety of habitats, from dry hillsides to wet meadows, often used in rock gardens.

1a. Stamens the same number as petals. **Crassula**
1b. Stamens usually twice as many as petals. (2)

Cruciferae

2a. Corolla of united petals, lobed to the middle. **Kalanchoe**
2b. Corolla of distinct petals or attached only at the base. (3)
3a. Floral parts in 4s or 5s. **Sedum**
3b. Floral parts in more than 5s. **Sempervivum**

Crassula L. Crassula
Succulent herbs; leaves opposite, thick, entire; cultivated. **C. argentea** Thunb.

Kalanchoe Adans.
Erect, branched succulents; leaves opposite, thick, crenate-dentate, producing young plants from axils of teeth; calyx 25–35 mm long; corolla longer than calyx, reddish. Air-Plant. **K. pinnata** Pers.

Sedum L. Stonecrop
1a. Plants 3–5.5 dm tall; leaves to 7.5 cm long and 5 cm wide; cultivated. **S. spectabile** Boreau.
1b. Plants less than 3.5 dm tall; leaves seldom over 3 cm long and usually under 1 cm wide; mostly native. (2)
2a. Flowers yellow; stems usually less than 20 cm tall. (3)
2b. Flowers rose-purple to whitish; stems usually over 20 cm tall. (5)
3a. Leaves opposite; petals united at base; rocky areas mostly above 8,000 ft. **S. debile** S. Wats.
3b. Leaves alternate; petals distinct. (4)
4a. Leaves linear, terete or nearly so, under 12 mm long, sessile and crowded; rocky slopes and ridges above 5,000 ft. **S. stenopetalum** Pursh
4b. Leaves ovate, sessile, densely overlapping, under 6 mm long; commonly cultivated, escaping. **S. acre** L.
5a. Flowers in spikelike racemes with leafy bracts, the petals light rose or whitish. Rose Crown. **S. rhodanthum** Gray
5b. Flowers in a dense cyme, 1-sided on short branches, the petals rose-purple to greenish purple. Kings Crown. **S. integrifolium** (Raf.) A. Nels. ex Coult. & Nels.

Sempervivum L.
Succulent herbs, multiplying mostly by offsets; leaves thick, in dense rosettes; commonly cultivated. Hen-and-Chickens. **S. tectorum** L.

CRUCIFERAE — MUSTARD FAMILY

Plants herbaceous (rarely suffrutescent) annuals, biennials or perennials; leaves alternate; inflorescence a raceme, spike, or corymb; flowers regular; sepals 4; petals 4; stamens 4 and 2; ovary superior, 2-loculed; placenta parietal; fruit a silique or silicle.

1a. Fruit stipitate (the base of the fruit prolonged into a stalk, with the receptacle evident at its lower end). (2)

Subclass DICOTYLEDONEAE

1b. Fruit sessile or at most substipitate, the stipe less than 2 mm long. (4)
2a. Flowers purple; fruit strongly flattened parallel to the partition, over 1 cm broad; cultivated ornamental, occasionally escaping. **Lunaria**
2b. Flowers yellow or greenish yellow; fruit terete or subterete, less than 5 mm broad. (3)
3a. Leaves hastately lobed, at least some; stipe less than 4 mm long; wooded areas. **Chlorocrambe**
3b. Leaves not hastately lobed; stipe more than 10 mm long; seleniferous soils in open sites. **Stanleya**
4a. Fruit indehiscent, 1- to several-seeded and breaking into segments at maturity. (5)
4b. Fruit 1- to several-seeded, opening by means of valves. (6)
5a. Fruit 1-seeded, winged and resembling a samara; flowers yellow; introduced weed of waste places. **Isatis**
5b. Fruit several-seeded, not winged, breaking into segments at maturity; flowers pink-purple; introduced weed of waste places. **Chorispora**
6a. Fruit didymous, with the valves inflated and bladdery or flattened contrary to the partition. (7)
6b. Fruit various, but not as above. (8)
7a. Fruit strongly flattened, resembling spectacles; seeds 1 in each locule. **Dithyrea**
7b. Fruit with inflated, bladdery valves; seeds 2 in each locule. **Physaria**
8a. Fruit strongly compressed at right angles to the septum, mostly less than twice as long as broad; petals never yellow. (9)
8b. Fruit compressed parallel to the septum, or quadrangular or terete, mostly more than twice as long as broad; petals often yellow. (14)
9a. Seeds 1 row in each locule; hairs, when present, simple. (10)
9b. Seeds 2 rows in each locule; hairs, when present, stellate. (12)
10a. Petals unequal in size, the outer 2 much larger. **Iberis**
10b. Petals equal in size. (11)
11a. Plants spreading by rhizomes; fruit more or less inflated. **Cardaria**
11b. Plants lacking rhizomes; fruit not inflated. **Lepidium**
12a. Stem leaves auriculate-clasping; fruit 4–15 mm long. (13)
12b. Stem leaves not clasping; fruit 3–4 mm long. **Hutchinsia**
13a. Plants pubescent, at least below; fruit cuneate, somewhat heart shaped. **Capsella**
13b. Plants glabrous; fruit obovate to orbicular, winged and samaralike. **Thlaspi**
14a. Fruit not more than twice as long as broad. (15)
14b. Fruit over twice as long as broad. (20)

Cruciferae

15a. Fruit definitely compressed parallel to the partition. (16)
15b. Fruit terete or quadrangular, slightly if at all flattened. (18)
16a. Fruit oval to orbicular in outline; seeds 2 (1–8) in each locule. (17)
16b. Fruit elliptic or long-elliptic in outline; seeds many in each locule. **Draba**
17a. Flowers white to purple; short stamens with 2 nectaries on each side at base, 1 gland much longer than the other; plants perennial (grown as annual), 1–2 dm tall. **Lobularia**
17b. Flowers yellow, rarely white or pink; short stamens with a single basal nectary; plants perennial or annual, to 6 dm tall. **Alyssum**
18a. Fruit obovoid; valves with a distinct central nerve extending to the apex; upper leaves sagittate-clasping. **Camelina**
18b. Fruit ovoid; valves nerveless or, if a nerve present, then not extending to apex; upper leaves not sagittate-clasping. (19)
19a. Plants glabrous or with simple pubescence. **Rorippa**
19b. Plants silvery-pubescent with stellate hairs. **Lesquerella**
20a. Plants at least somewhat hairy with some or all hairs stellate or branched (rarely glabrous in **Hesperis**). (21)
20b. Plants glabrous or with simple hairs only (glandular-pubescent in **Parrya**). (29)
21a. Fruit at maturity strongly compressed parallel to the partition, rarely subterete, but then pods erect. (22)
21b. Fruit terete or quadrangular, not strongly compressed, with pods rarely erect. (23)
22a. Fruit not over 2 cm long, often twisted, elliptic to oblong-linear; petals yellow to white. **Draba**
22b. Fruit over 2 cm long, not twisted, linear; petals white, yellowish, or pink-purple. **Arabis**
23a. Leaves of stem pinnate, 2-pinnate, or deeply pinnatifid, basal leaves occasionally entire. (24)
23b. Leaves of stem entire or sinuate-dentate to shallowly pinnatifid. (25)
24a. Plants perennial, caespitose, less than 20 cm tall; petals white or rose; alpine regions. **Smelowskia**
24b. Plants annual or biennial, not caespitose; petals yellow; middle and low elevations. **Descurainia**
25a. Petals yellow, orange, or maroon; native. **Erysimum**
25b. Petals white, pink-purple, or reddish purple (rarely yellowish to maroon in **Matthiola**). (26)
26a. Petals white; pods less than 4 cm long; native. **Halimolobos**
26b. Petals pink-purple or reddish purple, rarely white; pods usually more than 4 cm long; cultivated or introduced weeds. (27)
27a. Flowers sessile or nearly so on pedicels less than 1 mm long; introduced weed of waste places. **Malcolmia**

Subclass DICOTYLEDONEAE

27b. Flowers borne on long pedicels, these usually over 5 mm long; ornamentals, cultivated and escaping. (28)
28a. Stigma lobes distinct; leaves entire. **Matthiola**
28b. Stigma lobes connate into a beak or cone; leaves serrate. **Hesperis**
29a. Anthers sagittate at base; stigma usually with 2 lobes, these extending over the valves of the pod; key doubtful species both ways. (30)
29b. Anthers not sagittate at base; stigma entire or, if 2-lobed, the lobes extending over edges of septum. (35)
30a. Fruit strongly flattened parallel to the septum. (31)
30b. Fruit terete, or rarely slightly flattened. (33)
31a. Plants glandular-pubescent throughout; high elevations in the spruce-fir forests of the Uinta Mountains. **Parrya**
31b. Plants glabrous throughout; middle and lower elevations throughout much of Utah. (32)
32a. Fruit erect or ascending, 4–5 mm wide; stem leaves auriculate-clasping. **Streptanthus**
32b. Fruit reflexed, 1–2 mm wide; stem leaves not as above. **Streptanthella**
33a. Calyx urn-shaped, nearly closed at top in anthesis; petals linear and grooved; stems inflated. **Caulanthus**
33b. Calyx campanulate, open at top in anthesis; petals broader at apex than at base; stems not inflated. (34)
34a. Petals white to rose-purple. **Thelypodium**
34b. Petals yellowish to yellow. **Sisymbrium**
35a. Fruit strongly compressed parallel to the septum at maturity. (36)
35b. Fruit terete, quadrangular or slightly compressed. (38)
36a. Fruit less than 2 cm long, often twisted, elliptic to linear-oblong in outline; petals yellow to white. **Draba**
36b. Fruit more than 2 cm long, not twisted, linear to narrowly oblong in outline; petals white, pink-purple or yellow. (37)
37a. Petals yellow; fruit long beaked; introduced annual; weed of waste places. **Diplotaxus**
37b. Petals white to pink-purple (rarely yellowish); fruit not long beaked; native biennial or perennial herbs of wide distribution. **Arabis**
38a. Fruit with a stout, indehiscent beak. **Brassica**
38b. Fruit scarcely beaked, tipped by a slender style or by a sessile stigma. (39)
39a. Cauline leaves, at least the upper ones, auriculate-clasping. (40)
39b. Cauline leaves not auriculate-clasping. (41)
40a. Fruit quadrangular, the valves keeled, sessile. **Conringia**
40b. Fruit terete, sessile or stipitate. **Sisymbrium**
41a. Petals white or purple, not yellow. (42)
41b. Petals yellow. (43)

Cruciferae

42a. Plants aquatic or growing in moist places, rooting at the nodes; fruit terete. **Rorippa**
42b. Plants terrestrial, sometimes in moist soil, not rooting at the nodes. **Cardamine**
43a. Fruit quadrangular, the valves keeled; seeds flattened. **Barbarea**
43b. Fruit terete; seeds not flattened. (44)
44a. Fruit not over 14 mm long; seeds in 2 rows in each cell. **Rorippa**
44b. Fruit usually over 14 mm long, usually 2.5–10 cm long; seeds in 1 row in each cell. **Sisymbrium**

Alyssum L.   Madwort

1a. Plants annual; basal leaves 1–3 cm long; weedy species of the foothills. **A. alyssoides** L.
1b. Plants perennial; basal leaves 5–12 cm long; cultivated ornamental. Basket-of-Gold. **A. saxatile** L.

Arabis L.   Rockcress

1a. Fruit strictly erect at maturity, mostly appressed to the rachis. (2)
1b. Fruit spreading to reflexed at maturity, with some species ascending, but then not strictly erect and appressed to the rachis. (4)
2a. Stems hirsute at base; styles present and distinct. (3)
2b. Stems glabrous to sparsely pubescent at base, not hirsute; styles absent or short and indistinct. **A. drummondii** Gray
3a. Fruit subterete, slightly flattened; sepals not saccate at base; cauline leaves usually glaucous. **A. glabra** (L.) Bernh.
3b. Fruit flattened; sepals more or less saccate at base; cauline leaves not glaucous. **A. hirsuta** (L.) Scop.
4a. Fruiting pedicels ascending or seldom spreading to as much as right angles to the rachis. (5)
4b. Fruiting pedicels spreading at nearly right angles to the rachis or reflexed. (9)
5a. Basal leaves densely gray-canescent; stems densely pubescent below. **A. selbyi** Rydb.
5b. Basal leaves sparsely pubescent or glabrous; stems glabrous to sparsely pubescent below. (6)
6a. Basal leaves obovate to broadly oblanceolate, rounded at the apex, often forming a flat basal rosette. **A. nuttallii** Robins.
6b. Basal leaves linear to linear-lanceolate, acute or rarely obtuse, ascending, not forming a flat basal rosette. (7)
7a. Petals 4–6 mm long; leaves 5–20 mm long; plants glabrous to finely hirsute below. **A. microphylla** Nutt.
7b. Petals usually 7 mm long or more; leaves usually more than 2 cm long; plants coarsely pubescent at least below. (8)
8a. Plants less than 3 dm tall; stems several to many from a branching caudex; fruit and pedicels slightly spreading. **A. lyallii** S. Wats.

Subclass DICOTYLEDONEAE

- 8b. Plants 3–9 dm tall; stems usually single; fruit and pedicels ascending to widely spreading. **A. divaricarpa** A. Nels.
- 9a. Basal leaves hirsute, the marginal hairs longer, or glabrous; stems hirsute below. (10)
- 9b. Basal leaves pubescent, never hirsute; stems hirsute or not hirsute below. (12)
- 10a. Stems 25–60 cm tall. **A. fendleri** (S. Wats.) Greene
- 10b. Stems 10–30 cm tall. (11)
- 11a. Pubescence stellate; leaf margins not long-ciliate. **A. demissa** Greene
- 11b. Pubescence simple; leaf margins long-ciliate. **A. pendulina** Greene
- 12a. Petals 12–20 mm long; fruit densely pubescent; cauline leaves linear. **A. pulchra** M. E. Jones ex S. Wats.
- 12b. Petals 4–12 (14) mm long; fruit usually glabrous; cauline leaves lanceolate or broader. (13)
- 13a. Basal leaves not over 2 cm long; cauline leaves not over 1 cm long; petals 4–6 mm long; stems 6–20 cm tall. (14)
- 13b. Basal leaves 2–10 cm long (sometimes less in **A. holboellii**); cauline leaves 1–8 cm long; petals 5–14 mm long; stems usually over 20 cm tall. (15)
- 14a. Fruiting pedicels 2–5 mm long; fruit 2–3.5 mm wide; basal leaves spatulate; cauline leaves usually ovate. **A. lemonii** S. Wats.
- 14b. Fruiting pedicels 5–8 mm long; fruit 1–1.5 mm wide; basal leaves linear-oblanceolate; cauline leaves oblong. **A. gunnisoniana** Rollins
- 15a. Fruiting pedicels strictly reflexed and appressed to the rachis; the fruit likewise appressed. **A. holboellii** Hornem. var. **retrofracta** (Graham) Rydb.
- 15b. Fruiting pedicels spreading at right angles to the rachis or strongly descending, but not appressed to the rachis; fruit likewise not appressed. (16)
- 16a. Stems finely appressed-pubescent below; basal leaves linear-oblanceolate, always entire, finely pubescent, the hairs minute. **A. lignifera** A. Nels.
- 16b. Stems hirsute below with spreading hairs; basal leaves linear-oblanceolate to broader, often toothed, the hairs usually long and coarse. (17)
- 17a. Fruiting pedicels hirsute; fruit strongly curved, widely spreading, 6–12 cm long; petals 8–14 mm long. **A. sparsiflora** Nutt.
- 17b. Fruiting pedicels glabrous to pubescent, not hirsute; fruit straight to curved inward, not strongly curved, spreading to pendulous, 3–7 cm long; petals 6–12 mm long. (18)
- 18a. Stems several to many; fruiting pedicels 10–12 mm long, glabrous; fruit pendulous to widely spreading. **A. perennans** S. Wats.
- 18b. Stems 1 to several; fruiting pedicels 5–15 mm long, glabrous or pubescent; fruit pendulous, not widely spreading. **A. holboellii** Hornem. var. **pinetorum** (Tidestr.) Rollins

Cruciferae

Barbarea R. Br. Wintercress

Glabrous biennials; stems 30–50 cm tall, erect; basal leaves lyrate-pinnatifid with a large terminal division; stem leaves lyrate-pinnatifid; petals pale yellow; fruit 3.5–5 cm long, subterete, beak 0.5–1.5 mm long; a weed of waste places. **B. orthoceras** Ledeb.

Brassica L. Mustard

1a. Upper stem leaves clasping; weeds of cultivated fields and waste places. Field Mustard. **B. campestris** L.
1b. Upper stem leaves not clasping. (2)
2a. Pedicels erect, less than 5 mm long; fruit appressed to the rachis, 1–2 cm long, not over 2 mm wide. Black Mustard. **B. nigra** (L.) Koch
2b. Pedicels spreading, over 5 mm long; fruit not appressed, 3.5–5 cm long, 2–3.5 mm wide. Leaf Mustard. **B. juncea** (L.) Coss.

Camelina Crantz False Flax

Pubescent annuals, 30–80 cm tall; lower leaves lanceolate, entire; upper leaves sagittate-clasping; petals yellowish; fruit 4–7 mm long; adventive from Europe, weed of waste places. **C. microcarpa** Andrz.

Capsella Medic. Shepherdspurse

Plants annual, 10–40 cm tall; lower leaves dentate to pinnatifid; petals 1.5–2 mm long, white; fruit cuneate, heart shaped; a common weed of cultivated land and waste places, one of Utah's earliest flowering mustards. **C. bursa-pastoris** (L.) Medic.

Cardamine L. Bittercress

1a. Leaves all simple; widely distributed along stream banks at middle and higher elevations. **C. cordifolia** Gray
1b. Leaves pinnately compound, at least some. (2)
2a. Petals 5–6 mm long; beak of fruit to 2.5 mm long. **C. breweri** S. Wats.
2b. Petals 2–4 mm long; beak of fruit to 1 mm long. (3)
3a. Leaflets oblong to ovate, 3–5, lower leaves simple. **C. unijuga** Rydb.
3b. Leaflets linear to elliptic, 11–17, basal leaves pinnate. **C. pennsylvanica** Muhl. ex Willd.

Cardaria Desv. Whitetop

1a. Ovary and fruit pubescent. **C. pubescens** (C. A. Meyer) Rollins
1b. Ovary and fruit glabrous; introduced from Europe; this species and the above are pernicious weeds which spread both by means of seeds and rhizomes. **C. draba** (L.) Desv.

Caulanthus S. Wats. Wild Cabbage

Plants biennial or short-lived perennial, 30–100 cm tall; stems glabrous and glaucous, more or less inflated; sepals 10–15 mm long, pur-

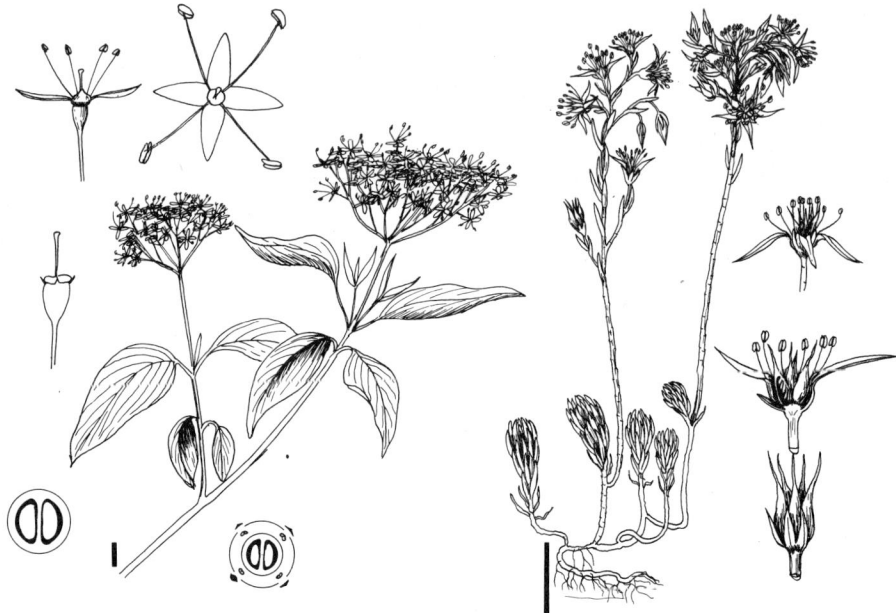

Figs. 37-38. 37. **Cornus stolonifera**, x .45. 38. **Sedum stenopetalum**, x 1

Figs. 39-40. 39. **Barbarea orthoceras**, x .21. 40. **Brassica campestris**, x .2

plish, hirsute; petals 15–20 mm long, purplish, linear and channeled; fruit 10–13 cm long, erect or ascending, subsessile; widely distributed, especially in eastern Utah. **C. crassicaulis** (Torr.) S. Wats.

Chlorocrambe Rydb.

Plants perennial, 6–10 dm tall; leaves petioled, mostly hastate, glabrous, 5–12 cm long; flowers spreading or reflexed; petals 5–6 mm long, greenish yellow; fruit terete, 6–9 cm long; stipe and style each 2 mm long. **C. hastatus** (S. Wats.) Rydb.

Chorispora DC.

Plants annual, 20–50 cm tall, glandular-viscid, but lacking forked hairs; sepals 4–6 mm long; petals 7–12 mm long, rose-purple; fruit 2.5–4 cm long, breaking into indehiscent joints, beak long; introduced from Eurasia, weeds of waste places. **C. tenella** DC.

Conringia Adans.   Haresear Mustard

Plants annual, 25–60 cm tall; leaves 4–10 cm long, oval to elliptic, deeply cordate-clasping; petals about 8 mm long. yellowish white; fruit 8–10 cm long; introduced from Europe, waste places. **C. orientalis** (L.) Dumort.

Descurainia Webb. & Berthel   Tansy Mustard

1a. Pods clavate or subclavate. **D. pinnata** (Walt.) Britt.

1b. Pods linear, not enlarged at the apex. (2)

2a. Leaves 2- to 3-pinnate; fruit more than 20-seeded. **D. sophia** (L.) Webb.

2b. Leaves simply pinnate with leaflets often deeply incised; fruit less than 20-seeded. (3)

3a. Fruits 3–7 mm long, attenuate at apex and tipped with a prominent style. **D. californica** (Gray) Schulz

3b. Fruits 8–15 mm long, not attenuate at apex; style short or lacking. **D. richardsonii** (Sweet) Schulz

Diplotaxis DC.   Sand Rocket

Plants annual or biennial, 30–50 cm tall, leafy only at base; leaves petiolate, 3–10 cm long; petals yellow, 5–8 mm long; fruit erect, 2–2.5 cm long, sessile. **D. muralis** (L.) DC.

Dithyrea Haw.   Spectacle Pod

Plants annual or possibly biennial, 20–60 cm tall; leaves dentate to entire; petals 5–8 mm long, white; fruit flattened contrary to the narrow partition, resembling a pair of eyeglasses; common in sandy situations in southern Utah. **D. wislizenii** Engelm. ex Wisliz.

Draba L.   Whitlowgrass

1a. Plants annual. (2)

1b. Plants perennial or biennial. (6)

Subclass DICOTYLEDONEAE

- 2a. Flowers white. (3)
- 2b. Flowers yellow. (4)
- 3a. Leaves usually entire; stems glabrous above; widely distributed. **D. reptans** (Lam.) Fern.
- 3b. Leaves usually denticulate; stems pubescent throughout; widely distributed. **D. cuneifolia** Nutt.
- 4a. Inflorescence and pedicels pubescent; Wasatch Mountains. **D. rectifructa** C. L. Hitchc.
- 4b. Inflorescence and pedicels glabrous. (5)
- 5a. Pedicles at least one and one-half times as long as the fruit; widely distributed. **D. nemorosa** L.
- 5b. Pedicels usually slightly longer than the fruit, but occasionally shorter; widely distributed. **D. stenoloba** Ledeb.
- 6a. Flowers white. (7)
- 6b. Flowers yellow or orange. (10)
- 7a. Flowering stems usually leafless. (8)
- 7b. Flowering stems usually with more than 1 leaf (1–10). (9)
- 8a. Leaves glabrous or with unforked hairs only; southwestern Utah, from Bryce Canyon and Cedar Breaks. **D. subalpina** Goodman & Hitchc.
- 8b. Leaves cinereous with fine, interwoven branched hairs; widespread at high elevations. **D. nivalis** Liljebl.
- 9a. Leaves usually denticulate; style 0.2–0.8 mm long; cauline leaves 1–10; widespread at high elevations. **D. lanceolata** Royle
- 9b. Leaves entire; style to 0.2 mm long or lacking; cauline leaves 1–2; La Sal Mountains. **D. fladnizensis** Wulfen ex Jacq.
- 10a. Flowers orange; southwestern and western Utah. **D. zionensis** C. L. Hitchc.
- 10b. Flowers lemon yellow to yellow; variously distributed. (11)
- 11a. Cauline leaves present, 1–15. (12)
- 11b. Cauline leaves lacking, all leaves basal. (15)
- 12a. Leaves glabrous or with a few cilia; talus and rocky cliffs, Uinta Mountains. **D. crassa** Rydb.
- 12b. Leaves more or less densely pubescent. (13)
- 13a. Petals 2–3 mm long, light yellow; moist rocks or banks above 7,000 ft., Wasatch Mountains. **D. brachystylis** Rydb.
- 13b. Petals 4–7 mm long, bright yellow. (14)
- 14a. Leaves bright green to slightly grayish, not long-ciliate, sparsely pubescent beneath with sessile or short-stalked hairs; southeastern and western Utah. **D. spectabilis** Greene
- 14b. Leaves grayish green or, if green, not as above; widely distributed. **D. aurea** Vahl. ex Hornem.
- 15a. Pedicels pubescent. (16)
- 15b. Pedicels glabrous. (18)

Cruciferae

16a. Styles less than 0.5 mm long; Wasatch Mountains. **D. densifolia** Nutt. ex T. & G.
16b. Styles more than 1 mm long. (17)
17a. Fruit glabrous or, if pubescent, the hairs simple or merely forked; known only from mountains between Panguitch Lake and Marysvale, Piute Co. **D. sobolifera** Rydb.
17b. Fruit densely pubescent with stellate hairs; Uinta Mountains. **D. ventosa** Gray
18a. Styles less than 0.5 mm long. (19)
18b. Styles usually more than 1 mm long. (20)
19a. Petals 2–3 mm long; style less than 0.2 mm long; widely distributed on rocks or talus near or above timberline. **D. crassifolia** R. Grah.
19b. Petals 4 mm long; style 0.2–0.5 mm long; talus and alpine meadows, Uinta and Wasatch mountains. **D. apiculata** C. L. Hitchc.
20a. Leaves oblanceolate; local endemic in the Bear River and Wellesville Mountain ranges. **D. maguirei** C. L. Hitchc.
20b. Leaves usually linear; widely distributed, possibly the most common perennial species. **D. oligosperma** Hook.

Erysimum L. Wallflower

1a. Flowers less than 10 mm long, pale yellow to yellow. (2)
1b. Flowers more than 10 mm long, and usually more than 15 mm long, pale yellow to orange or brown-purple; dry prairies and rocky or sandy bluffs, almost throughout Utah. Western Wallflower, Prairie Rocket. **E. asperum** DC.
2a. Flowers 3–4 mm long; pedicels slender, less than 1 mm wide, over 7 mm long; seeds and leaves are reported to be very bitter and, when consumed by cattle, to "taint" milk; moist waste places in valleys and canyons and in the Wasatch and Uinta mountains. Wormseed, Treacle Mustard. **E. cheiranthoides** L.
2b. Flowers 5–10 mm long; pedicels stocky, 1 mm wide or greater. (3)
3a. Fruits 4 cm long, erect, squarish; pedicels 7–7.5 mm long; plants simple or branched, but not divaricate; along riverbanks and sandy places, among sagebrush or even aspen-spruce to 9,000 ft., known from Garfield, Grand, San Juan and Salt Lake cos. **E. inconspicuum** (S. Wats.) Mac M.
3b. Fruits 5–7 cm long, straight or curved, divaricate; pedicels 4 mm long; leaves usually repand-dentate; plants often greatly branched, divaricate, rarely simple, annual (?); a naturalized plant from Europe, along roadsides, on empty lots or abandoned land. **E. repandum** L.

Halimolobos Tausch

Biennial or perennial herbs, 10–40 cm tall; basal leaves sinuate-dentate, petioled; cauline leaves sessile, auriculate; sepals 2–3 mm long;

Subclass DICOTYLEDONEAE

petals 3–4 mm long, white; fruit terete, erect, 2–4 cm long. **H. virgata** (Nutt.) Schulz

Hesperis L.   Rocket

Plants perennial or biennial, 3–10 dm tall; leaves 5–10 cm long, sessile, denticulate; flowers about 2 cm long, purplish (some races white flowered); fruit 5–10 cm long; cultivated ornamental. Damesviolet. **H. matronalis** L.

Hutchinsia R. Br.

Plants annual; stems low and spreading, 5–20 cm long; leaves short-petiolate below, sessile above; pedicels ascending or spreading; sepals and petals about 1 mm long; fruit 3–4 mm long; widely distributed. **H. procumbens** (L.) Desv.

Iberis L.   Candytuft

Evergreen perennials, woody at base, 1.5–4 dm tall; leaves 12-25 mm long; flowers white or tinged lilac; fruit about 5 mm long; cultivated border plant. Edging Candytuft. **I. sempervirens** L.

Isatis L.   Woad

Plants biennial or perennial, 3–10 dm tall; lower leaves oblanceolate, 7–10 cm long, upper reduced and sagittate-clasping; petals yellow, about 3 mm long; fruit 1-seeded, oblong, winged and samaroid, 12–15 mm long; waste places, introduced. Dyer's Woad. **I. tinctoria** L.

Lepidium L.   Peppergrass

1a. Upper cauline leaves perfoliate or sagittate-clasping. (2)
1b. Upper cauline leaves not perfoliate or clasping. (3)
2a. Plants with upper leaves perfoliate; basal leaves 2-pinnate; a common weed at lower elevations. **L. perfoliatum** L.
2b. Plants with upper leaves clasping; basal leaves entire to pinnatifid; weed at middle elevations. **L. campestre** (L.) R. Br. ex Ait.
3a. Plants perennial, the base frequently woody. (4)
3b. Plants annual. (5)
4a. Capsules 4–7 mm wide; southwestern Utah. **L. fremontii** S. Wats.
4b. Capsules less than 4 mm wide; widely distributed in sandy soils. **L. montanum** Nutt.
5a. Fruit prominently reticulate and long winged at apex; saline soils. **L. dictyotum** Gray
5b. Fruit not prominently reticulate or long winged at apex. (6)
6a. Pedicels much flattened; fruit widest above middle; waste places. **L. densiflorum** Schrad.
6b. Pedicels not much flattened; fruit widest at the middle; widespread. **L. virginicum** L.

Lesquerella S. Wats.   Bladderpod

1a. Fruit glabrous; southeastern Utah. **L. fendleri** (Gray) S. Wats.

Figs. 41-42. 41. **Camelina microcarpa**, x .2, 42. **Capsella bursa-pastoris**, x .2

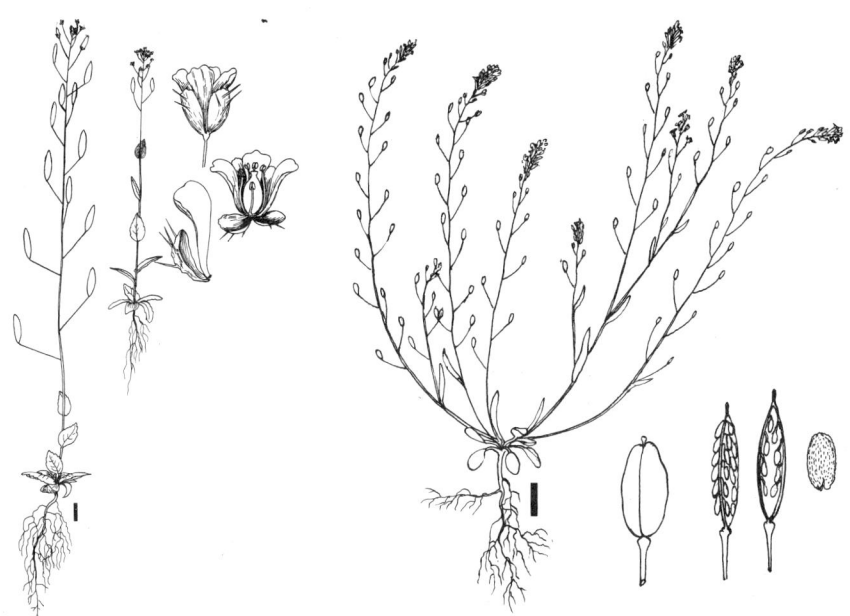

Figs. 43-44. 43. **Draba nemorosa**, x .25. 44. **Hutchinsia procumbens**, x .46

Subclass DICOTYLEDONEAE

1b. Fruit stellate pubescent; variously distributed. (2)
2a. Pedicels recurved, not straight or S-shaped; fruit globose or nearly so; eastern Utah. **L. ludoviciana** (Nutt.) S. Wats.
2b. Pedicels S-shaped, straight or uniformly curved upward. (3)
3a. Pedicels straight, not S-shaped. (4)
3b. Pedicels S-shaped. (5)
4a. Fruit 4–6 mm long, not compressed at apex; basal leaves linear; stems 5–18 cm tall; southern Utah. **L. intermedia** (S. Wats.) Heller
4b. Fruit 2.5–4 mm long, compressed near apex; basal leaves oblanceolate or wider; stems 1–5 cm tall; northeastern Utah. **L. subumbellata** Rollins
5a. Plants annual; southern Utah. **L. palmeri** S. Wats.
5b. Plants perennial; distribution various. (6)
6a. Basal leaves linear or narrowly oblanceolate. (7)
6b. Basal leaves oblanceolate or broader. (8)
7a. Pods sessile; northeastern Utah. **L. alpina** (Nutt.) S. Wats.
7b. Pods stipitate; central Wasatch Mountains. **L. garrettii** Payson
8a. Pods conspicuously elongated, when mature at least twice as long as wide; southern Utah. **L. wardii** S. Wats.
8b. Pods not conspicuously elongated. (9)
9a. Basal leaf blades narrowed gradually to the petiole; southeastern Utah. **L. rectipes** W. & S.
9b. Basal leaf blades abruptly narrowed to the petiole. (10)
10a. Pods acute at the apex; southern Utah. **L. wardii** S. Wats.
10b. Pods truncate at the apex; widely distributed in central and northern Utah (one of the segments of this complex has been designated as **L. multiceps** Maguire). **L. utahensis** Rydb.

Lobularia Desv. Sweet Alyssum

Plants perennial, although grown as annuals, 1.5–3 cm tall; leaves lanceolate or linear; flowers white or violet; fruit small, orbicular, beaked; cultivated border plant. **L. maritima** Desv.

Lunaria L. Moonwort

Plants annual or biennial, 4.5–10 dm tall; leaves coarsely toothed; flowers 2–2.5 cm long, mostly pink-purple, rarely white; pods thin and flat, 2.5 cm or more long and nearly as wide, stipitate; cultivated for showy flowers and for the lustrous septa of the pods. Dollar Plant. **L. annua** L.

Malcolmia R. Br. African Mustard

Plants annual, 10–40 cm tall, stellate-pubescent; leaves oblong or lanceolate; petals 6–8 mm long, pink-purple; fruit 4–6 cm long, stellate-pubescent; introduced weed of waste places. **M. africana** (L.) R. Br.

Matthiola R. Br. Stocks

Plants annual or biennial, stellate-pubescent; leaves lanceolate,

Cruciferae

short-petiolate, entire or serrulate; flowers sessile, 2 cm long, very fragrant, purplish; fruit 7.5–10 cm long, bifurcate apically; cultivated. Perfume-Plant. **M. bicornis** DC.

Parrya R. Br.

Acaulescent perennials from a caudex; leaves basal, glandular-hirsute, 6–8 cm long; scape glandular-hirsute; petals 15–18 mm long, purplish; pedicels 8–15 mm long, ascending; pods erect, glandular-hispid, 3–4 cm long, 4–7 mm wide; montane to alpine regions, Uinta Mountains. **P. platycarpa** Rydb.

Physaria (Nutt.) Gray Twinpod
1a. Style less than 3.5 mm long (1–2); septum lanceolate; basal sinus absent; southeastern Utah. **P. newberryi** Gray
1b. Style more than 4 mm long; septum variously shaped but not lanceolate. (2)
2a. Sinuses of pod equal above and below; valves nearly orbicular; northern and eastern Utah. **P. australis** (Payson) Rollins
2b. Sinuses of silique unequal-upper very deep, lower shallow or absent; valves variously shaped but not orbicular. (3)
3a. Silique highly inflated, 1.5–3 cm wide; valves membranous; basal leaves dentate, but not deeply lobed; northern and western Utah. **P. chambersii** Rollins
3b. Silique moderately inflated, less than 1.5 cm wide; valves coriaceous; basal leaves deeply lobed; known from the type locality Chandlar Canyon, Uintah Co. **P. grahamii** Morton

Rorippa Scop. Cress
1a. Leaves crenate or sinuate, simple; lower leaves at least 15 cm long; cultivated and escaping. Horseradish. **R. armoracia** (L.) A. S. Hitchc.
1b. Leaves pinnate or pinnatifid, seldom to 15 cm long. (2)
2a. Plants aquatic or semiaquatic; flowers white; leaves pinnately compound; introduced plant of springs, streams, and seeps. Watercress. **R. nasturtium-aquaticum** (L.) Schinz. & Thel.
2b. Plants of moist soil; flowers yellow; leaves pinnate or pinnatifid. (3)
3a. Stems pubescent, 3–10 dm tall. **R. islandica** (Oed.) Borbas
3b. Stems glabrous or glabrate. (4)
4a. Fruit globose, 2–3 mm long. **R. sphaerocarpa** (Gray) Greene
4b. Fruit elliptic to oblong or linear. (5)
5a. Plants perennial with horizontal rhizomes. **R. sinuata** (Nutt.) A. S. Hitchc.
5b. Plants annual or biennial, the rhizomes lacking. (6)
6a. Leaf segments acute. **R. curvisiliqua** (Hook.) Bessey
6b. Leaf segments obtuse. **R. obtusa** (Nutt.) Britt.

Sisymbrium L.

A difficult genus whose limits are uncertain; only 3 species treated

Subclass DICOTYLEDONEAE

herein, but more, assignable to this genus or to **Thelypodium**, are known from the region.

- 1a. Plants perennial with creeping rhizomes; at least middle and upper leaves entire and linear; native, common in the pinyon-juniper zone. **S. linifolium** Nutt.
- 1b. Plants annual or biennial, the rhizomes lacking; middle and upper leaves pinnatifid or hastate. (2)
- 2a. Fruit 1–1.5 cm long, erect; upper leaves hastate; introduced weed of waste places. **S. officinalis** (L.) Scop.
- 2b. Fruit 7–10 cm long, spreading; upper leaves pinnatifid; introduced weed of waste places. **S. altissimum** L.

Smelowskia C. A. Meyer

Caespitose perennials from a caudex; stems 5–20 cm tall, pubescent; basal leaves petioled, entire to pinnately parted; stem leaves pinnatifid; pedicels 5–10 mm long, ascending; petals 5–7 mm long, rose to white; fruit 5–12 mm long; alpine regions. **S. calycina** C. A. Meyer ex Ledeb.

Stanleya Nutt. Prince's Plume

- 1a. Middle and upper stem leaves sessile, sagittate-clasping. **S. viridiflora** Nutt. ex T. & G.
- 1b. Middle and upper stem leaves petiolate. (2)
- 2a. Plants woody at base; petals 1.5–3 mm wide; most common and most widespread of the **Stanleya** species. **S. pinata** (Pursh) Britt.
- 2b. Plants herbaceous throughout; petals 4–10 mm wide. **S. albescens** M. E. Jones

Streptanthella Rydb. Little Twistflower

Plants annual; stems 20–60 cm tall; lower leaves oblanceolate to spatulate, readily deciduous; stem leaves linear-lanceolate; sepals 3–5 mm long; petals white to yellowish; fruit 3–6 cm long; sandy sites throughout southern Utah. **S. longirostris** (S. Wats.) Rydb.

Streptanthus Nutt. Twistflower

Short-lived perennials; stems 30–90 cm tall; basal leaves spatulate, dentate apically; stem leaves sagittate; sepals 7–10 mm long, purplish; petals 10–15 mm long; fruit 5–10 cm long, 4–5 mm wide, flattened; locally common in pinyon-juniper community. **S. cordatus** Nutt. ex T. & G.

Thelypodium Endl.

- 1a. Stem leaves auriculate or clasping at base. **T. sagittatum** (Nutt.) Endl. ex Walp.
- 1b. Stem leaves not clasping, seldom slightly auriculate. (2)
- 2a. Pedicels 3–5 mm long, conspicuously flattened at base; stipe of fruit 1–3 mm long. **T. integrifolium** (Nutt.) Endl.

Figs. 45-46. 45. **Lepidium densiflorum,** x .21.   46. **Rorippa islandica,** x .2

Figs. 47-48.   47. **Sisymbrium altissimum,** x .22.   48. **Smelowskia calycina,** x .37

Subclass DICOTYLEDONEAE

2b. Pedicels 5–10 mm long, seldom flattened at base; stipe of fruit usually less than 1 mm long. **T. lilacinum** Greene

Thlaspi L. Pennycress

1a. Plants annual; fruit orbicular, broadly winged and deeply notched at apex; introduced weed of waste places. **T. arvensis** L.
1b. Plants perennial; fruit cuneate or obovate, not winged or slightly so, truncate or slightly notched at apex; native, high elevations. **T. alpestre** L.

CUCURBITACEAE — GOURD FAMILY

Trailing or climbing annual or perennial herbs with tendrils; leaves alternate, broad, usually simple but often deeply cut; tendrils simple or branched; flowers imperfect, the plants monoecious, sometimes dioecious; ovary inferior; petals 5, united; stamens 5, 2 pairs often united, mostly connate by anthers; carpels 3; fruit a pepo or a bladdery pod.

1a. Fruits fleshy, with hard or firm rind, indehiscent, often edible. (2)
1b. Fruits otherwise, berrylike, podlike, or bladdery, dry or fleshy, mostly inedible; mostly ornamental. (5)
2a. Corolla bell shaped and distinctly united, 5-lobed to middle or more. **Cucurbita**
2b. Corolla of 5 petals, or parted to the base, rotate or open bell shaped. (3)
3a. Tendrils simple. **Cucumis**
3b. Tendrils branched. (4)
4a. Leaves not lobed. **Lagenaria**
4b. Leaves pinnatifid. **Citrullus**
5a. Plants with tendrils, glabrous or hairy at nodes; ovules erect; filaments connate. **Echinocystis**
5b. Plants without tendrils, pubescent; ovules horizontal; filaments and stamens free. **Ecballium**

Citrullus Neck.

Monoecious annuals with long-running stems and branched tendrils; leaves mostly ovate, deeply pinnatifid; peduncles shorter than leaves; corolla about 3.5 cm broad, yellow, single in leaf axils, 5-lobed; fruit globular to oblong. Watermelon. **C. vulgaris** Schrad.

Cucumis L.

1a. Fruit prickly at maturity; cultivated. Cucumber. **C. sativus** L.
1b. Fruit net ribbed or rough at maturity; cultivated. Cantaloupe, Muskmelon. **C. melo** L.

Cucurbita L. Gourd

1a. Plants perennial from a greatly thickened root which is to 25 cm in diameter, ill smelling; fruit 5–10 cm in diameter, globose or

Dipsacaceae

nearly so, striped or mottled light and dark green; southwestern Utah. **C. foetidissima** H. B. K.
1b. Plants annual; cultivated. (2)
2a. Plants rough-hairy; leaves strongly lobed and triangularly pointed, dentate or serrate; peduncle expanded near fruit, not flared; seeds thick edged; buds pointed even preceding anthesis; calyx lobes of staminate flowers seldom foliaceous. Squash, Pumpkin, Gourd. **C. pepo** L.
2b. Plants soft, variously pilose; leaves shallowy if at all lobed, rounded or triangular, mostly not or only obscurely serrate. (3)
3a. Seeds thin with a hyaline ragged edge which often wears away; leaves broad-ovate to somewhat triangular; buds pointed in evening before anthesis; peduncle flared near fruit; calyx lobes of staminate flowers often foliaceous. Cushaw, Winter Crookneck. **C. moschata** Duchesne
3b. Seeds plump, margins obtuse and more or less elevated; leaves circular to reniform; buds not pointed in evening before anthesis; peduncle short and spongy, nearly cylindric, not flared near fruit; calyx lobes of staminate flowers not foliaceous. Fall and Winter Squash. **C. maxima** Duchesne

Ecballium A. Rich.   Squirting-Cucumber

Plants hairy-pubescent, trailing, without tendrils; pistillate flowers 1; staminate flowers in racemes; leaves triangular-ovate, 7.5–10 cm long, broadly cordate; fruit oblong, 3.5–5 cm long, rough-hairy, greenish, at maturity squirting brownish seeds; cultivated for odd fruits. **E. elaterium** A. Rich.

Echinocystis T. & G.

Annual vines with branched tendrils; leaves cordate-ovate, 7.5–12.5 cm long, shallowly 3- to 7-lobed; pistillate flowers usually 1; staminate flowers racemose or paniculate; fruit fleshy or dry, short-oblong, 3.5–5 cm long, puffy, with slender weak spines, opening at the end; seeds brownish; cultivated. Wild Cucumber. **E. lobata** T. & G.

Lagenaria Ser.   Bottle Gourd

Annual, musky-scented, soft, viscid-pubescent vines with branched tendrils; leaves cordate-ovate to reniform-ovate; flowers 5–10 cm broad, white, showy, 5-petaled, the staminate often surpassing the foliage; fruit 1–10 dm long, of many shapes; seeds oblong to nearly square, ridged on margins, tan to brown; cultivated. **L. siceraria** Standl.

DIPSACACEAE — TEASEL FAMILY

Herbs with opposite or whorled leaves; flowers ordinarily in dense, involucrate heads, perfect, with united petals, epigynous, more or less irregular; calyx small, usually deeply cut into 4 or 5 segments, or into more numerous teeth or hairs; corolla 4- to 5-lobed; stamens 4 or 2,

Subclass DICOTYLEDONEAE

alternate with the corolla lobes, exserted; ovary 1-loculed with a single ovule; fruit dry, indehiscent.

1a. Stems prickly, usually with leaves and involucre also prickly. **Dipsacus**
1b. Stems not prickly. (2)
2a. Receptacle densely hairy, without bracts. **Knautia**
2b. Receptacle with bracts. **Scabiosa**

Dipsacus L. Teasel

Biennial or perennial herbs; leaves lance-oblong, toothed and often prickly on the margin; an introduced weed, usually of moist situations. **D. sylvestris** Huds.

Knautia L. Bluebuttons

Perennial herbs, 3–10 dm tall; lowest leaves usually coarsely toothed, the others more or less deeply pinnatifid; heads 1.5–4 cm wide; involucral bracts 8–12 mm long; corolla lilac-purple; fruit 5–6 mm long; adventive from Europe, waste places. **K. arvensis** (L.) Duby

Scabiosa L. Pincushion-Flower

Annual herbs to 6 dm tall; basal leaves oblong-spatulate, coarsely dentate; stem leaves pinnately parted; flowers dark purple, rose, or white, in long-peduncled heads to 5 cm broad; involucels with ribs broadened toward apex; cultivated. Sweet Scabious. **S. atropurpurea** L.

ELAEAGNACEAE — OLEASTER FAMILY

Shrubs or small trees, with stellate pubescence; leaves alternate or opposite, simple, entire; flowers regular, perfect or imperfect; perianth 4-toothed in a single whorl from the apex of a hypanthium; stamens 4–8; pistils 1; ovary superior, though often appearing inferior, 1-loculed; fruit drupaceous.

1a. Leaves alternate; flowers perfect or polygamous; stamens 4. **Elaeagnus**
1b. Leaves opposite; flowers imperfect; stamens 8. **Shepherdia**

Elaeagnus L. Oleaster

1a. Branchlets and leaves with silvery scales only; leaves oblong to linear-oblong; introduced from Eurasia, cultivated and escaping, now well established. Russian Olive. **E. angustifolia** L.
1b. Branchlets and leaves with both brown and silvery scales; leaves oblong to elliptic or ovate; native, eastern Utah, rare in cultivation. Silverberry. **E. commutata** Bernh.

Shepherdia Nutt. Buffaloberry

1a. Leaves persistent, oval, ovate or suborbicular, often subcordate at base; fruit subglobose, scurfy; southern Utah. **S. rotundifolia** Parry

Ericaceae

1b. Leaves deciduous, elliptic-oblong to ovate, cuneate or rounded at base; fruit elongate, not scurfy. (2)
2a. Plants somewhat thorny, 1–6 m tall; leaves oblong, cuneate at base; fruit red, edible; common along stream courses at lower elevations. S. argentea (Pursh) Nutt.
2b. Plants not thorny, 1–2 m tall; leaves elliptic-oblong to ovate, usually rounded at base; fruit yellowish red, unpalatable; middle and higher elevations in mountains. S. canadensis (L.) Nutt.

Ericaceae — Heath Family

Shrubs or subshrubs with simple, alternate or opposite leaves; flowers perfect; calyx 4- to 7-parted or -cleft; corolla regular or slightly irregular, gamopetalous, 4- to 5-lobed; stamens as many, or twice as many, as petals; anthers dehiscent by longitudinal slits, terminal pores or tubes, sometimes appendaged; ovary superior or inferior, 2- to 10-loculed; fruit a capsule, berry, or drupe.

1a. Ovary inferior; fruit berrylike; leaves deciduous. **Vaccinium**
1b. Ovary superior; fruit capsular or berrylike; leaves evergreen. (2)
2a. Corolla rotate, with 10 pouches enclosing the anthers in bud; fruit capsular; leaves opposite or whorled. **Kalmia**
2b. Corolla urceolate to campanulate, no pouches present; fruit capsular or berrylike; leaves alternate. (3)
3a. Sepals definitely united at base, the tube becoming fleshy in fruit and enclosing the capsule; anthers not awned; flowers solitary in axils of leaves. **Gaultheria**
3b. Sepals nearly or quite distinct, not becoming fleshy, the ovary becoming a berry; anthers with a recurved dorsal awn; flowers terminal in short racemes or panicles. **Arctostaphylos**

Arctostaphylos Adans. Manzanita

1a. Leaves rounded at apex, not mucronate, bright green; fruit bright red, bitter astringent; prostrate shrub; widely distributed at middle and higher elevations. Bear-Berry **A. uva-ursi** (L.) Spreng.
1b. Leaves acute to rounded and mucronate or mucronulate at apex; fruit brown or red, acid. (2)
2a. Shrubs procumbent or prostrate, the lower branches rooting, glabrous; corolla 4–5 mm long; leaves broadest toward apex. **A. nevadensis** Gray
2b. Shrubs erect or ascending, the lower branches sometimes spreading and rooting. (3)
3a. Young twigs and branches of inflorescence resinous-glandular, not tomentulose; ovary and fruit glabrous; possibly Utah's most common large **Arctostaphylos**. **A. patula** Greene
3b. Young twigs and branches of inflorescence cinereous-tomentulose;

Subclass DICOTYLEDONEAE

fruit depressed-globose; nutlets irregularly separable; leaves seldom over 15 mm wide; plants not sprouting from base. **A. pungens** H. B. K.

Gaultheria L. Wintergreen

Evergreen perennials; leaves alternate; flowers perfect, solitary and axillary; corolla campanulate, white, the 10 stamens included; anthers not awned, opening by terminal pores; ovary superior, 5-loculed; calyx becoming fleshy, enclosing the capsule and forming a globose, berrylike, scarlet fruit. **G. humifusa** (Graham) Rydb.

Kalmia L.

Plants shrubs to 20 cm tall; leaves evergreen, opposite or whorled; flowers in terminal corymbs or umbels; petals united, the corolla rotate, with 10 pouches beneath the 5 corolla lobes and the 5 sinuses; anthers awnless, opening by terminal pores; fruit a capsule. Bog Kalmia. **K. polifolia** Wangh.

Vaccinium L. Blueberry

1a. Flowers in clusters of 2–4 or, if 1, then always from a bud on wood of the previous season. **V. occidentale** Gray
1b. Flowers solitary in the axils of leaves on branches of the current season. (2)
2a. Branches terete, not distinctly angled. (3)
2b. Branches strongly angled longitudinally. (4)
3a. Plants 0.5–2 dm tall. **V. caespitosum** Michx.
3b. Plants 2–4 dm tall. **V. arbuscula** Merriam
4a. Leaves 15 mm long or less; branches very numerous; fruit red at maturity, 4 mm in diameter or less. **V. scoparium** Leiberg
4b. Leaves more than 15 mm long; branches not very numerous; fruit blue, purple, or black at maturity. (5)
5a. Plants usually more than 5 dm tall. **V. globulare** Rydb.
5b. Plants usually less than 5 dm tall. (6)
6a. Leaves acute apically, usually widest above the middle. **V. myrtillus** L.
6b. Leaves obtuse to rounded apically, usually widest below the middle. **V. arbuscula** Merriam

EUPHORBIACEAE — SPURGE FAMILY

Herbaceous or woody annuals or perennials; sap milky or watery; leaves simple, alternate, opposite or whorled; flowers unisexual; calyx present or absent (a calyxlike involucre, a cyathium, present in **Euphorbia**); corolla lacking; stamens 1 to many; ovary superior, 3-loculed; ovules 1–2 per locule; fruit a capsule.

1a. Plants with milky juice; flowers borne in cyathia. **Euphorbia**

Euphorbiaceae

1b. Plants with watery juice; flowers with a calyx. (2)
2a. Plants 10–20 dm tall or more; leaves 2–4 dm broad, deeply lobed; cultivated. **Ricinus**
2b. Plants 2–6 dm tall; leaves 2–8 cm broad, not lobed. **Croton**

Croton L.

1a. Plants annual; leaves lanceolate to oblong. **C. texensis** (Klotz.) Muell. Arg. ex DC.
1b. Plants perennial; leaves elliptic to oblanceolate. **C. longipes** M. E. Jones

Euphorbia L. Spurge

1a. Plants woody; cultivated ornamental with brightly colored upper leaves, usually red. Poinsettia. **E. pulcherrima** Willd.
1b. Plants herbaceous; if cultivated then otherwise different from above. (2)
2a. Glands of cyathia subtended by petaloid appendages or, if absent, the leaves all opposite. (3)
2b. Glands of cyathia not subtended by petaloid appendages; leaves alternate or opposite. (7)
3a. Bracts with broad white margins; stipules minute and glandlike or none; cultivated and escaping. Snow-on-the-Mountain. **E. marginata** Pursh
3b. Bracts lacking broad white margins; stipules well developed. (4)
4a. Plants perennial; leaves ovate or broader, at least some. (5)
4b. Plants annual; leaves ovate or narrower. (6)
5a. Petaloid appendages large, white; stipules triangular; seeds smooth. **E. albomarginata** T. & G.
5b. Petaloid appendages rarely as wide as the gland, often wanting; stipules subulate; seeds wrinkled. **E. fendleri** T. & G.
6a. Leaves linear to narrowly linear, entire. **E. parryi** Engelm.
6b. Leaves oblong-linear to ovate, entire to serrate. **E. glyptosperma** Engelm. ex Emory
7a. Glands deeply cupped; leaves opposite throughout. **E. dentata** Michx.
7b. Glands flat or convex, not deeply cupped; stem leaves alternate (rarely opposite in or below the inflorescence). (8)
8a. Rhizomes present; seeds smooth. (9)
8b. Rhizomes absent; seeds smooth, pitted, or wrinkled; native. **E. robusta** (Engelm.) Small ex Britt. & Brown
9a. Leaves linear, to 3 mm wide; plants 12–30 cm tall; cultivated and escaping. **E. cyparissias** L.
9b. Leaves linear to oblong, 4–12 mm wide; plants 30–60 cm tall; a noxious weed. **E. esula** L.

Subclass DICOTYLEDONEAE

Ricinus L.

Plants annual in temperate zone, 1–5 m tall; leaves parted; panicle 3–6 dm tall; fruit a capsule 12–25 mm long, covered with soft spines; cultivated for foliage; seeds poisonous. Castor Bean. **R. communis** L.

FAGACEAE — BEECH FAMILY

Deciduous, cultivated and native trees and shrubs; leaves alternate, dentate to pinnately lobed; flowers monoecious, usually axillary on young shoots; staminate flowers with a 4- to 7-lobed perianth, in slender catkins or in peduncled heads; pistillate flowers in 1s or 3s with a 4- to 8-lobed perianth; ovary inferior, usually 3-loculed and with 3 styles, a single ovule maturing; fruit a 1-seeded nut.

1a. Staminate flowers in slender pendulous heads; pistillate flowers usually in pairs; buds long, slender, and sharp pointed; leaves 2-ranked, with prominent lateral veins; cultivated trees. **Fagus**
1b. Staminate flowers in slender catkins; pistillate flowers 1; buds usually not long and slender; leaves various, but not usually as above; cultivated or native. (2)
2a. Terminal bud lacking; the fruit, a nut, subtended by a burlike involucre covered with long, sharp spines. **Castanea**
2b. Terminal bud present; the fruit, an acorn, subtended by a cup. **Quercus**

Castanea Mill. Chestnut

Trees to 35 m tall; leaves oblong-lanceolate, acuminate, coarsely serrate, 12–25 cm long; bur 4–7 cm wide; nut 10–25 mm wide, usually more than 1; once an important eastern deciduous forest tree, now reduced by blight; rarely cultivated in Utah. American Chestnut. **C. dentata** Borkh.

Fagus L. Beech

1a. Leaves with 9–14 pairs of veins, serrate. American Beech. **F. grandifolia** Ehrh.
1b. Leaves with 5–9 pairs of veins, denticulate to undulate; a form having purple leaves is known as var. **atropunicea** West. European Beech. **F. sylvatica** L.

Quercus L. Oak

1a. Leaves grayish, evergreen, pinnatifid to dentate with acute, spinulose lobes or teeth; southwestern Utah. Native Live Oak. **Q. turbinella** Greene
1b. Leaves green, deciduous, serrate or lobed. (2)
2a. Leaves serrate, not lobed; occasionally cultivated. Chestnut Oak. **Q. variabilis** Bl.
2b. Leaves lobed, rarely serrate to dentate in native shrubs. (3)

Fumariaceae

3a. Leaves or lobes bristle tipped or acute; acorns maturing in 2 years. (4)
3b. Leaves or lobes rounded or obtuse, sometimes mucronate. (6)
4a. Leaves 7- to 11-lobed, the sinuses halfway to midrib, dull; branches becoming reddish; cultivated trees to 20 m tall. Red Oak. **Q. borealis** Michx.
4b. Leaves 5- to 9-lobed, the sinuses three-fourths the distance to midrib, lustrous. (5)
5a. Leaves with conspicuous tufts of hair beneath; buds glabrous; cultivated trees to 25 m tall or more. Pin Oak. **Q. palustris** Muench.
5b. Leaves with small axillary tufts only; buds whitish-pubescent toward the tip; cultivated trees to 25 m tall. Scarlet Oak. **Q. coccinea** Muench.
6a. Leaves shallowly lobed to sinuately dentate; petiole very short; bark exfoliating and coiling; acorns stalked. Swamp White Oak. **Q. bicolor** Willd.
6b. Leaves mostly deeply lobed; petiole mostly over 12 mm long; bark not as above; acorns not usually stalked (except in **Q. robur**). (7)
7a. Leaves 5–12 cm long or less, with 3–7 pairs of lobes. (8)
7b. Leaves 10–30 cm long, with 2–5 pairs of lobes. (10)
8a. Leaves glabrous; acorns stalked; cultivated trees to 25 m tall. English Oak. **Q. robur** L.
8b. Leaves soft-pubescent beneath; acorns not stalked. (9)
9a. Leaves sinuate-dentate to shallowly lobed; plants frequently low-growing shrubs; dune areas, common in southeastern Utah. Wavy Leaf Oak. **Q. x undulata** Torr.
9b. Leaves usually distinctly lobed; plants commonly shrubs or small trees to 5 m tall but occasionally to 15 m tall and 2.5 dm in diameter or more; usually not in dune areas, distributed throughout much of Utah. Gambel Oak. **Q. gambelii** Nutt.
10a. Leaves with 2–3 pairs of lobes, 1 sinus extending nearly to midrib, pubescent beneath; acorn cup with scales forming a fringe around the cup; cultivated trees to 25 m tall. Bur Oak. **Q. macrocarpa** Michx.
10b. Leaves with 3–5 pairs of lobes, sinuses more uniformly about halfway to midrib, glaucous beneath; acorn cup without a distinctive fringe; cultivated trees to 30 m tall. White Oak. **Q. alba** L.

FUMARIACEAE — FUMITORY FAMILY

Plants herbaceous; leaves alternate or basal; flowers perfect, irregular; sepals 2, bractlike; petals 4, the 2 outer ones spreading at the apex and 1 or both saccate or spurred at base, the 2 inner ones united over the stigma; stamens 6, diadelphous, 3 in each set; ovary superior, 1-loculed; fruit a capsule.

Subclass DICOTYLEDONEAE

1a. Corolla 2-spurred, or bigibbous; flowers heart shaped, solitary, racemose, or paniculate. **Dicentra**
1b. Corolla 1-spurred; flowers not heart shaped, usually several in a raceme. (2)
2a. Fruit a few- to several-seeded dehiscent slender capsule; flowers yellow, whitish, or purplish. **Corydalis**
2b. Fruit 1-seeded, globose, indehiscent; flowers purplish. **Fumaria**

Corydalis Medic.

1a. Corolla white, rose, or purplish; stems stout, usually over 50 cm tall; corolla about 2 cm long, the spur longer than the body; capsule reflexed. **C. caseana** Gray
1b. Corolla yellow or whitish; stems slender, less than 50 cm tall. (2)
2a. Plants annual or biennial; flowers bright yellow; native. Golden Corydalis. **C. aurea** Willd.
2b. Plants perennial; flowers cream colored; cultivated ornamental. **C. ochroleuca** Koch

Dicentra Bernh.

1a. Stems scapose, the scapes usually 1-flowered; flowers white or pink; native, mountains. Dutchman's Breeches. **D. uniflora** Kell.
1b. Stems leafy, the flowers in racemes; flowers rose-red; cultivated ornamental. Bleeding Heart. **D. spectabilis** Lem.

Fumaria L. Fumitory

Leafy stemmed annuals; leaves finely dissected, compound; flowers pink-purplish; occasional weed introduced from Eurasia, waste places. **F. officinalis** L.

GENTIANACEAE — GENTIAN FAMILY

Annual, biennial, or perennial herbs; juice colorless, bitter; leaves opposite or whorled, sessile; flowers solitary or in cymes, perfect, regular, gamopetalous; calyx of 2–5 sepals; corolla of 4–5 petals; stamens the same number and alternate with the corolla lobes; ovary superior, 1-loculed, with 2 parietal placentae; styles 1 or none; stigma 2-lobed; fruit a capsule.

1a. Corolla lobes with 1 or 2 conspicuous fringed glands and pits near base of each lobe. **Swertia**
1b. Corolla lobes without glands or pits, or these not fringed. (2)
2a. Corolla salverform; anthers becoming twisted after anthesis. **Centaurium**
2b. Corolla campanulate, funnelform or cylindric; anthers not twisted after anthesis. **Gentiana**

Centaurium Hill Centaury
1a. Corolla lobes 7–8 mm long, little shorter than the tube; anthers

linear; stems angled, 20–40 cm tall; basal leaves spatulate. **C. calycosum** (Buckl.) Fern.
1b. Corolla lobes 3–6 mm long, half as long as the tube; anthers oblong; stems slender, 30 cm tall or less; basal leaves oblong to linear. (2)
2a. Corolla lobes about 4 mm long, oblong, obtuse. **C. exaltatum** (Griseb.) Wight
2b. Corolla lobes about 6 mm long, ovate, acute. **C. nuttallii** (S. Wats.) Heller

Gentiana L. Gentian

1a. Corollas plicate and folded in the sinuses; plants perennial or biennial (rarely annual in **G. fremontii**). (2)
1b. Corollas not plicate or folded in the sinuses; plants annual (perennial in **G. barbellata**). (7)
2a. Flowers 1 and terminal, less than 18 mm long; plants seldom over 10 cm tall. **G. fremontii** Torr. ex Frem.
2b. Flowers usually more than 1, over 20 mm long; plants seldom less than 10 cm tall. (3)
3a. Corollas yellowish white, streaked with purple, over 3 cm long. **G. romanzovii** Ledeb.
3b. Corollas bluish or purplish, less than 3 cm long (except in **G. parryi**). (4)
4a. Calyx lobes broadly ovate, obtuse. **G. calycosa** Griseb.
4b. Calyx lobes linear, linear-lanceolate, or none. (5)
5a. Floral leaves ovate, somewhat scarious. **G. parryi** Engelm.
5b. Floral leaves narrow, not at all scarious. (6)
6a. Calyx very irregular, cleft on 1 or both sides, resembling a spathe, lobes poorly developed (much less than one-third as long as the tube) or none. **G. forwoodii** Gray
6b. Calyx nearly regular, not spathaceous, lobes well developed (at least one-third as long as the tube). **G. affinis** Griseb. ex Hook.
7a. Corolla lobes usually 4, with toothed or fringed margins; corollas usually over 2 cm long, not fringed in the throat; flowers solitary or terminating long peduncles. **G. barbellata** Engelm.
7b. Corolla lobes usually 5, the margins entire or nearly so; corollas not over 2 cm long, more or less fringed in the throat; flowers several to many on short pedicels (except in **G. tenella**). (8)
8a. Flowers solitary on a naked peduncle 2–10 cm long; plants not over 10 cm tall. **G. tenella** Rottb.
8b. Flowers several to many in short-peduncled clusters; plants usually over 10 cm tall. (9)
9a. Calyx lobes very unequal, the outer 2 large and foliaceous, enclosing the smaller inner ones. **G. heterosepala** Engelm.
9b. Calyx lobes equal or nearly so. (10)

Figs. 49-50.   49. **Thlaspi arvensis,** x .2.   50. **Shepherdia canadensis,** x .29

Figs. 51-52.   51. **Arctostaphylos uva-ursi,** x .46.   52. **Gentiana calycosa,** x .44

Geraniaceae

10a. Stem leaves linear or linear-lanceolate; corolla yellowish, about 10 mm long. **G. tortuosa** M. E. Jones
10b. Stem leaves lanceolate or broader. (11)
11a. Flowers numerous, crowded, short-pediceled, greenish yellow to white, the lobes blue; bristles in throat few; leaves usually equaling the internodes. **G. strictiflora** (Rydb.) A. Nels
11b. Flowers few, distinctly pediceled, blue or greenish yellow, the lobes blue; bristles in throat numerous; leaves usually much shorter than the internodes. **G. plebeia** Cham.

Swertia L.   Green-Gentian

1a. Corolla normally blue; lobes 4 or 5, each with a pair of sparsely fringed nectar pits at base; stem leaves alternate; usually higher elevations, in mountains. **S. perennis** L.
1b. Corolla not blue, usually greenish white or yellowish green, may be purple dotted; lobes normally 4, each with 1 or 2 large copiously fringed glands and pits; stem leaves opposite or whorled; usually middle and lower elevations. (2)
2a. Leaves without a distinct white margin; usually middle elevations. **S. radiata** (Kell.) Kuntze
2b. Leaves with a distinct white margin; usually lower elevations. (3)
3a. Sepals broadly ovate, acute; basal leaves lanceolate; stems mostly 1–2 m tall, usually simple. **S. utahensis** (M. E. Jones) St. John
3b. Sepals lanceolate, acuminate; basal leaves linear-oblanceolate; stems 3–10 dm tall, usually branched from the base. **S. albomarginata** (S. Wats.) Kuntze

GERANIACEAE — GERANIUM FAMILY

Herbaceous annuals or perennials; leaves opposite or alternate, stipulate; flowers perfect, mostly regular; sepals and petals 5, distinct; stamens 5 or 10, the filaments commonly more or less united at base; ovary superior, usually 5-loculed; fruit dry, 1-seeded in each locule, the valves separating from the base.

1a. Flowers with a spur which is adnate to the pedicel; cultivated ornamentals. **Pelargonium**
1b. Flowers lacking a spur; native or introduced weeds. (2)
2a. Leaves palmately lobed or divided; stamens all bearing anthers. **Geranium**
2b. Leaves pinnately dissected; only 5 stamens bearing anthers. **Erodium**

Erodium L'Her   Heronsbill

Plants annual with pinnately divided leaves; flowers purplish; stamens with anthers 5, alternating with 5 staminodes; introduced weed

Subclass DICOTYLEDONEAE

of waste places and overgrazed areas at lower elevations. **E. cicutarium** (L.) L'Her

Geranium L.   Cranesbill   (Contributed by Glen T. Nebeker)

1a. Plants annual; petals less than 1 cm long. (2)
1b. Plants perennial; petals more than 1 cm long. (3)
2a. Sepals awnless; fertile stamens 5. **G. pusillum** Burm.
2b. Sepals awned; fertile stamens 10. **G. carolinianum** L.
3a. Plants nonglandular (sometimes nonglandular in **G. richardsonii**). (4)
3b. Plants glandular in portions. (5)
4a. Petals reflexing at maturity, pilose one-third to one-half their length; extreme southern Utah. **G. atropurpureum** Heller
4b. Petals not reflexing at maturity, pilose one-fourth their length; Sevier, Wayne, and Garfield cos. **G. marginale** Rydb.
5a. Lower portions and pedicels of plant glandular; expected from eastern Utah. Parry's Geranium. **G. parryi** (Engelm.) Heller
5b. Lower portions of plant nonglandular, the pedicels glandular. (6)
6a. Petals white, pilose one-third to one-half their length; shady places. Richardson's Geranium. **G. richardsonii** Risch. & Trautv.
6b. Petals purple, pilose one-fourth their length. Fremont's Geranium. **G. fremontii** Torr. ex Gray

Pelargonium L'Her   Storksbill

Perennial herbs, indoors; leaves alternate, lobed; flowers slightly irregular, the 2 upper petals often larger and more highly colored than the lower 3; commonly cultivated. Geranium. **P. hortorum** Bailey

HALORAGACEAE — WATERMILFOIL FAMILY

Mostly aquatic perennial herbs with alternate or whorled leaves; stems wholly or partly submerged; flowers regular, minute, perfect or imperfect, sessile in axils of leaves or in terminal spikes; sepals 2–4 or lacking; petals 2–4 or lacking; stamens 1–8; ovary inferior, 1- to 4-loculed; styles 1–4; fruit indehiscent, drupelike or nutletlike.

1a. Submerged leaves pinnatifid; stems slender, not erect; flowers usually imperfect with 4–8 stamens in staminate flowers. **Myriophyllum**
1b. Submerged leaves not different from emergent leaves, simple, entire; stems stout and erect; flowers perfect, with 1 stamen. **Hippurus**

Hippurus L.   Marestail

Stems simple, 2–6 dm tall, glabrous, usually partly submerged; leaves 1–3 cm long, in whorls of 6 or more; flowers axillary; calyx lobes minute or lacking; petals lacking; fruit about 2 mm long; swamps and ponds. **H. vulgaris** L.

Hydrophyllaceae

Myriophyllum L. Watermilfoil
1a. Floral bracts pinnatifid, mostly much longer than the flowers; stems up to 200 cm long, not drying whitish; still water and slow streams. **M. verticillatum** L.
1b. Floral bracts serrate or entire, mostly shorter than the flowers; stems to 100 cm long, drying whitish; still water and slow streams. **M. spicatum** L. ssp. **exalbescens** (Fern.) Hult.

HIPPOCASTANACEAE — HORSECHESTNUT FAMILY

Trees with opposite, 3- to 9-foliolate leaves; flowers irregular in terminal panicles; calyx 4- to 5-lobed; petals 4–5; stamens 5–9; ovary superior, 3-loculed; fruit a leathery capsule.

Aesculus L.
1a. Winter buds resinous, sticky; leaflets 5–7, cuneate-obovate; flowers very showy. Horsechestnut. **A. hippocastanum** L.
1b. Winter buds not resinous or sticky; leaflets usually 5, elliptic to obovate or rarely cuneate-obovate; flowers not very showy. (2)
2a. Fruit prickly; stamens exserted from flower. Ohio Buckeye. **A. glabra** Willd.
2b. Fruit smooth; stamens included in flower. Yellow Buckeye. **A. octandra** Marsh

HYDROPHYLLACEAE — WATERLEAF FAMILY
(Contributed by Duane Atwood)

Perennial, biennial or annual herbs or shrubs; leaves alternate or sometimes opposite, entire to variously compound; flowers 5-merous, perfect, mostly in cymes, usually with a pair of scales attached at base of each filament; stamens 5, alternate with the corolla lobes; pistils 1, of 2 united carpels; ovary superior; styles 1, 2-cleft; fruit a capsule with 1–100+ seeds.

1a. Plants aromatic shrubs; leaves leathery, evergreen; Washington Co. **Eriodictylon**
1b. Plants annual or herbaceous perennials; leaves not leathery or evergreen. (2)
2a. Calyx lobes dimorphic, the 3 outer conspicuously enlarged, cordate and veiny in fruit, the 2 inner lobes linear; Washington Co. **Tricardia**
2b. Calyx lobes similar or, if somewhat unequal, then not with the above combination of characteristics. (3)
3a. Plants acaulescent; flowers solitary at the end of elongate, naked peduncles. **Hesperochiron**
3b. Plants mostly caulescent; stems more or less leafy; flowers in scorpioid cymes or solitary in leaf axils. (4)

Subclass DICOTYLEDONEAE

- 4a. Ovary 1-loculed. (5)
- 4b. Ovary partially or completely divided by the intrusion of the narrow parietal placentae. (7)
- 5a. Plants perennial; stamens exserted. **Hydrophyllum**
- 5b. Plants annual; stamens included. (6)
- 6a. Herbage glabrate; stems sharply angled and armed with minute, reflexed prickles; seeds usually 1. **Nemophila**
- 6b. Herbage viscid and scented; stems not as above; seeds 7-15. **Eucrypta**
- 7a. Stamens unequally inserted in corolla tube; flowers 1, axillary, in small, dense, leafy clusters. **Nama**
- 7b. Stamens equally inserted on corolla tube; flowers mostly in cymes. (8)
- 8a. Corolla pale yellow or cream colored, marcescent; flowers long-pedicellate, the pedicels and flowers about 1 cm long, pendulous. **Emmenanthe**
- 8b. Corolla blue, purple, or white, if yellow then less than 1 cm long. **Phacelia**

Emmenanthe Benth. Whisperingbells

Erect annuals, villous-pubescent and glandular-viscid, 1-5 cm tall; leaves linear to oblong, pinnatifid; corolla yellowish, 8-12 cm long; stamens included; capsule 1-loculed; ovules numerous; Washington Co. E. penduliflora Benth.

Eriodictylon Benth. Yerba-Santa, Mountain-balm

Aromatic evergreen shrubs; leaves alternate, leathery, entire to toothed; corolla violet or white; stamens included; capsule cartilaginous; ovules 2-4; Washington Co. E. angustifolium Nutt.

Eucrypta Nutt.

Annual, viscid, scented herbs; leaves pinnately divided; corolla purplish, blue, or white, the tube yellow, 2-4 mm long; stamens included; capsule 2-3 mm in diameter; seeds 7-15; Washington Co. E. micrantha (Torr.) Heller

Hesperochiron S. Wats. Dwarf Hesperochiron

Acaulescent perennials; leaves spatulate to oblong, entire; corolla lilac, purple, or white, rotate to campanulate; stamens hairy at base; capsule 1-loculed; Cache Co. south through the central part of the state to Washington Co. H. pumilus (Dougl.) T. C. Porter

Hydrophyllum L. Waterleaf

- 1a. Flowers in open clusters; peduncles longer than the petioles of the subtending leaves; anthers linear to oblong, 1-2 mm long. (2)
- 1b. Flowers in dense capitate clusters; peduncles shorter than the petioles of the subtending leaves; anthers short, oblong, 0.6-1 mm long. (3)

Hydrophyllaceae

2a. Leaflets acuminate, with 8–12 acuminate teeth; cymes lax in fruit; southeastern Utah. **H. fendleri** (Gray) Heller
2b. Leaflets obtuse to abruptly acute, with 3–6 obtuse to acute teeth; cymes compact in fruit; Salt Lake, Tooele, and Summit cos. south through the Wasatch Range to Washington Co. **H. occidentalis** (S. Wats.) Gray
3a. Cymes lax, at least in fruit; pedicels 7–19 mm long, reflexed in fruit; plants low, 2.5 dm tall or less, more or less acaulescent; western and northwestern Utah. **H. capitatum** Dougl. var. **alpinum** S. Wats.
3b. Cymes capitate even in fruit; pedicels 2–5 mm long, not reflexed; plants usually taller, 1–5 dm high, caulescent; northern Utah in Box Elder, Weber, and Summit cos. south to Sanpete, Emery, and Millard cos. **H. capitatum** Dougl. var. **capitatum**

Nama L. Nama

1a. Style shallowy 2-lobed at apex; corolla tubular, 3–5 mm long; Uintah, Grand, Wayne, and Garfield cos. **N. densum** Lem. var. **parviflorum** (Greenm.) A. S. Hitchc.
1b. Style divided to the base; corolla mostly 8–15 mm long or, if less, then the shorter stem hairs retrorse (corolla 4–7 mm long in **N. retrorsum**). (2)
2a. Leaves mostly in clusters at ends of branches and in a basal rosette; herbage hirsutulous or pilose; Washington, Millard, and Tooele cos. **N. demissum** Gray
2b. Leaves well distributed along the stem; herbage hirsute or hispid. (3)
3a. Stems erect, fastigiate; shorter stem hairs retrorse; corolla 4–7 mm long; Kane, Garfield, and Grand cos. **N. retrorsum** J. T. Howell
3b. Stems more or less spreading; stem hairs ascending; corolla 7–15 mm long. **N. hispidum** Gray

Nemophila Nutt. ex Barton Nemophila

1a. Leaves all alternate; seeds usually 1; calyx 3 mm long; style cleft only at apex; capsule shorter than the strongly accrescent calyx; Box Elder and Cache cos. south to Sanpete and Juab cos. **N. breviflora** Gray var. **breviflora**
1b. Leaves all opposite; seeds mostly 2–4; calyx 1–3 mm long; style cleft about one-half its length; capsule exceeding the calyx; presently known only from Weber Co. **N. breviflora** Dougl. ex Benth. var. **austinae** (Eastw.) Brand

Phacelia Juss. Scorpion Weed

1a. Corolla yellow, withering-persistent and enclosing the mature capsule. (2)

Figs. 53-54. 53. **Swertia perennis**, x .21. 54. **Hippurus vulgaris**, x .29

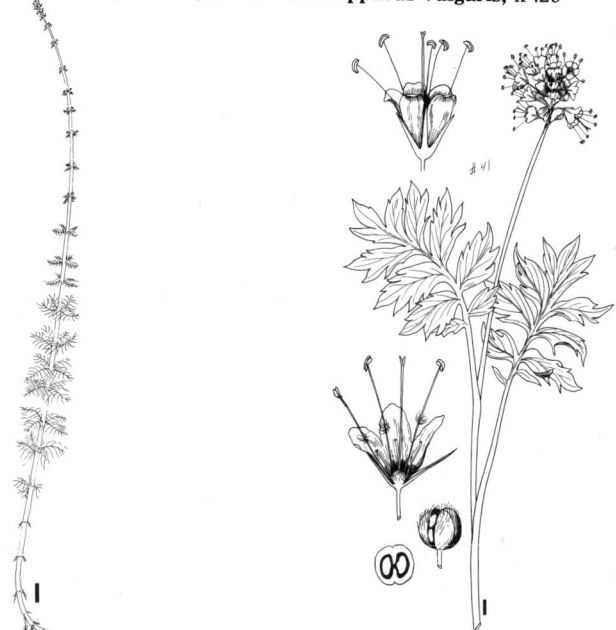

Figs. 55-56. 55. **Myriophyllum spicatum**, x .24. 56. **Hydrophyllum occidentalis**, x .21

Hydrophyllaceae

1b. Corolla blue, lavender, or violet to white, the tube sometimes yellow, deciduous. (3)
2a. Style and branches 1.25–2 mm long; ovules 10–15; Tooele Co. and probably elsewhere in western Utah. **P. scopulina** A. Nels.
2b. Style and branches 1 mm long; ovules 7–9; Tooele Co. and probably elsewhere in Utah. **P. salina** (A. Nels.) Howell
3a. Seeds transversely corrugated, numerous. (4)
3b. Seeds not transversely corrugated or, if so, the ventral surface of the seeds excavated on at least one side of a prominent ridge and with 2 ovules to each placenta. (6)
4a. Corolla 2–4.5 mm long, shorter than to nearly equaling the calyx. (5)
4b. Corolla 7–17 mm long, over twice the length of the calyx; Washington Co. **P. fremontii** Torr. in Ives
5a. Stems with black capitate glandular heads, at least on upper part of stem; calyx lobes spatulate; rare, disjunct distribution in Tooele, Beaver, Washington, Kane, San Juan, and Grand cos. **P. affinis** Gray
5b. Stems without black capitate glands; calyx lobes linear to oblanceolate; scattered throughout most of Utah, except the northwest portion. **P. ivesiana** Torr. in Ives
6a. Seeds excavated on ventral surface on one or more often both sides of a prominent ridge (Crenulatae group). (7)
6b. Seeds terete or angled and mostly foveate or reticulate, but not excavated ventrally. (21)
7a. Stamens and style included or nearly so. (8)
7b. Stamens and style exserted 2 mm or more. (9)
8a. Plants brittle, breaking easily; corolla 3–4 mm long, pale mauve to light blue; seeds dark brown when mature; to be expected in southwestern Utah. **P. coerulea** Greene
8b. Plants not brittle; corolla 6 mm long, lavender or white; seeds brown; Washington Co. **P. anelsonii** Macbr.
9a. Corolla small, 4 mm long or less, white, blue, or lavender, the lobes erose; Sevier Co. south to Wayne, Garfield, and Washington cos. **P. alba** Rydb.
9b. Corolla over 4 mm long, white or variously colored. (10)
10a. Corolla tubular, pale colored. (11)
10b. Corolla campanulate, purple, blue, lavender, or white (appearing tubular in some pressed specimens). (13)
11a. Seeds 3.5–4 mm long; cauline leaves sessile, or nearly so, auriculate; Emery, Wayne, and Washington cos. **P. rafaelensis** Atwood
11b. Seeds less than 3 mm long, black. (12)
12a. Inflorescence thyrsoid; stems solitary or, if branched, then near the base; Washington and Iron cos. **P. palmeri** Torr. ex S. Wats.
12b. Inflorescence open; stems branched throughout, especially at the base; Kane and San Juan cos. **P. constancei** Atwood

Subclass DICOTYLEDONEAE

13a. Leaves pinnately compound, finely dissected; mature seeds 2.4 mm long, excavated only on 1 side of the prominent ventral ridge; Utah Co. **P. glandulosa** Nutt. var. **argillacea** Atwood

13b. Leaves simple or, if compound, not finely so, the divisions broad, over 5 mm wide. (14)

14a. Corolla distinctly bicolored, the tube white or yellow, the lobes blue. (15)

14b. Corolla not distinctly bicolored, blue, purple, or white. (17)

15a. Cauline leaves sessile, auriculate; plants robust, 0.8–5.8 dm tall; endemic to Sanpete and Sevier cos. **P. utahensis** Voss

15b. Cauline leaves distinctly petiolate; plants not especially robust, less than 2.7 dm tall. (16)

16a. Stems branched at base; leaves simple, strigose and glandular; corolla tube white; seeds corrugated on the margins and ridge, dorsal surface smooth; Grand and San Juan cos. **P. howelliana** Atwood

16b. Stems simple or branched above; leaves essentially glabrous; corolla tube yellowish; seeds essentially lacking corrugations, dorsal surface deeply pitted; to be expected in Uintah and Grand cos. **P. splendens** Eastw.

17a. Corolla lavender; seeds lacking ventral corrugations; Kane and San Juan cos. **P. integrifolia** Torr. var. **integrifolia**

17b. Corolla blue or purple; seeds corrugated ventrally. (18)

18a. Mature seeds corrugated only on the ridge; pubescence of stems densely hispid, glandular above; Washington Co. **P. ambigua** M. E. Jones var. **ambigua**

18b. Seeds with margins and ridge corrugated; pubescence of stems mostly glandular, sometimes finely so. (19)

19a. Mature seeds dark brown; glandular pubescence long-stipitate; western Utah. **P. crenulata** Torr. ex S. Wats.

19b. Mature seeds light brown or reddish; glandular pubescence short-stipitate. (20)

20a. Anthers mostly branched throughout; seeds light brown; western Utah east through central Utah to Colorado, northeastern and southern Utah, except Washington Co. **P. corrugata** A. Nels.

20b. Anthers same color as filaments; corolla light blue, the lobes not widely spreading; stems mostly solitary or branched at base; seeds reddish brown; endemic to the Tropic Shale, Dakota Sandstone, and Kaiparowits formations in Kane and Garfield cos. **P. mammalariensis** Atwood

21a. Corolla campanulate to rotate or pelviform. (22)

21b. Corolla tubular or tubular-campanulate. (27)

22a. Plants biennial or perennial. (23)

22b. Plants annual. (25)

23a. Corolla marcescent, pelviform; seeds 8–18; filaments glabrous;

Hydrophyllaceae

throughout most of the high mountain ranges. **P. sericiea** (Graham) Gray

23b. Corolla deciduous, campanulate; stamens exserted about 6 mm; filaments hairy. (24)

24a. Plants perennial; basal leaves mostly entire, sometimes with 1–2 lateral lobes; corolla white to lavender; mountains and foothills. **P. hastata** Dougl. ex Lehm. ssp. **hastata**

24b. Plants biennial or weakly perennial; basal leaves pinnately dissected with 1–4 lobe pairs; corolla white to yellow-white; scattered in mountain areas. **P. heterophylla** Pursh ssp. **heterophylla**

25a. Filaments glabrous; seeds 4; leaves pinnately compound; Washington Co. **P. vallis-mortae** Voss

25b. Filaments sparsely hairy and sometimes minutely glandular; seeds 6–16. (26)

26a. Stems 0.3–1.5 dm tall; corolla 4–6 mm long, campanulate; style and branches 2–3 mm long; leaves mostly entire, sometimes few-toothed; Washington Co. **P. curvipes** Torr. ex S. Wats.

26b. Stems 1–5 dm tall; corolla 6–10 mm long, pelviform; style and branches 4.5–8 mm long; leaves entire or divided; Box Elder and Cache cos. south to Beaver and Sevier cos., also Kane Co. **P. linearis** (Pursh) Holz.

27a. Ovules 6–16 per ovary. (28)

27b. Ovules 20 or more per ovary. (30)

28a. Leaves oblong to elliptic; style, including branches, 1.5 mm long; filaments glabrous; flowers in dense sessile clusters; San Juan, Kane, and Washington cos. **P. cephalotes** Gray

28b. Leaves broadly ovate to orbicular; style, including branches, 1.5–4 mm long; filaments sparsely hairy; flowers in racemes, these 1–4 cm long; Kane Co. north to Carbon and Uintah cos. (29)

29a. Stems glandular-villous; style 2.5–4 mm long; Sevier, Piute, and Wayne cos. **P. demissa** Gray var. **heterotricha** Howell

29b. Stems glandular-puberulent; style 1.5–2 mm long; Uintah Co. south through Carbon Co. to Kane Co. **P. demissa** Gray var. **demissa**

30a. Corolla 8-14 mm long; style, including branches, 3.5–5 mm long; Garfield, Kane, and Washington cos. **P. pulchella** Gray

30b. Corolla 8 mm long or less, mostly less; style, including branches, 3 mm long or less. (31)

31a. Stem pubescence finely glandular-puberulent. (32)

31b. Stem pubescence glandular-villous or glandular-hirsutulous or, if glandular-puberulent, then the leaves dentate to crenate. (33)

32a. Filaments glabrous; style and branches 2.5–3 mm long; corolla 3–4.5 mm long; San Juan and Wayne cos. **P. indecora** Howell

32b. Filaments sparsely hairy below; style and branches 1.5–2 mm long; corolla 5–6 mm long; to be expected in western and southwestern Utah. **P. parshii** Gray

Subclass DICOTYLEDONEAE

33a. Leaves coarsely and radiately toothed; seeds 60–100, 0.5 mm long; Kane and Washington cos. **P. rotundifolia** Torr. ex S. Wats.
33b. Leaves entire to repand, crenate or dentate; seeds 60 or less (if 60, then less than 0.5 mm long). (34)
34a. Leaves dentate to crenate; style and branches 2–3 mm long; capsule 4–6 mm long; seeds 40–50, 1–1.3 mm long; to be expected in Washington Co. **P. peirsoniana** Howell
34b. Leaves entire; style and branches 1.5 mm long; capsule 2.5–4 mm long; seeds 22–37 or 60 per capsule, 1 mm long or less. (35)
35a. Corolla tubular, marcescent; seeds about 60, 0.3–0.4 mm long, reticulate; to be expected in Washington Co. **P. saxicola** Gray
35b. Corolla tubular-campanulate, deciduous; seeds 22–37, 0.6–1 mm long, pitted; western Utah, Uintah Co. **P. incana** Brand

Trichardia Torr. ex S. Wats.

Caulescent perennial herbs, 1–3 dm tall; leaves alternate, simple; flowers perfect in racemelike, terminal cymes; calyx lobes very unequal, 5; petals purplish, deciduous, broadly campanulate; stamens included, unequal; style included, 2-cleft; ovules 4–8, pendulous; Washington Co. **T. watsonii** Torr.

HYPERICACEAE — ST. JOHNSWORT FAMILY

Herbaceous perennials; leaves opposite, simple, punctate with glandular dots; flowers perfect, regular; sepals mostly 5; petals 5, yellow; stamens few to many, often in 3–5 clusters, the filaments united below; ovary superior, 1-loculed or 3- to 5-loculed; fruit a capsule.

Hypericum L. St. Johnswort

1a. Calyx lobes ovate, strongly black dotted along margins; stamens not united into groups; leaves oval to elliptic; native. **H. formosum** H. B. K.
1b. Calyx lobes lanceolate, punctate, slightly if at all black dotted; stamens united into 3–5 groups; leaves linear to oblong; an introduced weed to be expected in Utah. **H. perforatum** L.

JUGLANDACEAE — WALNUT FAMILY

Trees with alternate, pinnately compound leaves; flowers monoecious, the staminate in long drooping catkins, the pistillate solitary or few in a cluster; staminate perianth 3- to 5-lobed, the stamens 8–40; pistillate flower bracted and usually 2-bracteolate, the calyx 4-lobed and with 4 small petals; ovary inferior, 1-loculed or incompletely 2- to 4-loculed; styles 2; fruit drupaceous.

1a. Pith transversely partitioned; husk of fruit indehiscent; median lateral leaflets largest. **Juglans**

Labiatae

1b. Pith not partitioned; husk of fruit splitting into 4 valves; terminal leaflets largest. **Carya**

Carya Nutt. Hickory

1a. Scales of terminal bud 4–6, valvate; fruit with narrowly winged sutures; cultivated in southern Utah for edible nuts. Pecan. **C. illinoensis** (Wang.) Koch
1b. Scales of terminal bud more than 6, imbricate; fruit usually wingless; rare, cultivated tree. Shagbark Hickory. **C. ovata** (Mill.) Koch

Juglans L. Walnut

1a. Leaflets entire or nearly so, usually 7–9; bark whitish; cultivated for edible nuts. English Walnut. **J. regia** L.
1b. Leaflets serrate, 11–23; bark blackish. (2)
2a. Pith dark brown; bark with smooth ridges; fruit ovoid to short-cylindric, sticky pubescence; cultivated. Butternut. **J. cinerea** L.
2b. Pith light brown; bark with very rough ridges; fruit mostly subglobose, pubescent but not sticky; cultivated. Black Walnut. **J. nigra** L.

KRAMERIACEAE — KRAMERIA FAMILY

Low, woody, spinescent shrubs; leaves alternate, simple; flowers in racemes or more commonly solitary and axillary, irregular, purplish; petals 5, shorter than the sepals, the upper 3 petals well developed, the lower 2 reduced to glands; sepals 4 or 5; stamens 3 or 4, free or attached to the upper petals; fruit ovoid, indehiscent, 1-seeded, covered with spines.

Krameria (Loefl.) L. Ratany
1a. Pubescence of leaves of soft, matted, curly hairs; spines of fruit barbed only at apex; pedicels without stipitate glands; known from Beaver Dam Wash, Washington Co. **K. grayi** Rose & Painter
1b. Pubescence of leaves strigose, the hairs stiff and straight; spines of fruit barbed on sides below apex or sometimes barbless; pedicels with or without stipitate glands; rather widely distributed in southwestern Utah. **K. parvifolia** Benth.

LABIATAE — MINT FAMILY

Annual or perennial, usually aromatic herbs or shrubs; stems square, 4-angled; leaves simple, opposite; flowers perfect, irregular; calyx of 5 united sepals; corolla gamopetalous, 2-lipped; upper 2 petals connate, lower 3 connate; stamens 4, usually in 2 unequal pairs or only 2 by abortion; ovary superior, more or less 4-lobed, 2-loculed, 4-ovuled; styles 1, usually bifid, arising between the lobes or at apex; fruit of 4 nutlets.

Subclass DICOTYLEDONEAE

1a. Stamens with anthers 2 (2 staminodes sometimes also present). (2)
1b. Stamens with anthers 4. (8)
2a. Corolla regular or nearly so. **Lycopus**
2b. Corolla distinctly irregular, 2-lipped. (3)
3a. Plants herbaceous. (4)
3b. Plants woody shrubs. (6)
4a. Calyx 5-toothed, the teeth equal or nearly so. **Monarda**
4b. Calyx 2-lipped, the lips toothed or entire. (5)
5a. Flowers in axillary cymules; calyx teeth as long as the tube or nearly so. **Hedeoma**
5b. Flowers in terminal inflorescences; calyx various, but not as above. **Salvia**
6a. Calyx with more or less equal teeth, not 2-lipped; native. **Poliomintha**
6b. Calyx 2-lipped, the lips toothed or entire; native or cultivated. (7)
7a. Leaf margins revolute; leaves linear; flowers in short axillary racemes; cultivated. **Rosmarinus**
7b. Leaf margins not revolute or slightly so; leaves usually broader; flowers usually terminal; cultivated or native. **Salvia**
8a. Cultivated greenhouse ornamental with varicolored leaves (occasionally used in summer plantings). **Coleus**
8b. Cultivated or native plants, otherwise different from above. (9)
9a. Calyx teeth 10, hooked at apex; stems densely white-woolly. **Marrubium**
9b. Calyx teeth 5 or less, not hooked at apex; stems various. (10)
10a. Calyx 2-lipped, the lips entire, the upper lip bearing an erect crest. **Scutellaria**
10b. Calyx not 2-lipped or, if so, then at least 1 lip toothed or lobed, no crest present. (11)
11a. Upper lip of the corolla none, lower lip 5-lobed, 4 of the lobes minute. **Teucrium**
11b. Upper lip of corolla distinctly apparent, entire or 2- to 3-lobed. (12)
12a. Corolla regular or nearly so, mostly 4-lobed, the upper lobe sometimes larger than the others. (13)
12b. Corolla 2-lipped, the upper lip often 2-lobed and the lower lip 3-lobed. (14)
13a. Anther sacs parallel; flowers either in terminal spikes or in dense axillary clusters. **Mentha**
13b. Anther sacs divergent; flowers in terminal globose clusters. **Monardella**
14a. Plants shrubs or subshrubs; cultivated. (15)
14b. Plants herbaceous; cultivated or native. (16)
15a. Calyx 2-lipped, the lips entire or toothed. **Thymus**
15b. Calyx 5-toothed, the teeth equal or nearly so. **Lavandula**

Labiatae

16a. Upper lip of corolla obsolete, very short and reduced to 3 minute lobes; ovary shallowly 4-lobed. **Ajuga**
16b. Upper lip of corolla well developed, usually longer than the lower; ovary deeply 4-lobed. (17)
17a. Length of calyx about equaling corolla and always longer than corolla tube. (18)
17b. Length of calyx much less than corolla, rarely, if ever, as long as corolla tube. (19)
18a. Calyx broadly dilated (not split), shallowy 5-toothed, reticulately veined. **Molucella**
18b. Calyx not dilated, the teeth distinctly prominent, not conspicuously reticulately veined. **Ocimum**
19a. Bracts of inflorescence spinose toothed; upper calyx tooth ovate, twice as broad as the others. **Moldavica**
19b. Bracts of inflorescence, if present, not spinose toothed; calyx regular, slightly irregular or, if 2-lipped, otherwise different from above. (20)
20a. Anther sacs parallel or nearly so. (21)
20b. Anther sacs widely divergent or placed end to end. (23)
21a. Leaves palmately cleft with 3-5 lobes. **Leonurus**
21b. Leaves entire or toothed, not palmately cleft. (22)
22a. Leaves ovate-deltoid, definitely petioled; flowers many, densely crowded in a continuous or interrupted spike; upper (inner) pair of stamens longer than the lower. **Agastache**
22b. Leaves lanceolate to oblong-lanceolate, sessile; flowers not numerous, in a rather loose, spikelike raceme; upper (inner) pair of stamens shorter than the lower. **Dracocephalum**
23a. Flowers in clusters in axils of upper leaves (at least some), sometimes in only 1 terminal, leaf-subtended cluster. **Lamium**
23b. Flowers in dense or somewhat interrupted spikes, if leafy at all, only at the base. (24)
24a. Calyx distinctly 2-lipped; flowers in dense, uninterrupted spikes; bracts ovate to reniform, not at all leaflike. **Prunella**
24b. Calyx teeth equal or nearly so, not 2-lipped; flowers in rather interrupted spikes; bracts, when present, leaflike, except smaller. (25)
25a. Upper (inner) pair of stamens longer than the lower pair; plants distinctly aromatic; stems branched; leaves petioled. **Nepeta**
25b. Upper (inner) pair of stamens shorter than the lower pair; plants not distinctly aromatic; stems usually unbranched; leaves sessile or nearly so. **Stachys**

Agastache Clayton   Giant Hyssop

Perennial herbs, 6-20 dm tall; leaves 3.5-8 cm long, ovate or broader; verticils of flowers 4-15 cm long, 20-30 mm wide; calyx about 4-7 mm long, 5-toothed; corolla white to rose-purple, 2-lipped, the upper lip

Subclass DICOTYLEDONEAE

2-lobed. (Note: A second species, **A. pallidiflora** (Heller) Rydb., may be present in southeastern Utah.) **A. urticifolia** (Benth.) Kuntze

Ajuga L.   Bugleweed

Plants perennial, 10–30 cm tall, stoloniferous; leaves oblong-elliptic or obovate, petiolate; flowers usually violet or bluish; corolla well exserted; cultivated ornamental. **A. reptans** L.

Coleus Lour.

Perennial herbs (annual if planted out), 3–10 dm tall; leaves ovate, toothed or laciniate, acuminate, variously colored with yellow, red, purple; flowers blue or whitish; calyx 5-toothed; stamens united at base into a tube; greenhouse or house plants, occasionally planted outdoors. **C. blumei** Benth.

Dracocephalum L.   False Dragonhead
1a. Flowers usually less than 1.5 cm long; leaves thin in texture. **D. nuttallii** Britt. ex Britt. & Brown
1b. Flowers about 2 cm long; leaves firm in texture. **D. virginianum** L.

Hedeoma Pers.   Mock Pennyroyal

Plants perennial, 8–25 cm tall, from a woody caudex; leaves 1–2 cm long, linear to elliptic-oblong or ovate; cymules 1- to 6-flowered; calyx tube 5–7 mm long; corolla 6–12 mm long, rose to purple; southeastern Utah. **H. drummondii** Benth.

Lamium L.   Dead Nettle
1a. Upper leaves sessile or clasping; waste places. **L. amplexicaule** L.
1b. Upper leaves petioled; waste places. **L. purpureum** L.

Lavandula L.   Lavender

Perennial subshrubs to 10 dm tall; leaves oblong-linear or lanceolate; flowers lavender, 5–12 mm long, in 6- to 10-flowered whorls, forming interrupted spikes; cultivated. **L. officinalis** Chaix

Leonurus L.   Motherwort

Perennial herbs, 3–10 dm tall; leaves palmately 3- to 5-cleft; calyx about 6–8 mm long, the teeth spreading or reflexed; corolla 6–10 mm long, pale purple to rose or whitish, pubescent. **L. cardiaca** L.

Lycopus L.   Water Horehound
1a. Leaves pinnatifid, at least the lower; calyx teeth awn tipped; corolla equaling or slightly exceeding the calyx; plants lacking stolons. **L. americanus** Muhl. ex Bart.
1b. Leaves serrate or the lower rarely incised; calyx teeth acuminate; corolla slightly exceeding the calyx; plants stoloniferous. **L. lucidus** Turcz.

Labiatae

Marrubium L.   Horehound

Perennial herbs, 2–10 dm tall; leaves ovate or broader, petiolate; corolla whitish, 5–10 mm long; waste places. **M. vulgare** L.

Mentha L.   Mint
1a. Whorls of flowers axillary; native in most sites. **M. arvensis** L.
1b. Whorls of flowers in terminal spikes, or some in upper axils; cultivated and escaping. (2)
2a. Leaves petiolate; spikes more than 1 cm thick. (3)
2b. Leaves sessile; spikes less than 1 cm thick. Spearmint. **M. spicata** L.
3a. Leaves lanceolate or ovate-lanceolate, acute. Peppermint. **M. piperita** L.
3b. Leaves ovate or elliptic, obtuse. Bergamot Mint. **M. citrata** Ehrh.

Moldavica Adans.   Dragonhead

Annual or biennial (perennial) herbs, 20–60 cm tall; leaves lanceolate to oblong, coarsely serrate, upper with spinulose teeth; bracts awn toothed; calyx 4–15 mm long; corolla blue to pink, slightly longer than the calyx; waste places. **M. parviflora** (Nutt.) Britt. ex Britt. & Brown

Molucella L.   Shellflower

Annual herbs, 4–10 dm tall; leaves rounded to subcordate, 2–4 cm long, petiolate; flowers fragrant, in 6-flowered whorls; calyx with large, spreading, membranous border; corolla white, shorter than calyx; cultivated and escaping. **M. laevis** L.

Monarda L.   Horsemint
1a. Heads of flowers solitary; upper lip of corolla usually erect and straight; stamens exserted beyond corolla. **M. fistulosa** L.
1b. Heads 2 or more, forming an interrupted spike; upper lip of corolla curved; stamens usually not exserted. **M. pectinata** Nutt.

Monardella Benth.

Perennial herbs, woody at base, 20–30 cm tall; leaves 1–3 cm long, lanceolate or broader; flower clusters 1–3 cm wide; bracts ciliate; calyx 6–10 mm long; corolla 8–20 mm long, light purple. **M. odoratissima** Benth.

Nepeta L.   Catnip

Perennial aromatic herbs, 5–10 dm tall; leaves 2–8 cm long, ovate; calyx 4–7 mm long; corolla 7–12 mm long, white or purple spotted; cultivated and escaping. **N. cataria** L.

Ocimum L.   Basil

Annual herbs, 3–6 dm tall; leaves petiolate, ovate, 2.5–5 cm long; flowers white or purplish tinged; calyx to 5 mm long; corolla 6–12 mm long; stamens slightly exserted; cultivated. **O. basilicum** L.

Figs. 57-58. 57. **Phacelia heterophylla,** x .2. 58. **Phacelia linearis,** x .24

Figs. 59-60. 59. **Hypericum scouleri,** x .19. 60. **Marrubium vulgare,** x .2

Labiatae

Poliomintha Gray   Purple Sage

Plants shrubs to 1 m tall, often spreading and 2 m broad or more; leaves entire, linear or linear-oblong; flowers pale blue, rose, or purple, in small axillary cymules; calyx with subequal teeth; corolla 2-lipped; southern Utah. Immortalized by Zane Grey in his *Riders of the Purple Sage*. **P. incana** (Torr.) Gray

Prunella L.   Healall

Perennial herbs, 5–30 cm tall; leaves 2–7 cm long, ovate to lanceolate, cuneate; spikes 1.5–5 cm long or longer; calyx 5–10 mm long; corolla 8–15 mm long, purplish; widely distributed in moist places. **P. vulgaris** L.

Rosmarinus L.   Rosemary

Plants shrubs, 6–12 dm tall; leaves linear, to 3 cm long; flowers pale blue, about 12 mm long, in short, axillary racemes; stamens exserted; cultivated. **R. officinalis** L.

Salvia L.   Sage

1a. Flowers 4–6 cm long, bright red; cultivated herbs. **S. splendens** Sello
1b. Flowers less than 2.5 cm long, not bright red; cultivated or native herbs or shrubs. (2)
2a. Leaves white-woolly beneath; cultivated shrubs. Garden Sage. **S. officinalis** L.
2b. Leaves various but not white-woolly; not or rarely cultivated. (3)
3a. Plants shrubby perennials. Desert Sage. **S. carnosa** Dougl. ex Benth.
3b. Plants annuals. (4)
4a. Leaves dissected; lower anther sac fertile; southwestern Utah. **S. columbariae** Benth.
4b. Leaves simple, not dissected; lower anther sac abortive; to be expected in southeastern Utah. **S. reflexa** Hornem.

Scutellaria L.   Skullcap

1a. Leaves crenate, mostly 3–9 cm long; corolla dull blue. **S. galericulata** L.
1b. Leaves entire, usually under 3 cm long; corolla bright blue. (2)
2a. Corolla 20–30 mm long, the throat much expanded; leaves linear-oblong to oblong-ovate. **S. angustifolia** Pursh
2b. Corolla 12–20 mm long, the throat not much expanded; leaves oblong to elliptic, obtuse. **S. antirrhinoides** Benth.

Stachys L.   Betony

Perennial herbs, 15–80 cm tall; leaves 4–8 cm long; calyx 5–9 mm long; corolla 10–15 mm long, rose, mottled with darker spots; wet meadows. **S. palustris** L.

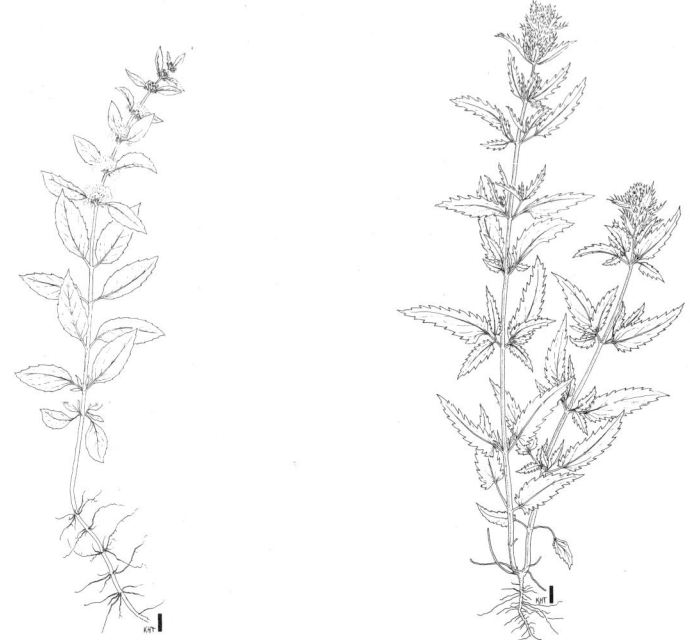

Figs. 61-62.   61. **Mentha arvensis,** x .25.   62. **Moldavica parviflora,** x .22

Figs. 63-64.   63. **Nepeta cataria,** x .26.   64. **Prunella vulgaris,** x .22

Leguminosae

Teucrium L. Germander
Perennial herbs, 30–60 cm tall; leaves 4–9 cm long, ovate to oval or lanceolate; calyx 5–7 mm long; corolla 7–15 mm long, rose or purplish, rarely cream colored; moist places. **T. canadense L.**

Thymus L. Thyme
Plants shrubs or subshrubs, 1.5–2 dm tall; leaves sessile, linear to ovate, about 5 mm long; flowers in whorls, several- to many-flowered, purplish; cultivated. **T. vulgaris L.**

LEGUMINOSAE — PEA FAMILY

Trees, shrubs, woody vines, or herbs; leaves alternate, compound or simple, mostly stipulate; inflorescence various but usually racemose; flowers papilionaceous, irregular or regular; calyx lobes 4 or 5; corolla with 0–5 petals; stamens diadelphous, monadelphous, or distinct, 5 to many; pistils 1, 1-loculed or 2-loculed (**Astragalus**); fruit a legume (pod) or a loment.

1a. Flowers regular, in dense heads or compact racemes; stamens 4 to many, usually long-exserted (Mimosoideae). (2)
1b. Flowers irregular, sometimes slightly so; stamens 10 or fewer, usually not exserted. (4)
2a. Plants small unarmed trees; flowers in dense heads to 5 cm in diameter. **Albizia**
2b. Plants shrubs or small trees bearing spines; flowers in dense, spicate racemes. (3)
3a. Spines recurved. **Acacia**
3b. Spines straight. **Prosopis**
4a. Corolla not papilionaceous, sometimes nearly regular, the upper petals enclosed by the others; stamens 10 or fewer, commonly distinct (Caesalpinoideae). (5)
4b. Corolla papilionaceous, much reduced (**Amorpha**) or missing (**Parryella, Petalostemon**), upper petal larger and enclosing the others in bud; stamens 10, rarely fewer (Papilionoideae). (8)
5a. Leaves simple, preceded by pink flowers. **Cercis**
5b. Leaves 1- or 2-pinnately compound, not preceded by pink flowers. (6)
6a. Plants herbaceous perennials; flowers yellow; native. **Hoffmanseggia**
6b. Plants trees; flowers white to greenish yellow; cultivated. (7)
7a. Leaves 1- and 2-pinnately compound on the same plant; flowers greenish yellow in spicate racemes; pods long and strap shaped. **Gleditsia**
7b. Leaves 2-pinnate; flowers white, long stalked, in open places. **Gymnocladus**
8a. Staminal filaments distinct to the base. (9)

Subclass DICOTYLEDONEAE

8b. Staminal filaments monadelphous or diadelphous. (12)
9a. Leaves pinnately compound; trees or herbs (Sophoreae). (10)
9b. Leaves palmately compound; herbs (Podalyrieae). (11)
10a. Trees (cultivated) or herbs (native) with solid leaf bases. **Sophora**
10b. Trees with hollow leaf bases covering superposed buds; cultivated. **Cladrastis**
11a. Pods inflated; flowers variously colored; cultivated. **Baptisia**
11b. Pods flat, linear to oblong; flowers yellow; native. **Thermopsis**
12a. Foliage glandular-dotted (Psoraleae). (13)
12b. Foliage not glandular-dotted. (18)
13a. Petals none; southeastern Utah. **Parryella**
13b. Petals 1–5; variously distributed. (14)
14a. Petals 1 (banner); cultivated. **Amorpha**
14b. Petals 5 (perianth of 5 petaloid stamens in **Petalostemon**); native, rare if at all in cultivation. (15)
15a. Stamens 5. **Petalostemon**
15b. Stamens 10. (16)
16a. Pods bearing uncinate appendages, several-seeded. **Glycyrrhiza**
16b. Pods lacking uncinate appendages, 1-seeded. (17)
17a. Leaves palmately compound; herbaceous. **Psoralea**
17b. Leaves odd-pinnately compound; woody or herbaceous. **Dalea**
18a. Rachis of leaves produced into a tendril of a bristlelike appendage (Vicieae). (19)
18b. Rachis of leaves not produced into a tendril or a bristlelike appendage or, if the latter, the plant a shrub. (21)
19a. Style strongly dilated; sepals foliaceous; cultivated. **Pisum**
19b. Style not strongly dilated; sepals not foliaceous. (20)
20a. Styles bearded down one side; plants usually somewhat coarse. **Lathyrus**
20b. Styles bearded at apex; plants usually slender. **Vicia**
21a. Filaments monadelphous; anthers of 2 kinds—5 large and 5 small; calyx 2-lipped (Genisteae). **Lupinus**
21b. Filaments diadelphous; anthers all alike; calyx not 2-lipped. (22)
22a. Leaflets toothed (rarely entire in **Trifolium**), usually 3 (Trifolieae). (23)
22b. Leaflets not toothed, entire. (25)
23a. Leaflets digitately compound, trifoliolate (rarely with more leaflets); flowers in very short, headlike racemes. **Trifolium**
23b. Leaflets pinnately trifoliolate; flowers in short to elongated racemes. (24)
24a. Leaflets toothed along the apical one-third or less; inflorescence a compact raceme. **Medicago**
24b. Leaflets toothed along the side from below the middle; inflorescence an elongate raceme. **Melilotus**

Leguminosae

25a. Flowers in umbels, loosely capitate or solitary in leaf axils (Loteae). **Lotus**
25b. Flowers in racemes, spikes, or spikelike heads. (26)
26a. Fruit a loment, detectable in the immature ovary (Hedysareae). (27)
26b. Fruit a legume. (28)
27a. Fruit 4- to several-seeded, usually not spiny or toothed; native. **Hedysarum**
27b. Fruit 1- to 2-seeded, more or less spiny toothed; adventive. **Onobrychis**
28a. Flowers bearing a rudimentary second staminal sheath; cultivated (Phaseoleae). (29)
28b. Flowers lacking a rudimentary second staminal sheath; cultivated or native (Galegeae). (30)
29a. Herbaceous vines or bushes; cultivated for edible pods. **Phaseolus**
29b. Woody vines; cultivated for foliage and flowers. **Wisteria**
30a. Plants trees, shrubs, or subshrubs; commonly cultivated (native and almost or quite herbaceous in **Peteria**). (31)
30b. Plants herbaceous; native or adventive, rare in cultivation. (34)
31a. Leaves even-pinnate, the terminal leaflet reduced to a bristle. **Caragana**
31b. Leaves odd-pinnate. (32)
32a. Shrubs without spines; flowers yellow; pods bladdery-inflated. **Colutea**
32b. Shrubs or trees with spines; flowers yellow, white, or pink; pods laterally compressed. (33)
33a. Plants trees or shrubs with white or pink flowers; leaflets with stipels. **Robinia**
33b. Plants subshrubs or herbs with yellowish flowers; leaflets without stipels. **Peteria**
34a. Plants with spiny stipules. **Peteria**
34b. Plants with stipules various but not spiny. (35)
35a. Keel bearing a porrect beak; ventral suture of pods produced internally, forming a partial or complete partition. **Oxytropis**
35b. Keel lacking a porrect beak; sutures variously produced or not at all produced internally. (36)
36a. Calyx closely subtended by a pair of caducous bractlets; corollas red-orange when fresh; adventive. **Swainsonia**
36b. Calyx without bractlets (except at pedicel base); corollas of various colors, but not red-orange. **Astragalus**

Acacia L.

Small trees or shrubs; branches armed with stout, curved spines; southwestern Utah. Catsclaw. **A. greggii** Gray

Subclass DICOTYLEDONEAE

Albizia Durazz.   Silk-Tree

Small, unarmed trees; flowers numerous in heads; numerous stamens long-exserted; filaments pink; cultivated ornamental. **A. julibrissin** Durazz.

Amorpha L.   False Indigo

Large shrubs with unarmed branches; flowers small, purplish; inflorescence a panicle; cultivated. **A. fruticosa** L.

Astragalus L.   Milkvetch

This large, complex genus has well over one hundred species in Utah. This comprehensive key to **Astragalus** represents an attempt to provide a means for the determination of all taxa known to occur in Utah. (For a more complete treatment, the serious student is referred to Barneby, R. C. 1964. Atlas of North American **Astragalus**. Mem. N. Y. Bot. Gard. 13:1–1188.)

1a. Leaflets awl shaped, spinulose tipped, confluent with the rachis; flowers few (1–3), very small; racemes sessile or nearly so. (2)
1b. Leaflets not awl shaped or spinulose tipped, jointed to the rachis or, if confluent with the rachis, then flowers more numerous, larger and on long peduncles. (4)
2a. Plants prostrate, mat forming; pubescence basifixed; usually of high elevations. **A. kentrophyta** Gray var. **implexus** (Canby) Barneby
2b. Plants erect or ascending (rarely prostrate); usually of middle and lower elevations. (3)
3a. Calyx 3.4–4.4 mm long, the teeth 1.5–2.4 mm long; pod 4–7 mm long, 1.5–2 mm wide; ovules 2–4. **A. kentrophyta** var. **elatus** S. Wats.
3b. Calyx at least 6 mm long, the teeth 3.4–5 mm long; pod 7–10 mm long, 2.8–4 mm in diameter; ovules 4–8; canyons of the Colorado River, southeastern Utah. **A. kentrophyta** var. **coloradoensis** M. E. Jones
4a. Plants annual, rarely biennial or perennial, but keyed both ways; banner not over 10 mm long. (5)
4b. Plants perennial or, if appearing annual, the flowers much larger. (11)
5a. Pods linear in outline, 2-loculed, or at least partially so. (6)
5b. Pods ovoid or ovoid-ellipsoid. (7)
6a. Keel obtusely rounded at tip; racemes elongating, the axis 0.5–3 cm long in fruit; southeastern Utah. **A. nuttallianus** A. DC. var. **micranthiformis** Barneby
6b. Keel triangular or narrowly deltoid at tip, acute; racemes not or scarcely elongating, to 1.2 cm long in fruit; Washington Co. **A. nuttallianus** var. **imperfectus** Barneby
7a. Racemes 2- to 10-flowered; eastern Utah, mostly north of the Tavaputs Plateau. **A. pubentissimus** T. & G.

Figs. 65-66. 65. **Stachys palustris,** x .21. 66. **Acacia greggii,** x .23

Figs. 67-68. 67. **Albizia julibrissin,** x .3. 68. **Amorpha fruticosa,** x .22

Subclass DICOTYLEDONEAE

- 7b. Racemes over 10-flowered; distribution various. (8)
- 8a. Pods appressed-strigose; annuals; widely distributed. **A. geyeri** Gray
- 8b. Pods strigillose, villous or hirsute; annuals, biennials, or short-lived perennials. (9)
- 9a. Flowers (7.5) 8.8–11.7 mm long; biennials or short-lived perennials; rare south of the Tavaputs escarpment, eastern Utah. **A. pubentissimus** T. & G.
- 9b. Flowers 5.2–8.2 mm long; annuals or short-lived perennials; eastern Utah, south of the Tavaputs escarpment. (10)
- 10a. Pod not or scarcely inflated, 9–17 mm long, 5–8 mm in diameter, curved-oblong; ovules 10–19; southeastern Utah. **A. sabulonum** Gray
- 10b. Pod bladdery-inflated, 15–20 mm long, 8–11 mm in diameter, ovoid-ellipsoid; ovules 20–28; east central Utah. **A. pardalinus** (Rydb.) Barneby
- 11a. Pubescence of herbage consisting largely or entirely of malpighian hairs (pick-shaped hairs attached near the middle). (12)
- 11b. Pubescence of herbage consisting of simple, basifixed hairs (attached at one end). (36)
- 12a. Terminal leaflet of at least some upper leaves continuous with the rachis; plants caulescent. (13)
- 12b. Terminal leaflets all articulated (jointed) to the rachis, the leaves all with several leaflets or all trifoliolate or, if some simple, then the plant acaulescent. (14)
- 13a. Stems slender, arising singly or few together from long rhizomatous caudex branches; stipules connate; pods stipitate, bladdery-inflated; widely distributed. **A. ceramicus** Sheld.
- 13b. Stems stout, arising several together from a hypogeous caudex; stipules free, large and foliaceous; pods sessile, narrowly oblong; known from Emery, Wayne, and Garfield cos. **A. woodruffii** M. E. Jones
- 14a. Stipules (at least those at lowest 1–3 nodes) connate. (15)
- 14b. Stipules all distinct. (26)
- 15a. Plants strictly acaulescent, forming tufts, mats, or mounds. (16)
- 15b. Plants caulescent. (19)
- 16a. Leaves all trifoliolate; stipules very large, hyaline; Summit Co. **A. gilviflorus** Sheld.
- 16b. Leaves simple (rarely with more leaflets). (17)
- 17a. Plants with longest leaves at least (7) 10 cm long; racemes 7- to 23-flowered, the fruiting rachis (4.5) 6–18 cm long; narrow endemic along Entrada Sandstone, vicinity of Dinosaur National Monument. **A. chloodes** Barneby
- 17b. Plants with longest leaves less than 8 (usually 3–4) cm long; racemes 1- to 9- (11-) flowered, the rachis not over 3.5 cm long in fruit. (18)

Leguminosae

18a. Flowers 10.5–20 mm long; pod 15–35 mm long, 2–3 mm wide, linear in outline; upper leaves frequently with several leaflets; Duchesne and Uintah cos. **A. detritalis** M. E. Jones
18b. Flowers 5–9.5 mm long; pod less than 15 mm long, 1.5–3 mm in diameter; upper leaves mostly simple; known from a wider range in northeastern Utah. **A. spatulatus** Sheld.
19a. Pod erect, oblong-cylindric, fully bilocular. (20)
19b. Pod variably orientated, unilocular. (22)
20a. Flowers erect at anthesis, stems from a caudex; Daggett Co. **A. adsurgens** Pallas ssp. **robustior** (Hook.) Welsh
20b. Flowers nodding at anthesis; stems from creeping rhizomes; widely distributed. (21)
21a. Pod terete, not sulcate dorsally; ovary and pod glabrous; central and northern Utah. **A. canadensis** L. var. **canadensis**
21b. Pod grooved dorsally; ovary and pod pubescent; extreme northern and southwestern Utah. **A. canadensis** var. **brevidens** (Gand.) Barneby
22a. Stems diffuse and prostrate, sometimes matted; pod dorsally compressed, not bisulcate ventrally. (23)
22b. Stems erect or ascending, not mat forming; pod bisulcate ventrally. (24)
23a. Racemes 7- to many-flowered; flowers mostly ochroleucous or dull purple; leaflets 11–17; southwestern Utah. **A. humistratus** Gray var. **humivagans** (Rydb.) Barneby
23b. Racemes 1- to 3-flowered; flowers purple; leaflets 7–11; Kane and San Juan cos. **A. sesquiflorus** S. Wats.
24a. Calyx teeth (3) 4.5–6 mm long, equaling or exceeding the length of the tube, hispid with long spreading hairs; flowers pink-purple, 9–10.5 mm long; leaflets silvery-strigillose on both sides; Emery, Grand, and Wayne cos. **A. flavus** Nutt. ex T. & G. var. **argillosus** (M. E. Jones) Barneby
24b. Calyx teeth shorter than the tube, not hispid with long spreading hairs; flowers white or yellowish or tinged with pale lilac or blue, 11–17.8 mm long; leaflets glabrous to glabrate above. (25)
25a. Flowers cream, straw, or yellow colored, the keel 8–10 mm long; eastern Utah. **A. flavus** var. **flavus**
25b. Flowers whitish, sometimes blue or lilac, the keel 6.5–8 mm long; central to southeastern and southwestern Utah. **A. flavus** var. **candicans** Gray
26a. Wing petals deeply cleft at apex; plants acaulescent; pods bilocular. **A. calycosus** Torr. ex S. Wats.
26b. Wing petals entire or obscurely emarginate; plants acaulescent or short-caulescent; pods unilocular. (27)
27a. Leaflets 3–5, or the pod hirsute with spreading hairs up to 2 mm long. (28)

Subclass DICOTYLEDONEAE

27b. Leaflets more numerous, except in seedling plants and earlier leaves; pod strigillose. (29)

28a. Pod densely hirsute with hairs 2–2.5 mm long; Sevier and western Wayne cos. **A. loanus** Barneby

28b. Pod strigillose, the hairs up to 2 mm long; Carbon, Emery, Grand, Wayne, and Garfield cos. **A. musiniensis** M. E. Jones

29a. Leaflets (1) 3–5, mostly 1–3.5 cm long; petioles persistent on root crown, forming a thatch. **A. musiniensis** M. E. Jones

29b. Leaflets 7 or more in mature leaves. (30)

30a. Pod narrowly oblong to oblong-ellipsoid, straight or nearly so, laterally compressed when ripe; Carbon, Emery, Grand, and San Juan cos. **A. cymboides** M. E. Jones

30b. Pod obliquely ovoid to narrowly ellipsoid, mostly incurved, if straight then dorsally compressed. (31)

31a. Pod hirsute with spreading hairs to 2 mm long; Sevier and western Wayne cos. **A. loanus** Barneby

31b. Pod strigillose, the hairs appressed and much less than 2 mm long. (32)

32a. Walls of pod at least 1 mm thick, the exocarp and endocarp separated by a thick mesocarp; petals mostly 2-colored; Uinta Basin (rare in Emery and Garfield cos.) **A. chamaeleuce** Gray ex Ives

32b. Walls of pod much less than 1 mm thick, becoming leathery or papery when ripe; petals 1- or 2-colored. (33)

33a. Pod ovate or lanceolate in outline, 9–22 mm long, laterally compressed only in the beak; flowers 11–18.5 mm long, whitish or purple tinged; Sevier, Piute, Wayne, and Garfield cos. **A. castaneiformis** S. Wats. var. **consobrinus** Barneby

33b. Pod variable, but commonly crescent shaped and mostly 20–50 mm long, laterally compressed; flowers variable in size, nearly always pink-purple. (34)

34a. Pod persistent (or tardily deciduous), mostly lance-ovoid in outline; San Juan and Grand cos. **A. missouriensis** Nutt. var. **amphibolus** Barneby

34b. Pod readily deciduous, mostly ellipsoid in outline; widespread in southern Utah. (35)

35a. Banner less than twice the length of calyx; keel 14–19 mm long; throughout southern Utah. **A. amphioxys** Gray var. **amphioxys**

35b. Banner 2–2.5 times the length of calyx; keel 19–23 mm long; throughout the range of var. **amphioxys** except for Washington Co. **A. amphioxys** var. **vespertinus** (Sheld.) M. E. Jones

36a. Leaves simple, oval to suborbicular; Duchesne, Uintah, Carbon, Sanpete, and Grand cos. **A. asclepiadoides** M. E. Jones

36b. Leaves pinnately compound or, if simple, then the blade linear-filiform. (37)

37a. Stipules connate into a 2-toothed sheath (at least the lower 3). (38)

Leguminosae

37b. Stipules all distinct. (72)
38a. Leaflets all (or at least the terminal one of some leaves) continuous with the rachis. (39)
38b. Leaflets all regularly pinnate, jointed to the rachis. (49)
39a. Flowers 14.5–30 mm long. (40)
39b. Flowers 7–11 mm long. (43)
40a. Flowers cream colored, yellowish, or white; leaves all regularly pinnate, the leaflets of upper leaves not reduced in size or in number; Daggett Co. **A. nelsonianus** Barneby
40b. Flowers mostly pink-purple, with white or pallid wing tips; leaves, at least the upper, with leaflets much reduced in size and number or both; distribution not as above. (41)
41a. Pod erect, or narrowly ascending; western and northwestern Utah. **A. toanus** M. E. Jones
41b. Pod declined or deflexed; Emery and Uintah cos. (42)
42a. Stems glabrous or nearly so, diffuse and ascending, 4–5.5 dm long; pod elliptic to oblong-elliptic, 1.2–2.5 cm long, 5–7.5 mm in diameter, glabrous; San Rafael Swell, Emery Co. **A. rafaelensis** M. E. Jones.
42b. Stems more or less strigillose, yellowish green, erect, 1.5–3 dm long; pod linear-oblong in outline, 2–3 cm long, 4–6 mm in diameter, thinly strigillose; Morrison, Carmel, and Moenkopi formations, Uintah Co. **A. saurinus** Barneby
43a. Pod ellipsoid to half-ellipsoid, 9–15 mm long, very flat; Carbon, Emery, Grand, and San Juan cos. **A. wingatanus** S. Wats.
43b. Pod linear or linear-oblanceolate, 12–50 mm long. (44)
44a. Root crown or caudex above ground; leafy spurs present on caudex forming a basal tuft. (45)
44b. Root crown or caudex below ground; no basal leafy tuft present. (46)
45a. Leaflets (3) 7–11; flowers 6–8 mm long; pod linear in outline; strigillose; ovules 8–12; high mountains of northern Utah. **A. miser** Dougl. ex Hook. var. **tenuifolius** (Nutt.) Barneby
45b. Leaflets 11–21, except on some early leaves; flowers 6.5–9.5 mm long; pod oblanceolate in outline, strigillose; ovules 14–19; common in aspen zone and higher, mountains. **A. miser** var. **oblongifolius** (Rydb.) Cronq.
46a. Banner recurved through about 50 degrees; ovules 4–8; Carbon, Emery, Grand, and San Juan cos. **A. wingatanus** S. Wats.
46b. Banner recurved through more than 90 degrees; ovules 10–26. (47)
47a. Pod narrowly oblong, mostly 1–1.7 cm long, 3–3.7 mm in diameter, 4–6 times longer than wide; ovules 10–16; leaflets and leaf rachis commonly expanded into flat, grasslike blades of thin texture; moist sites, Tooele and Juab cos. **A. diversifolius** Gray
47b. Pod linear-lanceolate, linear, or linear-oblanceolate in outline, 2.5–

Subclass DICOTYLEDONEAE

5 cm long, 2.5–3 mm in diameter, 8–18 times longer than wide, rarely shorter and broader but the ovules then 18–26; leaflets and leaf rachis usually very narrow, the leaflets, when present, usually folded and of stiff texture; dry soil, common and widespread. (48)

48a. Pod elongate, linear or linear-lanceolate, 2.5–5 cm long, 2.5–3 mm in diameter, 8–18 times longer than wide; widespread but not in extreme southwestern Utah. **A. convallarius** Greene var. **convallarius**

48b. Pod shorter and relatively broad, 1.3–3.4 cm long, 3.4–4 mm in diameter, 4–6 times longer than wide; southwestern Utah. **A. convallarius** var. **finitimus** Barneby

49a. Stems and leaves hirsute; plants coarse, with nodding white flowers and pendulous stipitate pods, the body linear, glabrous, 2-loculed; Sevier, Beaver, and Utah cos. **A. drummondii** Dougl. ex Hook.

49b. Stems and leaves glabrous, strigillose, or minutely villous; pod differing in some respects. (50)

50a. Calyx tube less than 4 mm long; banner less than 13 mm long. (51)
50b. Calyx tube more than 4 mm long; banner over 13 mm long. (67)

51a. Pod sessile, greatly inflated, fully 2-loculed, subglobose, papery. (52)
51b. Pod inflated or not but, if so, then 1-loculed. (53)

52a. Calyx 3.3–4.6 mm long, tube 2–2.7 mm long, teeth 1–2 mm long; high limestone ridges near timberline, mountains of western Utah. **A. platytropis** Gray

52b. Calyx 5.5–7 mm long, tube 3–4 mm long, teeth 1.8–3 mm long; dunes and sandy sites along the Zion escarpment, Kane and Washington cos. **A. striatiflorus** M. E. Jones

53a. Pod both sessile and bladdery-inflated. (54)
53b. Pod not greatly inflated, either sessile or stipitate. (59)

54a. Stems 5–7 cm long. (55)
54b. Stems 10–70 cm long. (57)

55a. Caudex subterranean; ovules about 20; mountains north of Bullion Creek, near Marysvale, Piute Co. **A. perianus** Barneby

55b. Caudex above ground; ovules 10–14; distribution not as above. (56)

56a. Terminal leaflet confluent with the rachis; petals pink-purple with white wing tips; foothills adjacent to Wyoming in Summit and Rich cos. **A. jejunus** S. Wats.

56b. Terminal leaflet jointed; petals ochroleucous, immaculate; shore of Navajo Lake, Garfield Co., and Cedar Breaks, Iron Co. **A. limnocharis** Barneby

57a. Pod 6–12 mm long, 5- to 9-ovuled, deciduous from the receptacle; moist rushy meadows and stream banks, rare, Wayne Co. **A. bodinii** Sheld.

57b. Pod 12–32 mm long, 12- to 32-ovuled, continuous with the receptacle, deciduous with the pedicel; dry situations. (58)

58a. Calyces and stems silvery-canescent with straight, appressed hairs; pod appressed-strigillose, 1.2–2.2 cm in diameter; keel tip

Leguminosae

triangular, beaklike; ovules 21–32; Garfield and San Juan cos. **A. fucatus** Barneby

58b. Calyces and stems minutely villous or loosely strigillose with sinuously incurved, widely spreading, or retrorse hairs; pod loosely strigillose or minutely villous, 6–13 mm in diameter; keel tip deltoid, obtuse; ovules 10–20; Garfield, Kane, and Washington cos. **A. subcinereus** Gray

59a. Pods linear, linear-oblanceolate, or narrowly elliptic in outline, laterally compressed, 2-sided. (60)

59b. Pods various in outline, but terete, or dorsally compressed, or 3-angled. (63)

60a. Caudex subterranean; Carbon, Emery, Grand, and San Juan cos. **A. wingatanus** S. Wats.

60b. Caudex above ground. (61)

61a. Pod stipitate; stipules turning black upon drying; northern Utah and down the central ranges at middle elevations. **A. tenellus** Pursh

61b. Pod sessile; stipules not blackening. (62)

62a. Leaflets 7–11 or less; flowers 6–8 mm long; pod linear in outline, strigillose; ovules 8–12; high mountains, northern Utah. **A. miser** Dougl. ex Hook. var. **tenuifolius** (Nutt.) Barneby

62b. Leaflets 11–21 (except on early leaves); flowers 6.5–9.5 mm long; pod oblanceolate in outline, strigillose; ovules 14–19; common in aspen zone and higher mountains. **A. miser** var. **oblongifolius** (Rydb.) Cronq.

63a. Pod nearly always erect, sessile, bisulcate ventrally; leaflets mostly linear; stems shorter than the inflorescences; Emery, Garfield, and San Juan cos. **A. moencoppensis** M. E. Jones

63b. Pod deflexed or pendulous, if bisulcate ventrally then stipitate and the leaflets of a broader type; stems longer than the inflorescences. (64)

64a. Pod bisulcate, the valves transversely corrugated; Grand, Uintah, Duchesne, Utah, and Carbon cos. **A. bisulcatus** (Hook.) Gray var. **haydenianus** (Gray) Barneby

64b. Pod not bisulcate, the valves not transversely corrugated. (65)

65a. Petals irregularly graduated, the broad and prominent keel a little longer and nearly twice as broad as the narrow whitish wings; pod pendulous, stipitate, 3-angled, semibilocular, pilose; Wasatch Mountains. **A. alpinus** L.

65b. Petals regularly graduated; pod either sessile or 1-loculed. (66)

66a. Calyx 3.5–5.8 mm long, tube 2.7–4.3 mm long, teeth 0.5–1.3 mm long; Abajo and La Sal mountains. **A. flexuosus** (Hook.) G. Don var. **flexuosus**

66b. Calyx 3.3–4.1 mm long, tube 2.4–2.7 mm long, teeth 0.7–1.1 mm long; pod sessile; Uintah, Carbon, Emery, and Grand cos. **A. flexuosus** var. **diehlii** (M. E. Jones) Barneby

Subclass DICOTYLEDONEAE

67a. Calyx tube campanulate; pod pendulous, stipitate, 1-loculed. (68)

67b. Calyx tube cylindric or subcylindric; pod stipitate, but either erect or 2-loculed. (70)

68a. Body of the pod 3-angled, the angles all acute or subacute, the 3 faces equal in width, or nearly so, all flat or slightly concave, the dorsal face sometimes a little narrower than the lateral ones and shallowly sulcate; Uinta Basin. **A. racemosus** Pursh var. **treleasei** C. L. Porter

68b. Body of the pod dorsally compressed, the dorsal surface convex, the ventral surface bisulcate. (69)

69a. Pod either smooth or faintly reticulate but, if reticulate, the body over 1 cm long and never really roughened; banner 10–17.5 mm long; Daggett, Uintah, and other counties from central to southern Utah. **A. bisulcatus** (Hook.) Gray var. **bisulcatus**

69b. Pod predominantly cross-reticulate, the lateral angles roughened, the body less than 1 cm long; banner 10 mm long or less; Uintah, Duchesne, Utah, Wasatch, Carbon, and Grand cos. **A. bisulcatus** var. **haydenianus** (Gray) Barneby

70a. Flowers crowded into ovoid heads; pod shortly stipitate, more or less covered by the calyx, long-villous, bilocular; higher and middle elevations, mostly throughout Utah. **A. agrestis** Dougl. ex G. Don

70b. Flowers loosely racemose; pod exserted far beyond the calyx on an elongate stipe, the body not long-villous. (71)

71a. Body of the pendulous pod narrowly oblong, 3-angled, grooved dorsally, 2-loculed; flowers nodding; higher elevations in San Juan, Grand, Sevier, Sanpete, Utah, Wasatch, and Summit cos. **A. scopulorum** T. C. Porter ex Port. & Coult.

71b. Body of the erect pod ovoid, inflated, uni- or subunilocular; flowers ascending; Kane, Washington, and Iron (?) cos. **A. ampullarius** S. Wats.

72a. Leaflets, at least the terminal one of some leaves, continuous with the rachis. (73)

72b. Leaflets all jointed in all leaves. (83)

73a. Pod thickly fleshy, becoming sharply 4-angled, with 4 subequal, shallowly concave faces; Kane, Washington, Iron, and Beaver cos. **A. tetrapterus** Gray

73b. Pod thinly or not at all fleshy, variously compressed or round but never 4-angled, the valves papery or leathery. (74)

74a. Lowest stipules 1–2.5 cm long; pod erect, curved, ascending, sessile, laterally compressed; Emery, Wayne, and Garfield cos. **A. woodruffii** M. E. Jones

74b. Lowest stipules much smaller; pod pendulous, if laterally compressed then stipitate. (75)

75a. Pod inflated, stipitate; Grand and Emery cos. **A. eastwoodiae** M. E. Jones

Leguminosae

75b. Pod not inflated, either sessile or stipitate. (76)
76a. Flowers nodding at full anthesis, the banner (12–24 mm long) recurved through about 35–50 degrees; keel tip obtuse; pod long-stipitate, the stipe (3) 4–15 mm long, the body either laterally or dorsally compressed, laterally so if the flowers are purple. (77)
76b. Flowers ascending or spreading at full anthesis, if not decidedly ascending then the pod sessile, the banner 8–12 mm long; keel tip beaklike; pod sessile or nearly so or, if stipitate, its body more or less dorsally compressed and petals purplish. (79)
77a. Petals pink-purple; pods laterally compressed, glabrous; Carbon, Emery, Sevier, Wayne, and Kane cos. **A. coltonii** M. E. Jones var. **coltonii**
77b. Petals white, cream colored, or lemon yellow; pod dorsally compressed. (78)
78a. Leaflets, both terminal and lateral, linear-oblanceolate, linear, or filiform, none over 3 mm wide; calyx tube cylindric, 2.7–4 mm in diameter, the teeth 0.6–2 mm long; flowers 13.5–19.5 mm long; widespread in Utah, but rare in Uinta Basin. **A. lonchocarpus** Torr.
78b. Leaflets all oblanceolate or oblong-lanceolate, the widest 4–6.5 mm wide; calyx tube 4–5.5 mm in diameter, the teeth 1.7–2.6 mm long; flowers 20–24 mm long; Uintah Co. **A. hamiltonii** C. L. Porter
79a. Pod stipitate, the stipe 3–6 mm long, the body more or less dorsally compressed, flattened, or shallowly grooved dorsally; southeastern Utah. (80)
79b. Pod sessile or nearly so, the stipe not over 1 mm long, the body variably compressed, but dorsally compressed only in **A. duchesnensis** of the Uinta Basin. (81)
80a. Stems 1.5–3.5 dm long, greenish-cinereous, strigillose; leaves mostly pinnate; racemes mostly 8- to 33-flowered, the axis 4–20 cm long in fruit; calyx (3.8) 4.8–7 mm long, the tube (3.3) 4–5.2 mm long; body of pod 2–3.2 mm long, 3.5–4.5 mm in diameter; ovules 20–24; White Canyon, San Juan Co. **A. nidularius** Barneby
80b. Stems 4–7 dm long, green, subglabrous; leaves, all but the lowest, reduced to the rachis; racemes 4- to 12-flowered, the axis greatly elongating and 1–4 dm long in fruit; calyx 2.7–3.4 mm long, the tube 1.5–2.9 mm long; body of pod 1.7–2.1 cm long, 3–3.4 mm in diameter; ovules 10–12; Capitol Reef, Wayne Co. **A. harrisonii** Barneby
81a. Pod dorsally compressed in the lower half, laterally compressed distally, hence the dorsal suture depressed proximally; Uinta Basin. **A. duchesnensis** M. E. Jones
81b. Pod more or less strongly compressed laterally its whole length, the dorsal suture as prominent toward the base as distally; south of the Uinta Basin. (82)
82a. Calyx 5.3–8.5 (9) mm long, the tube (3.8) 4.2–6.2 mm long, 1.9–2.8

Subclass DICOTYLEDONEAE

mm in diameter, the teeth (1) 1.2–3 mm long; ovules 16–24; Emery and Wayne cos. **A. episcopus** S. Wats.

82b. Calyx 3.8–4.8 mm long, the tube 3–3.5 mm long, 2.2–2.6 mm in diameter, the teeth 0.7–1.3 mm long; ovules 8–14; Kane and Washington cos. **A. lancearius** Gray

83a. Flowers small, the banner 12 mm long or less. (84)

83b. Flowers larger, the banner 12.5–28 mm long. (108)

84a. Peduncles of 2 forms, the early ones nearly basal and 1-flowered, the later ones elongate and remotely 2- to 7-flowered; pod deflexed, inversely boat shaped, semibilocular; Carbon, Emery, Wayne, and Garfield cos. **A. brandegei** T. C. Porter ex Port. and Coult.

84b. Peduncles all alike, none nearly basal or, if so, the pod otherwise; pod either ascending or 1-loculed or strongly incurved or stipitate. (85)

85a. Stems from a subterranean caudex; pod declined or pendulous, linear, or narrowly oblong in outline, uni- or subunilocular. (86)

85b. Stems from an above-ground caudex; pod usually ascending, always so if narrowly oblong. (89)

86a. Pod sessile and triangular in cross section, grooved dorsally, subunilocular, a narrow partial septum present; leaves all regularly pinnate with 3–7 pairs of leaflets; Comb Wash, west of Bluff, San Juan Co. **A. cronquistii** Barneby

86b. Pod variably compressed, stipitate or sessile but, if any part of the body is dorsally compressed, then either stipitate or at least the upper leaves reduced to the rachis or both; distribution not as above. (87)

87a. Flowers ascending or spreading at full anthesis, if not decidedly ascending then the pod sessile, the banner 8–12 mm long and the keel tip beaklike; pod sessile or nearly so, its body terete or slightly dorsally compressed; Juab and Beaver cos. **A. pinonis** M. E. Jones

87b. Flowers nodding at anthesis, the banner 6.5–8.5, or 12–24 mm long; keel obtuse; pod stipitate, the stipe 1.4–15 mm long, the body either laterally compressed or 3-angled. (88)

88a. Banner 12–24 mm long; stipe (3) 4–15 mm long; pod laterally compressed; Grand and San Juan cos. **A. coltonii** M. E. Jones var. **moabensis** M. E. Jones

88b. Banner 6.5–8.5 mm long; stipe 1.4–2 mm long; pod 3-angled; Washington and Millard cos. **A. straturensis** M. E. Jones

89a. Plants subacaulescent, the stems shorter than the peduncles. (90)

89b. Plants strongly caulescent, the stems much longer than the peduncles. (91)

90a. Pods declined or deflexed, incurved, dorsally compressed and sulcate dorsally below the beak, more or less inflated, uni- or subunilocular, the valves hirsute with widely spreading hairs which are

Leguminosae

minutely bulbous thickened at the base; San Juan, Grand, Emery, Wayne, Garfield, Kane, and Washington cos. **A. desperatus** M. E. Jones var. **desperatus**

90b. Pods ascending, straight or slightly curved, 3-angled, 2-loculed or semibilocular, strigillose with short, appressed hairs; San Juan Co. **A. monumentalis** Barneby

91a. Pods 1-loculed. (92)

91b. Pods semibilocular or 2-loculed. (98)

92a. Racemes 2- to 10-flowered. (93)

92b. Racemes more than 10-flowered. (97)

93a. Pod elevated on a stipe 1–2.5 mm long, the valves strigillose or short-minutely villous; local perennials of different distribution. (94)

93b. Pod sessile on the conical or flat receptacle, the valves villous with long spreading hairs; biennials or short-lived perennials. (95)

94a. Petals purple or purplish (ochroleucous in Iron Co. specimens); pod ovoid-ellipsoid, shortly or obscurely beaked, the valves brightly mottled; herbage loosely strigillose, the leaflets glabrescent above; Wayne and Iron cos. **A. serpens** M. E. Jones

94b. Petals whitish, with pink veins, or pinkish; pod ovoid-acuminate, strongly beaked, the valves not brightly mottled; herbage strigillose, the leaflets glabrous above; Grand Co. **A. wetherillii** M. E. Jones

95a. Pods villous or hirsute, the hairs 0.7–2.2 mm long; Uinta Basin, rare south of the Tavaputs escarpment. **A. pubentissimus** T. & G.

95b. Pod glabrous or strigillose, if loosely strigillose the hairs not over 0.7 mm long. (96)

96a. Ovary and pod glabrous; leaflets 17–23, mostly 3–8 mm long, thinly ciliate; Sevier Valley and Henry Mountains. **A. wardii** Gray

96b. Ovary and pod pubescent; leaflets 11–17, 4–20 mm long, pubescent on both sides; Emery, Wayne, and Garfield cos. **A. pardalinus** (Rydb.) Barneby

97a. Ovary and pod glabrous; flowers very small, the banner 5.2–7.4 mm long, the keel 3.9–4.6 mm long; Sevier Valley and Henry Mountains. **A. wardii** Gray

97b. Ovary and pod pubescent; flowers larger, the banner and keel at least 9 mm long; Uinta Basin and area adjacent to Tavaputs escarpment. **A. pubentissimus** T. & G.

98a. Pod narrowly lanceolate or lance-elliptic in outline, never inflated, semibilocular; extreme northwestern Utah. **A. iodanthus** S. Wats.

98b. Pod greatly to not at all inflated, but always greatly swollen and 2-loculed in the range of the preceding; distribution various. (99)

99a. Pod scarcely or not inflated, the body slenderly ellipsoid or linear-ellipsoid, mostly under 5 mm in diameter, several times longer than wide. (100)

99b. Pod globose or ovoid, either at least 7 mm in diameter, or at least half as broad as long. (101)

Subclass DICOTYLEDONEAE

- 100a. Flowers large, the keel 10–15 mm long, the calyx 6.3–9.5 mm long; San Juan, Grand, Emery, Wayne, Kane, and Washington cos. **A. lentiginosus** Dougl. ex Hook. var. **palans** (M. E. Jones) M. E. Jones
- 100b. Flowers small, the keel about 8.5 mm long, the calyx about 5 mm long; Bear Valley, south central Utah. **A. lentiginosus** var. **ursinus** (Gray) Barneby
- 101a. Racemes in fruit (and often in early anthesis) lax and open, becoming 4–18 cm long; Washington Co. (102)
- 101b. Racemes in fruit short and compact, the axis not over 4 (5) cm long; mostly other than Washington Co. (103)
- 102a. Herbage green, the stems glabrous or nearly so; pod glabrous. **A. lentiginosus** var. **vitreus** Barneby
- 102b. Herbage cinereous, the stems canescent; pod pubescent. **A. lentiginosus** var. **stramineus** (Rydb.) Barneby
- 103a. Flowers relatively small, the keel 6–8.5 (9) mm long. (104)
- 103b. Flowers larger, the keel (8.5) 9–16.5 mm long. (105)
- 104a. Pods thinly papery, commonly transparent or translucent; Beaver and Iron cos. **A. lentiginosus** var. **salinus** (Howell) Barneby
- 104b. Pods stiffly papery, opaque or nearly so; western Juab Co. **A. lentiginosus** var. **scorpionis** M. E. Jones
- 105a. Petals white (the keel tip sometimes maculate); northern Utah. **A. lentiginosus** var. **platyphyllidius** (Rydb.) Peck
- 105b. Petals purple; distribution various. (106)
- 106a. Leaflets 9–15; peduncles 1–2 cm long; Sanpete Co. **A. lentiginosus** var. **chartaceous** M. E. Jones
- 106b. Leaflets 15–23; peduncles mostly 2.5–8.5 cm long; distribution various. (107)
- 107a. Pod typically very strongly incurved, the lance-acuminate or narrow beak 6–15 mm long; western Wayne, western Garfield, Iron, Beaver, Millard, Piute, and Sevier cos. **A. lentiginosus** var. **araneosus** (Sheld.) Barneby
- 107b. Pod typically little incurved, the broadly triangular or deltoid beak 3–10 mm long; Garfield Co. **A. lentiginosus** var. **diphysus** (Gray) M. E. Jones
- 108a. Plants subacaulescent or truly acaulescent, the proper stems not longer than the longest leaves and inflorescence. (109)
- 108b. Plants strongly caulescent, the peduncles much shorter than the stems and obviously lateral. (124)
- 109a. Leaves very densely hirsute-tomentose, with the longer hairs straight and spirally twisted; pod fully 2-loculed; throughout eastern and southern Utah. **A. mollissimus** Torr. var. **thompsonae** (S. Wats.) Barneby
- 109b. Leaves variably pubescent to subglabrous but, if densely tomen-

tose, the hairs all extremely fine, sinuous and cottony, none straight and spirally twisted. (110)
110a. Pods bladdery-inflated, papery, 1-loculed. (111)
110b. Pods not inflated or, if somewhat so, then the valves leathery or pithy, never bladdery. (112)
111a. Pod 1–2 cm long, hirsute; Garfield Co. **A. desperatus** M. E. Jones var. **conspectus** Barneby
111b. Pod 2.5–6 cm long, strigillose or subglabrous; Duchesne, Juab, Sanpete, Emery, Garfield, Iron, and Kane cos. **A. megacarpus** (Nutt.) Gray
112a. Pubescence of leaves (and commonly of the entire plant) softly villous-tomentose, composed of extremely fine, cottony, contorted or entangled hairs; pod both villous-tomentose and hirsute. (113)
112b. Pubescence of leaves various, composed either of straight, appressed or narrowly ascending hairs or of spreading-incurved and sometimes sinuous or contorted hairs, but not of extremely fine, entangled hairs, never entirely cottony-tomentose; pod strigillose, glabrous, minutely villous, simply hirsute (in **A. eurekensis**), or both villous-hirsute and tomentose (in **A. marianus** and **A. newberryi**). (115)
113a. Leaflets mostly obovate and obtuse; flowers bright pink-purple; Beaver, Piute, and western Wayne cos. northward except in Rich, Daggett, Uintah, Grand, and Emery cos. **A. utahensis** (Torr.) T. & G.
113b. Leaflets various, but where the range of this and the preceding overlap, in Box Elder Co., they are either elliptic or the petals are whitish. (114)
114a. Petals whitish or ochroleucous except for the purple keel tip; Box Elder, Daggett, and Uintah cos. **A. purshii** Dougl. ex Hook. var. **purshii**
114b. Petals purple or pink-purple; Box Elder and Rich cos. **A. purshii** var. **glareosus** (Dougl.) Barneby
115a. Plants strictly and obligately acaulescent; either the leaflets 3–11, or the racemes 1- to 6-flowered, usually both (but leaflets sometimes to 17 and flowers to 8 in **A. newberryi**, which has densely hirsute and tomentose pods). (116)
115b. Plants either caulescent or acaulescent, but if acaulescent then the leaflets more than 11 in some leaves, or the flowers more than 8 in some racemes, or both, normally caulescent if the pod densely villous-hirsute. (118)
116a. Pod strigillose or hirsutulous, more rarely villous, the hairs to 2 mm long, the valves thickly fleshy, and at least 1.5 mm thick when ripe; Carbon, Emery, Wayne, Garfield, and Grand cos. **A. musiniensis** M. E. Jones
116b. Pod densely long-hirsute with stiff hairs from (2) 2.5–4.5 mm long,

Subclass DICOTYLEDONEAE

sometimes densely tomentulose, the valves less than 1 mm thick when ripe; rare east of the Wasatch Mountains. (117)

117a. Petals purple, sometimes pale; pod obliquely ovoid, the valves concealed by the tomentulose undervesture; Tooele Co. southward in the Great Basin, Kane and Washington cos. in the Colorado drainage. **A. newberryi** Gray var. **newberryi**

117b. Petals ochroleucous, faintly tinged with lavender, exceptionally purple; pod lance-ovoid, the stiffly leathery valves not or scarcely concealed by an undervesture of shorter curly hairs; western Garfield and Iron cos. northward to Tooele and Wasatch cos. **A. eurekensis** M. E. Jones

118a. Leaflets more than 21 in some mature leaves; Washington and Millard cos. **A. tephrodes** Gray var. **brachylobus** (Gray) Barneby

118b. Leaflets not more than 21 in some leaves; distribution various. (119)

119a. Pod strigillose or minutely villous, the hairs not over 1.6 mm long. (120)

119b. Pod either simply hirsute or both hirsute and tomentose, the hairs (1.7) 2–3.5 mm long. (121)

120a. Vesture of the leaflets appressed or nearly so; banner 17–21 mm and the keel 15–19 mm long; pod 2–3.5 cm long; ovules 27–36; Sevier, Piute, Millard, Beaver, Iron, and Juab cos. **A. marianus** (Rydb.) Barneby

120b. Vesture of the leaflets mostly ascending; banner 18–22.5 mm and the keel 12–13.5 mm long; pod 1–1.2 cm long; ovules 14–16; Sanpete Co. **A. desereticus** Barneby

121a. Pod brightly mottled red, ventrally sulcate toward the base; rocky ledges and talus of sandstone canyons or escarpments in Washington, Kane, and San Juan cos. **A. zionis** M. E. Jones

121b. Pod, at least beneath the vesture, green, turning brownish, not or scarcely ventrally sulcate; widespread. (122)

122a. Banner 15–17.5 mm and keel 12–15.2 mm long; pod curved-ellipsoid, 3–4 times as long as greatest width, densely silky with appressed hairs; Garfield, Iron, and Kane cos. **A. argophyllus** Nutt. ex T. & G. var. **panguicensis** (M. E. Jones) M. E. Jones

122b. Banner 18–24 mm and keel 15.9–20.3 mm long; pod ovoid to lance- or oblong-ellipsoid, mostly 2–3 times longer than wide, variously pubescent. (123)

123a. Petals bright pink-purple, the banner 22–24 mm long, 9–13.7 mm wide; moist meadows, stream banks, or lake shores, from Sanpete Co. northward. **A. argophyllus** var. **argophyllus**

123b. Petals tinged with lilac or dull purple, the banner 18–21.5 (22.5) mm long, 7–9.4 mm in diameter; dry, gravelly hills, often in sagebrush, mostly from Sanpete Co. northward. **A. argophyllus** var. **martinii** M. E. Jones

Leguminosae

124a. Pod papery, bladdery-inflated, 1-loculed, sessile, or elevated on a stipelike gynophore (a stalk of receptacular origin, usually jointed to the pod). (125)
124b. Pod not inflated or, if so, either truly stipitate or 2-loculed. (131)
125a. Stems from a hypogeous caudex; leaflets 15–27, closely crowded together and folded; shale barrens, eastern Uintah Co. **A. lutosus** M. E. Jones
125b. Stems from an above-ground caudex; leaflets fewer, well spaced; distribution otherwise. (126)
126a. Pod 1–2 cm long, hirsute; Garfield Co. **A. desperatus** M. E. Jones var. **conspectus** Barneby
126b. Pod 2.5–6 cm long, strigillose or subglabrous; more widely distributed. (127)
127a. Pod 1.5–3 cm long, solid or somewhat turgid but never bladdery, the valves fleshy, becoming stiffly leathery, nearly always inflexed dorsally as a partial though narrow and obscure septum; banner at least 16.5 mm long. (128)
127b. Pod 3–6 cm long, bladdery-inflated, 1-loculed, the valves papery (if some pods less than 3 cm long, the flowers less than 1.5 cm long). (129)
128a. Petals ochroleucous; southwestern and northern Utah. **A. beckwethii** T. & G. var. **beckwethii**
128b. Petals purple or bicolored, the wings then with pale or white tips; extreme western Utah. **A. beckwethii** var. **purpureus** M. E. Jones
129a. Plants subacaulescent, the stems 1–5 cm long, the internodes mostly concealed by stipules and the leaves gathered into a basal tuft; peduncles 0.5–2.5 (7) cm long; racemes 3- to 5- (8-) flowered; ovary and pod strigillose; widely distributed. **A. megacarpus** (Nutt.) Gray
129b. Plants caulescent, the stems 5–20 cm long; peduncles from median and upper axils, 4–13 cm long; racemes 4- to 13-flowered; ovary and pod glabrous; widely distributed. (130)
130a. Calyx tube cylindric, 7.8–8.5 mm long; gynophore 10–11 mm long; petals 2-colored; Iron Co. **A. oophorus** S. Wats. var. **lonchocalyx** Barneby
130b. Calyx tube broadly campanulate, 4–6.5 mm long; gynophore 3.5–8 (10) mm long; petals ochroleucous; Utah, Juab, Sanpete, Sevier, Piute, western Wayne, Kane, Iron, and Washington cos. **A. oophorus** var. **caulescens** (M. E. Jones) M. E. Jones
131a. Pod sessile, 1-loculed, when ripe sharply 4-angled and 4-sided; stems from a buried caudex; Beaver, Iron, Washington, and Kane cos. **A. tetrapterus** Gray
131b. Pod not 4-angled; stems mostly from a superficial caudex, if from a buried one then the pod either stipitate or 2-loculed. (132)

Subclass DICOTYLEDONEAE

132a. Pod stipitate, the stipe at least 2 mm long. (133)
132b. Pod sessile or subsessile, the stipe, if present, less than 2 mm long. (149)
133a. Pod pendulous. (143)
133b. Pod erect or ascending, never pendulous. (134)
134a. Stems procumbent, diffuse; pod ascending, the 2-loculed body laterally compressed; Kaiparowits Plateau, Kane Co. **A. malacoides** Barneby
134b. Stems erect and ascending; pod erect, subterete or 3-angled, never flattened. (135)
135a. Body of pod 3-angled, grooved dorsally, 2-loculed or nearly so; Washington and Kane cos. **A. eremiticus** Sheld.
135b. Body of pod terete or nearly so, uni- or subunilocular; distribution mostly not as above. (136)
136a. Leaflets (1) 3-11; racemes 6- to 10-flowered; flowers yellowish, very large, the calyx 15-17.5 mm long; Grand Co. **A. sabulosus** M. E. Jones
136b. Leaflets mostly 9-27; if flowers white or ochroleucous then the racemes over 10-flowered; flowers of moderate size, the calyx 6-14 mm long; widespread. (137)
137a. Flowers white or ochroleucous, nodding at full anthesis, the calyx dorsally gibbous; pod fleshy, becoming leathery or woody. (138)
137b. Flowers purple or purplish, ascending at full anthesis, the calyx little oblique at base; pod mostly thinly leathery or stiffly papery. (141)
138a. Calyx tube cylindric or subcylindric, (6) 6.5-8.8 mm long, the teeth (2.3-6.8 mm long) commonly setaceous and divergent or decurved in age; calyx and petals pure white; ovules 22-38; Uintah Co. **A. pattersonii** Gray ex Brand
138b. Calyx tube broadly campanulate, 4.4-6.5 (7.5) mm long, the teeth (0.3-6.6 mm long) erect or nearly so; calyx and petals ochroleucous; ovules (40) 44-84; southeastern and southern Utah. (139)
139a. Pod long-stipitate, the stipe 4.5-8 mm long; San Juan Co. **A. praelongus** Sheld. var. **lonchopus** Barneby
139b. Pod sessile or shortly stipitate, the stipe not over 3 times its diameter or over 2.5 mm long. (140)
140a. Body of pod (9) 10-15 mm in diameter; Sevier, Wayne, Garfield, Kane, and Washington cos. **A. praelongus** var. **praelongus**
140b. Body of pod 6-10 (11) mm in diameter; Carbon, Emery, Wayne, and Grand cos. **A. praelongus** var. **ellisiae** (Rydb.) Barneby ex Turner
141a. Pod horizontally spreading or declined; stems short, relatively slender, loosely to quite densely tufted, mostly 2-10 (14) cm long; Emery, Grand, and San Juan cos. **A. eastwoodiae** M. E. Jones

Leguminosae

141b. Pod erect or ascending; stems stout, erect and ascending in clumps, mostly 10–35 cm long; southern Utah. (142)

142a. Pod stipitate, the stipe 2 (3–7) mm long; racemes 1–7 (9) cm long in fruit; flowers large, the calyx (8.9) 9.6–12.3 mm, the keel 13.4–19 mm long; Emery, Grand, Wayne, and San Juan cos. **A. preussii** Gray var. **preussii**

142b. Pod sessile or nearly so, the stipe reduced to a thick, obconic neck hardly longer than wide; racemes 4–23 cm long in fruit; flowers smaller, the calyx 6.5–9.4 mm long, the keel 11.1–12.8 mm long; southwestern Washington Co. **A. preussii** var. **laxiflorus** Gray

143a. Body of pod plumply ovoid or subglobose, greatly inflated, not or slightly laterally compressed (see 137a). **A. eastwoodiae** M. E. Jones

143b. Body of pod linear or linear-lanceolate. (144)

144a. Flowers ascending at anthesis; pod sessile or stipitate. (145)

144b. Flowers nodding at anthesis; pod long-stipitate, the stipe (3) 4–15 mm long. (146)

145a. Pod sessile or nearly so, the stipe not over 1 mm long, the body laterally compressed; Emery and Wayne cos. **A. episcopus** S. Wats.

145b. Pod stipitate, the stipe 3–6 mm long, the body more or less dorsally compressed; San Juan Co. **A. nidularius** Barneby

146a. Petals purple; pod laterally compressed, glabrous. (147)

146b. Petals white, cream colored, or lemon yellow; pod dorsally compressed. (148)

147a. Stems (1) 2–4 dm long, long relative to height of whole plant; leaves with 8–18 leaflets, the terminal one jointed on all leaves; Grand and San Juan cos. **A. coltonii** M. E. Jones var. **moabensis** M. E. Jones

147b. Stems 1–2.5 dm long, short relative to height of whole plant; lateral leaflets 2–8 (10), those of the upper leaves reduced in size or wanting, the terminal one continuous with the rachis; Carbon, Emery, Sevier, Wayne, and Kane cos. **A. coltonii** var. **coltonii**

148a. Leaflets linear or filiform, none over 2.5 (3) mm wide; calyx tube 2.7–4 mm in diameter, the teeth 0.6–2 mm long, the banner 13.5–19.5 mm long, the keel 10.5–14 mm long; widespread, mostly not in the Uinta Basin. **A. lonchocarpus** Torr.

148b. Leaflets oblanceolate or oblong-oblanceolate, the widest 4–6.5 mm wide; calyx tube 4–5.5 mm wide, the teeth 1.7–2.6 mm long, the banner 20–24 mm long, the keel 13.7–16.6 mm long; Uintah Co. **A. hamiltonii** C. L. Porter

149a. Pod broadly ovoid or subglobose, inflated, fully 2-loculed. **A. lentiginosus** Dougl. ex Hook. (Note: For varietal determination return to couplet 95a/95b.)

149b. Pod narrowly elliptic, lanceolate, linear-oblong, or oblong in outline, not inflated or, if broadly ovoid to subglobose, then 1-loculed or nearly so. (150)

Subclass DICOTYLEDONEAE

150a. Pod subuni- or unilocular, the septum narrow, often obscure, less than half as wide as the cavity. (151)
150b. Pod fully 2-loculed, the septum complete (or nearly so), more than half as wide as the cavity. (158)
151a. Plants malodorous selenophytes; lower stipules narrow, not obtuse and several-nerved; southeastern and southern Utah. (153)
151b. Plants not of seleniferous soils, not malodorous; lower stipules ovate, obtuse and several-nerved; mostly in the Great Basin. (152)
152a. Petals ochroleucous, immaculate; stems and pods erect; northwestern Utah and Juab cos. **A. adanus** A. Nels.
152b. Petals pink-purple or 2-colored; stems diffuse or prostrate, the pods ascending from procumbent peduncles. **A. cibarius** Sheld.
153a. Leaflets 3–11; racemes 6- to 10-flowered; flowers yellowish; calyx 15–17.5 mm long; Grand Co. **A. sabulosus** M. E. Jones
153b. Leaflets 9–27; if flowers white or ochroleucous, then racemes over 10-flowered; calyx 6–14 mm long. (154)
154a. Flowers purple or purplish, ascending at full anthesis; pod mostly thinly leathery or stiffly papery. (155)
154b. Flowers white or ochroleucous, nodding at full anthesis; pod fleshy, becoming leathery or woody. (156)
155a. Pod erect or narrowly ascending from erect peduncles; stems commonly stout, erect or ascending in clumps, mostly 1–3.5 dm long; Washington Co. **A. preussii** Gray var. **laxiflorus** Gray
155b. Pod horizontally spreading or declined, borne on weakly ascending or reclining peduncles; stems short, loosely to densely tufted, mostly 2–10 (14) cm long; Emery, Grand, and San Juan cos. **A. eastwoodiae** M. E. Jones
156a. Calyx tube (6) 6.5–8.8 mm long, the teeth 2.3–6.8 mm long; calyx and petals pure white; ovules 22–38; Uintah Co. **A. pattersonii** Gray
156b. Calyx tube 4.4–6.5 (7.5) mm long, the teeth 0.3–6.6 mm long; calyx and petals ochroleucous; ovules (40) 44–84. (157)
157a. Body of pod (9) 10–15 mm in diameter; Sevier, Wayne, Garfield, Kane, and Washington cos. **A. praelongus** Sheld. var. **praelongus**
157b. Body of pod 6–10 (11) mm in diameter; Carbon, Emery, Wayne, and Grand cos. **A. praelongus** var. **ellisiae** (Rydb.) Barneby
158a. Pod erect or ascending, laterally compressed, incurved through nearly one-fourth of a circle; Washington Co. **A. minthorniae** (Rydb.) Jepson var. **gracilior** (Barneby) Barneby
158b. Pod spreading or slightly ascending, either terete or obscurely 3-angled; mostly other than Washington Co. (159)
159a. Pod sharply compressed, 3-angled, grooved dorsally; Glen Canyon. **A. bryantii** Barneby
159b. Pod commonly subterete, rarely sulcate dorsally or, if so, the angles all rounded. **A. lentiginosus** var. **palans** (M. E. Jones) M. E. Jones

Leguminosae

Baptisia Vent. Wild Indigo
Plants tall with palmately compound leaves and inflated, somewhat woody or leathery pods; cultivated ornamental. **B. leucantha** T. & G.

Caragana Lam. Pea Tree
Small to large shrubs or small trees with bright yellow flowers and even-pinnate leaves. **C. arborescens** Lam.

Cercis L. Redbud
1a. Leaves rounded or emarginate at apex, much broader than long; native, southwestern and southeastern Utah. **C. occidentalis** Torr.
1b. Leaves abruptly short-acuminate, not much broader than long; common in cultivation. Eastern Redbud. **C. canadensis** L.

Cladrastris Raf. Yellow Wood
Trees to 15 m tall; flowers white, in many-flowered, drooping panicles; occasionally cultivated. **C. lutea** (Michx.) Koch

Colutea L. Bladder Senna
Shrubs with yellow flowers in axillary, long-peduncled, few-flowered racemes; occasionally cultivated. **C. arborescens** L.

Dalea Juss. Indigo Bush
1a. Plants decumbent, herbaceous, or woody only at the base; San Juan Co. **D. lanata** Spreng.
1b. Plants erect or ascending shrubs. (2)
2a. Branches lacking reflexed hairs, little if at all glandular; leaflets 4–15 mm long. **D. fremontii** Torr. ex Gray
2b. Branches when young with reflexed hairs, conspicuously punctate with orange glands; leaflets 2–4 mm long. (3)
3a. Calyx copiously pubescent externally, the ribs not prominent. **D. polyadenia** Torr. ex S. Wats.
3b. Calyx glabrous externally, the ribs rather prominent. **D. thompsonae** (Vail) L. O. Williams

Gleditsia L. Honey Locust
Trees, armed with stout thorns (thornless in numerous cultivated forms); flowers in spicate racemes, regular, inconspicuous. **G. triacanthos** L.

Glycyrrhiza L. Licorice
Herbaceous perennials; flowers whitish to yellowish, in axillary racemes; pods armed with uncinate appendages, several-seeded; widely distributed. **G. lepidota** Pursh

Gymnocladus L. Kentucky Coffee-Tree
Unarmed trees with 2-pinnate leaves; flowers greenish white, in

Subclass DICOTYLEDONEAE

panicles to 25 cm long; pods 10–15 cm long, thick; seeds large; cultivated. **G. dioica** (L.) Koch

Hedysarum L. Sweet Vetch
Perennial herbs with racemes of bright pink-purple flowers; fruit a loment; widely distributed. **H. boreale** Nutt.

Hoffmanseggia Cav. Rushpea
Perennial herbs with 2-pinnate leaves; flowers in racemes, yellow; pods flat; southeastern Utah. **H. repens** (Eastw.) Cockerell

Lathyrus L. Sweet Pea, Wild Pea
1a. Leaflets 2; stems winged; introduced perennials or annuals. (2)
1b. Leaflets 4 or more; stems angled but not winged; plants perennial, native. (4)
2a. Plants annual, pubescent; flowers 25–30 mm long. Sweet Pea. **L. odoratus** L.
2b. Plants perennial, glabrous; flowers up to 20 mm long. (3)
3a. Stipules linear-lanceolate, 1–2 cm long; calyx 7–9 mm long, the teeth considerably shorter than the tube. **L. sylvestris** L.
3b. Stipules lanceolate to ovate, 2–4 cm long; calyx 8–12 mm long, the teeth subequal to the tube. Perennial Sweet Pea. **L. latifolius** L.
4a. Keel conspicuously shorter than the wings; calyx glabrous, or the teeth ciliate (lower tooth usually longer than the tube); stipules large, foliaceous; flowers pink-purple, occasionally white; mountain brush and aspen woodlands in central and northern Utah. **L. pauciflorus** Fern. var. **utahensis** (M. E. Jones) Peck
4b. Keel usually subequal to the wings; calyx sometimes hairy, the lower tooth often shorter than the tube; stipules not foliaceous; flowers pink-purple, pale lavender, pinkish violet, cream, or white. (5)
5a. Flowers 8–16 mm long, pale lavender tinged to pinkish violet, cream, or white, often polychrome in a population; common at middle elevations, especially in aspen. **L. lanzwertii** Kellogg
5b. Flowers usually more than 17 mm long, usually bright pink- or blue-purple; widely distributed, usually at lower elevations. (6)
6a. Plants villous-pubescent; leaflets mostly 1–2.5 cm long; flowers 18–25 mm long, the banner not deeply cordate, the blade as long as broad; calyx tube 3.5–5.5 mm long, the teeth 2.2–3.8 mm long; western and northern Utah. **L. brachycalyx** Rydb. ssp. **brachycalyx**
6b. Plants glabrous or sparsely pubescent; leaflets mostly more than 2.5 cm long; flowers often over 25 mm long, or blade of banner often deeply cordate and mostly broader than long; calyx tube various, but mostly shorter or longer than above, the teeth various. (7)
7a. Flowers (15) 17–25 mm long, the banner often deeply cordate, the sinus to 4 mm deep, the blade varying in width compared to total

length of petal; calyx tube 4–5 mm long, the teeth 1.5–2.3 mm long; southern and southeastern Utah. **L. brachycalyx** Dydb. ssp. **zionis** (C. L. Hitchc.) Welsh

7b. Flowers 20–30 mm long, the banner but shallowly retuse, the sinus to 1.5 mm deep, the blade not more than half as broad as total length of petal; lower calyx teeth but little shorter than tube; central and east central Utah. **L. brachycalyx** Rydb. ssp. **eucosmus** (Butters & St. John) Welsh

Lotus L. Trefoil

1a. Leaf rachis elongated between lower and upper leaflets (the lower pair of leaflets often considered foliaceous stipules); cultivated and escaping. **L. corniculatus** L.
1b. Leaf rachis not elongated between the leaflets or, if so, the stipules represented by glands or obsolete. (2)
2a. Stems rigid, somewhat woody below, the internodes commonly more than twice as long as the leaves; southwestern Utah. **L. rigidus** (Benth.) Greene
2b. Stems not rigid, not woody below, the internodes usually much less than twice as long as the leaves; southern Utah. (3)
3a. Peduncles usually shorter than the leaves, or obsolete. **L. wrightii** (Gray) Greene
3b. Peduncles much longer than the leaves. **L. utahensis** Ottley

Lupinus L. Lupine (Contributed by D. B. Dunn)

1a. Plants perennial. (10)
1b. Plants annual. (2)
2a. Flowers solitary, axillary, equaling foliage, which forms a dense tuft like a pincushion, commonly less than 2.5 cm tall and 5 cm in diameter; rare, to be looked for in western Utah. **L. uncialis** S. Wats.
2b. Flowers in terminal racemes; plants usually much larger and not a compact tuft, 2.5 cm tall. (3)
3a. Ovules 4–6; cotyledons petioled; leaflets densely pubescent on upper surface as well as lower; Sonoran habitat, to be looked for in southern and western Utah. Bajada Lupine. **L. concinnus** Agardh. var. **orcuttii** (S. Wats.) C. P. Sm.
3b. Ovules 2, rarely 3; cotyledons sessile; leaflets glabrous on upper surface, or with only a few hairs near the margins. (4)
4a. Flowers borne in elongate racemes, 3 cm long or more. (7)
4b. Flowers borne in pedunculate, headlike racemes (sessile in variety of **kingii**). (5)
5a. Plants essentially acaulescent, the stems, at most, only 1–2 cm long; upper lip of calyx 2 mm long or less, the lobes obsolete. Short-stem Lupine. **L. brevicaulis** S. Wats.
5b. Plants caulescent, 1–2 dm tall, with distinct stems even when young; upper lip of calyx 3–6 mm long with distinct lobes. (6)

Subclass DICOTYLEDONEAE

6a. Peduncles mostly elongate, extending well above the foliage; widely distributed, sagebrush to clearings in ponderosa pine. King's Lupine. **L. kingii** S. Wats.
6b. Peduncles poorly developed, often obsolete or only 1–2 cm long; inflorescence contained within the foliage; plants branched. **L. kingii** var. **argillaceus** (Woot. & Standl.) C. P. Sm.
7a. Peduncles obsolete, rarely as much as 1 cm long; inflorescence shorter than the leaves; plants coarsely spreading-hairy; banner less than 5 mm wide; western and southwestern Utah. Rusty Lupine. **L. pusillus** ssp. **intermontanus** (Heller) Dunn
7b. Peduncles evident; inflorescence exceeding the foliage, sometimes only slightly; banner 6–10 mm wide. (8)
8a. Peduncles 3–6 cm long, decumbent or erect; racemes dense; mature pods ovoid, 5–6 mm wide, margins flat, not constricted between the seeds, sparsely pubescent on the sides; generally interpreted as acaulescent. **L. flavoculatus** Heller
8b. Peduncles 1–3.5 cm long, erect; racemes lax; mature pods 6–7 mm wide, torulose, abundantly pubescent; distinctly caulescent. (9)
9a. Calyx cup and pedicel glabrous, only the teeth with some hairs; corolla commonly deep blue; wings often appear inflated and stand wide apart along the top side; Sonoran habitat, southern and southwestern Utah. Yelloweye Lupine. **L. pusillus** ssp. **rubens** (Rydb.) Dunn
9b. Calyx cup and pedicels pilose to hirsute; corolla blue to white; wings appear to lie rather close to the keel, the upper margins generally touching most of their length; eastern Utah. Rusty Lupine. **L. pusillus** Pursh ssp. **pusillus**
10a. Leaflets permanently pubescent above (sometimes a lens is needed). (24)
10b. Leaflets glabrous above (a few hairs near the margins above in **L. ammophilus**). (11)
11a. Banner reflexed at or below the midpoint, glabrous or pubescent distally along the dorsal crest; lower petioles elongate; largest leaflets generally over 10 mm wide. (12)
11b. Banner reflexed above the midpoint, glabrous or with sparse pubescence beneath upper lip of calyx; lower petioles shorter (except in **L. ammophilus**). (15)
12a. Banner pubescent along the dorsal crest; known only from type material at Marysvale, probably only a hybrid of **L. sericeus**. **L. marianus** Rydb.
12b. Banner glabrous. (13)
13a. Stems not fistulous, but some hollow, caespitose, generally about 3–3.5 dm tall, or less; lower petioles 6–12 cm long; to be sought in the Uinta Mountains, on dry slopes in spruce zone, about 9,000 ft. **L. arcticus** S. Wats. var. **tetonensis** (E. Nels.) C. P. Sm.

Leguminosae

13b. Stems fistulous; lower petioles 10–25 cm long; largest leaflets over 10 mm wide; wet or dry habitats. (14)

14a. Flowers 8–12 mm long; bracts persistent or subpersistent; keel glabrate to sparsely ciliate above near the acumen; stems appear glabrous; leaflets 9–10; wet habitats, not known from Utah, but to be sought in the Uinta Mountains. Burke's Lupine. **L. burkei** S. Wats.

14b. Flowers 12 mm long or more; bracts caducous; keel ciliate above near the acumen; stems hirsute to strigose, to 10 dm tall; leaflets 6–10, sharply acute, elliptic-oblanceolate; pubescence appressed or with spreading villi; basal leaves gone by fruiting time; slopes of coniferous zones. **L. amplus** Greene

15a. Leaves mostly basal; petioles 8–13 cm long; pubescence coarsely spreading-hirsute; stems from an extensively branched rootstock, subrhizomatous; calyx very gibbous above; sandy desert habitat. Sand Lupine. **L. ammophilus** Greene

15b. Leaves well distributed along the stems; petioles often less than twice the length of the leaflets; pubescence minute to strigose or sericeous. (16)

16a. Leaflets linear-oblanceolate, conduplicate on drying, often 4–6 mm wide or less. (20)

16b. Leaflets broad, oblanceolate, the largest often over 10 mm wide, generally flat but the younger folded. (17)

17a. Flowers 5–7 mm long, in a lax raceme; keel densely ciliate. Lodge Pole Lupine. **L. parviflorus** Nutt.

17b. Flowers 8–12 mm long; keel sparsely ciliate to glabrous; banner generally glabrous under the calyx lip. (18)

18a. Flowers horizontal at anthesis; upper lip of calyx very gibbous; racemes 10–15 cm long; slender stems become red if nights are cold enough; commonly found from 10,000 to 11,500 ft. in spruce zone. Redstem Lupine. **L. rubricaulis** Greene

18b. Flowers pendant at anthesis; upper lip of calyx slightly gibbous; Wasatch Mountains only. (19)

19a. Leaflets with obtuse or rounded tips; racemes 25–30 cm long; plants 7–10 dm tall. Spathulate Lupine. **L. spathulatus** Rydb.

19b. Leaflets with acute to obtuse tips; racemes 7–12 cm long; plants 4–6 dm tall. **L. maculatus** Rydb.

20a. Flowers 5–7 mm long; racemes dense, 10–20 cm long; meadow to sagebrush habitat. **L. floribundus** Greene

20b. Flowers 8–12 mm long; racemes dense or more lax (long and dense in white-flowered **L. ingratus**). (21)

21a. Banner glabrous; foliage generally quite green; keel glabrous or only a few cilia near the acumen. (22)

21b. Banner pubescent beneath the lip of calyx; foliage silvery due to

Subclass DICOTYLEDONEAE

conduplicate leaves, sericeous on the bottom; keel ciliate near the acumen; sagebrush zone. (23)

22a. Flowers blue, lacking an eyespot; clearings in the spruce zone, generally above 10,000 ft. **L. rubricaulis** Greene

22b. Flowers off-white, with a brown eyespot on drying; mountains of northern New Mexico, possibly not reaching Utah. **L. ingratus** Greene

23a. Wings 4.5–6 mm wide; flowers 10–12 mm long, orbicular in outline, viewed laterally. Silvery Lupine. **L. argenteus** Pursh ssp. **argenteus**

23b. Wings 4.5 mm wide or less; flowers 7–10 mm long, narrow as viewed laterally; upper lip of calyx gibbous at base or protruding backward as a slight spur. **L. argenteus** var. **tenellus** (Dougl. ex G. Don) Dunn

24a. Banner glabrous dorsally. (37)

24b. Banner pubescent on the back in the distal portion near the crest or beneath upper lip of calyx. (25)

25a. Calyx at most gibbous at base of upper lip; wings and lower edge of keel glabrous. (28)

25b. Calyx with a distinct spur at base of upper lip (see also **L. argenteus** var. **stenophyllus**); wings pubescent or keel ciliate below the claws. (26)

26a. Wings and/or keel ciliate below near the claws and also with villi laterally ahead of the claws, of the keel particularly. (27)

26b. Wings generally glabrous except for a tuft of pubescence laterally near the tip of the wings; banner sometimes glabrate; flowers often largely yellow but varicolored strains known; northwestern Utah. Longspur Lupine. **L. arbustus** Dougl. ex Lindl. ssp. **calcaratus** (Kell.) Dunn

27a. Lower leaves longer petioled; plants densely sericeous throughout; northern and western Utah. Tailcup Lupine. **L. caudatus** Kell. ssp. **caudatus**

27b. Lower leaves as well as upper with very short petioles, only slightly longer than the leaflets; upper leaf surface glabrate on occasion (probably of hybrid origin from **L. argenteus**). **L. caudatus** ssp. **argophyllus** (Gray) Phillips

28a. Banner reflexed at or below the midpoint, sericeous over the distal portion, near crest. (29)

28b. Banner reflexed above the midpoint, pubescent on the back beneath calyx lobes or over greater part of back. (31)

29a. Stems stout, fistulous, clumped, 7–10 dm tall, spreading hispidulous-hairy below; montane areas, Colorado and Utah. **L. bakeri** Greene

29b. Stems solitary or few, branching above, solid or hollow but not fistulous, 4–7 dm tall. (30)

30a. Stem pubescence spreading to retrorse; southern Utah and Arizona. **L. barbiger** S. Wats.

Leguminosae

30b. Stem pubescence appressed; widespread. Silky Lupine. **L. sericeus** Pursh
31a. Banner pubescent over most of back; pedicels only 1–3 mm long; stems velutinous to woolly. (32)
31b. Banner pubescent only under calyx lobes. (33)
32a. Flowers 5–7 mm long; plants 3–5 dm tall. **L. leucophyllus** var. **tenuispicus** (A. Nels.) C. P. Sm.
32b. Flowers 8–10 mm long; plants generally more robust. Velvet Lupine. **L. leucophyllus** Dougl.
33a. Flowers 6–7 mm long; foliage sericeous, 12–15 cm tall; northern Utah and adjacent Idaho. **L. evermannii** Rydb.
33b. Flowers 8 mm long or more; foliage over 15 cm tall. (34)
34a. Stems with spreading to retrorse hairs. Palmer's Lupine. **L. palmeri** S. Wats.
34b. Stems with appressed or, at most, well-ascending longer hairs. (35)
35a. Upper surface of leaflets densely sericeous. **L. holosericeus** Nutt. ex T. & G.
35b. Upper surface of leaflets sparsely strigose to thinly sericeous, appears glabrous at times to the unaided eye. (36)
36a. Base of calyx gibbous, narrow; leaflets thinly sericeous to strigose above; flowers 8–10 mm long. **L. argenteus** var. **stenophyllus** (Nutt. ex Rydb.) Davis
36b. Base of calyx not obviously gibbous; leaflets thinly strigose to glabrate in appearance above; flowers 10–12 mm long. Mountain Lupine. **L. alpestris** A. Nels.
37a. Flowers orbicular, large, 12–15 mm long; plants 2.5–4 dm tall; banner reflexed at midpoint; Idaho, Montana, Wyoming, and Utah. Wyeth's Lupine. **L. wyethii** S. Wats.
37b. Flowers small, orbicular, less than 8 mm long or 11–13 mm long and slender. (38)
38a. Flowers orbicular, 5–8 mm long or less; banner reflexed above midpoint. (39)
38b. Flowers slender, in dense or lax racemes; banner not reflexed above midpoint; plants dwarf, tufted. (41)
39a. Foliage to 10 cm tall; alpine areas above 10,000 ft. in Tetons, to be sought in Uinta Mountains. **L. roseolus** Rydb.
39b. Foliage normally well over 15 cm tall; racemes 5–10 cm long; sagebrush to ponderosa pine zones. (40)
40a. Leaflets densely sericeous; appressed hairs on stems, densely sericeous (the spreading hispidulous condition was named **L. osterhautianus** C. P. Sm. but banner was pubescent). **L. hillii** Greene
40b. Leaflets thinly sericeous to strigose with appressed hairs, glabrate or glabrous above; plants green in appearance; racemes dense, 8–20 cm long. Many Flowered Lupine. **L. floribundus** Greene
41a. Peduncles shorter than foliage, arising near the base; stems often

Subclass DICOTYLEDONEAE

poorly developed; petioles of the tuft of basal leaves 6–12 cm long. (42)

41b. Peduncles rising well beyond the foliage; plants covered with coarse scaberulous ringed hairs; northeastern Nevada and southern Idaho, should be expected in northwestern Utah. **L. lyallii** Gray ssp. **subpandens** C. P. Sm. ex Dunn

42a. Flowers 11–13 mm long; leafy part generally 10–20 cm tall, densely villous-hairy; racemes (10–15 cm long) exceed the foliage. **L. volutans** Greene

42b. Flowers 9–11 mm long; peduncles obsolete, the entire raceme included in the foliage; stems obsolete; petioles 6–10 cm long. Stemless Lupine. **L. caespitosus** Nutt.

Medicago L.   Alfalfa

1a. Flowers 7–11 mm long, purplish; plants erect, perennial; common in cultivation, escaping and persisting. Alfalfa. **M. sativa** L.
1b. Flowers 3–5 mm long, yellow; plants prostrate to decumbent, annual; a weed of lawns and waste places. Black Medic. **M. lupulina** L.

Melilotus Adans.   Sweetclover

1a. Flowers white, 4–5 mm long; pods reticulately nerved. White Sweetclover. **M. alba** Descr.
1b. Flowers yellow, 4.5–7 mm long; pods cross-ribbed. Yellow Sweetclover. **M. officinalis** (L.) Lam.

Onobrychis Scop.   Sainfoin

Perennial herbs, 20–50 cm tall; leaves odd-pinnate; flowers pink-purple; occasional weed, introduced. **O. viciaefolia** Scop.

Oxytropis DC.   Locoweed

1a. Plants usually caulescent; stipules only slightly adnate to the petioles; pods pendulous. **O. deflexa** (Pall.) DC.
1b. Plants scapose or subscapose; stipules adnate to the petioles through half their length or more; pods erect or spreading. (2)
2a. Inflorescence, pod, and sometimes other plant parts glandular-viscid. **O. viscida** Nutt.
2b. Inflorescence and pod variously pubescent, but not glandular-viscid. (3)
3a. Racemes 1- to 5-flowered, subcapitate. (4)
3b. Racemes 6- to many-flowered, usually not subcapitate. (7)
4a. Calyx turgid at full anthesis, becoming inflated and finally enclosing legume, which rarely exceeds 1 cm in length; extreme northeastern Utah. **O. multiceps** T. &. G.
4b. Calyx campanulate, not turgid, not becoming inflated or enclosing fruit, at length ruptured along one side; pod usually over 1 cm in length. (5)

Leguminosae

5a. Pod ellipsoid, leathery in texture; widely distributed, usually at upper-middle elevations. **O. parryi** Gray
5b. Pod ovoid, inflated, the valves papery in texture; southern and eastern Utah. (6)
6a. Flowers 6–12 mm long; leaflets 7–17; widely distributed in southern Utah. **O. oreophila** Gray
6b. Flowers 15–17 mm long; leaflets 3–7; Garfield and Uintah cos. **O. jonesii** Barneby
7a. Flowers bright pink-purple. (8)
7b. Flowers whitish or yellowish. (9)
8a. Pubescence composed of simple, basifixed hairs; northeastern Utah. **O. obnapiformis** C. L. Porter
8b. Pubescence composed of malpighian hairs (attached in the middle, pick shaped); widely distributed. **O. lambertii** Pursh
9a. Wings dilated apically, at least 5 mm wide; widely distributed. **O. sericea** Nutt.
9b. Wings less than 5 mm wide; known from the north slope of the Uinta Mountains. **O. campestris** (L.) DC.

Parryella T. & G.   Dunebroom
Shrubs with odd-pinnate leaves and long spikelike racemes; Grand and San Juan cos. **P. filifolia** T. & G.

Petalostemon Michx.   Prairie Clover
1a. Corolla rose colored or purplish. **P. searlsiae** Gray
1b. Corolla white. (2)
2a. Stems and leaves glabrous; calyx tube glabrous or sparsely pubescent. **P. occidentale** (Heller) Fern.
2b. Stems and leaves usually pilose; calyx tube pilose or villous. **P. flavescens** S. Wats.

Peteria Gray
Perennial herbs from a woody caudex; leaves odd-pinnate with many leaflets; stipules spiny; flowers ochroleucous to whitish; southern Utah. **P. thompsonae** S. Wats.

Phaseolus L.   Bean
The pole and bush beans of gardens. **P. vulgaris** L.

Pisum L.   Pea
The common, white-flowered pea of gardens. **P. sativum** L.

Prosopis L.   Mesquite
1a. Fruit tightly coiled into a cylinder; petals united; stipules developing into spines, adnate to the petiole. **P. pubescens** Benth. ex Hook.
1b. Fruit flattened, straight; petals distinct; stipules not developing into

Figs. 69-70.   69. **Astragalus cibarius**, x .29.   70. **Caragana arborescens**, x .34

Figs. 71-72.   71. **Melilotus officinalis**, x .22.   72. **Pisum sativum**, x .26

Leguminosae

spines, the thorns modified branches from axillary buds. **P. glandulosa** Torr.

Psoralea L. Scurfpea

1a. Flowering stems from thick, tuberous, rounded or fusiform rootstocks; subcaulescent; main stem not more than 10 cm long; plants conspicuously pubescent; leaves prevailing 5-foliolate; leaflets broadly obovate to nearly orbicular; inflorescence conspicuously bracteate; pods regularly or irregularly circumscissile near middle, with surface not glandular-warty and with a beak equal to or longer than pod. (2)

1b. Flowering stems from branching rootstocks; strongly caulescent; main stem seldom less than 20 cm long; plants not conspicuously pubescent; leaves mostly 3-foliolate; leaflets narrowly obovate, oblanceolate, or linear; inflorescence inconspicuously bracteate; pods indehiscent, with a glandular-warty surface and with a beak much shorter than body of pod. (6)

2a. Lowermost calyx lobes more than twice as wide as others; seeds reticulate; sandy flats and washes in desert areas, 1,500–4,000 ft., in creosote bush–Joshua tree community, southern (or southwestern?) Utah. **P. castorea** S. Wats.

2b. Lowermost calyx lobes about twice as wide as others; seeds smooth. (3)

3a. Petiole pubescence predominantly appressed or ascending. (4)

3b. Petiole pubescence spreading or retrorse. (5)

4a. Flowers 5–11 mm long; corolla projecting well beyond calyx lobes; red clay on sandy slopes (?), 4,000–5,000 ft., juniper community, central eastern Utah. **P. aromatica** Payson

4b. Flowers 15–20 mm long; corolla barely projecting beyond calyx lobes (by 2 mm); dry clay or red sand slopes, benches, or ridges, 4,500–6,000 ft., with juniper, Duchesne Co. and eastern tier of counties. **P. megalantha** Woot. & Standl.

5a. Plants acaulescent; leaflets grayish green above, inconspicuously punctate, densely or sparsely pubescent on both sides; roots used as food by Paiutes; dry, rocky, red-sandy or calcareous soils, with sagebrush, scrub oak, at 3,500–5,500 ft., southwestern Utah. **P. mephitica** S. Wats.

5b. Plants short-caulescent; leaves bright green, glabrous, and conspicuously punctate above; red clay mesas among juniper at about 5,500 ft., Kane Co. **P. epipsila** Barneby

6a. Corolla whitish, though tips of keel often purple; pods subglobose, rounded, and abruptly beaked at apex; branch leaves all 3-foliolate. (7)

6b. Corolla pale to deep purple; pods ovoid-elliptic, somewhat tapering into the beak; lower leaves often 4- to 5-foliolate. (8)

Subclass DICOTYLEDONEAE

- 7a. Leaflets, at least of lower leaves, broadly obovate to broadly oblanceolate; pods densely white-pilose; sandy soils, to about 5,000 ft., questionably in northwestern or western Utah. **P. lanceolata** Pursh var. **purshii** (Vail) Piper
- 7b. Leaflets all linear to narrowly oblanceolate; pods sparsely strigose to nearly glabrous; in sand, to 5,000 ft., throughout Utah with possible exception of extreme northwest. **P. lanceolata** Pursh var. **lanceolata**
- 8a. Leaves, except those near base of flowering stem, reduced to small subulate scales; pods white-sericeous; plants shrubby; sandy areas in southeastern Utah. **P. juncea** Eastw.
- 8b. Leaves well developed, mostly 3-foliolate; pods glabrous; plants herbaceous. (9)
- 9a. Leaflets of stem leaves oblanceolate, oblong, or linear; racemes elongate, 4–10 cm long, many-flowered; primarily plains and prairie areas, in diverse soil types, and often in open pine lands, southwestern Utah (?). **P. tenuiflora** Pursh var. **floribunda** Nutt.
- 9b. Leaflets of stem leaves obovate or broadly oblanceolate; racemes short, 3–5 cm long; southwestern Utah. **P. tenuiflora** Pursh var. **bigelovii** Rydb.

Robinia L.

- 1a. Upper 2 calyx teeth connate, forming a lip; branchlets and peduncles not hispid. (2)
- 1b. Upper 2 calyx teeth deeply cleft, the lobes triangular-acuminate; branchlets or peduncles (or both) glandular-hispid. (3)
- 2a. Branchlets and peduncles glandular-viscid; flowers rose-pink; not definitely known from Utah, but to be expected. Clammy Locust. **R. viscosa** Vent. ex Vauq.
- 2b. Branchlets and peduncles glabrous; flowers white (some horticultural forms have pink flowers). Black Locust. **R. pseudoacacia** L.
- 3a. Branchlets and peduncles glandular-hispid; shrubs (sometimes grafted to **R. pseudoacacia** and appearing treelike). Rose-acacia. **R. hispida** L.
- 3b. Branchlets glabrous; peduncles glandular-hispid or glandular-pubescent; small trees. New Mexico Locust. **R. neomexicana** Gray

Sophora L.

- 1a. Plants trees; flowers white, in panicles, these 15–30 cm long; cultivated. Japanese Pagoda Tree. **S. japonica** L.
- 1b. Plants herbaceous; flowers whitish or bright pink-purple, in racemes; native in southern Utah. (2)
- 2a. Flowers bright pink-purple; plants velvety-tomentose; leaves narrowly linear. **S. stenophylla** Gray ex Ives
- 2b. Flowers whitish; plants sericeous; leaves oblong or oblong-obovate. **S. nuttalliana** Turner

Leguminosae

Swainsonia Salisb.
Rhizomatous perennial herbs, 30–100 cm tall; corolla 12–15 mm long, orange-red; introduced. **S. salsula** (Pall.) DC.

Thermopsis R. Br. Golden Pea
Rhizomatous perennial herbs, 30–60 cm long; leaves 3-foliolate; flowers 15–20 mm long, bright yellow; pods erect; common in pasture lands. **T. montana** Nutt.

Trifolium L. Clover
1a. Plants annual. **T. variegatum** Nutt.
1b. Plants perennial. (2)
2a. Flowers subtended by an involucre of distinct or connate bracts. (3)
2b. Flowers not subtended by an involucre; sometimes stipules of upper leaves somewhat involucral. (6)
3a. Heads 1- to 4-flowered; plants 1–3 cm tall; high elevations. **T. nanum** Torr.
3b. Heads several-flowered; plants over 3 cm tall; high or low elevations. (4)
4a. Plants glabrous, at least the calyx; middle and higher elevations. **T. parryi** Gray
4b. Plants plainly pubescent, at least on the calyx. (5)
5a. Calyx bladdery-inflated, enclosing the corolla at maturity; plants moderately pubescent; weeds of waste places. **T. fragiferum** L.
5b. Calyx not bladdery-inflated; plants rather densely pubescent; montane and alpine regions. **T. dasyphyllum** T. &. G.
6a. Inflorescence sessile or nearly so; stipules of uppermost leaf forming a pseudo-involucre beneath the inflorescence; cultivated. Red Clover. **T. pratense** L.
6b. Inflorescence pedunculate, the peduncles usually several times the length of inflorescence. (7)
7a. Calyx glabrous, or sparsely pubescent with scattered hairs. (8)
7b. Calyx densely pubescent to villous or plumose. (10)
8a. Flowers 5–9 mm long; inflorescence axillary. (9)
8b. Flowers 10 mm long or longer; heads often terminal. **T. longipes** Nutt.
9a. Corolla usually pinkish; plants stoloniferous; calyx glabrous; cultivated. White Dutch Clover. **T. repens** L.
9b. Corolla usually pink or reddish; plants usually not stoloniferous; calyx hairy at base of teeth; cultivated. **T. hybridum** L.
10a. Pedicels over 1.5 mm long at maturity; inflorescence less than 15-flowered. **T. gymnocarpon** Nutt.
10b. Pedicels less than 1.5 mm long at maturity; inflorescence usually more than 15-flowered. (11)
11a. Plants glabrous; flowers reflexed. **T. kingii** S. Wats.
11b. Plants sparsely to densely pubescent; flowers reflexed or erect. (12)

Subclass DICOTYLEDONEAE

- 12a. Flowers straight at base. **T. longipes** Nutt.
- 12b. Flowers curved at base. (13)
- 13a. Calyx plumose-villous. **T. eriocephalum** Nutt.
- 13b. Calyx sparsely villous. **T. kingii** S. Wats.

Vicia L. Vetch

- 1a. Peduncle, if present, much shorter than the leaflets; flowers few (often solitary or paired), sessile or subsessile in upper leaf axils; cultivated as a green manure, introduced from Europe. **V. sativa** L.
- 1b. Peduncle well developed, nearly equaling to much exceeding the leaflets; flowers 2 to many. (2)
- 2a. Corolla 5–9 mm long; flowers yellowish white to purple; southern Utah. **V. exigua** Nutt.
- 2b. Corolla 12–28 mm long; flowers purplish to violet; variously distributed. (3)
- 3a. Mature inflorescence shorter than the subtending leaves, 2- to 12-flowered; flowers 15–28 mm long; racemes glabrous or short-pubescent; native. **V. americana** Muhl.
- 3b. Mature inflorescence equaling or exceeding the subtending leaves, many-flowered; flowers 12–20 mm long; racemes spreading-villous; introduced. **V. villosa** Roth

Wisteria Nutt.

- 1a. Leaflets 13–19. **W. floribunda** DC.
- 1b. Leaflets 7–13 (usually 11). **W. sinensis** Sweet

LENTIBULARIACEAE — BLADDERWORT FAMILY

Aquatic or semiaquatic herbs; leaves divided, usually bearing small inflated bladders that float the plant and trap small water organisms; flowers perfect, irregular, solitary or racemose; calyx 2-lipped, lips entire; corolla united, yellow, 2-lipped, spur at base; stamens 2; ovary superior, 1-loculed; styles 1; stigma 2-cleft; capsule 2-valved.

Utricularia L. Bladderwort

- 1a. Corolla about 12 mm broad, spur hornlike, curved and conspicuous; leaves usually pinnately divided, the ultimate divisions usually over 3 mm long on some; shallow water of ponds and streams. **U. vulgaris** L.
- 1b. Corolla about 6 mm broad, spur reduced to a protuberance, almost lacking; leaves dichotomously divided, the ultimate segments usually less than 3 mm long; shallow water. **U. minor** L.

LIMNANTHACEAE — FALSEMERMAID FAMILY

Low annual herbs of wet places; leaves alternate, 1- to 3-pinnately dissected; flowers tiny and inconspicuous, regular, solitary; sepals and petals 3; stamens 6; ovary superior, 2- or 3-loculed; fruit indehiscent.

Loasaceae

Floerkea Willd. Falsemermaid

A single genus and species treated, with the characteristics of the family; plants to 3 dm long; leaves 12–50 mm long; flowers white; moist places. **F. proserpinacoides** Willd.

LINACEAE — FLAX FAMILY

Annual or perennial herbs; leaves alternate, simple; flowers perfect, regular; sepals 5, persistent; petals 5, usually blue or yellow, soon falling; stamens 5, the filaments united at their bases; ovary superior, of 5 united carpels; styles distinct or united below; capsule 10-loculed, 10-seeded.

Linum L. Flax

1a. Petals blue (rarely white); stigmas elongate or at least longer than wide; sepals glandless, may be toothed. (2)
1b. Petals yellow or orange-yellow; stigmas capitate; sepals with marginal glands, at least inner ones. (3)
2a. Plants annuals; inner sepals ciliated or toothed on margins; stigmas much elongated; cultivated, may escape around fields and waste places. Flax. **L. usitatissimum** L.
2b. Plants perennials, inner sepals entire; stigmas only slightly longer than wide; native. **L. perenne** L.
3a. Styles distinct; sepals not or scarcely aristate, the outer ones entire, or with a few glandular teeth. **L. kingii** S. Wats.
3b. Styles united nearly to the apex; sepals spinose-aristate, the outer ones with numerous glandular teeth. (4)
4a. Pedicels and stems densely puberulent; angles of stem not winged. **L. puberulum** (Engelm.) Heller
4b. Pedicels and stems glabrous, or sparsely and obscurely puberulent; angles of stem narrowly winged. **L. aristatum** Engelm. ex Wisliz.

LOASACEAE — LOASA FAMILY

Annual, biennial, or perennial herbs; herbage stiff-hairy; stems whitish and shining, the bark exfoliating; leaves alternate, simple but often pinnatifid; flowers perfect, regular; calyx 5-lobed; petals 5–10, separate; ovary inferior, 1-loculed; placentae parietal; ovules in 1 or 2 rows on each placenta; fruit a capsule.

Mentzelia L. Blazing Star

1a. Petals 2–6 (8) mm long, 5 present; plants annual; stems usually less than 4 mm in diameter. (2)
1b. Petals 8–80 mm long or more, 5–10 present; plants perennial or biennial or, if annual, otherwise different from above; stems usually well over 4 mm in diameter. (3)

Subclass DICOTYLEDONEAE

- 2a. Seeds irregularly angled, not grooved, or only slightly so on one angle, tuberculate; leaves linear-oblanceolate. **M. albicaulis** Dougl. ex Hook.
- 2b. Seeds regularly and sharply angled, grooved on the angles, minutely roughened; leaves usually ovate or ovate-oblong, at least the upper ones. **M. dispersa** S. Wats.
- 3a. Staminal filaments broadened, cuspidate at apex; southern Utah. **M. tricuspis** Gray
- 3b. Staminal filaments narrowed, not cuspidate at apex; distribution various. (4)
- 4a. Seeds pendulous on narrowly filiform placentae; extreme southern Utah. **M. nitens** Greene
- 4b. Seeds horizontal on broad horizontal placentae. (5)
- 5a. Petals 50–80 mm long; calyx lobes 20–40 mm long; common blazing star of northern Utah. **M. laevicaulis** (Dougl.) T. & G.
- 5b. Petals 6–20 mm long; calyx lobes 2–15 mm long. (6)
- 6a. Leaves entire or slightly dentate; south central and southwestern Utah. (7)
- 6b. Leaves sinuate-dentate to pinnately toothed or lobed; eastern, east central, and southeastern Utah. (8)
- 7a. Stem pubescent; petals 12–18 mm long; southwestern Utah. **M. integra** (M. E. Jones) Tidestr.
- 7b. Stem smooth, polished; petals 6–8 mm long; south central Utah. **M. argillosa** Darlington
- 8a. Petals obtuse; east central or southeastern Utah. (9)
- 8b. Petals acute; northeastern Utah. (11)
- 9a. Capsule 15–20 mm long, base acute. **M. multiflora** (Nutt.) Gray
- 9b. Capsule 8–15 mm long, base rounded. (10)
- 10a. Anthers minutely roughened. **M. pterosperma** Eastw.
- 10b. Anthers glabrous. **M. longiloba** Darlington
- 11a. Capsules 7–10 mm long. **M. humilis** (Gray) Darlington
- 11b. Capsules 10–20 mm long. **M. pumila** (Nutt.) T. & G.

LOGANIACEAE — LOGANIA FAMILY

Plants shrubs with simple, opposite or whorled leaves; flowers regular, usually perfect, 4- to 5-merous; corolla of united petals; stamens as many as the corolla lobes, alternate with them; ovary superior, 2-loculed; style usually simple; stigma capitate or 2-lobed; fruit a capsule.

Buddleja Houst.

- 1a. Low shrubs, 2–3 dm tall; leaves 1.5–3 cm long, linear to linear-oblong; axils usually with fascicles of very small leaves; inflorescence of globose clusters or cymules, forming 2–4 heads, about 10–15 mm

Loranthaceae

thick and about 10–15 mm apart, at ends of branches; corolla purple; southern Utah. **B. utahensis** Coville

1b. Tall shrubs, 1–3 m tall; leaves 10–25 cm long, ovate-lanceolate to lanceolate; inflorescence a dense terminal panicle 10–15 cm long; corolla lilac or deeper in color; cultivated. **B. davidii** Franch.

LORANTHACEAE — MISTLETOE FAMILY

Plants parasitic, attached to host plants by specialized roots; stems with swollen joints; leaves opposite, scalelike; flowers dioecious or monoecious; perianth 2- to 5-lobed; stamens as many as lobes of the perianth; ovary inferior, 1-loculed; styles and stigmas 1; fruit a berry.

1a. Anthers 1-celled; perianth of pistillate flowers 2-lobed; fruits on short, often curved pedicels, longer than wide, greenish to purplish; flowers not in spikes. **Arceuthobium**

1b. Anthers 2-celled; perianth of pistillate flowers normally 3-lobed; fruits sessile, globose or nearly so, whitish or reddish; flowers in axillary spikes. **Phoradendron**

Arceuthobium Bieb. Dwarf Mistletoe

1a. Staminate flowers paniculate, nearly all terminal; accessory branches arising in such a manner to form a whorled arrangement; host plant almost exclusively **Pinus contorta**. **A. americanum** Nutt. ex Engelm.

1b. Staminate flowers mostly axillary, in simple or clustered spikes; accessory branches in such a manner to form fan-shaped clusters; host plant usually not **Pinus contorta**. (2)

2a. Stems slender, about 1 mm in diameter at base, commonly scattered along stem of host plant (**Pseudotsuga**). **A. douglasii** Engelm. ex Wheeler

2b. Stems stout, 2–5 mm in diameter at base, commonly clustered along stem of host plant (not **Pseudotsuga**). (3)

3a. Plants yellowish, flowering normally in early summer; stems seldom less than 3 mm in diameter at base; parasitic on **Pinus ponderosa**. **A. vaginatum**. (H. B. K.) Eichler ex Mort.

3b. Plants olive green, brown, or sometimes yellowish, flowering in late summer; stems commonly about 2 mm in diameter at base; parasitic on **Pinus flexilis, P. aristata, P. edulis, P. monophylla, P. contorta; Abies lasiocarpa, A. concolor; Picea engelmannii, P. pungens. A. campylopodium** Engelm. ex Gray

Phoradendron Nutt. Mistletoe

1a. Stems not crowded, the branches relatively slender, terete, flexuous; berries usually red; parasitic on leguminous shrubs or trees. **P. californicum** Nutt.

Subclass DICOTYLEDONEAE

1b. Stems crowded, the branches stout, obscurely quadrangular, rather rigid; berries whitish; parasitic on juniper. **P. juniperinum** Engelm.

MAGNOLIACEAE — MAGNOLIA FAMILY

Trees or large shrubs; leaves alternate, mostly entire; flowers large and showy, regular, perfect, hypogynous; sepals 3, often petaloid; petals 6–12; stamens and carpels numerous, spirally arranged; fruit a follicle or samara.

1a. Leaves lobed, truncate at apex; flowers appearing after leaves. **Liriodendron**
1b. Leaves entire, acute or acuminate; flowers appearing before leaves or, if after, then plants evergreen. **Magnolia**

Liriodendron L. Tulip Tree

Deciduous, cultivated trees to 35 m tall; leaves to 12 cm long, suborbicular with broad lobes on either side of a truncate apex; flowers solitary, terminal, tulip shaped, yellow-green. **L. tulipifera** L.

Magnolia L.

1a. Flowers appearing before leaves; fruit cylindric, asymmetric, twisted. (2)
1b. Flowers appearing with or after leaves; fruit symmetric. (3)
2a. Leaves 4–10 cm long, dark green; flowers white, fragrant; shrubs or small trees; cultivated for early showy blooms. Star Magnolia. **M. stellata** Maxim.
2b. Leaves usually over 10 cm long, light green; flowers pink to rose tinged; shrubs or small trees; cultivated for early showy blooms. Showy Magnolia. **M. soulangeana** Soul.
3a. Leaves 18–25 cm long; flowers greenish yellow; cultivated trees to 30 m tall. Cucumber Tree. **M. acuminata** L.
3b. Leaves 10–20 cm long; flowers white. (4)
4a. Leaves deciduous, mostly glabrous beneath; petiole thin; flowers 5–8 cm long; large shrubs or trees to 20 m tall; cultivated. Sweet Bay. **M. virginiana** L.
4b. Leaves evergreen, leathery, pubescent beneath; petiole thick; flowers 15–20 cm long; trees to 30 m tall; cultivated mostly in warmer regions for showy blossoms and lustrous leaves. Bull Bay. **M. grandiflora** L.

MALVACEAE — MALLOW FAMILY

Annual or perennial herbs or shrubs with more or less mucilaginous juice; leaves alternate, simple, generally palmately veined or lobed; flowers regular, perfect; calyx of 5 more or less united sepals; petals 5, hypogynous, more or less united at base; stamens many, monadelphous;

Malvaceae

ovary superior, of several carpels, each 1-loculed, usually united to each other; styles united below; fruit a capsule or the carpels separating at maturity.

1a. Fruit a loculicidal capsule; carpels 3–5; flowers mostly solitary in the axils; ovules and seeds several in each carpel. (2)
1b. Fruit a schizocarp, of several or many carpels or, if fruit dehiscent, then the carpels more or less separate. (3)
2a. Style branches 5; calyx 5-lobed; petals yellow, purple, red, pink, or white; plants woody or herbaceous. **Hibiscus**
2b. Style unbranched; calyx entire or shallowly dentate; petals white or yellow; plants shrubby. **Gossypium**
3a. Style branches filiform, stigmatic surface lateral. (4)
3b. Style branches ending in a capitate or truncate stigma. (7)
4a. Flowers in elongate racemes; involucel none; stamens in 2 whorls— 1 apical, the other subapical, the filaments more or less united in groups. **Sidalcea**
4b. Flowers axillary, solitary or in small clusters, or in loose terminal cymes; involucel present, the bractlets narrow; stamens not in 2 whorls, the filaments not united in groups. (5)
5a. Plants perennial, the taproot thick; leaves palmately 5-cleft, the lobes also deeply cleft. **Callirrhoe**
5b. Plants annual or biennial, the roots not thickened. (6)
6a. Calyx bractlets 3; plants procumbent; leaves crenate, often shallowly cleft with broad round lobes; peduncles short; petals not more than 15 mm long; filaments all apical or subapical on the staminal column. **Malva**
6b. Calyx bractlets 6–9; plants erect, the stems strict, hairy; leaves large and rough, rounded cordate; flowers large, 10 cm broad or more. **Althaea**
7a. Carpels differentiated into a reticulate indehiscent basal portion and a smooth dehiscent apical portion, the 2 portions separated by a ventral notch; petals orange. **Sphaeralcea**
7b. Carpels not sharply differentiated into 2 portions as above or, if so (as in species of **Sida**), then the portions not separated by a distinct notch; petals mostly other than orange. (8)
8a. Ovules and usually the seeds 2 or more per carpel or, if 1 ovule, then the carpels membranous and much inflated; carpels dehiscent to base or nearly so, not reticulate. (9)
8b. Ovule 1; carpels not membranous or much inflated, indehiscent or dehiscent only toward apex. (10)
9a. Involucel of 3 narrow bractlets; column antheriferous to far below apex; petals pink or lavender; carpels hirsute with long, simple hairs. **Iliamna**
9b. Involucel none; column antheriferous only at or near apex; petals mostly yellow, red, or orange; fruit cylindric or ovoid. **Abutilon**

Subclass DICOTYLEDONEAE

10a. Involucel present; petals yellow-orange, purple, or purplish; ovules erect or ascending; carpels indehiscent. **Malvastrum**

10b. Involucral none or, if present, then petals yellowish white; ovules usually pendulous; carpels indehiscent or apically dehiscent. **Sida**

Abutilon Mill.   Flowering Maple

1a. Plants annual; herbage velvety-pubescent; leaves ovate-orbicular, cordate, entire or slightly serrate; flowers bright yellow; petals 6–9 mm long; introduced weed. Velvet-Leaf. **A. theophrasti** Medic.

1b. Plants perennial (at least indoors); herbage glabrous or thinly pubescent; leaves 3-lobed, side lobes often small, green or variegated; flowers yellow or orange; corolla to 3.5 cm long, often veined crimson; cultivated. **A. pictum** Walp.

Althaea L.   Hollyhock

Tomentose or pilose herbs with lobed or divided leaves and solitary or racemose flowers; calyx 5-cleft; petals 5; stamens petaloid in many forms; color highly variable; gardens and roadsides. **A. rosea** Cav.

Callirrhoe Nutt.   Poppymallow

Perennial herbs with procumbent stems; leaves 5- to 7-parted; bractlets of involucel 3, linear; petals 2–3 cm long, crimson; style branches filiform, longitudinally stigmatic; carpels 10–20, each 1-loculed and 1-seeded. **C. involucrata** (T. & G.) Gray

Gossypium L.   Cotton

Plants mostly 6–12 dm tall; leaves cordate or subcordate; flowers white or light yellow; fruit 4–6 cm long, usually 4- to 5-loculed; cultivated. Upland Cotton. **G. hirsutum** L.

Hibiscus L.   Rosemallow

1a. Calyx herbaceous, not inflated, closely applied to or filled by the capsule; perennial. (2)

1b. Calyx bladdery-inflated, soon scarious, closed over the globular capsule; annual weed of fields and waste places. Flower-of-an-Hour. **H. trionum** L.

2a. Shrubs with rhombic-ovate, 3-lobed, glabrous leaves; cultivated. Rose of Sharon. **H. syriacus** L.

2b. Perennial herbs or tall growing annuals. (3)

3a. Plants annual with hairy leaves; corolla greenish yellow; cultivated for the edible unripe capsules. Okra. **H. esculentus** L.

3b. Plants perennial; stem and often hastate median and upper leaves glabrous; flowers commonly flesh colored with a purple center; cultivated. (Note: This plant has served as a parent for several hybrids varying in color from white to deep red.) **H. militaris** Cav.

Iliamna Greene   Wild Hollyhock

Perennial plants, 6–20 cm tall; leaves cordate to reniform; flowers

Malvaceae

large, pinkish white to rose-purple, the petals 18–30 mm long; ovules and seeds 2–4 per carpel; meadows and along streams, widely distributed. **I. rivularis** (Dougl.) Greene

Malva L. Mallow

Annual or biennial herbs; stems procumbent, ascending or erect; leaves orbicular or reniform; flowers solitary or in small clusters; involucel present; petals white or pinkish; styles filiform, longitudinally stigmatic; fruit of several carpels, separating at maturity; weed of gardens, fields, and waste places. **M. neglecta** Wallr.

Malvastrum Gray Falsemallow

Annual herbs; stems decumbent or prostrate, 1–4 cm long; leaves suborbicular; bractlets slender; calyx 3–5 mm long; petals whitish or pinkish, 5–6 mm long; carpels indehiscent, not more than 15; southwestern Utah. **M. exile** Gray

Sida L.

Perennial herbs; flowers axillary, usually solitary, white or yellowish; bractlets 1–3, or apparently none; carpels 5 to many; styles filiform with capitate stigmas; carpels 1-seeded, indehiscent or tardily so near apex. **S. hederacea** (Dougl.) Torr. ex Gray

Sidalcea Gray

1a. Corolla yellowish white or white, 10–15 mm long; stems usually glabrous to the inflorescence; wet meadows or along streams. **S. candida** Gray
1b. Corolla pink, rose, or lilac, rarely white; stems usually not glabrous to the inflorescence. (2)
2a. Flowers nearly sessile; calyx stellate-pubescent to bristly with a mixture of stellate hairs; corolla pinkish; moist places in canyons and on mountainsides. **S. oregana** (Nutt.) Gray
2b. Flowers with conspicuous pedicels; calyx hirsute; corolla rose-purple; meadows and mountainsides. **S. neomexicana** Gray

Sphaeralcea St. Hil. Globe Mallow (Contributed by J. A. M. Jefferies)

1a. Inflorescence racemose, rarely with more than 1 flower at each node; involucral bracts present or wanting. (2)
1b. Inflorescence subthyrsoid to thyrsoid-glomerate, more than 1 flower at most nodes; involucral bracts present. (4)
2a. Plants caespitose, the herbage densely tomentose; hairs with rays radiating in more than a single plane; leaf blades flat, sparingly 3- to 5-lobed if at all, margins irregularly dentate or crenate; involucral bracts present; more than one-half of carpel face smooth; known only from Millard and Beaver cos. **S. caespitosa** M. E. Jones
2b. Plants not caespitose, the internodes elongate, the herbage moderately pubescent; hairs appressed; leaf blades folded, some 3- or

Figs. 73-74. 73. **Trifolium pratense**, x .22. 74. **Utricularia vulgaris**, x .25

Figs. 75-76. 75. **Linum perenne**, x .21. 76. **Malva neglecta**, x .27

Malvaceae

5-parted or -divided; involucral bracts present or wanting; more than one-half of carpel face reticulate. (3)

3a. Herbage silvery-scaly; hairs with rays more than one-fourth united; bractlets of involucel present; lower leaves 3-divided, upper leaves with a single lobe, the lobes all entire; Emery, Wayne, Garfield, Kane, Grand, and San Juan cos. **S. leptophylla** (Gray) Rydb.

3b. Herbage grayish- to whitish-pubescent; hairs with rays only sparingly united; bractlets of involucel wanting; leaves 3- or 5-parted or -divided, the lobes coarsely and often regularly toothed; almost throughout Utah. **S. coccinea** (Nutt.) Rydb.

4a. Plants sparsely pubescent, the herbage bright green. (5)

4b. Plants moderately to densely pubescent or canescent, the herbage (especially on younger stems) appearing yellowish, whitish, or grayish. (7)

5a. Leaves pedately 3- or 5-parted or -divided, the lobes with narrow, regularly pinnatifid, toothed margins, the teeth nearly at right angles to the vein; carpels with transparent areolae; known from Washington, Garfield, and San Juan cos. **S. rusbyi** Gray

5b. Leaves sparingly 3- to 5-lobed to pedately parted or divided, the lobes broader, with margins irregularly dentate or variously lobed, not as above; carpels with opaque areolae. (6)

6a. Leaves faintly lobed with coarsely and unevenly dentate margins or deeply parted to divided with margins coarsely and irregularly lobed, the base subcordate to cuneate; known from Emery, Duchesne, Utah, Salt Lake, Summit, Cache, and Box Elder cos. **S. munroana** (Dougl.) Spach

6b. Leaves 3- or 5-parted or -divided, the margins more regularly cleft, lobed, or toothed, the base subcordate to deeply cordate; known from most of Utah, except for the Uinta Basin. **S. grossulariifolia** (H. & A.) Rydb.

7a. Inflorescence loosely thyrsoid (may appear paniculate), leafy; flowers not numerous at each node; peduncles generally elongate; calyx long, always surpassing the fruit; carpels tall, with heavy, well-defined reticulae which extend onto back of carpel; known from Washington and San Juan cos. **S. ambigua** Gray

7b. Inflorescence thyrsoid-glomerate; flowers generally numerous at each node, not especially leafy; calyx small, often equaled or surpassed by the fruit; carpels with moderately well-defined to extremely faint reticulae, the reticulae confined to the lateral face of the carpel. (8)

8a. Leaves 3- or 5-cleft, -parted, or -divided; herbage moderately pubescent to white-canescent or subtomentose; carpels with moderately well-defined reticulae on less than one-half of carpel face; widespread in western and southern Utah. **S. grossulariifolia** (H. & A.) Rydb.

8b. Leaves shallowly 3- or 5-lobed; herbage moderately pubescent to

229

Subclass DICOTYLEDONEAE

densely white- or yellowish-canescent; carpels with finely defined or nearly obscure reticulae on lower one-third of carpel; common in southern and eastern Utah, less common in northern counties. **S. parvifolia** A. Nels.

MARTYNIACEAE — MARTYNIA FAMILY

Stout, annual, viscid-pubescent herbs; leaves opposite, at least below; inflorescence terminal, racemose; flowers perfect, zygomorphic; calyx of 5 separate sepals; corolla of 5 united petals; stamens 2, the other 2 represented by staminodes; pistils 1; the ovary superior, 2-carpelled, 1-loculed; fruit a horned capsule.

Proboscidea Keller ex Schmid.

A single genus and species treated, with the characteristics of the family; southwestern Utah. **P. louisianica** (Mill.) Thell.

MELIACEAE — MAHOGANY FAMILY

Plants trees; leaves alternate, 1- to 3-pinnate; flowers in panicles; sepals 5–6; corolla of 5–6 petals, white or purple; stamens mostly 10–12, the filaments connate and forming a tube; ovary superior, 3- to 8-celled; fruit a drupe.

Melia L. Chinaberry

A single genus and species treated, with the characteristics of the family; cultivated in southwestern Utah. **M. azedarach** L.

MORACEAE — MULBERRY FAMILY

Trees, shrubs, or herbs, often with milky juice; leaves alternate, entire, serrate or lobed; flowers monoecious or dioecious, small, regular, usually in heads; fruit a small achene, or drupe usually enveloped by a fleshy perianth and forming a multiple fruit.

1a. Plants woody, trees or shrubs. (2)
1b. Plants erect or twining herbs. (3)
2a. Leaves serrate or dentate, often lobed; branches not spiny. **Morus**
2b. Leaves entire; branches spiny. **Maclura**
3a. Plants erect herbs; leaves palmately compound. **Cannabis**
3b. Plants twining vines; leaves palmately lobed. **Humulus**

Cannabis L. Marijuana

Stout, rough herbs, with 5–11 digitate leaflets; flowers greenish, dioecious, axillary, staminate panicled, pistillate spicate; occasional weed, especially in northern Utah. **C. sativa** L.

Nyctaginaceae

Humulus L. Hop

Perennial twining herbs; stems rough-hairy; leaves ovate, deeply 3- to 5-lobed, dentate; flowers in spikes in pairs with leafy bracts; fruit in papery clusters conspicuous in fall; canyons and foothills. American Hop. **H. americanus** Nutt.

Maclura Nutt. Osage Orange

Deciduous trees or shrubs to 10 m tall; leaves ovate-oblanceolate, 8–12 cm long; fruit drupaceous in large, globose multiples the size and appearance of a green orange. **M. pomifera** Schneid.

Morus L. Mulberry

1a. Leaves glabrous above, glossy, dark green; fruit white or violet; bark light gray; trees to 15 m tall; weeping and fruitless forms also cultivated. White Mulberry. **M. alba** L.
1b. Leaves dull green, mostly pubescent. (2)
2a. Leaves variously lobed; fruit red, turning dark; trees to 20 m tall. Red Mulberry. **M. rubra** L.
2b. Leaves usually not lobed but tapering to a point; fruit dark; trees shrubby, to 10 m tall. Black Mulberry. **M. nigra** L.

Nyctaginaceae — Four O'clock Family

Annual or perennial herbs; stems usually swollen at nodes; leaves simple, opposite or subopposite; flowers perfect, regular, reduced to a petaloid calyx, subtended by calyxlike bracts; stamens 1 to many; pistils 1; ovary superior, 1-loculed, 1-ovuled; fruit indehiscent.

1a. Flowers 1–10 in a cluster, surrounded by united bracts or subtended by 3 bracts which are united at the base (distinct or nearly so in **Hermidium** and **Selinocarpus**); stigmas capitate or hemispheric. (2)
1b. Flowers many in a cluster, surrounded by 4–6 distinct bracts; stigmas linear. (6)
2a. Bracts subtending flowers distinct, 1 at base of each flower. (3)
2b. Bracts subtending flowers united, at least below. (4)
3a. Perianth purplish red; fruit elliptic, smooth. **Hermidium**
3b. Perianth greenish; fruit with 3–5 papery wings. **Selinocarpus**
4a. Fruit strongly compressed, oval to obovate in outline with dentate margins inrolled over the dorsal face. **Allionia**
4b. Fruit terete or angled, not compressed, lacking inrolled, dentate margins. (5)
5a. Fruit strongly 5-angled longitudinally, constricted at the base; subtending bracts greatly enlarged in fruit. **Oxybaphus**
5b. Fruit almost smooth (may have 10 faint lines); subtending bracts not enlarged in fruit. **Mirabilis**

Subclass DICOTYLEDONEAE

6a. Perianth 5-lobed; wings of fruit thickish, opaque, not continuous around the fruit. **Abronia**
6b. Perianth 4- or 5-lobed; wings of fruit thin, translucent, continuous around the fruit. **Tripterocalyx**

Abronia Juss. Sand Verbena

1a. Plants caespitose from caudices; stems short, erect or nearly so; flowers white. **A. nana** S. Wats.
1b. Plants not caespitose from caudices; stems long, commonly decumbent or trailing; flowers white, reddish or purplish. (2)
2a. Bracts subtending flowers small, lanceolate, acute or acuminate; flowers reddish or purplish. (3)
2b. Bracts subtending flowers large and broad, ovate or obovate, acutish; flowers white, greenish white or pink tinged. (4)
3a. Flowers less than 12 mm long, purplish red; leaves broadly lanceolate; stems glabrous or merely viscid; southern Utah. **A. pumila** Rydb.
3b. Flowers 15–20 mm long, purple or purplish red; leaves broadly ovate to oblong; stems densely villous; southwestern Utah. **A. villosa** S. Wats.
4a. Stems viscid-villous above. **A. salsa** Rydb.
4b. Stems glabrous or viscid-pubescent above, not at all villous. **A. fragrans** Nutt. ex Hook.

Allionia L. Umbrellawort

1a. Plants perennial; outer margin of fruits with 3 triangular, nonglandular teeth on each side, usually incurved and covering nearly the entire surface; southeastern and southwestern Utah. **A. incarnata** L.
1b. Plants annual; outer margin of fruits with several gland-tipped teeth on each side, not covering the surface; southeastern Utah. **A. choisyi** Standl.

Hermidium S. Wats.

Perennial herbs, 1.5–4 dm tall; leaves opposite, thick, short-petiolate, obovate to ovate; subtending bracts united; perianth about 2 cm long, purplish red, stamens included; fruit 1- to 10-nerved, not compressed; western Utah and Uinta Basin. **H. alipes** S. Wats.

Mirabilis L. Four O'Clock

1a. Involucre 1-flowered; southwestern Utah. **M. bigelovii** Gray
1b. Involucre 3- to several-flowered (often 1-flowered in cultivated **M. jalapa**). (2)
2a. Plants cultivated, rarely persisting, common. Four O'Clock. **M. jalapa** L.
2b. Plants native, rare in cultivation. (3)

Nymphaeaceae

3a. Perianth 3–6 cm long; involucre campanulate, 3- to 10-flowered; stamens 5, the filaments connate at base; southeastern and southwestern Utah. **M. multiflora** (Torr.) Gray ex Torr.
3b. Perianth less than 1 cm long; involucre broadly spreading, usually 3-flowered; stamens 3, the filaments distinct. **M. oxybaphoides** Gray ex Torr.

Oxybaphus L'Her ex Willd.

1a. Leaf blades narrowly linear, 5 or more times longer than wide, acute or attenuate at base. (2)
1b. Leaf blades broader, not more than 4 times as long as broad, short-cuneate, truncate or cordate at base; not definitely known from Utah, to be expected in extreme southern Utah. **O. pumilus** Standl.
2a. Stems, involucre, perianth, and fruit glabrous or sparsely strigose; southern Utah. **O. glaber** S. Wats.
2b. Stems usually glandular and pilose or villous above; involucre pilose or villous, often glandular; perianth more or less pubescent; fruit copiously villous; widely distributed, Utah's most common species. **O. linearis** (Pursh) Robins.

Selinocarpus Gray

Perennial herbs or subshrubs; herbage pubescent with inflated white hairs; leaves opposite, oval to ovate, fleshy; flowers few per involucre, yellowish green, 3.5–4 cm long; fruit about 7 mm long, with 5 papery wings; southwestern Utah. **S. diffusus** Gray

Tripterocalyx (Torr.) Hook. Sandpuff

1a. Stems glabrous or nearly so; peduncles often longer than the subtending leaves. **T. pedunculatus** (M. E. Jones) Standl.
1b. Stems densely pubescent; peduncles shorter than the subtending leaves. **T. micranthus** (Torr.) Hook.

NYMPHAEACEAE — WATERLILY FAMILY

Aquatic perennials; leaves long-petiolate, the blades floating on the surface, peltate or cordate; flowers perfect, regular; sepals usually 4; petals 3 to many; pistil various, superior.

1a. Plants with floating oval leaves; sepals about 2.5 cm long, numerous, yellow, often tinged with red; native, usually in lakes at higher elevations. **Nuphar**
1b. Plants with floating cordate leaves; flowers showy, of many colors; sepals 4; petals and stamens many; cultivated. **Nymphaea**

Nuphar Smith   Yellow Pondlily

The common pondlily of Utah mountains, usually in lakes and ponds at higher elevations. **N. polysepalum** Engelm.

Subclass DICOTYLEDONEAE

Nymphaea L. Waterlily

Numerous horticultural forms, mostly hybrids involving several species, are known, cultivated. **N. odorata** Ait.

OLEACEAE — OLIVE FAMILY

Trees, shrubs or subshrubs, with opposite, rarely alternate, simple or compound-pinnate leaves; flowers perfect or imperfect, regular; calyx usually 4-lobed, sometimes obsolete; corolla of 4 or more distinct or united petals, or none; stamens usually 2; ovary superior, 2-loculed; fruit a drupe, berry, or capsule.

- 1a. Fruit a winged samara; leaves compound-pinnate (often simple in **F. anomala**). **Fraxinus**
- 1b. Fruit not a winged samara; leaves usually simple (sometimes with 3 leaflets in **Forsythia**). (2)
- 2a. Corolla lacking, or rarely of 1 or 2 small petals; calyx minute or none; native, southeastern Utah. **Forestiera**
- 2b. Corolla present; calyx well developed. (3)
- 3a. Flowers bright yellow, appearing with or before the foliage; cultivated. **Forsythia**
- 3b. Flowers usually other than yellow or, if so, then native plants of southern Utah. (4)
- 4a. Flowers bright yellow; plants subshrubs, woody only at very base; native of southern Utah. **Menodora**
- 4b. Flowers not bright yellow; plants shrubs, woody the length of stems; cultivated. (5)
- 5a. Fruit a loculicidal capsule, usually persistent on the plant. **Syringa**
- 5b. Fruit a drupe or berry. (6)
- 6a. Corolla with linear, elongate, nearly separate petals; panicles drooping. **Chionanthus**
- 6b. Corolla funnelform, with broad lobes; panicles erect or merely spreading. (7)
- 7a. Panicles scaly bracted, from the old axils; flowers dioecious or polygamous; stigma capitate; leaves leathery, evergreen; drupe 1-loculed. **Osmanthus**
- 7b. Panicles open, terminal, erect; flowers perfect; stigma 2-lobed; leaves tardily deciduous; drupelike berry 2-loculed. **Ligustrum**

Chionanthus L. Fringe-Tree

Low trees or shrubs with deciduous, entire leaves; flowers in loose, drooping panicles from lateral buds; calyx 4-parted, persistent; petals barely united at base; stamens 2 (rarely more); stigma notched; drupe fleshy, 1-loculed; cultivated. **C. virginicus** L.

Forestiera Poir.

Shrubs to 3.5 m tall; leaves opposite, simple; flowers inconspicuous,

Oleaceae

dioecious or polygamo-dioecious, in small fascicles or panicles, appearing before the leaves; calyx minute or none; corolla lacking or with 1 or 2 small petals; stamens usually 2; fruit a 1-seeded black or purplish drupe; southern or southeastern Utah. F. neomexicana Gray

Forsythia Vahl.
1a. Branches hollow between nodes; leaves occasionally compound; cultivated. F. suspensa (Thunb.) Vahl.
1b. Branches with chambered or solid pith. (2)
2a. Leaves often with 3 leaflets; pith solid at nodes, chambered between; branches erect; cultivated. F. intermedia Zab.
2b. Leaves always simple; pith chambered throughout. (3)
3a. Leaves ovate to broad-ovate; cultivated. F. ovata Nakai
3b. Leaves ovate to oblong-lanceolate; cultivated. F. viridissima Lindl.

Fraxinus L. Ash

1a. Leaves simple, rarely 3-foliolate; small trees or shrubs to 10 m tall; native, eastern and southern Utah. Single-leaf Ash. F. anomala Torr. ex S. Wats.
1b. Leaves compound, usually with 5 or more leaflets; trees. (2)
2a. Buds black or brownish black. (3)
2b. Buds gray or brown, but not dark brown. (4)
3a. Buds black; leaves medium green; cultivated. European Ash. F. excelsior L.
3b. Buds brownish black; leaves dark glossy green; cultivated. German Ash. F. holotricha Koehne
4a. Branchlets 4-angled; leaflets 7–11; cultivated. Blue Ash. F. quadrangulata Michx.
4b. Branchlets terete; leaflets 3–9. (5)
5a. Leaves glabrous, 20–38 cm long; leaflets usually 7; cultivated. White Ash. F. americana L.
5b. Leaves pubescent, at least on lower midrib. (6)
6a. Leaflets 3–5, 2–5 cm long; native, southwestern Utah, also cultivated. Modesto Ash. F. velutina Torr.
6b. Leaflets 5–11, 5–15 cm long; cultivated. (7)
7a. Leaflets 5–9, stalked; petiolules 3–6 mm long; calyx persistent; cultivated, possibly Utah's most common ash. (Note: The var. lanceolata (Borkh.) Sarg. differs in having glabrous leaflets and petioles and narrower leaflets.) Red Ash. F. pennsylvanica Marsh
7b. Leaflets 7–11, sessile; calyx lacking on the fruit; cultivated. Black Ash. F. nigra Marsh

Ligustrum L. Privet

1a. Branchlets glabrous or barely puberulent; leaves glabrous. (2)
1b. Branchlets strongly pubescent; leaves pubescent beneath, at least on the midrib. (3)

Subclass DICOTYLEDONEAE

2a. Leaves tardily deciduous; branchlets puberulent or glabrous; corolla tube scarcely as long as the limb; anthers much shorter than the corolla lobes; cultivated and escaping, possibly Utah's most common privet. Common Privet. **L. vulgare** L.
2b. Leaves semievergreen; branchlets glabrous; corolla tube about 3 times the length of lobes; anthers as long as corolla lobes; cultivated. California Privet. **L. ovalifolium** Hassk.
3a. Calyx glabrous; anthers much shorter than corolla lobes; branching upright; berries in vertical clusters; cultivated. **L. amurense** Carr.
3b. Calyx pubescent; anthers and corolla lobes subequal; branching horizontal; berries in horizontal clusters. **L. obtusifolium** Sieb. & Zucc.

Menodora H. & B.

Plants subshrubs, woody only at very base, nearly herbaceous; stems 1–3.5 dm tall; leaves 1–3 cm long, opposite or alternate, especially above; calyx deeply cleft, with 5–7 lobes; corolla usually 5-lobed, subrotate, bright yellow, 7–12 mm long; capsule 5–7 mm long, circumscissile; southern Utah. **M. scabra** (Engelm.) Gray

Osmanthus Lour.

Evergreen shrubs or small trees with leathery leaves; flowers very fragrant, dioecious, in scaly-bracted panicles from old axils; calyx 4-cleft, persistent; corolla white to yellowish, 4-cleft; stamens usually 2; ovary with 2 ovules in each locule; drupe 1-seeded; cultivated greenhouse plant, probably the "jasmine" used in Chinese jasmine tea. **O. fragrans** (Thunb.) Lour.

Syringa L. Lilac
1a. Plants trees to 10 m tall, with one main trunk; corolla tube scarcely longer than calyx; flowers yellowish white; cultivated. Tree Lilac. **S. amurensis** Rupr.
1b. Plants shrubs to 5 m tall, usually with many stems; corolla tube much longer than calyx; flowers variously colored. (2)
2a. Panicles from true terminal buds. (3)
2b. Panicles from 2 lateral buds. (4)
3a. Leaves 5–11 cm long; lenticels few; corolla lobes upright; cultivated. Hungarian Lilac. **S. josikaea** Jacq.
3b. Leaves 10–16 cm long, cuneate at base, pilose along veins beneath; flowers lilac to whitish; cultivated. Late Lilac. **S. villosa** Vahl.
4a. Leaves truncate or subcordate at base; flowers very fragrant, variously colored; cultivated, Utah's most common lilac. Common Lilac. **S. vulgaris** L.
4b. Leaves cuneate at base. (5)
5a. Panicle 5–8 cm long; leaves lanceolate; shrubs to 2 m tall; cultivated. Persian Lilac. **S. persica** L.

Onagraceae

5b. Panicle 8–16 cm long; leaves ovate-lanceolate; shrubs to 5 m tall; cultivated. Chinese Lilac. **S. chinensis** Willd.

ONAGRACEAE — EVENING PRIMROSE FAMILY

Caulescent or acaulescent herbs, or rarely woody plants; leaves alternate or basal; flowers perfect; hypanthium adnate to the ovary and usually prolonged beyond; sepals 4 or 2; petals distinct, 4 or 2, inserted on the hypanthium; stamens as many or twice as many as the petals; ovary inferior, usually 4-loculed; styles 1; stigma capitate, 4-lobed or discoid; fruit a capsule, nut, or berry.

1a. Sepals, petals, and stamens 2; fruit indehiscent, usually with hooked hairs. **Circaea**
1b. Sepals and petals 4; stamens usually 8; fruit lacking hooked hairs. (2)
2a. Seeds with a tuft of hair at one end. (3)
2b. Seeds without a tuft of hair. (4)
3a. Hypanthium less than 1 cm long, or lacking, no scales within; flowers white, pinkish, or purplish. **Epilobium**
3b. Hypanthium 2–3 cm long, with a row of 8 scales within; flowers scarlet. **Zauchneria**
4a. Fruit indehiscent, a nut or berry. (5)
4b. Fruit dehiscent, a capsule. (6)
5a. Hypanthium colored and showy; shrubs with berries; cultivated greenhouse ornamentals, occasional in outside plantings. **Fuchsia**
5b. Hypanthium not colored and showy; biennial or annual herbs with nutlike fruit; native weedy plants. **Gaura**
6a. Ovary 2-loculed; hypanthium not prolonged beyond apex of ovary; flowers minute (petals not over 1.5 mm long). **Gayophytum**
6b. Ovary 4-loculed; hypanthium prolonged beyond ovary; flowers large or small (petals over 1.5 mm long). (7)
7a. Anthers usually versatile (attached near the middle); petals white or yellow, often pink or red with age. **Oenothera**
7b. Anthers innate (attached near one end); petals pink to lavender, rarely whitish, but not yellow. (8)
8a. Sepals reflexed, or the apices remaining united and turning to one side at flowering; petals clawed. **Clarkia**
8b. Sepals erect; petals small or wanting. **Boisduvalia**

Boisduvalia Spach

Annual caulescent herbs, 2–30 cm tall; leaves lanceolate to lance-ovate or oblong, 4–15 mm long; flowers solitary in leaf axils; hypanthium about 0.5 mm long; petals pinkish or purplish, 2–4 mm long; capsule subterete, 5–8 mm long. **B. glabella** (Nutt.) Walp.

Subclass DICOTYLEDONEAE

Circaea L.   Enchanters Nightshade
Perennial herbs, 2–6 dm tall; leaf blades 2–6 cm long; calyx and corolla about 1 mm long; capsule 2 mm long. **C. pacifica** Asch. & Mag.

Clarkia Pursh
Annual caulescent herbs, 2–10 dm tall; leaves subopposite, the blades 2–7 cm long; flowers in elongated spikes; sepals green; petals 5–10 mm long, rose-purple; capsules 1–3 cm long. **C. rhomboidea** Dougl.

Epilobium L.   Willowherb
1a. Hypanthium not prolonged beyond ovary; flowers large, the petals 1–2 cm long, spreading; stigmas 4-lobed. (2)
1b. Hypanthium prolonged beyond ovary; flowers with petals less than 1 cm long, ascending; stigmas not lobed. (3)
2a. Leaves 5–15 cm long, membranous, lateral veins definite, confluent in marginal loops; racemes many-flowered, elongate, not leafy. Fireweed. **E. angustifolium** L.
2b. Leaves 2–6 cm long, fleshy, not prominently veined; racemes few-flowered, short and leafy. **E. latifolium** L.
3a. Plants annual in dry situations; stems with exfoliating epidermis. **E. paniculatum** Nutt. ex T. & G.
3b. Plants perennial, usually in moist situations; stems not exfoliating. (4)
4a. Leaves linear or linear-oblong, margins more or less revolute, sessile or nearly so. **E. palustre** L.
4b. Leaves linear-lanceolate, ovate or oval, margins not revolute, sessile or petioled. (5)
5a. Rhizomes bearing globose or ovoid winter buds with imbricate fleshy leaves. (6)
5b. Rhizomes without globose or ovoid winter buds. (8)
6a. Stems glabrous below, pubescent or glandular above, but not with decurrent, longitudinal lines of hair from leaf bases. **E. brevistylum** Barbey ex Brewer & Wats.
6b. Stems with decurrent, longitudinal lines of hair from leaf bases. (7)
7a. Leaves linear-lanceolate, often decurrent at base, not crowded, the margins irregularly dentate. **E. halleanum** Hausskn.
7b. Leaves, at least the basal, ovate with rounded bases, often crowded, the margins entire. **E. saximontanum** Hausskn.
8a. Stems 30–100 cm tall, freely branched above; leaves sessile or short-petiolate; capsules 4–6 cm long; tuft of hair on seeds white. **E. adenocaulon** Hausskn.
8b. Stems 5–30 cm tall, simple or nearly so; leaves indistinctly petiolate; capsules 2–5 cm long; tuft of hair on seeds white to dingy. (9)
9a. Stems not densely caespitose, erect, 10–30 cm tall; leaves 1.5–5 cm long; capsules 3–5 cm long. **E. hornemanii** Reichenb.

Onagraceae

9b. Stems densely caespitose, sigmoidally curved, 5–15 cm tall; leaves 1–2 cm long; capsules 2–4 cm long. (10)
10a. Capsules linear, slender, not over 1 mm thick and 2–4 cm long; leaves oblong-ovate to lanceolate. **E. alpinum** L.
10b. Capsules subclavate, 1.5–2 mm thick and 2–2.5 cm long; leaves broadly ovate. **E. clavatum** Trel.

Fuchsia L.

Shrubby perennials, 3–12 dm tall or more; leaves ovate, 3–10 cm long, toothed; hypanthium twice or more as long as ovary; sepals usually crimson or pink; petals rose, purple, or white, 10–25 mm long; stamens long-exserted. **F. hybrida** Voss

Gaura L. Butterflyweed

Winter annuals or biennials, 2–20 dm tall; leaves 3–10 cm long; cauline leaves almost sessile; flowers in terminal, slender spikes; hypanthium 1.5–3 mm long; sepals 1.5–3 mm long; petals 1.5–2 mm long, pink or rose; fruit indehiscent, nutlike, 6–10 mm long. **G. parviflora** Dougl. ex Hook.

Gayophytum A. Juss.
1a. Capsule pedicellate; plants freely branched above base, repeatedly dichotomous; upper leaves bractlike. (2)
1b. Capsule subsessile; plants mostly branched at base, little branched above; upper leaves quite well developed. **G. racemosum** T. & G.
2a. Petals 0.5 mm long; capsule 2–5 mm long, shorter than the deflexed pedicel; plants glabrous. **G. ramosissimum** T. & G.
2b. Petals 1–1.5 mm long; capsule 5–12 mm long, exceeding the pedicel. **G. nuttallii** T. & G.

Oenothera L.   Evening Primrose

1a. Stigmas with 4 linear lobes; flowers mostly opening in evening. (2)
1b. Stigmas capitate, discoid or slightly 4-lobed or 4-toothed; flowers mostly opening in daytime. (13)
2a. Capsule winged, at least in the upper part, ovoid and rather short for its thickness; petals yellow when young; plants acaulescent or nearly so (subgenus Lavauxia). **O. flava** (A. Nels.) Garrett
2b. Capsule terete or round angled, but never winged, usually elongated; petals white or yellow when young; plants acaulescent or caulescent. (3)
3a. Flowers yellow when young; seeds sharply angled, in 2 rows in each locule; plants caulescent (subgenus Onagra). (4)
3b. Flowers white when young (yellow in the annual **O. primiveris**); seeds not sharply angled, in 1 or 2 rows; plants caulescent or acaulescent. (6)
4a. Hypanthium tube 8–15 cm long. **O. longissima** Rydb.

Subclass DICOTYLEDONEAE

- 4b. Hypanthium tube 3–4.5 cm long. (5)
- 5a. Petals and sepals 25–40 mm long. **O. hookeri** T. & G.
- 5b. Petals and sepals less than 20 mm long. **O. rydbergii** House
- 6a. Plants acaulescent or nearly so; seeds with a deep furrow along the raphe; capsule ovoid or ovoid-lanceolate (subgenus Pachylophus). (7)
- 6b. Plants caulescent; seeds lacking a deep furrow along the raphe; capsule oblong, cylindric, or subcylindric. (8)
- 7a. Flowers yellow, red with age; plants annual; southwestern Utah. **O. primiveris** Gray
- 7b. Flowers white, usually pink with age; plants perennial; widely distributed, highly variable. **O. caespitosa** Nutt.
- 8a. Capsule membranous, somewhat enlarged above the base; seeds in 2 rows in each locule, with shallow pits in regular rows on the surface (subgenus Raimannia). (9)
- 8b. Capsule woody, somewhat narrowed toward apex; seeds in 1 row in each locule, not pitted (subgenus Anagra). (10)
- 9a. Plants perennial from slender underground rhizomes; throat of hypanthium with long, conspicuous white hairs; petals 7–15 mm long; capsule 8–20 mm long. **O. coronopifolia** T. & G.
- 9b. Plants annual (or winter annual); throat of hypanthium lacking long hairs; petals 15–40 mm long; capsule 20–40 mm long. **O. albicaulis** Pursh
- 10a. Plants spring or winter annuals, coarse; basal leaf blades rhombic, 2–8 cm long; capsules woody; southwestern Utah. **O. deltoides** Torr. & Frem.
- 10b. Plants perennial or occasionally biennial; basal leaf blades smaller and narrower. (11)
- 11a. Plants glabrous or glabrate with possibly a few appressed hairs or some glandular pubescence above; widely distributed. **O. pallida** Lindl.
- 11b. Plants canescent to gray-strigillose or villous, especially in the upper parts. (12)
- 12a. Capsules contorted; buds with long hairs as well as short appressed ones; petals 10–18 mm long; eastern to west central Utah. **O. trichocalyx** Nutt. ex T. & G.
- 12b. Capsules spreading, usually not contorted; buds lacking long hairs; petals 15–25 mm long; north central Utah. **O. latifolia** (Rydb.) Munz
- 13a. Ovary with a long, narrow, upper sterile part grading into the basal, enlarged fertile part, the upper part simulating a hypanthium; plants perennial and acaulescent (subgenus Taraxia). (14)
- 13b. Ovary without such an elongate sterile part; plants annual or perennial, caulescent or, if acaulescent, then annual. (15)
- 14a. Leaves deeply pinnatifid; capsule densely pubescent. **O. breviflora** T. & G.

Onagraceae

14b. Leaves entire or with a few teeth; capsule glabrous. **O. heteranthera** Nutt.
15a. Hypanthium tube 25–50 mm long or longer; stamens all nearly equal in length (subgenus Salpingia). **O. lavandulaefolia** T. & G.
15b. Hypanthium tube 1–15 mm long; stamens of 2 lengths. (16)
16a. Capsule distinctly pedicellate, cylindric to clavate; leaves mostly near base of plants; petals yellow when young (subgenus Chylismia). (17)
16b. Capsule sessile or very nearly so, cylindric; leaves scattered; petals yellow to white (subgenus Sphaerostigma). (22)
17a. Flowers axillary, 4–5 mm broad; seeds oblong, deeply concave on the inner side and somewhat wing margined; southwestern and western Utah. **O. pterosperma** S. Wats.
17b. Flowers in terminal racemes; seeds obovoid, rounded or angled, not winged. (18)
18a. Capsules linear, elongate, usually over 2 cm long. (19)
18b. Capsules somewhat enlarged at one end, usually less than 2 cm long. (20)
19a. Stems coarse, commonly branched only at base; pedicels 3–15 mm long; capsules widely spreading, 5–9 cm long; anthers pubescent; southwestern Utah. **O. pallidula** (Munz) Munz
19b. Stems slender, commonly freely branched above; pedicels 10–25 mm long; anthers glabrous; widespread in southern Utah. **O. multijuga** S. Wats.
20a. Branches in well-developed plants capillary and arising freely throughout the plant; anthers oblong to linear-oblong, glabrous; style not longer than petals; southwestern Utah. **O. parryi** S. Wats.
20b. Branches in well-developed plants few to several and arising at base of plant only, not capillary; anthers linear, beset with scattered white hairs; style longer than petals. (21)
21a. Stems fairly coarse; flowers crowded in close terminal clusters; leaves frequently with supplementary pinnules on petioles; petals 4–7 mm long; southwestern Utah. **O. clavaeformis** Torr. & Frem.
21b. Stems slender; flowers few, not congested; leaves ovate, subentire; petals usually less than 4 mm long; widely distributed in Utah. **O. scapoidea** Nutt. ex T. & G.
22a. Flowers yellow, often drying greenish, borne in axils of foliage leaves. (23)
22b. Flowers white (yellowish in **O. minor** and **O. decorticans**), often drying pinkish, borne in terminal spikes. (24)
23a. Plants with several naked, fine, often capillary stems, each bearing a leafy inflorescence at apex; capsule subfusiform, almost straight, 5–8 mm long; north central Utah. **O. andina** Nutt.
23b. Plants with stems leafy from base; capsules terete, straight or

241

Subclass DICOTYLEDONEAE

coiled, 15–40 mm long; central, western, and southwestern Utah. **O. contorta** Dougl. ex Hook.

24a. Capsules cylindric, terete, linear, not thickened basally, scarcely if at all coiled, not attenuate apically. (25)

24b. Capsules not strictly cylindric, somewhat enlarged basally, and attenuate at the tip. (26)

25a. Petals 5–7 mm long, suborbicular; style exceeding the corolla, 10–13 mm long; hypanthium 4–6 mm long; capsules spreading, occasionally coiled; southern Utah. **O. refracta** S. Wats.

25b. Petals up to 3 mm long, spatulate; style shorter than the corolla, 3.5–4 mm long; hypanthium 2.5–3 mm long; capsules widely spreading; southern and western Utah. **O. chamaenerioides** Gray

26a. Mature capsules usually distinctly contorted and coiled, not merely bent or curved, quite slender, not subfusiform in shape. (27)

26b. Mature capsules merely curved or bent, not contorted or twisted, subfusiform in shape. (28)

27a. Flowers minute; petals 1–2 mm long, narrowly obovate; style 1.5–3 mm long; hypanthium 1–2 mm long; longer filaments twice as long as shorter filaments; central and eastern Utah. **O. minor** (A. Nels.) Munz

27b. Flowers larger; petals 3.5–5 mm long, orbicular-ovate; style 6–12 mm long; hypanthium 3–8 mm long; filaments subequal; northern and western Utah. **O. alyssoides** H. & A.

28a. Leaves mostly near the base, glabrate, lance-ovate to oblanceolate; stems glabrous or glabrate; epidermis promptly exfoliating; capsule 15–25 mm long; seeds ash colored; southwestern Utah. **O. decorticans** (H. & A.) Greene

28b. Leaves well distributed, glandular-pubescent to glandular-villous, ovate to oblong-ovate; stems glandular-pubescent to glandular-villous; epidermis exfoliating tardily if at all; capsule 10–15 mm long; seeds brownish; northeastern Utah. **O. boothii** Dougl. ex Hook.

Zauschneria Presl.

Perennial herbs, 1–6 dm tall; leaves opposite, sessile, 2–3 cm long; inflorescence spicate; flowers scarlet, large and showy; hypanthium 1.8–2 cm long, inflated above the ovary, then narrowed into a long tube; petals 8–10 mm long; capsule linear, 1.5–2 cm long. **Z. garrettii** A. Nels.

Orobanchaceae — Broomrape Family

Root parasites without green foliage; stems erect, usually yellowish or purplish; leaves reduced to alternate scales; flowers perfect, irregular; calyx 4- to 5-toothed or -cleft; corolla united; stamens 4, didynamous; filaments slender; ovary superior, 1-loculed; stigma discoid or lobed; capsule 1-loculed, 2- to 4-valved.

Papaveraceae

Orobanche L. Broomrape
1a. Flowers without floral bracts, solitary or several on a stem, on long slender pedicels. (2)
1b. Flowers with 2, sometimes 3, floral bracts; inflorescence a spike, a corymb, or a panicle; flowers sessile or pedicellate. (3)
2a. Stems slender, very short, bearing 1 or sometimes 2–3 pedicels many times the length of stem; calyx lobes awl shaped with a long tip, longer than the tube; parasitic on various hosts. **O. uniflora** L.
2b. Stems stout, each bearing 3–12 pedicels equaling or shorter than tube; parasitic on various hosts. **O. fasciculata** Nutt.
3a. Flowers pedicellate, lower pedicels 8–25 mm long, sometimes shorter above; inflorescence compactly corymbose; anthers woolly; corolla lips about 7–8 mm long; lobes of upper corolla lips obtuse or rounded; parasitic on **Artemisia** and associated shrubs. **O. corymbosa** (Rydb.) Ferris
3b. Flowers sessile, lower ones sometimes short-pedicellate; inflorescence essentially spicate, axis much elongated with age. (4)
4a. Corolla lobes, especially of lower lip, narrowly acute; calyx lobes 1.5–2 times the length of tube; parasitic on **Franseria** and other shrubs; southern Utah. **O. cooperi** (Gray) Heller
4b. Corolla lobes obtuse or rounded; calyx lobes 2 or more times the length of tube; parasitic on Compositae; western and southern Utah. **O. multiflora** Nutt.

Oxalidaceae — Woodsorrel Family

Plants herbs; leaves palmately 3-foliolate, alternate or basal; flowers in cymose or umbellate inflorescences, or solitary on axillary peduncles; flowers perfect, regular; sepals 5; petals 5; stamens 10, united at base; ovary superior, 5-loculed, with 5 styles; fruit a capsule; sap sour.

Oxalis L. Woodsorrel
1a. New leafy flowering stems repent, rooting at nodes, from slender taproots; common greenhouse weeds. **O. corniculata** L.
1b. New leafy flowering stems ascending, not rooting at sometimes persistent and decumbent bases; occasional weed in Utah. **O. stricta** L.

Papaveraceae — Poppy Family

Annual or perennial herbs, usually with milky yellowish juice; leaves mostly alternate, entire, lobed or divided; flowers solitary or in clusters, perfect, regular; sepals 2 (3), caducous; petals 6, rarely more; stamens numerous, in whorls; pistils 1, superior, 1-loculed; fruit a capsule.

1a. Leaves mainly opposite. **Platystemon**
1b. Leaves mostly alternate or basal. (2)

Figs. 77-78.   77. **Sphaeralcea coccinea**, x .29.   78. **Nuphar polysepalum**, x .26

Figs. 79-80.   79. **Syringa villosa**, x .25.   80. **Orobanche fasciculata**, x .28

Papaveraceae

2a. Flowers brick red. **Roemeria**
2b. Flowers white, yellowish, yellow, or orange. (3)
3a. Herbage, sepals, and capsules prickly; leaf blades large, sinuate-dentate or sinuate-pinnatifid; sepals with hornlike appendages. **Argemone**
3b. Herbage, sepals, and capsules not prickly (somewhat so in **Arctomecon**); leaf blades not sinuate or pinnatifid; sepals not appendaged. (4)
4a. Leaves entire or merely toothed near apex. **Arctomecon**
4b. Leaves pinnately parted, pinnately compound, or ternately dissected. (5)
5a. Stigmas expanded and disklike, extending over openings of broad poricidal capsule. **Papaver**
5b. Stigmas not expanded, the capsule elongate, 2-valved. **Eschscholtzia**

Arctomecon Torr. & Frem.   Desert Poppy

Perennial herbs; leaves mostly basal, long-hirsute, cuneate-obovate, mostly apically toothed; flowers large, 1 to several at the ends of long peduncles; petals 4 or 6, white or pale yellow; capsule oblong, ovoid or obovoid, 3- to 6-valved; southwestern Utah. **A. humilis** Coville

Argemone L.   Prickly Poppy

1a. Prickles of sepals and, when present, those of sepal horns perpendicular; lower cauline leaves mostly lobed to four-fifths or more the distance to midrib; flowers 7–12 mm broad; western and southwestern Utah. **A. munita** Dur. & Hilg.
1b. Prickles of sepals and sepal horns, when present, ascending, or sepals smooth; lower cauline leaves often lobed to four-fifths the distance to midrib; flowers 4–8 cm broad; southeastern Utah. **A. corymbosa** Greene

Eschscholtzia Cham. ex Nees   California Poppy

1a. Receptacle with 2 rims—the inner erect and hyaline, the outer spreading; annual or occasionally perennial. (2)
1b. Receptacle with only an erect hyaline rim, the outer rim absent or rudimentary. (3)
2a. Outer rim of receptacle 0.5–2 mm wide; petals 1.5–3 cm long; native, southwestern Utah. **E. mexicana** Greene
2b. Outer rim of receptacle 2–4 mm wide; petals 2–6 cm long; cultivated. **E. californica** Cham.
3a. Stems scapose, the leaves mostly in a basal tuft; petals 10–25 mm long; southwestern Utah. **E. glyptosperma** Greene
3b. Stems leafy; petals less than 10 mm long. **E. minutiflora** S. Wats.

Papaver L.   Poppy

1a. Plants scapose, the leaves basal; cultivated and native, the latter usually above 10,000 ft. in mountains. **P. nudicaule** L.

Subclass DICOTYLEDONEAE

1b. Plants with leafy stems. (2)
2a. Peduncles with coarse, appressed hairs; showy cultivated plant. Oriental Poppy. **P. orientale** L.
2b. Peduncles with weak, spreading hairs; cultivated and escaping. Corn Poppy. **P. rhoeas** L.

Platystemon Benth. Cream Cups

Plants annual, with opposite entire leaves; flowers subscapose, solitary on peduncles 1–2 dm long; sepals 6–10 mm long; petals cream, 8–16 mm long; carpels 6–25, united at first, separate in fruit; southwestern Utah. **P. californicus** Benth.

Roemeria Medic.

Plants annual, with leafy stems branching near base; a noxious weed in southwestern Utah. Field Poppy. **R. refracta** DC.

PLANTAGINACEAE — PLANTAIN FAMILY

Annual or perennial, acaulescent or short-stemmed herbs; flowers small, perfect or imperfect, regular, borne in bracteate spikes or heads; calyx 4-parted, persistent; corolla 4-lobed; stamens 4 or 2, alternate with the corolla lobes; ovary superior, 1- to 2-loculed; fruit a circumscissile capsule.

Plantago L.
1a. Leaves copiously villous, linear or lanceolate. (2)
1b. Leaves glabrous or nearly so, occasionally pubescent at base. (3)
2a. Floral bracts subulate or narrowly lanceolate, not scarious margined or only indistinctly so, at least lower ones commonly longer than calyx; widespread. **P. purshii** R. & S.
2b. Floral bracts broadly lanceolate to nearly orbicular, conspicuously scarious margined to apex; southern Utah. **P. insularis** Eastw.
3a. Stamens 2; leaves linear; corolla lobes in some flowers erect and closing over the capsule; flowers more or less dioecious or polygamous; moist, saline soil. **P. elongata** Pursh
3b. Stamens 4; leaves lanceolate to ovate. (4)
4a. Leaves narrowly oblong-lanceolate to narrowly lanceolate; adventive weed of lawns, fields, and waste places. **P. lanceolata** L.
4b. Leaves oblong-lanceolate to ovate. (5)
5a. Crown of plant conspicuously reddish silky-hairy; moist saline meadows. **P. eriopoda** Torr.
5b. Crown of plant not conspicuously reddish silky-hairy. (6)
6a. Leaves broad, abruptly contracted into the petiole; capsule breaking at the middle; adventive, a common lawn weed. **P. major** L.
6b. Leaves lanceolate, gradually tapering into the petiole; capsule breaking to near base or well below the middle; slopes and mountains, mostly at high elevations. **P. tweedyi** Gray

POLEMONIACEAE — PHLOX FAMILY

Annual or perennial herbs, rarely suffrutescent; leaves simple and entire to compound; flowers perfect, regular, solitary or variously clustered; corolla usually of 5 united petals; stamens usually 5, alternate with corolla lobes; ovary superior, 3-loculed; styles 1; stigmas 3-lobed; fruit a capsule.

1a. Foliage leaves represented by connate, entire cotyledons and bracts. **Gymnosteris**
1b. Foliage leaves proper present. (2)
2a. Leaves prevailingly opposite. (3)
2b. Leaves prevailingly alternate (opposite in some **Leptodactylon**). (5)
3a. Stamens equally inserted on corolla tube. **Linanthus**
3b. Stamens unequally inserted on corolla tube. (4)
4a. Plants perennial, or rarely annual; corolla tube over 6 mm long, salverform, with a narrow throat. **Phlox**
4b. Plants annual; corolla tube less than 6 mm long, flaring. **Microsteris**
5a. Stamens unequally inserted on corolla tube; calyx enlarging in fruit; annuals with entire or pinnatifid leaves. **Collomia**
5b. Stamens equally inserted on corolla tube, or in throat of corolla; plants various, but not usually as above. (6)
6a. Plants perennial or biennial. (7)
6b. Plants annual. (9)
7a. Leaves palmately parted, spinulose; plants often woody at base. **Leptodactylon**
7b. Leaves simple, pinnatifid, or pinnately compound. (8)
8a. Calyx green; flowers solitary or clustered; corolla campanulate to rotate-funnelform, blue or white; leaves pinnately parted, at least below. **Polemonium**
8b. Calyx more or less scarious in the sinuses; flowers mostly cymose; corolla salverform or trumpet shaped, with an open throat. **Gilia**
9a. Calyx lobes subequal; flowers solitary, cymose, or in heads. (10)
9b. Calyx lobes unequal; flowers in dense bracteate heads. (11)
10a. Corolla lobes essentially alike; leaf lobes mostly not setose or spine-tipped. **Gilia**
10b. Corolla lobes more or less unlike; leaf lobes setose or spine tipped. **Langloisia**
11a. Plants cobwebby-pubescent, at least the inflorescence with a feltlike mass of interlaced hairs; capsule dehiscent from top; leaves and bracts rarely with rigid, spinose lobes. **Eriastrum**
11b. Plants not having a feltlike mat of interlaced hairs; capsule dehiscent mostly from below, or indehiscent; leaves and bracts usually with rigid, spinose lobes. **Navarretia**

Subclass DICOTYLEDONEAE

Collomia Nutt.

1a. Plants often exceeding 3 dm in height; flowers in few- to many-flowered clusters, white to salmon pink or apricot. **C. grandiflora** Dougl.
1b. Plants mostly under 3 dm in height. (2)
2a. Plants 5–30 cm tall, pubescent; flowers in few- to many-flowered clusters, rose-purple to nearly white. **C. linearis** Nutt.
2b. Plants 1–2 cm tall, glabrous; flowers solitary or rarely in pairs, purplish or pinkish white. **C. tenella** Gray

Eriastrum Woot. & Standl.

1a. Corolla lobes nearly equaling tube, violet-blue to white; southern Utah. **E. sapphirinum** (Eastw.) Mason
1b. Corolla lobes much shorter than tube, blue. (2)
2a. Anthers barely exserted; heads of flowers broad, with a rounded base; bracts spreading; stems low, diffuse, 1–10 cm tall. **E. diffusum** (Gray) Mason
2b. Anthers included; heads of flowers with an acute base; bracts ascending; stems 10–20 cm tall. **E. filifolium** (Nutt.) Woot. & Standl.

Gilia R. & P.

1a. Inflorescence capitate or spicate-glomerate. (2)
1b. Inflorescence openly paniculate or thyrsoid-paniculate, with evident pedicels. (8)
2a. Leaves entire or toothed. (3)
2b. Leaves pinnatifid or 3-cleft, at least some of them. (5)
3a. Plants perennial, frequently suffrutescent. **G. congesta** Hook.
3b. Plants annual. (4)
4a. Leaves linear-filiform; flowers in small terminal heads; calyx teeth lanceolate. **G. gunnisonii** T. & G.
4b. Leaves oblanceolate; calyx teeth subulate. **G. depressa** M. E. Jones
5a. Plants annual. (6)
5b. Plants perennial. (7)
6a. Leaves with linear lobes; calyx 5 mm long, the teeth short, subulate; corolla 9–10 mm long, white, the lobes oval. **G. pumila** Nutt.
6b. Leaves with short, oblong lobes; calyx 4–5 mm long, the teeth lanceolate; corolla 4–5 mm long, white, the lobes minute. **G. polycladon** Torr.
7a. Heads of flowers in a more or less dense spikelike inflorescence. **G. spicata** Nutt.
7b. Heads of flowers single or in corymbs. **G. congesta** Hook.
8a. Leaves entire, rarely few-toothed, the basal ones sometimes pinnatifid. (9)
8b. Leaves distinctly toothed or pinnatifid, at least some. (10)

Polemoniaceae

9a. Calyx glabrous; corolla white or yellowish, 4–5 mm long; glabrous annuals to 25 cm tall; leaves filiform. **G. filiformis** Parry
9b. Calyx, pedicels, and branches glandular; corolla white or bluish, about 3 mm long; glandular-puberulent annuals to 15 cm tall; leaves linear. **G. tenerrima** Gray
10a. Corolla 20–40 mm long, pink, scarlet, or white. (11)
10b. Corolla less than 20 mm long, variously colored. (14)
11a. Inflorescence more or less flat topped, corymbose. (12)
11b. Inflorescence thyrsoid, narrow. (13)
12a. Corolla 30–40 mm long, white, the lobes obtuse, 10 mm long. **G. longiflora** (Torr.) G. Don
12b. Corolla 20–30 mm long, bluish white, the lobes acute, 4–5 mm long. **G. laxiflora** (Coult.) Osterh.
13a. Tube and throat of corolla 20–40 mm long, the lobes long-acuminate; stamens exserted. **G. aggregata** (Pursh) Spreng.
13b. Tube and throat of corolla about 20 mm long (often somewhat less), the lobes acute or short-acuminate; stamens included. **G. arizonica** (Greene) Rydb.
14a. Stamens exserted. (15)
14b. Stamens included. (17)
15a. Inflorescence narrow; corolla bluish white, about 12 mm long; stems simple. **G. stenothyrsa** Gray
15b. Inflorescence broad, the branches numerous; stems branching from the base or nearly so. (16)
16a. Corolla tube 4–6 mm long, bluish; plants more or less glandular or glabrate; leaf segments linear. **G. calcarea** M. E. Jones
16b. Corolla tube 7–10 mm long, blue; plants glabrous, at least below; leaf segments oblong or obovate. **G. mcvickerae** M. E. Jones
17a. Leaves orbicular to oblong or obovate in outline; southwestern Utah. **G. latifolia** S. Wats. ex Parry
17b. Leaves various but not as above. (18)
18a. Corolla 4–6 mm long. **G. leptomeria** Gray
18b. Corolla 6–18 mm long. (19)
19a. Limb of corolla 8–12 mm broad. **G. subnuda** Torr. ex Gray
19b. Limb of corolla 8 mm broad or less. (20)
20a. Calyx one-fourth as long as corolla tube. **G. scopulorum** M. E. Jones
20b. Calyx one-half to two-thirds as long as corolla tube. (21)
21a. Corolla lobes obtuse; leaf segments oblong, mostly entire. **G. sinuata** Dougl. ex DC.
21b. Corolla lobes acute; leaf segments oblong, mostly spinulose toothed. **G. hutchinsifolia** Rydb.

Gymnosteris Greene

Small annuals, with basal cotyledons, then naked except for the

Subclass DICOTYLEDONEAE

terminal heads of few flowers subtended by an involucre of a few leafy bracts; calyx 3–4 mm long; corolla tube about 5 mm long, limb about 1.5–3 mm wide, pale yellow. **G. parvula** (Rydb.) Heller

Langloisia Greene
- 1a. Corolla distinctly 2-lipped; leaves pinnatifid with a ligulate or spatulate rachis and single marginal bristles. **L. schottii** (Torr.) Greene
- 1b. Corolla almost regular; leaves abruptly dilated toward apex, the marginal bristles mostly in pairs. **L. setosissima** (T. & G.) Greene

Leptodactylon Nutt. (includes **Linanthastrum** Ewan)
- 1a. Leaves opposite or nearly so. (2)
- 1b. Leaves alternate. (3)
- 2a. Leaves glabrous to hispidulous, the segments linear, 10–18 mm long. **L. nuttallii** (Gray) Rydb.
- 2b. Leaves glandular, the segments awl shaped, acerose, 10–16 mm long. **L. watsonii** (Gray) Rydb.
- 3a. Plants low, densely caespitose, the stems short, densely clothed with persistent crowded leaves; corolla about 12 mm long, 4-merous. **L. caespitosum** Nutt.
- 3b. Plants 10–60 cm tall, suffrutescent, the branches elongate, not densely clothed with leaves; corolla 15–20 mm long, 5-merous. **L. pungens** (Torr.) Nutt.

Linanthus Benth.
- 1a. Flowers sessile or on pedicels less than 5 mm long. (2)
- 1b. Flowers on pedicels 5 mm long or more. (3)
- 2a. Calyx lobes free to base, membrane margined but not with an obvious hyaline membrane below the sinuses; corolla campanulate. **L. demissus** (Gray) Greene
- 2b. Calyx lobes united by bordering membranes; corolla funnelform to salverform. **L. bigelovii** (Gray) Greene
- 3a. Corolla scarcely exceeding calyx, glabrous within; filaments glabrous. **L. harknessii** (Curran) Greene
- 3b. Corolla mostly 2–5 times as long as calyx, with a hairy ring within or filaments hairy at the base. (4)
- 4a. Corolla 2–4 mm long; hairs on throat at base of stamens. **L. septentrionalis** Mason
- 4b. Corolla 10–30 mm long; hairs on base of filaments. **L. liniflorus** (Benth.) Greene

Microsteris Greene

Small annuals; leaves entire, the lower opposite and the upper alternate; flowers small, pedicellate, usually in pairs in upper axils, white to pinkish purple, 8–12 mm long; calyx 5-lobed, lobes subequal; stamens included, unequally inserted in corolla tube; ovary globose, 3-loculed; stigma 3-lobed. **M. gracilis** (Dougl. ex Hook.) Greene

Polemoniaceae

Navarretia R. & P.

1a. Main stems finely retrorse-pubescent with crisped white hairs; mature capsule thin-membranous; corolla white, usually 4-lobed; style 2-cleft; herbage not glandular. **N. minima** Nutt.
1b. Main stems with spreading hairs to almost glabrous or, if hairs retrorse, not crisped and appressed against the stems; mature capsule with thicker wall; corolla yellow, usually 5-lobed; style 3-cleft; herbage glandular-puberulent. **N. breweri** (Gray) Greene

Phlox L.

1a. Plants loosely tufted; shoots usually over 10 cm long, little branched; flowers in cymes. (2)
1b. Plants more or less caespitose; shoots tending to be short and branched, mostly less than 10 cm long; flowers solitary or in few-flowered clusters. (4)
2a. Corolla tubes 20–25 mm long. **P. stansburyi** (Torr.) Heller
2b. Corolla tubes less than 20 mm long. (3)
3a. Plants often with glandular hairs; membrane at junction of calyx teeth gibbous keeled; widely distributed. **P. longifolia** Nutt.
3b. Plants lacking glandular hairs; membrane at junction of calyx teeth flat or nearly so; northeastern Utah. **P. grahamii** Wherry
4a. Leaves flat or nearly so, 1–2 mm broad or more. (5)
4b. Leaves more or less keel shaped, up to 1 mm broad. (7)
5a. Leaves glabrous; corolla about 2 cm long, the lobes equaling the tube. **P. multiflora** A. Nels.
5b. Leaves more or less glandular; corolla lobes 5–8 mm long. (6)
6a. Corolla tube pilose externally. **P. gladiformis** (M. E. Jones) E. Nels.
6b. Corolla tube glabrous externally. **P. caespitosa** Nutt.
7a. Leaves beset with woolly hairs; corolla tube twice as long as calyx. (8)
7b. Leaves ciliate, glandular or pubescent, not woolly; corolla tube exceeding calyx. (9)
8a. Leaves imbricate, at length spreading, 5–10 mm long. **P. hoodii** Richards.
8b. Leaves densely imbricate, appressed, up to 3 mm long, **P. bryoides** Nutt.
9a. Calyx glandular. **P. rigida** Benth. ex DC.
9b. Calyx not glandular, more or less pubescent. **P. austromontana** Coville

Polemonium L.

1a. Corolla limb definitely shorter than tube; corolla tubular to funnelform; high elevations in mountains. **P. viscosum** Nutt.
1b. Corolla limb exceeding or subequal to tube; corolla campanulate to rotate from a short funnelform tube. (2)

Figs. 81-82.　81. **Plantago lanceolata**, x .21.　82. **Collomia linearis**, x .29

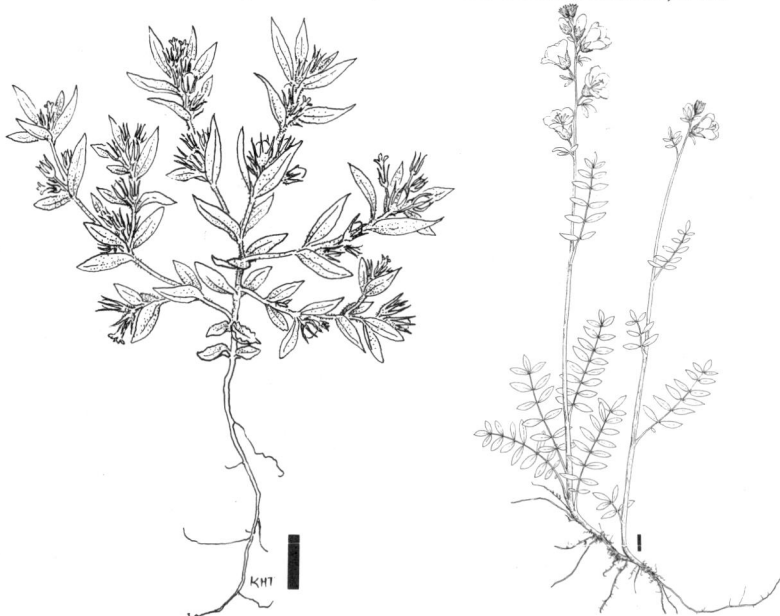

Figs. 83-84.　83. **Microsteris gracilis**, x .74.　84. **Polemonium coeruleum**, x .22

Polygonaceae

2a. Leaves predominantly cauline. (3)
2b. Leaves predominantly basal. (5)
3a. Corolla slightly exceeding or shorter than calyx. **P. micranthum** Benth.
3b. Corolla greatly exceeding calyx. (4)
4a. Terminal leaflets discrete. **P. coeruleum** L.
4b. Terminal leaflets confluent. **P. foliosissimum** Gray
5a. Plants erect; peduncles strict. **P. coeruleum** L.
5b. Plants caespitose; peduncles branching. (6)
6a. Calyx lobes about one and one-half times the length of calyx tube. **P. delicatum** Rydb.
6b. Calyx lobes equaling calyx tube. **P. pulcherrimum** Hook.

POLYGALACEAE — MILKWORT FAMILY

Perennial subshrubs; leaves simple, entire; flowers perfect, irregular; sepals 5; petals 3, the lower 2 forming a keel; stamens 6 or 8, the filaments united into a tube; anthers opening by apical pores; ovary superior; style simple; fruit a 2-loculed capsule.

Polygala L. Milkwort
1a. Plants intricately branched shrubs, usually much over 15 cm tall; flowers 4–5 mm long; leaves rarely over 3 mm wide. **P. acanthoclada** Gray
1b. Plants subshrubs, woody only at the base, usually less than 15 cm tall; flowers 7–10 mm long; leaves 3–6 mm wide. **P. subspinosa** S. Wats.

POLYGONACEAE — BUCKWHEAT FAMILY

Plants herbs, subshrubs, or shrubs, erect or climbing; leaves alternate, opposite or whorled; stipules sheathing or absent; flowers perfect or polygamo-dioecious, regular; perianth 2- to 6-parted or -cleft; stamens 2–9; styles 2 or 3; ovary superior, 1-loculed, 1-ovuled; fruit an achene.

1a. Sheathing stipules lacking; flowers subtended by a campanulate, turbinate, or cylindric involucre, or by a folded, 2-toothed bract. (2)
1b. Sheathing stipules (ocreae) present; flowers not subtended by an involucre or by a folded, 2-toothed bract. (5)
2a. Flowers solitary, subtended by a single, folded, 2-toothed bract. **Pterostegia**
2b. Flowers several (solitary in **Chorizanthe**), subtended by a campanulate, turbinate, or cylindric involucre. (3)
3a. Teeth or lobes of involucre not bristly or spiny apically. **Eriogonum**
3b. Teeth or lobes of involucre ending in bristles or spines. (4)

Subclass DICOTYLEDONEAE

4a. Involucre subtending 2 or more flowers, its teeth tipped with straight spines or bristles; upper bracts connate-perfoliate, forming a cup-shaped disk. **Oxytheca**
4b. Involucre usually subtending 1 flower, its teeth tipped with hooked or straight spines; bracts not connate-perfoliate; southwestern Utah. **Chorizanthe**
5a. Sepals 4; styles 2; leaf blades basal and reniform. **Oxyria**
5b. Sepals 5 or 6; styles 2 or 3; leaf blades various but not basal and reniform. (6)
6a. Sepals 5 (rarely 4), all similar and erect in fruit. **Polygonum**
6b. Sepals 6, in 2 sets—inner ones erect and winged in fruit, or wings from the achenes, outer ones reflexed and often smaller. (7)
7a. Sheaths (ocreae) large and prominent; stamens 8–10; leaf blades ovate to orbicular; cultivated and persisting. **Rheum**
7b. Sheaths (ocreae) not prominent, evanescent; stamens 6; leaf blades narrower; native or introduced weeds, not cultivated. **Rumex**

Chorizanthe R. Br.

1a. Bracts 2-lobed; involucres small, with 3 broad, spreading spurs at the base. **C. thurberi** (Gray) S. Wats.
1b. Bracts entire; involucres not spurred. (2)
2a. Involucres 6-toothed, the tube 6-ribbed (angled), the teeth strongly hooked apically, less than 2 mm long; stems very brittle and soon falling apart; foliage leaves all basal; stem leaves reduced to subulate bracts. **C. brevicornu** Torr.
2b. Involucres with fewer than 6 teeth, these to 1 cm long or more, the tube either 3-ribbed (angled) or not ribbed; stems not very brittle; lower stem leaves like the basal ones. **C. rigida** (Torr.) Torr.

Eriogonum Michx. Wild Buckwheat (Contributed by James L. Reveal)

1a. Flowers with stipelike base, mostly yellow to reddish yellow or rarely cream, glabrous or pubescent; low, spreading, caespitose to shrubby perennials. KEY II, p. 267.
1b. Flowers not attenuated into a stipelike base, various; annuals or perennials. (2)
2a. Plants perennial. (3)
2b. Plants annual or, if perennial, with distinctly inflated stems and long-peduncled involucres bearing yellow, hirsute flowers. (4)
3a. Plants caespitose to large shrubs, not monocarpic; flowers white to yellow, usually enlarging in fruit; stems glabrous to floccose or tomentose; achenes not winged, usually hidden by flowers, brown to black. KEY I, p. 255.
3b. Plants tall, strict and erect, monocarpic perennials; flowers yellow, not enlarging in fruit; stems strigose; achenes distinctly winged and obviously exserted beyond flowers, usually yellowish. KEY III, p. 269.
4a. Involucres smooth, not ribbed or angled, usually distinctly pedun-

Polygonaceae

cled or, if sessile, then not pressed to the stems; annuals or perennials. KEY IV, p. 270.

4b. Involucres angled to strongly ribbed, strongly appressed to the stems and always sessile; strictly annuals. KEY V, p. 274.

KEY I. Plants caespitose to shrubby perennials.

1a. Plants shrubby or subshrubby to herbaceous perennials, not pulvinate or caespitose. (2)
1b. Plants pulvinate or caespitose. (52)
2a. Plants distinctly shrubby or subshrubby, woody above the caudex and not dying back completely to the ground after each year. (3)
2b. Plants herbaceous perennials, dying back completely after each growing season, the new growth arising from woody underground caudex branches. (32)
3a. Flowers pubescent, 2.5–3 mm long, white to pink; involucres 2.5–3.5 mm long, usually congested into tight ball-like clusters atop elongated branches; leaves narrowly linear or nearly so, often revolute, 6–18 mm long, (1) 2–6 mm wide, usually strongly fasciculate; shrubs and subshrubs up to 5 dm high and 8 dm across, flowering in spring and early summer; restricted to Washington Co. E. fasciculatum Benth. var. **polifolium** (Benth. in DC.) T. & G.
3b. Flowers glabrous; involucres usually not in densely congested heads, mostly solitary; leaves not in fascicles or, if so, then plants not from Washington Co. (4)
4a. Stems smooth, glabrous to tomentose; leaves usually persistent (at least some), the blades usually more than 1 cm long or, if less, then plants of eastern Utah; inflorescences with involucres arranged in loose to compact terminal cymes; widespread. (5)
4b. Stems angled or ribbed or, if smooth, then obviously scabrous, or, if smooth and tomentose, then inflorescence not as above; leaves usually not persistent and less than 1 cm long; inflorescences with involucres arranged in divaricately branched panicles; restricted to Washington Co. (30)
5a. Inflorescence large with numerous branches and branchlets bearing racemosely arranged involucres at their tips; stems glabrous to floccose or rarely densely tomentose; leaves usually narrowly elliptic to elliptic or, if broader, then densely tomentose on both surfaces; typically of sandy habitats. (6)
5b. Inflorescence small and compact, cymose, with involucres dichotomously arranged even at tips of branches; stems usually tomentose, rarely floccose or glabrous; leaves elliptic or orbicular, rarely linear to linear-oblanceolate; various habitats. (13)
6a. Leaves pubescent, at least below. (7)
6b. Leaves glabrous on both surfaces except for microscopic hairs on midveins in some. (12)

Subclass DICOTYLEDONEAE

7a. Leaves linear-lanceolate to oblanceolate or elliptic, 1.5–4.5 cm long; involucres sessile or, if peduncled, then stems glabrous. (8)

7b. Leaves rounded or nearly so, 0.5–1 cm in diameter, densely white-tomentose on both surfaces; stems densely lanate; involucres short-pedunculate; low, spreading subshrubs, 1–2 (2.5) dm high and 2–5 dm across, with white flowers 1.5–2 mm long; western Tooele and Juab cos. **E. nummulare** M. E. Jones

8a. Leaves linear-lanceolate to oblanceolate or narrowly elliptic, 1.5–4.5 cm long, 2–8 mm wide; branches floccose to glabrous; eastern Utah. (9)

8b. Leaves oblanceolate to elliptic, 1–2.5 (3) cm long, (5) 10–20 mm wide; branches glabrous or tomentose; western Utah. (11)

9a. Flowers yellowish; stems floccose to glabrate; large, spreading shrubs, 4–12 dm high and up to 2 m across, flowering in late summer; leaves 1.5–4 cm long and only 2–4 mm wide; Emery and Grand cos. southward to eastern Garfield and central San Juan cos. **E. leptocladon** T. & G. var. **leptocladon**

9b. Flowers white. (10)

10a. Stems thinly tomentose to floccose; large, spreading shrubs, 2–8 dm high and up to 2 m across, flowering in late summer; leaves 1.5–3 cm long and 3–8 (10) mm wide; common from southern Emery Co. southward into Kane and San Juan cos. **E. leptocladon** T. & G. var. **ramosissimum** (Eastw.) Reveal

10b. Stems glabrous; large, spreading shrubs, 0.5–1.3 m high and up to 2 m across, flowering in late summer; leaves 2–4.5 cm long and 2–8 (10) mm wide; infrequent in central Garfield and eastern Kane cos. **E. leptocladon** T. & G. var. **papiliunculum** Reveal

11a. Stems and branches tomentose; leaves scattered along lower stems, the blades 1–3 cm long, 6–15 mm wide; involucres sessile, 2–2.5 mm long; large to medium-sized, diffuse shrubs, 2–8 dm high and up to 1.5 m across, flowering in summer; scattered locations in Tooele, Juab, Millard, and Washington cos. **E. kearneyi** Tidestr.

11b. Stems and branches glabrous; leaves restricted to base of plant, the blades 1–2.5 (2.8) cm long, 8–17 mm wide; involucres on peduncles (2) 5–10 (12) mm long; small subshrubs, 2–4 dm high and up to 5 dm across, flowering in midsummer; known only from central Millard Co. **E. ammophilum** Reveal

12a. Stems and branches bright green; inflorescences composed of stout, firm branches divided 3–6 times; flowers bright yellow, 3–4 mm long; involucres (2.5) 3–3.5 mm long; leaves revolute, narrowly elliptic, 2.5–4.5 cm long, 6–10 mm wide; large, spreading shrubs, 4–10 dm high and up to 2 m across, flowering in late summer and early fall; endemic to San Rafael Desert, Emery Co. **E. smithii** Reveal

12b. Stems and branches yellowish green; inflorescences composed of slender, fragile branches divided 10–20 times; flowers pale yellow

Polygonaceae

to white, (2) 2.5–3 mm long; involucres 2–2.5 mm long; leaves flat, elliptic, 1.5–4.5 cm long, 6–12 mm wide; large, erect shrubs, 4–8 (10) dm high and 0.5–1 m across, flowering in late summer; known currently only from clay hills just southwest of Fredonia, Mohave Co., Arizona, but to be expected in Kane Co., Utah. **E. mortonianum** Reveal

13a. Leaves flat, not revolute or with margins rolled or thickened. (14)
13b. Leaves revolute. (27)
14a. Leaf blades 0.2–4 cm long. (15)
14b. Leaf blades 3–7 cm long. (23)
15a. Leaf apices sharply acute, the blades mostly narrowly elliptic or narrower, 1–8 mm wide; widespread low subshrubs. (16)
15b. Leaf apices slightly acute to rounded, the blades oblanceolate to elliptic or even orbicular, (0.5) 1–3 cm wide; obviously woody shrubs of eastern and southern Utah. (17)
16a. Leaves flat, 2.5–6 (8) mm wide; stems glabrous to floccose; low subshrubs, 2–4 dm high, flowering in summer; common in northern and western Utah. **E. microthecum** Nutt. var. **laxiflorum** Hook.
16b. Leaves, at least some, revolute, 1–2.5 (3) mm wide; stems tomentose to lanate; low subshrubs to rather tall and erect shrubs up to 1 m high, flowering in summer; rather common in eastern and southern Utah. **E. microthecum** Nutt. var. **foliosum** (T. & G.) Reveal
17a. Flowers white or brownish white. (18)
17b. Flowers yellowish. (22).
18a. Leaves oblanceolate to elliptic, 1–3 (4.5) cm long, 1–2 cm wide. (19)
18b. Leaves elliptic-oblong to ovate-orbicular, 1–3 (4) cm long, 1–3 (3.5) cm wide. (21)
19a. Involucres 1.5–2.5 mm long, 1–1.5 mm wide; stems and branches spreading into subglobose crowns; branches whitish-tomentose; leaves lanceolate to oblanceolate or elliptic, spreading outwardly from stem, the blades 1–3 (4.5) cm long, 0.5–1 (1.5) cm wide, on petioles 2–6 mm long; flowers white, 2–3 (3.5) mm long; low subshrubs or more frequently large, roundish shrubs to 8 dm high and 1.5 m across, flowering mainly in mid- and late summer (including E. revealianum Welsh); common in eastern Utah southward into western Kane Co. **E. corymbosum** Benth. in DC. var. **corymbosum**
19b. Involucres 2.5–3.5 mm long, 1.5–2 mm wide (see **E. thompsonae** var. **albiflorum** if plants with bright green, glabrous stems forming low, scraggly subshrubs on red clay hills near Zion National Park [46b]). (20)
20a. Stems and crown open and erect, the branches brownish-tomentose; leaves 2–3.5 cm long, 0.5–1.5 cm wide, brownish-tomentose, more or less erect and somewhat appressed to stem; erect shrubs, (3) 6–10 dm high and up to 1.3 m across, flowering in summer; flowers brownish white, 2.5–3 mm long; margin of Uinta Basin in mountains

Subclass DICOTYLEDONEAE

above 6,000 ft. elevation, in Duchesne, Uintah, and Utah cos. **E. corymbosum** Benth. in DC. var. **erectum** Reveal & Brotherson
- 20b. Stems and crown spreading, the branches silvery-tomentose; leaves 3–4 cm long, (0.5) 1–2 cm wide, silvery-tomentose, especially below, spreading away from stem; large shrubs, 8–12 dm high and up to 2 m across, flowering in summer and fall; flowers brownish white, 2–2.5 mm long; endemic to clay hills near Wellington, Carbon Co. **E. corymbosum** Benth. in DC. var. **davidsei** Reveal
- 21a. Plants greenish, thinly floccose; leaves ovate-orbicular, densely tomentose below, subglabrous to glabrous and greenish above, the blade 1–3.5 cm long and wide; flowers white, 2.5–3 mm long; large, massive shrubs, 3–12 dm high and up to 2.5 m across, forming a large, compact, densely branched crown of rigid interlocking branchlets, flowering in late summer and fall; eastern Utah from Emery and Grand cos. southward into San Juan Co. **E. corymbosum** Benth. in DC. var. **orbiculatum** (Stokes) Reveal & Brotherson
- 21b. Plants brownish white, usually tomentose; leaves elliptic-oblong, densely white-tomentose below, less so and often brownish-floccose above, (1.5) 2–3 cm long, 1–2.5 cm wide; flowers brownish white, 2–2.5 mm long; large, open shrubs, 5–10 dm high and up to 1.3 m across, with an open, moderately branched crown of stout branchlets, flowering in summer and early fall; eastern San Juan Co. **E. corymbosum** Benth. in DC. var. **velutinum** Reveal
- 22a. Involucres 1–2 mm long; flowers bright yellow, 1.5–2.5 mm long; stems floccose to glabrous; leaves elliptic, 1–4 cm long, 0.5–1.5 cm wide, tomentose below, thinly pubescent to glabrous and green above; shrubs or subshrubs to 1 m high and 1.3 m across, flowering mainly in late summer and early fall; southern Utah in Garfield and Beaver cos. southward. **E. corymbosum** Benth. in DC. var. **glutinosum** (M. E. Jones) M. E. Jones
- 22b. Involucres 2.5–3.5 mm long; flowers yellow to pale yellow, 2.5–3 mm long; stems tomentose to floccose; leaves lanceolate to elliptic, (1) 2–4 cm long, 0.5–1 cm wide, tomentose below, subglabrous to floccose or sparsely tomentose above; highly variable subshrubs up to 3.5 dm high and 4 dm across, flowering mainly in late summer; Duchesne and Utah cos. (including **E. corymbosum** var. **albogilvum** Reveal); a hybrid species resulting from interaction between varieties of **E. corymbosum** and **E. brevicaule**. **E. x duchesnense** Reveal
- 23a. Leaf apices mostly rounded, the blades oblanceolate to elliptic, 1–3 (5) cm long; various long-leaved phases of the species (see couplets 19–22). **E. corymbosum** Benth. in DC.
- 23b. Leaf apices acute, usually sharply so, the blades mostly lanceolate, usually more than 3 cm long. (24)
- 24a. Stems and branches subglabrous to tomentose; involucres tomentose without. (25)
- 24b. Stems and branches glabrous; involucres 2–3 mm long; flowers

Polygonaceae

cream to pale yellowish white, 2–3 mm long; leaves lanceolate, 3–6 cm long, 4–8 mm wide, densely white-tomentose below, subglabrous and greenish above, on petioles 5–10 mm long; erect subshrubs to nearly herbaceous perennials, 3–5 dm high, flowering in summer and early fall; restricted to Mowry Shale outcrops near and in Dinosaur National Monument. **E. saurinum** Reveal

25a. Involucres 2.5–3 mm long; inflorescences with several short branches, these subglabrous; flowers white, 3–3.5 mm long; leaves lanceolate, 3–5 cm long, 0.5–1 cm wide, on petioles 3–6 mm long; low, scraggly shrubs up to 5 dm high and 8 dm across, flowering in fall; restricted to Mancos Shale hills near Wellington, Carbon Co. **E. lancifolium** Reveal & Brotherson

25b. Involucres 3–4 mm long; inflorescences open with only a few long branches, these usually tomentose. (26)

26a. Involucres 3.5–4 mm long; flowers white, 3.5–4.5 mm long; leaves lanceolate, 3.5–7 cm long, 3–6 (8) mm wide, densely white-tomentose below, less so and greenish above, on petioles 5–10 (18) mm long; low, spreading subshrubs up to 4 dm high and 5 dm across, flowering in late summer; restricted to Bad Land Cliffs, Duchesne Co. **E. hylophilum** Reveal & Brotherson

26b. Involucres 3–3.5 mm long; flowers yellow, 2.5–3 mm long; leaves lanceolate to elliptic, (1) 2–4 cm long, 0.5–1 cm wide, tomentose below, subglabrous to floccose and green above; variable subshrubs up to 3.5 dm high and 4 dm across, flowering mainly in late summer; scattered locations in Duchesne and Utah cos.; a hybrid species resulting from interaction between varieties of **E. corymbosum** and **E. brevicaule**. **E. x duchesnense** Reveal

27a. Leaf blades 2–6 cm long. (28)

27b. Leaf blades 0.5–2 cm long. (29)

28a. Inflorescences densely cymose, 0.2–1.2 (1.5) dm long, glabrous and bright green; plants forming round, compact shrubs, 2–6 (8) dm high and 3–12 dm across, flowering in late summer; involucres 2–3 mm long; flowers white, 2.5–4 mm long; known from the Four Corners area of Colorado and to be expected in San Juan Co. **E. leptophyllum** (T. & G.) Woot. & Standl.

28b. Inflorescences open to densely cymose, 0.1–0.6 dm long, floccose or rarely glabrate and green; plants forming open, diffuse shrubs, 0.5–10 dm high and up to 5 dm across, flowering in summer and early fall; involucres 2–3 mm long; flowers white, 2–3 mm long; common throughout eastern and southern Utah. **E. microthecum** Nutt. var. **foliosum** (T. & G.) Reveal

29a. Plants low subshrubs, 1–2 dm high and 3–8 dm across, with thinly floccose to glabrous stems and branches, flowering in spring; leaves oblanceolate, 5–12 (15) mm long, 0.8–1.7 (2) mm wide, tomentose below, thinly pubescent and green above; inflorescences umbellate-cymose, more or less glabrous; involucres (3.5) 4–4.5 mm long;

Subclass DICOTYLEDONEAE

flowers white, 3–3.5 mm long; endemic to the Comb Wash area of San Juan Co. **E. clavellatum** Small

29b. Plants low, matted subshrubs, less than 1 dm high and up to 3 dm across, with tomentose branches, flowering in spring; leaves linear-oblanceolate to narrowly elliptic, 5–12 (15) mm long, 1–2 (3) mm wide, tomentose below, floccose to tomentose above; inflorescences umbellate-cymose, tomentose; involucres 2–4 mm long; flowers white, 2.5–4 mm long; Carbon Co. southward to southeastern Garfield Co. eastward into western Colorado. **E. bicolor** M. E. Jones

30a. Outer whorl of tepals obovate, narrowed at base; branches grayish-tomentose; inflorescence composed of mostly horizontal tiers of zigzag branchlets; leaves oblanceolate to oblong-lanceolate, 6–10 (15) mm long, 2–4 mm wide, tomentose; involucres 2–2.5 mm long; flowers pale yellowish white to white, 2–2.5 mm long; known from a single collection made in the 1870s supposedly from Washington Co. **E. plumatella** Dur. & Hilg.

30b. Outer whorl of tepals nearly orbicular, subcordate to truncate at base; branches green, glabrous, smooth or angled, scabrous in some; inflorescence composed of mostly rigid or spinescent, slender to stout branchlets; leaves linear-lanceolate to elliptic, 4–10 (12) mm long, 1.5–5 mm wide, tomentose below, less so or more commonly glabrous and green above; involucres 0.7–1.5 mm long; flowers yellowish white, 1.5–2 mm long. (31)

31a. Stems scabrous, not angled or ribbed, with racemosely arranged involucres 1–1.5 mm long at tips of often spinescent branchlets; plants low, stout shrubs, 3–6 dm high and up to 8 dm across, flowering in fall; leaves lanceolate, 5–8 mm long, 1.5–2 mm wide; Virgin Narrows, Washington Co. **E. heermannii** Dur. & Hilg. var. **subracemosum** (Stokes) Reveal

31b. Stems angled and finely scabrous, with numerous, highly congested, rigid branchlets bearing small, closely arranged clusters of involucres near their tips, the involucres 0.7–1.3 mm long; plants low subshrubs, 1–4 dm high and up to 6 dm across, flowering in summer; leaves linear-lanceolate to narrowly elliptic, 4–12 mm long, 2–5 mm wide; Washington Co. **E. heermannii** Dur. & Hilg. var. **sulcatum** (S. Wats.) Munz & Reveal

32a. Involucres not racemosely arranged along elongated branches. (33)
32b. Involucres racemosely arranged along elongated branches. (51)
33a. Stems and branches pubescent. (34)
33b. Stems and branches glabrous. (39)
34a. Plants tall, more than 1 dm high, not forming mats. (35)
34b. Plants short, less than 1 dm high, forming spreading mats. (38)
35a. Leaves linear, revolute or nearly so, (3) 5–10 cm long, 1–4 mm wide; stems up to 2 dm long; inflorescences capitate or more commonly subcapitate or umbellate; flowers yellow or cream, 2.5–3.5 mm long; widely scattered and highly variable herbaceous perennials up to 2

Polygonaceae

dm high, flowering in summer and early fall; typically of the foothills and lower mountains of central and northeastern Utah. **E. brevicaule** Nutt. var. **laxifolium** (T. & G.) Reveal

35b. Leaves oblanceolate to narrowly elliptic, not linear. (36)
36a. Inflorescences cymose; central Utah. (37)
36b. Inflorescences capitate; leaves oblanceolate to narrowly elliptic, 2–5 (7) cm long, 3–7 mm wide; stems 1–2.5 dm long, tomentose; involucres clustered, 3.5–4.5 mm long, tomentose; flowers cream to yellow, 2.5–3.5 (4) mm long; herbaceous perennials forming mats 1.5–3 dm across, flowering in spring; endemic to Cache Valley. **E. loganum** A. Nels.
37a. Inflorescences cymose, 3–7 cm long; leaves oblanceolate to elliptic, the blade 3–7 cm long, 3–5 mm wide; involucres 2–3 mm long; flowers yellow, 3–4 mm long; perennials, 1–1.5 dm high, flowering in midsummer; infrequent in Utah, Tooele, and Juab cos. **E. brevicaule** Nutt. var. **cottamii** (Stokes) Reveal
37b. Inflorescences compoundly divided, 3–10 cm long; leaves lanceolate to elliptic, the blade 1–4 (6) cm long, 3–10 mm wide; involucres 2–3.5 mm long; flowers cream to pale yellow, 2.5–3.5 mm long; herbaceous perennials, 1.5–4 dm high, flowering in late summer; restricted to clay hills in Sevier and Piute cos. southward into Iron Co. **E. spathulatum** Gray
38a. Inflorescences capitate to subcapitate or umbellate, with densely tomentose stems; leaves linear to narrowly oblanceolate, the blades 3–7 cm long, 1.5–5 mm wide, densely tomentose, at least below; involucres usually clustered, 3–4 mm long; flowers yellow or rarely cream, 2.5–3.5 mm long; low, matted plants, flowering in summer (including **E. chrysocephalum** Gray); low desert ranges and mountains up to 10,000 ft. in northern and northeastern Utah. **E. brevicaule** Nutt. var. **laxifolium** (T. & G.) Reveal
38b. Inflorescences cymose, 1–3 cm long, with floccose and green stems, rarely glabrous on a single individual; leaves linear to linear-oblanceolate, (0.5) 1–2 cm long, 1.5–2 (2.5) mm wide, sparsely tomentose below, floccose above; involucres solitary, 1.5–2 (2.5) mm long; flowers bright yellow, 1.5–2.5 mm long; low herbaceous perennials up to 1 dm high, flowering in summer; restricted to Grand Valley in Grand and Garfield cos. **E. contortum** Small ex Rydb.
39a. Leaves linear to narrowly lanceolate or oblanceolate, not broadly elliptic, spatulate, or rotund. (40)
39b. Leaves broadly elliptic to spatulate or rotund. (44)
40a. Leaves 3–10 cm long, 1–7 mm wide, linear to narrowly oblanceolate, occasionally revolute; stems 1–2.5 dm long, pale green; inflorescences cymose, (3) 5–25 cm long; involucres 2–4 mm long; flowers yellow, 2.5–3 mm long; common perennials, flowering in summer; foothills and lower mountains, northern Utah. **E. brevicaule** Nutt. var. **brevicaule**

Subclass DICOTYLEDONEAE

40b. Leaves 1–5 cm long, linear and tightly revolute or lanceolate to oblanceolate and flat; eastern and southern Utah. (41)

41a. Leaves 1–4 cm long, linear and tightly revolute; stems erect, 5–12 cm long, bright green; inflorescences densely cymose with numerous branches, 3–15 cm long; involucres 2–3 mm long; flowers bright yellow, 1.5–2 mm long; perennials up to 3.5 dm high, flowering in late summer; endemic to the Uinta Basin of Duchesne and Uintah cos. **E. viridulum** Reveal

41b. Leaves 1–5 cm long, lanceolate to oblanceolate and flat; flowers white or pale yellow. (42)

42a. Inflorescences narrowly cymose, 1.5–2.5 dm long, glabrous and pale green; leaves lanceolate, 1.5–2.5 cm long; stems erect, 1–2 dm long; involucres 2–2.5 mm long; flowers cream to pale yellow, 2–2.5 mm long; perennials up to 3.5 dm high, flowering in late summer; endemic to shale hills in eastern Uintah Co. and adjacent Colorado. **E. ephedroides** Reveal

42b. Inflorescences open and spreading, usually grayish or reddish; flowers white. (43)

43a. Leaves 1.5–3 cm long, 2–5 (7) mm wide, oblanceolate, on petioles 5–10 (12) mm long; stems somewhat spreading, 8–15 cm long, glabrous and grayish; involucres 1, turbinate, 3–4 mm long; flowers 3–3.5 mm long; perennials up to 3 dm high, flowering in late summer; endemic to low clay hills east of Monticello, San Juan Co. **E. humivagans** Reveal

43b. Leaves (2) 3–5 cm long, 2–4 mm wide, lanceolate, on petioles 1–2 cm long; stems erect, 1–1.5 dm long, glabrous and often reddish; involucres clustered, turbinate-campanulate, 2.5–3.5 (4) mm long; flowers 2–3 mm long; perennials up to 3 dm high, flowering in midsummer; endemic to the Roan Cliffs area, Grand Co. **E. intermontanum** Reveal

44a. Leaves elliptic to spatulate, densely white-tomentose below, floccose to glabrous and green above or, if tomentose on both surfaces, then the leaf blade tapering to the petiole; eastern Utah. (45)

44b. Leaves nearly rotund to ovate, densely white-tomentose on both surfaces. (50)

45a. Stems and branches bright green and glabrous; leaves oblong to elliptic, (2) 3–5 cm long, 8–15 mm wide, tomentose below, green and glabrous above, on petioles 3–7 cm long; stems erect, 12–25 cm long; inflorescences cymose, 1–3 dm long; involucres sessile, 2–3.5 mm long; flowers yellow or white, 3–3.5 mm long; erect herbaceous perennials or rarely subshrubs; southern Utah. (46)

45b. Stems and branches grayish; northeastern, central, or western Utah. (47)

46a. Flowers yellow; herbaceous perennials, 2–4 dm high, flowering in late summer and early fall; endemic to the Kanab area of Kane Co. and adjacent Arizona. **E. thompsonae** S. Wats. var. **thompsonae**

46b. Flowers white; herbaceous perennials or rarely subshrubs, 2–5 dm high, flowering in late summer and fall; Washington Co. **E. thompsonae** S. Wats. var. **albiflorum** Reveal
47a. Involucres not clustered; central Utah and along the Wasatch Front. (48)
47b. Involucres usually clustered, occasionally with some solitary involucres on some individual plants; eastern or western Utah. (49)
48a. Leaves crenulate, narrowly elliptic to elliptic, 1.5–4 cm long, (3) 4–7 mm wide, tomentose below, floccose above; stems erect, 1–3 dm long; inflorescences cymose, (8) 10–15 cm long; involucres 3–4 mm long; flowers white, 2–2.5 mm long; low, spreading herbaceous perennials, flowering in late summer; along the Wasatch Front in the mountains from Sanpete Co. northward to Salt Lake Co. **E. brevicaule** Nutt. var. **wasatchense** (M. E. Jones) Reveal
48b. Leaves not crenulate, elliptic to spatulate, 1–3 cm long, 5–8 (10) mm wide, tomentose below, slightly less so to floccose above; stems erect, 8–15 cm long; inflorescences cymose, 5–25 cm long; involucres 2–2.5 mm long; flowers white, 1.5–2.5 mm long; erect herbaceous perennials, flowering in mid- and late summer; endemic to clay hills in Piute and Sevier cos. (including **E. spathuiforme** Rydb.). **E. ostlundii** M. E. Jones
49a. Leaves not crenulate, the leaf blades elliptic, (1.5) 2–4 cm long; stems erect, 1–2 dm long; inflorescences cymose, 5–15 cm long; involucres 2–4 mm long; flowers white, 1.5–3 mm long; erect herbaceous perennials up to 4 dm high, flowering throughout the summer; common in scattered locations in eastern Utah from Uintah Co. southward to Garfield Co. **E. batemanii** M. E. Jones
49b. Leaves crenulate, the leaf blades elliptic, 0.5–2 cm long; stems spreading to weakly erect, 5–10 cm long; inflorescences subcapitate to cymose, up to 7 cm long; involucres 2.5–3 mm long; flowers white, (1.5) 2–3 mm long; low, spreading herbaceous perennials up to 2 dm high, flowering in mid- and late summer; endemic to Bull Mountain, Henry Mountains, Garfield Co. **E. cronquistii** Reveal
50a. Stems and branches tomentose; leaves orbicular to rotund, 0.5–1 cm in diameter; flowering stems 3–7 cm long; involucres 1.5–2 mm long, tomentose, on slender peduncles 2–8 (10) mm long; flowers white, 1.5–2 mm long; low subshrubs (but appearing to be merely herbaceous at times) up to 2.5 dm high; endemic to sandy areas in western Tooele and Juab cos. **E. nummulare** M. E. Jones
50b. Stems and branches glabrous; leaves ovate to round, 1.2–2 cm long, 1–1.7 cm wide; flowering stems 5–20 cm long; involucres 2.5–4 mm long, sessile; flowers white, 2.5–3 mm long; erect herbaceous perennials up to 4.5 dm high; endemic to clay hills in southwestern Millard Co. **E. eremicum** Reveal
51a. Stems tomentose, or, if glabrous, then stems not fistulous or green, up to 3 dm long, thinly floccose in some; leaves elliptic to ovate

Subclass DICOTYLEDONEAE

or oval, (1.5) 2–6 (10) cm long, 1–2.5 (3.5) cm wide, lanate to tomentose below, floccose to glabrous above; inflorescences bearing 5–20 or more racemosely arranged involucres, these (2) 3–5 mm long; flowers white to cream, (2) 2.5–5 mm long; tall, erect perennials up to 10 dm high, flowering mainly in midsummer; common in foothills and mountains throughout most of Utah. **E. racemosum** Nutt.

51b. Stems glabrous, usually fistulous, bright green, up to 2.5 dm long; leaves elliptic or oblong-ovate to ovate, 2–4.5 cm long, 1.5–2.5 (3) cm wide, lanate to tomentose below, thinly floccose to glabrous above; inflorescences bearing 8–15 racemosely arranged involucres, these 1.5–3 mm long; flowers white, 2–3.5 (5) mm long; weakly erect herbaceous perennials up to 6 dm high, flowering in fall; in deep shade or in protected areas, endemic to Zion National Park and adjacent parts of western Kane Co. **E. zionis** J. T. Howell

52a. Involucres ebracteate, the bracts (1) 3–5 mm below the base of involucre; leaves linear-oblanceolate to narrowly elliptic, 5–12 (15) mm long, 1–2 (3) mm wide; flowers white, 2.5–4 mm long, the tepals dissimilar, the outer whorl broadly obovate to nearly orbicular; clay hills of eastern Utah. **E. bicolor** M. E. Jones

52b. Involucres with bracts immediately below the involucre or, if ebracteate, then plants densely pulvinate and with pilose flowers. (53)

53a. Tepals similar or nearly so, glabrous or pubescent. (54)
53b. Tepals distinctly dissimilar, the outer whorl of tepals much larger and wider than the narrower, but as long to slightly longer inner whorl, glabrous. (66)

54a. Plants 1.5–3 dm high; leaves oblanceolate to narrowly elliptic, 2–5 (7) cm long, 3–7 mm wide; flowers white or yellow; spreading, matted perennials with tomentose stems; restricted to Cache Valley. **E. loganum** A. Nels.

54b. Plants less than 1.5 dm high or, if taller, then leaves not as above; desert ranges and mountains of Utah. (55)

55a. Leaves linear to linear-oblanceolate, (0.5) 1–2 cm long, 1.5–2 (2.5) mm wide, usually revolute; stems and branches floccose or rarely glabrous, bright green; flowers bright yellow, glabrous; inflorescences cymose; clay hills in Grand and Garfield cos. **E. contortum** Small ex Rydb.

55b. Leaves not linear or, if so, flowers not yellow or glabrous and plants not of clay hills in eastern Utah; inflorescences capitate or, if branched or divided, then flowers pilose without. (56)

56a. Flowers glabrous. (57)
56b. Flowers villous or pilose without. (62)

57a. Flowers yellow or pale yellowish or, if whitish, then plants of Box Elder Co. (58)
57b. Flowers white; southern Utah. (61)

Polygonaceae

58a. Stems glabrous to floccose; leaves plane and usually crenulate. (59)
58b. Stems densely tomentose or, if floccose, then plants of the Wasatch Plateau; leaves usually revolute or, if flat, then margins not crenulate. (60)
59a. Flowers bright yellow, 1.5–2.5 mm long; leaves narrowly oblanceolate to narrowly elliptic, the blades 0.5–1.5 (2) cm long, 2–4 mm wide, tomentose below, glabrous and green above, the margin crenulate; stems scapose, 0.4–15 cm long, glabrous or sparsely floccose; involucres 2–3 mm long, floccose; flowering in summer; endemic restricted to the Wasatch Mountains mostly above 10,000 ft. from Mt. Nebo northward to southern Box Elder Co. **E. grayi** Reveal
59b. Flowers greenish white to pale yellow, 2–3 mm long; leaves broadly elliptic, the blades 3–10 mm long, 2–4 (5) mm wide, tomentose below, subglabrous to glabrous above, the margin crenulate; stems scapose, 6–12 cm long, glabrous; involucres 1.5–2.5 mm long, glabrous; flowering in summer; endemic known only from Willard Peak, Box Elder Co. **E. nanum** Reveal
60a. Leaves plane and flat, not revolute, elliptic to broadly elliptic, 1–2.5 cm long, (3) 5–8 mm wide, tomentose on both surfaces; stems scapose, (2) 3–6 cm long, densely tomentose; involucres 2.5–4 mm long, tomentose; flowers bright yellow, 2.5–3 mm long; low, matted perennials, flowering in early summer; just entering Utah in extreme western Box Elder and Tooele cos. **E. desertorum** (Maguire) R. J. Davis
60b. Leaves revolute, linear to narrowly oblanceolate, 2.5–5 cm long, 1.5–3 (5) mm wide, tomentose below, floccose to subglabrous above; stems scapose, 0.5–1.5 (2) dm long, tomentose to floccose or glabrous; involucres 3–4 mm long, tomentose or rarely glabrous; flowers yellow, 2.5–3.5 mm long; low, matted perennials, flowering in summer; desert ranges and in foothills and mountains of the Wasatch Range and especially on the Wasatch Plateau to 10,000 ft. **E. brevicaule** Nutt. var. **laxifolium** (T. & G.) Reveal
61a. Leaves 1.5–6 (7) cm long, 2–5 mm wide, narrowly oblanceolate, tomentose below, floccose to subglabrous and green above, the margin entire or crenulate, flat or slightly inrolled in some; stems scapose, 8–25 (30) cm long, glabrous; involucres (2) 2.5–3 mm long, glabrous; flowers 2–2.5 mm long, white; low caespitose perennials forming loose mats up to 2 dm across, flowering in late spring; restricted to clay hills in southern Utah from Sevier Co. southward to Washington and Kane cos. **E. panguicense** (M. E. Jones) Reveal var. **panguicense**
61b. Leaves 0.5–1.5 cm long, the margin always crenulate; stems 2–7 cm long; flowers (2) 2.5–3 mm long; caespitose perennials forming compact mats, flowering in midsummer; restricted to the Cedar Breaks area of Iron Co. **E. panguicense** (M. E. Jones) Reveal var. **alpestre** (Stokes) Reveal

Subclass DICOTYLEDONEAE

62a. Ovaries and achenes glabrous. (63)

62b. Ovaries and achenes pubescent. (65)

63a. Flowers white, becoming rustic to rose or red with age, 3–4.5 mm long; involucres 6- to 8-lobed with 5–12 flowers. (64)

63b. Flowers yellow, 1.8–2.3 mm long, densely pilose without; involucres 4-lobed, 2.8–3.2 mm long, villous, with 2–4 flowers; stems lacking, the involucres in the center of the leaves; leaves forming tuftlike rosettes, the blades oblanceolate, 1–3.5 mm long, 0.9–1.2 mm wide, silky-pilose, revolute; low caespitose perennials with 20–50 rosettes of leaves forming a low, rounded mound 7–14 cm across, flowering in spring to early summer; rare, known only from the Red Canyon area and near Widtsoe, Garfield Co. **E. aretioides** Barneby

64a. Leaves in loose tuftlike rosettes, the blades narrowly elliptic, 4–10 mm long, 1–2 mm wide, silky-tomentose; stems 2–5 (8) cm long, villous, prostrate; inflorescences subcapitate to cymose-umbellate; involucres 4–5 mm long, villous, 6- to 10-lobed; flowers 3–4.5 mm long, densely pilose without; plants loosely caespitose with 10–20 rosettes, forming an indistinct mat 1–6 cm across, flowering in spring and early summer; rare, restricted to the desert ranges of western Utah from Millard Co. southward to Kane Co. **E. villiflorum** Gray

64b. Leaves in tight tuftlike rosettes, the blades oblanceolate to elliptic, 3–4 mm long, 0.7–1 mm wide, silky-tomentose; stems 1–9 mm long, villous, erect; inflorescences capitate; involucres 2–4 mm long, villous, 7- to 10-lobed; flowers 3–4 mm long, pilose without; plants densely pulvinate and forming hemispheric cushions of several hundred rosettes, forming distinct mounds or mats 1–4 dm across, flowering in spring and early summer; rather common, restricted to pinyon-juniper woodlands in Duchesne and Emery cos. **E. tumulosum** (Barneby) Reveal

65a. Leaves 2–5 (6) mm long, 2–4 mm wide, elliptic, tomentose; stems scapose, up to 2 cm long, erect, floccose to tomentose; involucres 2–3.5 mm long, the lobes 0.5–2 mm long, floccose to tomentose; flowers white or yellow, 2.5–3.5 (4) mm long, densely pilose without; achenes sparsely pubescent; low, spreading perennials, forming pulvinate mats up to 2.5 dm across, flowering in spring and early summer; foothills and low ranges in the deserts of western Utah. **E. shockleyi** S. Wats. var. **shockleyi**

65b. Leaves (4) 5–12 mm long, 2–6 mm wide, oblanceolate to spatulate; stems up to 3 cm long; involucres (3) 4–6 mm long, the lobes (1) 2–3 mm long; flowers white, 3–4 mm long; achenes usually densely pubescent; perennials, forming mats up to 4 dm across, flowering in spring and early summer; mainly pinyon-juniper woodlands of eastern Utah from the Uinta Basin southward to Kane and San Juan cos. **E. shockleyi** S. Wats. var. **longilobum** (M. E. Jones) Reveal

266

Polygonaceae

66a. Leaves mostly over 1 cm long; scapes mostly 5–30 cm long; involucres (3.5) 4–7 mm long, turbinate; flowers yellow or white, (3) 4–7 mm long; common, lower elevations. (67)

66b. Leaves 2–8 mm long and wide, broadly elliptic to rotund, densely lanate and white on both surfaces; scapes 0.3–5 cm long, floccose to lanate; involucres turbinate-campanulate, 3–4.5 mm long; flowers white, 2–3 mm long, the tepals not as strongly dissimilar; low, matted caespitose plants forming mats up to 3 dm across; restricted to alpine regions mostly above 9,000 ft. in the Deep Creek Mountains of Juab and Tooele cos., and the Raft River Range of northwestern Box Elder Co. E. ovalifolium Nutt. var. nivale (Canby in Cov.) M. E. Jones

67a. Flowers white or cream, maturing purplish in some, 4–5 mm long; leaves obovate to oval and round, 0.5–2 (3) cm long, rarely longer and narrower; stems scapose, (4) 5–20 cm long; involucres 4–5 (6.5) mm long; perennials, flowering in spring and early summer; common, throughout most of Utah. E. ovalifolium Nutt. var. ovalifolium

67b. Flowers yellow, 4–5 mm long; leaves elliptic to spatulate, (2) 3–6 cm long, rarely smaller and broader; stems scapose, 5–30 cm long; involucres 4–6 (7) mm long; flowering in spring and early summer; common, mainly in northern Utah. E. ovalifolium Nutt. var. multiscapum Gand.

KEY II. Perennial plants with stipitate flowers.

1a. Flowers glabrous without. (2)
1b. Flowers pubescent without. (8)
2a. Stems without a whorl of bracts about the middle. (3)
2b. Stems with a whorl of leaflike bracts about halfway up the stem; leaves mostly oblanceolate, 2–5 cm long, 4–10 (15) mm wide, lanate to tomentose below, thinly floccose to glabrous above; stems erect, 1–3 dm long; inflorescences simple or compoundly umbellate, tomentose to floccose; involucres 2–4.5 mm long with long reflexed lobes; flowers white to cream, 4–9 mm long including the 1.5–3 mm long stipe; achenes sparsely pubescent at the beak; common, loosely matted herbaceous perennials up to 5 dm high, arising from spreading, woody caudex branches and flowering mainly in late spring and summer; throughout mountains of northern Utah. E. heracleoides Nutt.

3a. Primary branches of inflorescence simple, not divided or compoundly umbellate but, if so, then flowers cream colored and plants of northern Utah. (4)

3b. Primary branches of the inflorescence branched 1 to several times, forming a compound umbel; leaves elliptic, 1–3 cm long, sparsely pubescent on both surfaces or occasionally glabrous; stems 0.5–2 dm long, floccose to glabrous; involucres 2–3 (3.5) mm long with reflexed lobes 1–3 mm long; flowers bright yellow, rarely cream

Subclass DICOTYLEDONEAE

        colored in Washington Co., 6–7 mm long, including the stipe; tall herbaceous perennials up to 7 dm high, arising from more or less erect, woody caudex branches; restricted to southern half of Utah. **E. umbellatum** Torr. var. **subaridum** Stokes

4a. Flowers bright yellow. (5)
4b. Flowers mostly whitish to cream or pale yellowish. (7)
5a. Leaves tomentose below, floccose to glabrous above, elliptic or nearly so, 1–2 cm long; stems mostly 1–2 dm long, tomentose to floccose; involucres 2–3.5 mm long with reflexed lobes 1.5–3 mm long; flowers 4–7 mm long, including the stipe; common herbaceous perennials up to 5 dm high, arising from spreading, woody caudex branches, flowering throughout summer; mountainous areas throughout most of Utah. **E. umbellatum** Torr. var. **umbellatum**
5b. Leaves glabrous on both surfaces. (6)
6a. Inflorescences umbellate to subcapitate, up to 3 cm long; leaves elliptic, 1–2 cm long, glabrous or with scattered hairs on the midrib or margin; stems erect, 0.7–2 dm long, thinly floccose to glabrous; flowers 4–7 mm long, including the stipe; common herbaceous perennials up to 5 dm high, arising from spreading to weakly erect, woody caudex branches and flowering throughout summer; foothills and middle elevations in the mountains of central and northern Utah. **E. umbellatum** Torr. var. **aureum** (Gand.) Reveal
6b. Inflorescences capitate or nearly so, up to 1 cm long; leaves elliptic to spatulate, 0.5–1.5 cm long, glabrous or merely thinly floccose below on immature blades; stems erect or prostrate, up to 5 cm long, thinly floccose to glabrous; flowers 3–6 mm long, including the stipe; infrequent herbaceous perennials up to 1 dm high, arising from compact, spreading, woody caudex branches and flowering mainly in midsummer; alpine regions of northern and central Utah. **E. umbellatum** Torr. var. **porteri** (Small) Stokes
7a. Leaves thinly tomentose to tomentose below, floccose to subglabrous or rarely glabrous above, the tomentum whitish, the blades elliptic to ovate-elliptic, 1–2 (2.5) cm long; stems 1–2 dm long, mostly floccose; inflorescences umbellate or rarely compoundly umbellate, up to 4 cm long; involucres 2–3 mm long with reflexed lobes 1–2.5 mm long; flowers pale yellow, cream or whitish, 4–8 mm long, including the stipe; plants up to 5 dm high, arising from more or less erect, woody caudex branches, flowering mainly in midsummer; infrequent in desert ranges and along foothills of the western Wasatch Mountains of northern and western Utah. **E. umbellatum** Torr. var. **dichrocephalum** Gand.
7b. Leaves densely matted, whitish or more frequently brownish, lanate below, green to olive green and thinly floccose to glabrous and shiny above, the blades oblanceolate to elliptic, (0.3) 0.5–2 (4) cm long; stems 1.5–3 dm long, brownish-floccose; inflorescences umbellate, up to 9 cm long; involucres 2–3.5 mm long with reflexed lobes

Polygonaceae

1–4 mm long; flowers cream colored, 3–6 mm long, including the stipe; commonly densely matted, highly spreading herbaceous perennials up to 5 dm high, arising from prostrate, spreading, woody caudex branches, flowering throughout the summer (including **E. subalpinum** Greene); common in foothills and mountains of northern Utah. **E. umbellatum** Torr. var. **majus** Hook.

8a. Stems ebracteate and scapose, (1) 3–8 (10) dm long, floccose to glabrous; leaves elliptic to obovate or oblong-spatulate to nearly oval, 2–10 (15) mm long, 1.5–4 (5) mm wide, tomentose, rarely less so above; inflorescences capitate; involucres 2–3.5 mm long with reflexed lobes 2–3.5 mm long; flowers yellow, becoming reddish in some, 2.5–5 mm long in anthesis, up to 10 mm long in fruit; perennials forming mats up to 4 dm across, flowering in spring; desert ranges of western Utah. **E. caespitosum** Nutt.

8b. Stems bracteate and not scapose, 0.2–2 dm long, tomentose to floccose; leaves oblanceolate to elliptic, 1–3 cm long, 0.3–1.5 cm wide, tomentose below, floccose above; inflorescences subcapitate to umbellate, up to 2 dm long, floccose; involucres 3–7 mm long with erect lobes less than 1 mm long; flowers yellow, 3–8 mm long, including the stipe. (9)

9a. Leaves oblanceolate to elliptic, 1–3 cm long, 0.5–1.5 cm wide; inflorescences up to 2 dm long, divided 1–3 times or more; involucres 3–7 mm long; flowers 5–8 mm long; spreading to erect plants forming mats up to 5 dm across, flowering from midsummer to early fall; eastern Utah from Duchesne and Uintah cos. southward to San Juan Co. **E. jamesii** Benth. in DC. var. **flavescens** S. Wats.

9b. Leaves elliptic, 1–1.5 cm long, 0.3–0.8 cm wide; inflorescences up to 6 cm long, if divided, then only once; involucres 3–4 mm long; flowers 3–6 mm long; densely matted perennials forming mats up to 5 dm across, flowering in midsummer; restricted to sandstone outcrops and blow sand in Zion National Park, Washington and Kane cos. **E. jamesii** Benth. in DC. var. **rupicola** Reveal

KEY III. Tall monocarpic perennials with winged fruits.

Only a single species in the Utah area; plants 5–15 dm high, arising from a soft, chambered taproot; leaves linear-lanceolate to lanceolate or oblanceolate, 5–15 cm long, 0.3–1.5 cm wide, strigose below, becoming glabrous above in most; stems erect, 2–10 dm long, strigose to nearly glabrous; inflorescences open paniculate cymes, 2–8 dm long, sparsely strigose to glabrous; peduncles erect, slender, 0.5–3 cm long; involucres 2–4 mm long; flowers yellow, 1.5–2.5 mm long in early anthesis, becoming 3–6 mm long in late anthesis or early fruit; achenes yellowish, 5–9 mm long, glabrous, distinctly winged; flowering from late spring to late fall (including **E. triste** S. Wats.); common throughout southern and eastern Utah. **E. alatum** Torr. in Sitgr.

Subclass DICOTYLEDONEAE

KEY IV. Annuals (rarely perennials) with smooth involucres.

1a. Leaves glabrous, pilose, hispid or villous on one or both surfaces. (2)
1b. Leaves tomentose to lanate, at least below. (8)
2a. Flowers pubescent without, yellow, with pilose to hirsute hairs. (3)
2b. Flowers glabrous, white, 1–2.5 mm long; leaves basal, the blades obovate to round or reniform, 1–5 cm long, 1–5 cm wide, sparsely villous to hirsute or glabrous; stems 5–15 cm long, glabrous to sparsely hispid; inflorescences open, 5–30 cm long, glabrous to hispid; peduncles slender, erect, 0.5–2 cm long, glabrous; involucres 0.6–1.3 mm long; annuals, 1–4 dm high, flowering in summer; common, eastern Utah. E. gordonii Benth. in DC.
3a. Involucres forming distinct tubes. (4)
3b. Involucres composed of 2 distinct whorls of 3 foliaceous bractlike lobes. (7)
4a. Plants glabrous or, if glandular, glands small and infrequent and restricted to base of stems. (5)
4b. Plants glandular; leaves basal, the blades broadly elliptic to oval, 7–25 mm long, 5–20 mm wide, pilose-hirsutulous; stems 3–10 cm long, glandular throughout; peduncles slender, straight or curved, ascending to erect, 3–10 mm long, glandular only on the lower half; involucres 1.3–2 mm long, glabrous, 5- (rarely 4-) lobed; flowers yellow, 1–1.5 (2) mm long, pilose; annuals, 0.5–3 dm high, flowering in summer; rare, restricted to scattered locations in desert ranges of western Utah. E. howellianum Reveal
5a. Involucres 5-lobed; plants annual or perennial with open inflorescences, the lower nodes with 3–5 branchlets or major branches, the upper nodes dichotomous or trichotomous; flowers yellow, 1–3 mm long; southern and eastern Utah. (6)
5b. Involucres 4-lobed; plants strictly annual with verticillate whorls of branches at each node, especially at lower nodes, often with 5–20 radiating secondary branchlets at lower ones; leaves basal, the blades round-oblong to rounded, 1–2.5 cm long, 1–2 cm wide, hirsute; stems erect, slender or rarely slightly fistulous, 0.5–1.5 (2) dm long; peduncles capillary, 5–15 mm long, glabrous; involucres 0.7–1 mm long; flowers 1–2 (2.5) mm long, short-hirsute; weedy annuals, flowering from spring to fall; common, Washington Co. E. trichopes Torr. in Emory
6a. Plants perennials or first-year flowering perennials, (0.5) 2–10 dm high; leaves basal, the blades oblong to round or reniform, 1–2.5 cm long, 1–2 cm wide, short-hirsute; stems erect, grayish, usually inflated; peduncles capillary, erect, 5–20 mm long; involucres 1–1.5 mm long; flowers (1) 2–2.5 (3) mm long; flowering from spring to summer; common across southern Utah and becoming less frequent in eastern Utah as far north as Uintah Co. E. inflatum Torr. & Frem. var. inflatum

Polygonaceae

6b. Plants strictly annuals, 0.5–4 dm high; leaves basal, the blades round to reniform, 0.5–2 cm long and wide, short-hirsute; stems erect, bright green, strongly inflated; peduncles filiform, erect, (5) 10–20 mm long; involucres 1–1.5 mm long; flowers 2–2.5 mm long; flowering in spring and early summer; common, Uinta Basin southward, and becoming less frequent to Kane and San Juan cos. **E. inflatum** Torr. & Frem. var. **fusiforme** (Small) Reveal

7a. Leaves strictly basal, the blades orbicular, 0.5–2 cm long and wide, strigose when young, becoming glabrous; stems erect, 3–7 cm long, minutely glandular; inflorescences erect to spreading and open, 0.5–2.5 dm long, glabrous except for some glands at nodes; peduncles fifilorm, 1–3 cm long, flexed to an acute angle about three-fourths its length, glandular to the middle; involucres 2–3 mm long, usually glabrous; flowers yellow, 1.5–3.5 mm long, pilose; annuals, flowering in spring and early summer; infrequent, clay hills in eastern Utah. **E. flexum** M. E. Jones

7b. Leaves basal and cauline, the basal blades spatulate, (1) 2–4 cm long, (0.5) 1–2.5 cm wide, the cauline blades linear-lanceolate to oblanceolate, 0.5–4.5 cm long, 2–10 mm wide, glabrous; stems prostrate to suberect, 1–3 cm long, glabrous; inflorescences open, 0.5–2 dm long, glabrous; peduncles, when present, straight and erect, up to 4 cm long, glabrous; involucres 2–3 mm long; flowers yellow, 1.5–3 mm long, pilose; annuals, flowering from spring to early fall; infrequent, clay hills and flats in eastern Utah. **E. salsuginosum** (Nutt.) Hook.

8a. Leaves strictly basal or, if not, then leaves sheathing up stems and not at nodes. (9)

8b. Leaves basal and cauline. (21)

9a. Involucres 1–3 mm long or, if shorter, then flowers with saccate-dilated bases with the 2 pouches conspicuous in fruiting material. (10)

9b. Involucres (0.3) 0.5–1 mm long. (20)

10a. Flowers glabrous without. (11)

10b. Flowers glandular or sparsely pubescent without. (19)

11a. Outer tepals cordate at base, mostly oblong to orbicular or, if obtuse, then plants distinctly scaberulous or margins of the tepals entire and not crisped. (12)

11b. Outer tepals truncate to obtuse at base. (17)

12a. Involucres deflexed, sessile or peduncled. (13)

12b. Involucres erect or arising from side of stem and remaining horizontal. (16)

13a. Plants glabrous. (14)

13b. Plants glandular; leaves basal, the blades orbicular to cordate, 1–3 (4) cm long, (1.5) 2–4 (5) cm wide; stems stout, 2–7 cm long; inflorescences flat-topped to spreading; peduncles up to 15 mm long, strongly deflexed; involucres 1–2.5 mm long; flowers white to red-

Subclass DICOTYLEDONEAE

dish, 1–2.5 mm long, the tepals dissimilar, the outer whorl ovate to oblong; erect to spreading annuals, 5–30 (40) cm high, flowering from spring to fall; Washington Co. (including **E. parryi** Gray). **E. brachypodum** T. & G.

14a. Involucres narrowly turbinate to turbinate, 1.5–2 mm long; leaves basal, the blades cordate to reniform or nearly orbicular, 1–2.5 (4) cm long, 2–4 (5) cm wide; stems slender, 3–30 cm long; inflorescences erect or spreading, open to diffuse, 1–5 dm long; peduncles up to 15 mm long, strongly deflexed; flowers white to pink, 1–2.5 (3) mm long, the tepals dissimilar; erect to spreading annuals, (0.5) 1–7 dm high, flowering from spring to fall; western and southern Utah. (15)

14b. Involucres broadly campanulate to hemispheric, 1–2 mm long; leaves basal, the blades cordate to subreniform, (1) 2–5 cm long, 2–6 cm wide; stems slender, 0.5–3 cm long; inflorescences spreading to flat-topped; peduncles lacking; involucres strongly deflexed; flowers yellow to reddish yellow, 1.5–2 mm long, the tepals dissimilar, the outer whorl orbicular to hastate; erect to spreading annuals, 1–6 dm high, flowering from summer to early fall; throughout Utah. **E. hookeri** S. Wats.

15a. Tepals broadly cordate at base, obovate, the flowers (1.5) 2–2.5 (3) mm long; involucres mostly 2 mm long, sessile or peduncled; mainly in southern Utah, but sporadically elsewhere. **E. deflexum** Torr. in Ives var. **deflexum**

15b. Tepals truncate at base, oblong, not cristate (compare with 17a), the flowers 1–2 mm long; involucres mostly 1.5 mm long, sessile; mainly in western Utah, but sporadically elsewhere. **E. deflexum** Torr. in Ives var. **nevadense** Reveal

16a. Stems and branches glabrous, (0.5) 2–20 cm long; leaves basal, the blades subcordate to orbicular, 2–5 cm long and wide; inflorescences narrow and strict with long whiplike branches, (0.5) 1–8 dm long; peduncles erect, up to 2 mm long; involucres 2–2.5 (3) mm long; flowers white, 1.5–2 mm long, the tepals dissimilar, the outer whorl oblong; annuals up to 1 m high, flowering in summer and fall; rare, known only from Iron and Washington cos. **E. insigne** S. Wats.

16b. Stems and branches scaberulous, 5–15 cm long; leaves basal, the blades cordate, 1–3 cm long and wide; inflorescences spreading and flat-topped, 0.5–4 dm long; peduncles lacking; involucres 1.5–2.5 mm long, horizontally arranged on stems, bending downward in fruit; flowers white to rose or red, 1–1.5 mm long, the tepals dissimilar, the outer whorl obovate; annuals, flowering in fall; infrequent to rare, clay hills in eastern Utah from Grand Co. southward to Kane and San Juan cos. **E. scabrellum** Reveal

17a. Outer whorl of tepals pandurate, crisped along margins (compare with 15b); leaves basal or sheathing up stems, the blades round-

Polygonaceae

ovate to orbicular, (0.5) 1–2 cm long and wide; stems 0.3–2 dm long, glabrous; inflorescences open, erect to spreading, 0.5–5 dm long, glabrous; peduncles lacking in some, otherwise cernuous, spreading or ascending, 1–25 mm long, glabrous; involucres turbinate, (1) 1.5–2 mm long; flowers white to pinkish, 1–2 mm long; annuals up to 6 dm high, flowering from spring to fall; common throughout Utah. (18)

17b. Outer whorl of tepals oblong to oval, entire margined; leaves strictly basal, the blades round to broadly reniform, 5–25 mm long and wide; stems 3–15 cm long, sparsely glandular; inflorescences mostly open and spreading, 5–20 cm long, glandular; peduncles curving downward, 3–10 mm long, glandular; involucres campanulate, 2–3 mm long; flowers white to rose or red, rarely pale yellowish, 2–3 mm long; annuals, flowering in summer; infrequent, widely scattered across northern Utah. **E. nutans** T. & G.

18a. Peduncles present; throughout Utah. **E. cernuum** Nutt. var. **cernuum**

18b. Peduncles lacking and thus involucres sessile throughout inflorescense; mainly desert valley floors of western Utah. **E. cernuum** Nutt. var. **viminale** (Stokes) Reveal in Munz

19a. Flowers short-hispidulous, the tepals dissimilar, the outer whorl of tepals saccate on each side of cordate base, yellow and 0.8–1 mm long in early anthesis, becoming 1.2–2 mm long and white to rose in fruit; leaves basal, the blades round to reniform, 5–20 mm long and wide; stems 2–10 cm long, glabrous except for glands at base; inflorescences open and spreading, 0.5–2.5 dm long, glabrous; peduncles capillary, 5–20 mm long, glabrous; involucres 0.6–1.2 mm long, glabrous; annuals, 0.5–3 dm high, flowering in spring; Washington Co. **E. thomasii** Torr.

19b. Flowers glandular, the tepals dissimilar, the outer whorl not saccate, yellow and 1–1.7 mm long in anthesis, becoming yellowish red and 2–2.5 mm long in fruit; leaves basal, the blades oblong-ovate to rounded, 0.5–2 (3) cm long, 0.4–2 (2.5) cm wide; stems 1–8 cm long, glabrous except for scattered glands at base; inflorescences open and spreading, 0.5–2.5 dm long, glabrous; peduncles slender, 1–3.5 (4) cm long, glabrous; involucres 1–1.5 mm long, glandular; locally common annuals, 0.5–3 dm high, flowering in spring; Washington Co. **E. pusillum** T. & G.

20a. Flowers yellow to red, 0.5–1.5 mm long, the tepals elliptic to obovate, glabrous; leaves basal, the blades oblong to orbicular, (0.5) 1–4 cm long, (0.5) 1–3 cm wide, floccose to glabrate above; stems 1–5 cm long, glabrous except for villous bases; inflorescences compact and dense, 0.5–2 dm long, glabrous; peduncles filiform, erect, (3) 5–10 mm long, glabrous; involucres (0.3) 0.5–1 mm long, 4-lobed; achenes 0.6–1 mm long, dark brown to black; low, spreading annuals, 0.5–2.5 dm high, flowering in summer; common, sandy areas in eastern Utah. **E. wetherillii** Eastw.

Subclass DICOTYLEDONEAE

20b. Flowers white to pink or rose, 0.8–2 mm long, the tepals lanceolate to ovate, glabrous or sparsely hirsute; leaves basal or sheathing up stem, the blades orbicular to reniform, (0.5) 1–3.5 cm long, 0.5–4 cm wide, hirsute above; stems 2–15 dm long, glabrous except for hispid bases; inflorescences open to diffuse, 0.5–4 dm long, glabrous; peduncles filiform, 0.5–2.5 cm long, glabrous; involucres 0.5–1 mm long, 5-lobed; achenes 1.7–2 mm long, light brown; erect annuals, 1–5 dm high, flowering from spring to summer; infrequent in southern Utah. **E. subreniforme** S. Wats.

21a. Flowers yellow, 1–3 mm long, glabrous, the outer whorl oblong-ovate with a large saccate base on each side of truncate to cordate base; leaves basal and cauline, linear-lanceolate to linear-oblanceolate, 1–2 (4) cm long, 2–4 mm wide, lanate below; stems villous; peduncles erect, (1) 2–5 cm long, sparsely villous to glabrous; involucres 1–2 mm long with 5 lanceolate lobes 1–3 mm long; rare annuals, flowering from late summer to fall, 1–2 dm high; restricted to desert ranges in southwestern Utah (an as yet undescribed variety of). **E. pharnaceoides** Torr. in Sitgr.

21b. Flowers white to pale yellowish or pink to red with a large, conspicuous, rose to purple midrib, 1–2.5 mm long, glandular, the outer whorl elliptic to roundish or obovate with an enlarged inflated area including most of surface below middle of tepal; leaves basal and cauline, the blades lanceolate to obovate, 1–3 (4) cm long, 1–1.5 (2) cm wide, tomentose below; stems tomentose; peduncles spreading, (5) 10–20 mm long, often glandular-puberulent; involucres 1–1.5 (2) mm long with 5 short, erect teeth; common annuals, flowering in spring and summer; desert ranges of western Utah. **E. maculatum** Heller

KEY v. Annuals with ribbed or angled involucres.

1a. Leaves tomentose, at least below. (2)
1b. Leaves puberulent to villous. (5)
2a. Stems glabrous, 0.5–1.5 (2) dm long; leaves basal, the blades round to reniform, 1–2 cm long and wide, tomentose below, floccose to glabrate above; inflorescences erect and rather strict, 0.5–2 dm long; involucres cylindric-turbinate, (2.5) 3–4 (5) mm long, glabrous; flowers white to pink, 1.5–2 mm long, glabrous, the tepals oblong-obovate; rare annuals, flowering in summer; scattered locations in Washington and Kane cos. **E. davidsonii** Greene
2b. Stems tomentose to floccose. (3)
3a. Leaves basal, not sheathing along stems; plants low and spreading. (4)
3b. Leaves cauline, the blades narrowly oblanceolate to elliptic, (0.5) 1–3 cm long, tomentose on both surfaces; inflorescences erect, narrow and strict, 1–4 (5) dm long, tomentose; involucres turbinate, 1.5–2.5 mm long, floccose; flowers white to pink, 1.5–2 mm long,

Polygonaceae

glabrous, the outer whorl of tepals broadly fan shaped, the inner whorl narrower; infrequent annuals, 1–5 dm high, flowering in summer and early fall; southern Washington and Kane cos. **E. polycladon** Benth. in DC.

4a. Flowers yellow to red, 1.5–2 (3) mm long, glabrous, the outer whorl of tepals broadly fan shaped, the inner whorl narrower; leaves basal, the blades rounded, 0.5–2 cm long and wide, tomentose below, floccose to tomentose above; stems spreading, 0.3–0.8 dm long, floccose; inflorescences dense, forming compact masses of numerous floccose branches with their tips often curving inward, 0.3–2.5 dm long; involucres turbinate, 1 mm long, floccose; rather common annuals, (0.5) 1–3 dm high, flowering in spring and summer; desert regions of western Utah. **E. nidularium** Cov.

4b. Flowers white, 1.5–2 mm long, glabrous, the outer whorl of tepals narrowly fan shaped, the inner whorl slightly narrower; leaves basal, the blades subcordate to cordate, 0.5–1.5 cm long, 0.5–2 cm wide, tomentose, less so to glabrate above; stems spreading, 0.3–0.8 dm long, floccose to tomentose; inflorescences open, forming a spreading mass of few floccose to tomentose branches with straight tips, 0.5–2.5 dm long; involucres campanulate, 1.5–2 mm long, floccose to tomentose; widely scattered and infrequent annuals, 1–3 dm high, flowering in summer; across southern Utah and occasionally in desert valleys and ranges of western Utah. **E. palmerianum** Reveal in Munz

5a. Stem leaves foliaceous at lower nodes, the blades elliptic-oblong to orbicular, 1–3 cm long, 1–2 cm wide, puberulent to short-pilose; stems spreading to decumbent, 3–5 cm long, puberulent; inflorescences spreading, 0.5–2.5 dm long, puberulent; involucres turbinate, 1–2 mm long, pilose, 5-lobed; flowers yellowish, 1.5–2 mm long, hispidulous or glandular, the tepals similar, oblong; infrequent annuals, 1–2 (3) dm high, flowering in summer and early fall; clay hills and flats in eastern Utah. **E. divaricatum** Hook.

5b. Stem leaves bractlike, the basal blades obovate to rounded, 0.5–1.5 cm long, 0.5–1.5 cm wide, sparsely villous, those of stem highly reduced; stems erect or spreading, 3–8 cm long, silky-puberulent; inflorescences open and spreading, 0.5–2.5 dm long, silky-puberulent; involucres turbinate, 1–1.5 mm long, villous, 4-lobed; flowers white to red, 1–1.5 mm long, glabrous to hispidulous, the tepals slightly dissimilar, the outer whorl abcordate, the inner whorl narrower; infrequent annuals, 0.5–3 dm high, flowering in spring and summer; central and southwestern Utah. **E. puberulum** S. Wats.

Oxyria Hill   Mountain Sorrel

Perennial herbs, 5–30 cm tall; leaves 1–3.5 cm in diameter; perianth 1.5–2 mm long, red or greenish; achenes broadly winged; high elevations in mountains. **O. digyna** (L.) Hill

Subclass DICOTYLEDONEAE

Oxytheca Nutt.

Plants annual with dichotomously branched stems, 1–3 dm tall; basal leaves oblong-oblanceolate, 1.5–4 cm long; bracts of upper nodes 3, connate-perfoliate into a single cuplike disk, 1–2 cm broad; southwestern Utah. **O. perfoliata** T. & G.

Polygonum L. Knotweed

- 1a. Plants twining. (2)
- 1b. Plants erect or spreading, not twining. (3)
- 2a. Plants annual; weed of cultivated and waste places. **P. convolvulus** L.
- 2b. Plants perennial; cultivated, semiwoody ornamental vine. Silver Lace-Vine. **P. aubertii** L. Henry
- 3a. Leaves with a hingelike joint at the attachment of blade and sheath; flowers in axillary clusters or, if in rather spikelike terminal clusters, then in the axils of bracts; ocreae 2-lobed or deeply cut. (4)
- 3b. Leaves without a hingelike joint; flowers in terminal, sometimes also axillary, spikelike clusters, the bracts of the inflorescence scarious and reduced to sheaths; ocreae not commonly 2-lobed or cut, may be ciliate. (11)
- 4a. Flowers crowded toward ends of branches, appearing to be terminal; bracts leaflike, usually broader than leaves. (5)
- 4b. Flowers scattered along the stems in small axillary clusters. (6)
- 5a. Leaves 4–10 mm long; styles lacking. **P. kelloggii** Greene
- 5b. Leaves 10–40 mm long; styles short but evident. **P. watsonii** Small
- 6a. Fruit reflexed. (7)
- 6b. Fruit erect or nearly so. (8)
- 7a. Perianth and achenes 1.5–2.6 mm long; leaves linear to linear-lanceolate. **P. engelmannii** Greene
- 7b. Perianth and achenes over 2.6 mm long; leaves linear-lanceolate or wider. **P. douglasii** Greene
- 8a. Stems decumbent or prostrate; achenes dark brown or reddish brown at maturity. **P. aviculare** L.
- 8b. Stems erect or ascending from a branching base; achenes black at maturity. (9)
- 9a. Leaves elliptic-lanceolate, ovate-lanceolate or broader, the upper leaves not reduced, rather crowded. **P. minimum** S. Wats.
- 9b. Leaves lanceolate, oblanceolate or narrower, the upper leaves reduced in length and width, rather scattered. (10)
- 10a. Upper bracts subulate; achenes smooth and shining; stems branching at base, seldom throughout. **P. sawatchense** Small
- 10b. Upper bracts linear-oblong, not subulate; achenes minutely roughened, not shining; stems usually branching throughout. **P. ramosissimum** Michx.
- 11a. Plants of high altitudes; ocreae open, oblique, lobed or lacerate;

Polygonaceae

      basal leaves long-petiolate, the cauline sessile or short-petiolate; inflorescence 1. (12)
11b. Plants of low and middle elevations; ocreae closed, truncate, ciliate or entire; leaves all cauline and similar. (13)
12a. Racemes bearing bulblets below, linear, 5–8 mm in diameter; basal leaves cordate or subcordate. **P. viviparum** L.
12b. Racemes lacking bulblets, oblong, over 1 cm in diameter; basal leaves cuneate. **P. bistortoides** Pursh
13a. Inflorescences usually 1, all terminal; plants perennial, usually aquatic or semiaquatic; flowers bright pink; stamens 5. (14)
13b. Inflorescences several to many, axillary as well as terminal; plants mostly annual, not really aquatic; flowers bright rose to white or green; stamens usually more than 5. (15)
14a. Inflorescence 1–3 cm long, over 1 cm wide. **P. amphibium** L.
14b. Inflorescence 3–10 cm long, less than 1 cm wide. **P. coccineum** Muhl. ex Willd.
15a. Sheaths (ocreae) without marginal bristles. **P. lapathifolium** L.
15b. Sheaths (ocreae) with marginal bristles. (16)
16a. Perianth not glandular-punctate, or obscurely so, pink to purple; inflorescence rather densely flowered, not interrupted (except occasionally near base). **P. persicaria** L.
16b. Perianth glandular-punctate, pale green or whitish; inflorescence lax and interrupted. (17)
17a. Racemes erect; achenes smooth and shining; leaves lanceolate to linear-lanceolate. **P. punctatum** Ell.
17b. Racemes nodding, at least in fruit; achenes granular to dull; leaves lanceolate to ovate-lanceolate. **P. hydropiper** L.

Pterostegia Fisch. & Mey.

      Plants annual; dichotomously branched stems 1–5 dm long; lower leaves fan shaped, usually 2-lobed; upper leaves entire or slightly toothed; involucre 2-lobed, 2–3 mm long in fruit, laciniate toothed; sepals usually 6; southwestern Utah. **P. drymaroides** Fisch. & Mey.

Rheum L.   Rhubarb

      Perennial herbs, 12–20 dm tall in flower; leaves mostly basal, with thick, edible petioles and large cordate-ovate leaves; cultivated and persisting. **R. rhaponticum** L.

Rumex L.   Dock   (Contributed by J. B. Karren)
1a. Leaves, at least some, hastate; achenes naked, exceeding the small sepals; herbage sour or acid to the taste; introduced weed, waste places. **R. acetosella** L.
1b. Leaves various but not hastate; achenes enclosed in valves (enlarged sepals) of various shapes and sizes; herbage sour or not sour. (2)

Subclass DICOTYLEDONEAE

- 2a. Flowers tending to be dioecious; leaves mostly basal; valves lacking grains (small raised tubercles); herbage sour or acid to the taste; high elevations in mountains. **R. paucifolius** Nutt. ex S. Wats.
- 2b. Flowers perfect or monoecious; leaves basal or cauline and alternate; valves with or without grains; herbage not sour or acid. (3)
- 3a. Valves very large, 5–30 mm wide, without grains. (4)
- 3b. Valves less than 5 mm wide, with or without grains. (6)
- 4a. Valves 20–30 mm wide; plants rhizomatous; occasional in sandy or saline soils. **R. venosus** Pursh
- 4b. Valves 5–12 mm wide; plants not rhizomatous. (5)
- 5a. Valves 6–12 mm wide; leaves not cordate at base; occasional in sandy soils at lower elevations. **R. hymenosepalus** Torr.
- 5b. Valves 5–6 mm wide; leaves cordate at base; occasional in moist situations. **R. occidentalis** S. Wats.
- 6a. Valves denticulate, with spines. (7)
- 6b. Valves entire, without spines. (8)
- 7a. Plants perennial; basal leaves broadly ovate and cordate; only 1 valve with a grain; introduced weed, waste places. **R. obtusifolius** L.
- 7b. Plants annual; basal leaves linear-lanceolate, usually cuneate at base and undulate on margins; each valve with a grain; occasional, widely distributed. **R. fueginus** Phil.
- 8a. Valves all possessing grains; stems usually several from 1 root; widely distributed, usually at higher elevations. **R. triangulivalvis** Rech.
- 8b. Valves not all possessing grains, usually only 1; stems usually single. (9)
- 9a. Leaves linear-lanceolate, usually less than 5 cm wide; grains on valves 1–3, large and swollen, quite variable; widely distributed introduced weed. **R. crispus** L.
- 9b. Leaves broadly lanceolate, usually more than 5 cm wide; grains on valves 1 (the other 2 may be slightly enlarged); a poorly understood species in Utah. **R. patientia** L.

PORTULACACEAE — PURSLANE FAMILY

Perennial or annual, more or less succulent (rarely somewhat woody) herbs; leaves alternate, opposite or basal, entire; flowers perfect, regular or nearly so; sepals commonly 2; petals 3–16, often more or less united at base; stamens few to many, opposite the petals; ovary commonly superior, 1-loculed; styles 2- to 8-cleft or distinct; fruit a circumscissile capsule or dehiscent by 2–3 valves.

- 1a. Capsule 2- to 3-valved, dehiscing from the apex. (2)
- 1b. Capsule splitting by means of a cap. (5)
- 2a. Sepals deciduous. **Talinum**
- 2b. Sepals persistent. (3)

Portulacaceae

3a. Capsule dehiscent by 2 valves; inflorescence secund (1-sided); style long, filiform; petals upon drying twist about the style. **Spraguea**
3b. Capsule dehiscent by 3 valves; inflorescence not secund or rarely inconspicuously so. (4)
4a. Plants with deep-seated corms or fleshy roots; ovules 6. **Claytonia**
4b. Plants fibrous rooted or reproducing by runners or bulblets; ovules 3. **Montia**
5a. Ovary completely superior; capsule splitting upward from line of dehiscence. **Lewisia**
5b. Ovary partly inferior; capsule not splitting in upper part. **Portulaca**

Claytonia L.   Spring Beauty

1a. Plants with fusiform, fleshy, purplish red taproots; basal leaves numerous, petioles winged; petals 5–10 cm long, white to pink, clawed; rock crevices and rock slides of higher mountains. **C. megarrhiza** (Gray) Parry ex S. Wats.
1b. Plants with globose corms, with 1 to several stems from each corm; basal leaves few or none; petals 7–12 mm long, white or pink, rounded to emarginate at apex; moist areas on mountain slopes. **C. lanceolata** Pursh

Lewisia Pursh   Bitterroot

1a. Sepals 2; bracts 1–2; pedicels not jointed to the peduncles. (2)
1b. Sepals 4–9; bracts 3–7; pedicels jointed to the peduncles; root fleshy; dry, rocky ridges and slopes in the mountains of north eastern Utah, along central mountain mass to Iron Co. **L. rediviva** Pursh
2a. Basal leaves numerous; roots thick, lacking corms. (3)
2b. Basal leaves absent or 1, the leaves subtending the inflorescence; corms globose; moist, bare mountain slopes, 6,000–10,000 ft., Salt Lake and Summit cos. **L. triphylla** (S. Wats.) Robins.
3a. Floral bracts similar to sepals and subtending them; sepals not denticulate; wet montane meadows, southwestern Utah. **L. brachycalyx** Engelm.
3b. Floral bracts dissimilar to sepals and remote from them; sepals denticulate, obtuse; meadows or moist open places on ridges, from 7,000–12,000 ft., throughout Utah. **L. pygmaea** (Gray) Robins.

Montia L.   Indian Lettuce, Miners Lettuce

1a. Stems with 2 to several pairs of leaves; inflorescence terminal or axillary; moist or wet ground, widespread. **M. chamissoi** (Ledeb.) Dur. & Jackson
1b. Stems with 1 pair of leaves subtending terminal inflorescence. (2)
2a. Pedicels of the inflorescence bracteate; bracts 4–10 mm long; plants perennial; stem leaves sessile; plant reproducing by bulb scales or

Figs. 85-86.   85. **Oxyria digyna**, x.29.   86. **Polygonum aviculare**, x .21

Figs. 87-88.   87. **Rumex acetosella**, x .21.   88. **Claytonia lanceolata**, x .2

Portulacaceae

from a persistent root crown; bogs and moist woods in mountains, northern Utah. **M. sibirica** (L.) Howell

2b. Pedicels of the inflorescence bractless or with a single bract at lowest branch. (3)

3a. Plants perennial with a horizontal rootstock; petals 8.5–12 mm long; shaded streams and marshy areas, northern Utah. **M. cordifolia** (S. Wats.) Pax & K. Hoffm.

3b. Plants annual; petals 2–7 mm long; seeds under lens more or less minutely studded with low tubercles. (4)

4a. Stem leaves united on both sides, forming a rounded, though often angled, disk; plants more or less fleshy, often reddish; petals 5, clawed; common in more or less shaded places, valleys and hills, widespread. **M. perfoliata** (Donn) Howell

4b. Stem leaves united on one side only, free above the base or forming a 2-lobed disk; plants tufted, glaucous; petals 2.5–5 mm long; racemes 5–25 mm long, subsessile or with short peduncle; open grassy or gravelly hillsides, western Utah. **M. spathulata** (Dougl.) Howell

Portulaca L.

1a. Petals variously colored, showy; plants with crinkled hairs in leaf axils; leaves terete or nearly so; common ornamental, cultivated. **P. grandiflora** Hook.

1b. Petals yellow; plants glabrous or with a few inconspicuous hairs in leaf axils; leaves spatulate or broader, fleshy but flattened. (2)

2a. Seeds under lens conspicuously and sharply tuberculate, almost spiny, 0.9–1 mm wide; leaves often retuse or emarginate at apex; southern Utah. **P. retusa** Engelm.

2b. Seeds under lens minutely and not sharply tuberculate, seldom more than 0.8 mm wide; leaves rounded or truncate at apex; common weed of cultivated areas, also used as a pot herb, widespread. Purslane, Pusley. **P. oleracea** L.

Spraguea Torr. Pussy Paws

Plants annual or perennial from a taproot; leaves mostly basal, stem leaves reduced; inflorescence of 1-sided spikes arranged in umbels, heads or panicles; flowers perfect; sepals 2, scarious, persistent; petals 4, withering around the style; stamens 3; styles united, long-filiform; stigmas 2; capsule membranous; high mountains. **S. umbellata** Torr.

Talinum Adans. Fame Flower

Herbs, or sometimes suffrutescent, often from a tuberous root; leaves usually alternate, blades flat or terete; flowers in cymes or solitary in leaf axils; sepals 2, deciduous; petals usually 5; stamens as many as, or more than, petals; style 3-lobed or 3-cleft; capsule 3-valved; southern Utah. **T. brevifolium** Torr.

Subclass DICOTYLEDONEAE

Primulaceae — Primrose Family

Plants herbaceous, annual or perennial; leaves simple, alternate, opposite or whorled; flowers perfect, regular; sepals commonly 5, more or less united; petals usually 5, united but sometimes cleft to base or wanting entirely; stamens as many as petals and opposite them, inserted on corolla tube; ovary superior or partly inferior, 1-loculed, with free central placentation; styles and stigmas 1; fruit a capsule.

- 1a. Corolla lacking. **Glaux**
- 1b. Corolla present. (2)
- 2a. Flowers not in true umbels; plants with leafy stems. (3)
- 2b. Flowers in true umbels; plants scapose. (4)
- 3a. Corolla lobes broad, each lobe curved around its stamen. **Steironema**
- 3b. Corolla lobes linear, 5- to 7-parted. **Lysimachia**
- 4a. Corolla lobes strongly reflexed. **Dodecatheon**
- 4b. Corolla lobes erect or spreading. (5)
- 5a. Plants perennial; flowers conspicuous and showy. **Primula**
- 5b. Plants annual or perennial; flowers inconspicuous, not showy. **Androsace**

Androsace L.   Rock-Jasmine

- 1a. Plants caespitose perennials; umbels subcapitate, pedicels not over 5 mm long; corolla showy, the lobes 2–4 mm long, longer than sepals; leaves dimorphic, inner ones broader; capsule few-seeded; high mountains. **A. carinata** Torr.
- 1b. Plants annuals or short-lived perennials; umbels never capitate, pedicels over 5 mm long; corolla not showy, the lobes seldom over 2 mm long, subequal to sepals (except in fruit); leaves nearly alike; capsule 5- to many-seeded. (2)
- 2a. Sepals broadly triangular, 3-nerved, flat; calyx tube not 5-angled; capsule at maturity much exceeding calyx; wet margins of lakes and streams. **A. filiformis** Retz.
- 2b. Sepals narrowly triangular to subulate, not 3-nerved, but keeled or inrolled from the sides; calyx tube 5-angled; capsule included in calyx. (3)
- 3a. Involucral bracts ovate to lanceolate-ovate; scapes 1–25; umbels 1- to 4-flowered when scapes 1, 2- to 10-flowered if more than 1; dry ground in mountains. **A. occidentalis** Pursh
- 3b. Involucral bracts narrowly lanceolate to subulate; scape number, puberulence, and presence of glands various; mountains. **A. septentrionalis** L.

Dodecatheon L.   Shooting Star

- 1a. Leaves abruptly contracted to petioles about as long as blades, the blades rounded to cordate-based, sinuate-dentate; petals white, dry-

Primulaceae

ing and persistent with stamens; capsule protruding beyond petals; moist situations, northern Utah. **D. dentatum** Hook.

1b. Leaves usually gradually attenuate to winged petioles, the blades seldom conspicuously toothed; petals usually colored, forced off by growing capsule and deciduous with stamens. (2)

2a. Stigma conspicuously capitate, usually at least twice as broad as style at midlength; filaments usually not much more than 1 mm long; flowers 4-merous; leaves linear-oblanceolate, usually less than 1 (1.5) cm broad, glabrous; inflorescence usually glabrous; capsule dehiscent by valves; mountain meadows and along streams. **D. alpinum** (Gray) Greene

2b. Stigma not conspicuously capitate, less than twice as thick as style; filaments usually united to form a tube more than 2 mm long, usually yellowish or orange, if purplish then filament above tube smooth or longitudinally rather than transversely wrinkled, or plants glabrous, or leaves entire; saline swamps and mountain meadows. **D. pauciflorum** (Dur.) Greene

Glaux L. Saltwort

Caulescent, succulent, glabrous herbs; leaves opposite, entire, sessile, oval to linear-oblong; calyx 5-lobed, campanulate, petaloid, white or pinkish, 3–5 mm long; corolla lacking; stamens 5, inserted at base of calyx, alternate to its lobes; ovary superior; capsule 5-valved at top; saline or subsaline, moist soil. **G. maritima** L.

Lysimachia L.

Perennial herbs with opposite or whorled leaves; flowers pedicellate, single and axillary or in terminal or axillary racemes, yellow, often purple dotted or purple streaked, mostly 5-merous; calyx lobed nearly full length; stamens exserted; capsule valvate, few-seeded; moist situations. **L. thyrsiflora** L.

Primula L. Primrose

1a. Corolla tube equal in length to, or only slightly longer than, calyx. (2)

1b. Corolla tube 1.5–2 times as long as calyx. (3)

2a. Pedicels in flower short, little if at all exceeding the bracts; limb of corolla deeply 2-lobed, purple; leaves mostly in flat rosettes and less than 10 cm long, white-farinose beneath; Garfield Co. (near Tropic) in 1894 and has not been collected since in that area, known also from near Manila, Daggett Co. **P. incana** M. E. Jones

2b. Pedicels in flower long, much exceeding the bracts; limb of corolla not deeply 2-lobed, at most emarginate, deep red or purple, rarely white, with a strong odor; leaves mostly erect, 10–25 cm long, green, not white-farinose, beneath; widespread above 8,000 ft. **P. parryi** Gray

3a. Scape bearing 1–3 flowers and usually 2 unequal involucral bracts;

Subclass DICOTYLEDONEAE

leaves mostly entire; corolla red or purple when dry; known only from Logan Canyon, Cache Co. **P. maguirei** L. O. Williams

3b. Scape bearing 10–20 flowers and numerous equal involucral bracts; leaves mostly denticulate; corolla tube yellowish, limb violet; often found in hanging gardens on the sheer canyon walls of southeastern Utah, reported from Kane, Garfield, Grand, San Juan, and Wayne cos. Cave Primrose. **P. specuicola** Rydb.

Steironema Raf. Loosestrife

Perennial herbs with opposite or whorled leaves; flowers axillary, solitary, usually 5-merous; calyx divided to near base; corolla rotate, yellow, lobed to near base, each petal enveloping its opposed stamen in bud; ovary superior; capsule valvate; damp meadows and along streams. **S. ciliatum** (L.) Raf.

Pyrolaceae — Wintergreen Family

Suffrutescent or herbaceous perennials; leaves simple, alternate, opposite, or appearing whorled, evergreen or much reduced; flowers usually perfect, regular, symmetric or nearly so; calyx mostly with 5 more or less distinct sepals; corolla mostly with 5 more or less distinct petals; stamens twice as many as petals; anthers usually opening by terminal or terminal-appearing pores or slits, frequently bearing 2 awnlike appendages; ovary superior, 4- to 5-celled; styles 1; stigmas 1, entire or lobed; fruit a capsule.

1a. Plants lacking chlorophyll; leaves reduced and scalelike, not evergreen; petals united; anthers with deflexed awns. **Pterospora**
1b. Plants with chlorophyll; leaves not reduced to scales, evergreen; petals distinct or nearly so; plants herbaceous or somewhat woody at very base only. (2)
2a. Stems 1-flowered. **Moneses**
2b. Stems more than 1-flowered. (3)
3a. Stems leafy though short; flowers corymbose; filaments dilated near base; styles very short, or lacking. **Chimaphila**
3b. Stems leafy at base only; flowers in elongated racemes; filaments not especially dilated at base; styles in most species over 3 mm long. **Pyrola**

Chimaphila Pursh   Pipsissewa

Herbaceous or suffruticose perennials with creeping rootstocks, 10–30 cm tall; leaves evergreen, leathery, often whorled; flowers perfect; sepals 5, 5–6 mm long; petals 5, distinct, 5–6 mm long; stamens 10, the filaments expanded near base; anthers opening by pores at basal (appearing apical) 2-horned end; ovary superior, 5-loculed, 5-lobed; style very short; stigma peltate; fruit a loculicidal capsule; high mountains. **C. umbellata** (L.) Bart.

Figs. 89-90. 89. **Montia chamissoi**, x .31. 90. **Androsace septentrionalis**, x .27

Figs. 91-92. 91. **Lysimachia thyrsiflora**, x .2. 92. **Chimaphila umbellata**, x .25

Subclass DICOTYLEDONEAE

Moneses Salisb.   Woodnymph

Perennial herbs with rootstocks; stems 1-3 cm above ground, scapose, the scapes 1-flowered, 5-10 cm tall; sepals mostly 5; petals mostly 5 distinct; stamens mostly 10; anthers opening by pores, 2-horned; ovary superior; stigma peltate; fruit a 4- to 5-loculed capsule. **M. uniflora** (L.) Gray

Pterospora Nutt.   Pinedrops

Plants purplish brown, lacking chlorophyll, 20-80 cm tall; leaves much reduced, 1-3.5 cm long; flowers perfect; sepals 5, slightly united; corolla pitcher shaped; stamens 10, included; anthers opening by slits, bearing awns; ovary superior, 5-loculed; capsule 5-lobed; rich woods in high mountains. **P. andromedea** Nutt.

Pyrola L.   Shinleaf

1a. Style straight, narrowing out to much wider peltate stigma; stamens not declined but filaments straight or nearly so. (2)
1b. Style deflexed at base then curved upward, gradually thickened to a truncate collar which is thicker than stigma; stamens curved upward. (3)
2a. Style short, 1-2 mm long; raceme not 1-sided; hypogynous disk not present; woods in high mountains. **P. minor** L.
2b. Style 4-5 mm long; raceme 1-sided; hypogynous disk present at base of ovary; woods in high mountains. **P. secunda** L.
3a. Petals pink to rose-purple; sepals often definitely longer than wide; wet woods and swamps in high mountains. **P. asarifolia** Michx.
3b. Petals white or greenish white; sepals about as wide as long. (4)
4a. Leaves mottled with white along veins above, usually acute at apex, lower surface pale or tinged with red; woods in high mountains. **P. picta** Smith ex Rees
4b. Leaves not mottled above, lower surface not pale or tinged with red, but usually rounded or obtuse at apex, 1-3 cm long, usually shorter than petioles; wet woods in high mountains. **P. chlorantha** Swartz

RANUNCULACEAE — BUTTERCUP FAMILY

Herbs, rarely vinelike or woody; leaves simple, deeply divided or variously compound; flowers perfect, rarely dioecious; all parts mostly free and distinct; sepals 3 to many, often petaloid; petals 3 to many or lacking; stamens and pistils often many; ovary superior; fruit an achene, follicle, or berry.

1a. Flowers distinctly irregular. (2)
1b. Flowers regular; petals may be spurred. (3)
2a. Upper sepal spurred at base. **Delphinium**
2b. Upper sepal hooded at apex. **Acontium**

Ranunculaceae

3a. Petals or sepals spurred. (4)
3b. Petals and sepals not spurred; flowers symmetric. (5)
4a. Leaves ternately compound; petals spurred. **Aquilegia**
4b. Leaves simple, filiform; sepals spurred. **Myosurus**
5a. Petals lacking, but sepals often look like petals, in which case there is only 1 whorl of colorful parts. (6)
5b. Petals and sepals both present but the latter readily deciduous. (10)
6a. Leaves simple, reniform or cordate. **Caltha**
6b. Leaves compound or deeply lobed. (7)
7a. Leaves opposite or whorled. (8)
7b. Leaves alternate. (9)
8a. Petioles on basal leaves much longer than those on upper leaves. **Anemone**
8b. Petioles all the same length. **Clematis**
9a. Leaves simple to lobed; flowers perfect. **Trautvetteria**
9b. Leaves more than once compound; flowers usually dioecious. **Thalictrum**
10a. Flowers in dense terminal racemes; pistils 1 per flower; fruit a fleshy, red berry. **Actaea**
10b. Flowers not in dense terminal racemes; pistils more than 1 per flower; fruit not a berry. (11)
11a. Flowers large; sepals more than 1 cm long; petals brownish red or whitish; fruit a follicle. (12)
11b. Flowers smaller; sepals less than 1 cm long; petals mostly white or yellow; fruit an achene. (13)
12a. Petals reduced to small linear organs much smaller than the petaloid sepals. **Trollius**
12b. Petals not as above, showy. **Paeonia**
13a. Petals with a nectary at base; mostly native. **Ranunculus**
13b. Petals lacking a nectary; introduced weed. **Adonis**

Aconitum L. Monkshood
   Perennial herbs; leaves palmately lobed or divided; flowers large, irregular, in terminal racemes; sepals blue or purple; moist mountain sites. **A. columbianum** Nutt.

Actaea L. Baneberry
   Perennial herbs; leaves 2–3 times compound; flowers tiny, white, racemose; sepals 3–6, petallike, falling off when flowers open; petals narrow, clawed, smaller than the sepals; rich soil in moist mountain sites. **A. rubra** (Ait.) Willd.

Adonis L.
   Annual herbs; leaves alternate, dissected; flowers yellow or red, solitary; introduced weed. Pheasants Eye. **A. annua** L.

Subclass DICOTYLEDONEAE

Anemone L. Windflower
1a. Plants cultivated, not found growing wild; flowers yellow to blue. **A. japonica** Sieb. & Zucc.
1b. Plants native, rarely cultivated; flowers whitish to purplish. (2)
2a. Leaves 2–4 times ternately cleft; sepals 5–8; stems from a woody crown; meadows and rocky areas at high elevations. **A. globosa** Nutt.
2b. Leaves 1- or 2-ternately compound; sepals 8–10; stems from short tubers; rocky areas at middle and lower elevations. Desert Anemone. **A. tuberosa** Rydb.

Aguilegia L. Columbine
1a. Stems and leaves conspicuously glandular-pubescent, 3–6 dm tall; sepals white, yellowish, blue, or reddish; canyons and hanging gardens, southeastern Utah. **A. micrantha** Eastw.
1b. Stems and leaves glabrous to faintly pubescent. (2)
2a. Sepals and petals blue; leaflets thick, closely clustered; 5–20 cm tall; gravelly, subalpine areas. **A. scopulorum** Tidestr.
2b. Sepals and petals not blue. (3)
3a. Flowers nodding; sepals and spurs scarlet, sometimes fading yellow. (4)
3b. Flowers erect; sepals and spurs not scarlet or fading yellow. (5)
4a. Stems 5–10 dm tall; leaflets 2–4 cm long; open woods. Crimson Columbine. **A. formosa** Fisch.
4b. Stems 1–4 dm tall; leaflets small; southern mountain areas. **A. elegantula** Greene
5a. Flowers blue or white, large; aspen areas. **A. coerulea** James
5b. Flowers yellow. (6)
6a. Spur hooked at tip; flowers may be reddish tinged; stems 2–6 dm tall; basal leaves 2-ternate; leaflets 1–4 cm long, thin, broadly obovate; open woods mostly below 8,000 ft. Yellow Columbine. **A. flavescens** S. Wats.
6b. Spur not hooked at tip; flowers golden yellow throughout; stems 3–10 dm tall; basal leaves mostly 3-ternate; mountains in moist situations. Golden Columbine. **A. chrysantha** Gray

Caltha L. Marsh Marigold
Fleshy perennial herbs; leaves mostly basal; sepals 5–15, petallike, showy; stamens and pistils many; flowers white or bluish outside; wet meadows, mountains. **C. leptosepala** DC.

Clematis L. Virgin's Bower
1a. Plants cultivated, sometimes escaping but not native. (2)
1b. Plants native, rare in cultivation. (3)
2a. Leaves pinnate; flowers blue or purple, over 25 mm broad. Purple Clematis. **C. jackmanii** Moore

Ranunculaceae

2b. Leaves 1- to 3-ternate; flowers yellow, under 25 mm broad; locally established and spreading. **C. orientalis** L.
3a. Leaves pinnately 5- to 7-foliolate; flowers blue or white. (4)
3b. Leaves 3-foliolate or 2-ternate; flowers blue or purple. (5)
4a. Plants woody vines; leaflets lanceolate to ovate, nearly glabrous; flowers white in cymes; stream banks growing over bushes. Western Virgin's Bower. **C. ligusticifolia** Nutt.
4b. Plants with stems erect; leaflets lanceolate to narrowly linear, densely pubescent; flowers solitary, blue to purple. Hairy Leather Flower. **C. hirsutissima** Pursh
5a. Leaves 3-foliolate; leaflets broadly ovate, entire or sparsely toothed; usually in deep forests. **C. columbiana** (Nutt.) T. & G.
5b. Leaves 2-ternate; leaflets lanceolate, deeply toothed to cleft; usually in forests. **C. pseudoalpina** (Kuntze) A. Nels.

Delphinium L.   Larkspur

1a. Plants annual; leaves finely divided into over 5 segments; petals united into 1; cultivated and escaping. Rocket Larkspur. **D. ajacis** L.
1b. Plants perennial; leaves less finely divided into less than 5 segments; petals distinct; native, rare in cultivation. (2)
2a. Stems scapose, the leaves primarily basal, sometimes with very much reduced stem leaves; primary leaf segments obovate; flowers dark blue; southern Utah. **D. scaposum** Greene
2b. Stems with well-developed leaves. (3)
3a. Leaves over 4 cm broad, broadly cleft; plants over 5 dm tall; flowers blue; meadows, mostly in northern Utah. Tall Larkspur. **D. occidentale** S. Wats.
3b. Leaves less than 4 cm broad, narrowly divided; plants under 5 dm tall. (4)
4a. Stems narrowing at ground level; flowers dark blue, occasionally white; widely distributed. Low Larkspur. **D. nelsonii** Greene
4b. Stems not narrowing at ground level; flowers light blue; mostly in southern Utah. Desert Larkspur. **D. amabile** Tidestr.

Myosurus L.   Mousetail

Small annuals; leaves basal, entire and linear; flowers 1, small; sepals 5; petals 5 or lacking; pistils many; wet areas, usually at middle elevations. **M. minimus** L.

Paeonia L.   Peony

1a. Flowers brownish red; native. Western Peony. **P. brownii** Dougl.
1b. Flowers variously colored, large and showy; cultivated. (2)
2a. Flowers white. **P. albiflora** Pallas
2b. Flowers red, most reddish varieties included. **P. anomala** L.

Subclass DICOTYLEDONEAE

Ranunculus L. Buttercup

1a. Achenes roughly transversely ridged; petals not glossy, white, the claws sometimes yellow; aquatic (subgenus Batrachium). (2)
1b. Achenes not transversely ridged (except in **R. sceleratus**); petals usually glossy, yellow or rarely red, white, or green. (4)
2a. Style persistent after flowering, the achene beak 0.7–1.0 mm long; dissected leaves once or sometimes twice trichotomous, then dichotomous; aquatic in sluggish water, central and northern Utah. **R. longirostris** Godr.
2b. Style largely deciduous after flowering, the achene beak 0.3–0.5 mm long. (3)
3a. Pedicels not recurved at fruiting time; submersed dissected leaves usually petioled, the first divisions arising usually, but not always, well above the nondilated stipular leaf bases (the ends of these not free), usually collapsing when withdrawn from the water, not circinate, usually about equaling or a little shorter than the internodes; achenes mostly 10–20 (40); dissected leaves usually repeatedly trichotomous; ponds, ditches, streams, mostly at lower elevations. **R. aquatilis** L.
3b. Pedicels recurved at fruiting time; submersed dissected leaves usually sessile, the first divisions arising within the usually dilated stipular leaf bases (the ends of these often free), usually not collapsing when withdrawn from the water, circinate, much shorter than the internodes; achenes mostly 30–45 (80); dissected leaves usually once or twice trichotomous; ponds, lakes, streams, and pools, at moderate elevations. **R. circinatus** Sibth.
4a. Sepals persistent in fruit; fruits either utricular or 3-chambered. (5)
4b. Sepals deciduous during, or soon after, anthesis; fruits 1-chambered. (6)
5a. Plants annual; stamens usually less than 10 (3–7); hypocotyl simulating a taproot; fruit with 1 basal seed chamber and 2 lateral empty vesicles, the beak lanceolate (subgenus Ceratocephalus); locally abundant in waste places. Bur Buttercup. **R. testiculatus** Crantz
5b. Plants perennial; stamens usually 10 or more; hypocotyl not simulating a taproot; fruit 1-chambered, utricular, the beak not lanceolate; petals red, rarely white; western and southwestern Utah. **R. juniperinus** M. E. Jones
6a. Pericarp striate, the nerves 3 or more on each face, these sometimes branched, the ovary wall thin and usually fragile at fruiting time (subgenus Cyrtorhyncha). (7)
6b. Pericarp not striate or nerved, thick and firm (subgenus Ranunculus). (8)
7a. Fruiting receptacle enlarged to several times its size in anthesis, cylindroid or long-ovoid; nectary scale overarching the nectary, truncate, the margins free from the blade of the petal; stolons

present; leaves simple; widely distributed, especially in moist situations. **R. cymbalaria** Pursh
7b. Fruiting receptacle but slightly enlarged from its size in anthesis; nectary scale not overhanging the nectary, consisting of a mere transverse ridge below the gland; stolons lacking; leaves compound; rare, dry mountainsides, northeastern Utah. **R. ranunculinus** (Nutt.) Rydb.
8a. Achenes covered with spines, hooks, or papillae or with papillae produced into hooked hairs; dorsiventral measurement of achene 3–6 times the lateral; receptacle in fruit 1–3 times its length in anthesis; northern Utah. **R. arvensis** L.
8b. Achenes smooth, sometimes hairy, but not as above. (9)
9a. Style and achene beak practically lacking or, if otherwise, the achene with a corky thickening on the margin of the body; nectary scale either with the gland in a pocket on its ventral surface or else the scale forked and prolonged anteriorly on the surface of the petal or surrounding the gland; aquatic or growing on mud. (10)
9b. Style and achene beak present, the body neither corky keeled nor with corky thickening on the margin of the body; nectary scale ventral to the nectary, covering it, apically truncate or rounded. (11)
10a. Achenes each with corky thickening beside the inconspicuous keel; leaves 1- or 2-parted or -lobed, pentagonal, 1–2 cm long, 1.5–2.5 cm broad, rarely dissected but not 3-ternately so; anthers elliptic, 0.5–1 mm long; petals 4–7 mm long; mostly in northern Utah. **R. gmelinii** M. E. Jones
10b. Achenes each with a conspicuous corky keel; leaves of aquatic specimens finely 3-ternately dissected into ribbonlike segments 1–2 mm broad, the complete blades 1.5–10 cm long, 2–12 cm broad; anthers oblong, 1–1.5 mm long; petals 7–15 mm long; northern Utah. **R. flabellaris** Raf. ex Bigel.
11a. Nectary scale attached to the petal laterally and forming a pocket; dorsiventral measurement of achene 1–2.5 times the lateral; receptacle in fruit mostly 3–15 times its length in anthesis; sepals usually tinged dorsally with purple or lavender. (12)
11b. Nectary scale free laterally for at least two-thirds its length, not forming a pocket; dorsiventral measurement of achene 3–15 times the lateral; receptacle in fruit mostly 1–3 times its length in anthesis; sepals usually not lavender or purple tinged, but sometimes markedly so. (18)
12a. Head of achenes globose, 10–20 mm in diameter; nectary scale usually apically ciliate; herbage glabrous; roots large and fleshy, 2–3 mm in diameter, the cluster conspicuous and dense; stems prostrate or ascending; rare, northern Utah. **R. glaberrimus** Hook.
12b. Head of achenes cylindric, ovoid, or globose, if globose, 2.5–7 mm in diameter, the horizontal diameter never in any case more than 9 mm. (13)

Subclass DICOTYLEDONEAE

13a. Roots markedly thickened into storage organs, these large for the relative size of the plant, 1.5–2.5 mm in diameter; locally abundant in northern Utah. Hillside Buttercup. **R. jovis** A. Nels.
13b. Roots not thickened markedly into storage organs, sometimes of varying diameter. (14)
14a. Achenes swollen or broadened at bases and therefore obovoid-oblong. (15)
14b. Achenes obovoid, flattened-obovoid, or discoid, not swollen or flat stiped at bases. (16)
15a. The 3 primary divisions of the basal leaves 1-lobed or the middle one entire; head of achenes cylindroid or ovoid-cylindroid; high peaks, La Sal Mountains. **R. eschscholtzii** Schlect.
15b. The 3 primary divisions of the basal leaves 2-divided into linear divisions; head of achenes ovoid; high peaks, Wasatch and Uinta mountains. **R. adoneus** Gray
16a. Achenes glabrous; high mountains, mostly in southern Utah. **R. pedatifidus** J. E. Smith ex Rees
16b. Achenes canescent. (17)
17a. Nectary scale and petal glabrous; moist slopes in mountains at higher elevations. **R. inamoenus** Greene
17b. Nectary scale ciliate, the surface of the petal sometimes hairy also; mountain meadows, frequently at higher elevations. **R. cardiophyllus** Hook.
18a. Stems rooting at the nodes, some of them usually stoloniferous; receptacle usually hispid; meadows and marshes at lower elevations. **R. repens**. L.
18b. Stems never rooting; receptacle glabrous. (19)
19a. Achene beaks 0.3–0.6 mm long; introduced, extreme northern Utah. **R. acris** L.
19b. Achene beaks 1–2 mm long; known from Orton's Ranch, Garfield Co. **R. acriformis** Gray

Thalictrum L.   Meadow Rue

Erect perennials to 1 m tall; leaves 3-ternate; flowers panicled or corymbed, perfect or more usually dioecious; sepals 4–7, greenish white, petallike; shady areas in mountains. **T. fendleri** Engelm.

Trautvetteria Fisch. & Mey.

Perennial herbs; leaves palmately cleft; flowers perfect, numerous; plants 3–10 dm tall, stout; leaves 5- to 11-parted, the lobes coarsely toothed; sepals 4–5, petaloid, the petals lacking; subalpine regions. **T. carolinensis** Vail

Trollius L.   Globeflower

Perennial herbs; leaves palmately divided; flowers solitary; sepals 5–15; whitish petaloid; petals 5–8, small; mostly subalpine regions. **T. laxus** Salisb.

RESEDACEAE — MIGNONETTE FAMILY

Annual or perennial herbs; leaves 5–10 cm long, oblong and deeply lobed to pinnatifid; flowers perfect, irregular, small, in narrow racemes; calyx 4–7 parted; petals 4–7; stamens 8–30, inserted on one side of the flower; ovary superior, 3- to 6-carpelled, 1-loculed with 3–6 placentae; styles or stigmas 3–6; fruit a 3- to 6-lobed capsule, horned at apex, opening at top before seeds mature.

Reseda L. Mignonette

A single genus and species treated, with the characteristics of the family; plants to 6 dm tall; flowers greenish yellow; introduced species, frequently cultivated. **R. lutea** L.

RHAMNACEAE — BUCKTHORN FAMILY

Shrubs or small trees with simple, alternate, opposite, or fascicled leaves; flowers small, perfect or polygamous, regular or nearly so; sepals 4 or 5, tube lined with a disk; petals distinct, 4 or 5; stamens equal in number to and opposite the petals; ovary free or coalescent to the fleshy disk, 2- to 3-loculed, each cell with 1 ovule; fruit a capsule or a berrylike drupe.

1a. Fruit dry, capsular; petals 5, long clawed, 2 mm long or more; ovary adnate to disk for lower part, partly inferior. **Ceanothus**
1b. Fruit fleshy, drupelike; petals 4, rarely 5, short clawed, about 1–1.5 mm long; ovary slightly if at all adnate to disk, essentially superior. **Rhamnus**

Ceanothus L. New Jersey Tea, Mountain Lilac

1a. Leaves opposite, pinnately veined with several broad veins; stipules thick, commonly persistent (at least lower portion); leaves not more than 2.5 cm long, thick, pilose or tomentulose when young, at least beneath; inflorescence small, surpassing leaves little if at all; southern Utah. **C. greggii** Gray
1b. Leaves alternate, often palmately 3-nerved; stipules thin, usually soon deciduous except for thickened basal scar. (2)
2a. Branches spinescent; leaves both entire and permanently pubescent beneath; woods and slopes, probably more common in eastern half of Utah. **C. fendleri** Gray
2b. Branches unarmed; leaves toothed or, if entire, then glabrate below. (3)
3a. Leaves entire, seldom over 2 cm long, commonly rounded or subcordate at base, usually less than twice as long as wide; southern Utah. **C. martinii** M. E. Jones
3b. Leaves toothed, over 2 cm long, thick evergreen, glossy varnished above and strongly aromatic; hillsides and mountain slopes, northern half of Utah. Deerbrush. **C. velutinus** Dougl. ex Hook.

Subclass DICOTYLEDONEAE

Rhamnus L. Buckthorn

- 1a. Winter buds without bud scales; flowers in pedunculate umbels, perfect, 5-merous; southern Utah. **R. betulaefolia** Greene
- 1b. Winter buds scaly; flowers solitary or in sessile umbels, imperfect, 4-merous. (2)
- 2a. Plants cultivated, rarely if ever escaping, ornamental shrubs or low trees, bark of which has purgative properties; branches, at least some, usually spiny; leaves about 2 times as long as wide. **R. cathartica** L.
- 2b. Plants native shrubs to 3 m tall; branches never spiny; leaves about 3 times as long as wide; southeastern Utah. **R. smithii** Greene

ROSACEAE — ROSE FAMILY

Plants herbs, shrubs, or trees; leaves usually alternate or fascicled, and stipulate; flowers perfect, regular, often showy; stamens often numerous, borne on the 5-sepaled hypanthium; petals usually 5, borne on the hypanthium; carpels 1 to many; pistils 1 to many; ovary superior, inferior or half-inferior; fruit a drupe, pome, aggregate, accessory, or achene.

- 1a. Plants herbaceous. (2)
- 1b. Plants woody, decumbent or upright shrubs or trees. (9)
- 2a. Plants low growing, usually under 10 cm tall and spreading. (3)
- 2b. Plants taller, more upright. (6)
- 3a. Plants not stoloniferous, runners absent. **Potentilla**
- 3b. Plants stoloniferous, runners medium to long. (4)
- 4a. Flowers white; leaves 3-foliolate. **Fragaria**
- 4b. Flowers yellowish; leaves 3-foliolate or otherwise. (5)
- 5a. Leaflets 3-notched at apex; plants caespitose, matlike in growth habit. **Sibbaldia**
- 5b. Leaflets deeply serrate, densely whitish-pubescent beneath; plants not matlike. **Potentilla**
- 6a. Petals lacking; pistils 1–3; sepals 4, petallike, greenish. **Sanguisorba**
- 6b. Petals present, 5 or more, usually brightly colored, not greenish; pistils 1 to many. (7)
- 7a. Pistils 1–6; stamens 5; flowers in dense cymes, capitate; leaves pinnately compound. **Ivesia**
- 7b. Pistils many; stamens more than 10; flowers single or in few-flowered cymes; leaves pinnate or palmate. (8)
- 8a. Bractlets alternating with sepals; style becoming plumose and elongated in fruit; leaves pinnate. **Geum**
- 8b. Bractlets absent; style not elongating; leaves may be pinnate but usually 5-foliolate. **Potentilla**
- 9a. Leaves compound. (10)
- 9b. Leaves simple. (16)

Rosaceae

10a. Leaves 2-pinnately compound into minute divisions; plants aromatic. **Chamaebatiaria**
10b. Leaves 1-pinnately or palmately compound, if 2-pinnate, the divisions large and the plants not aromatic. (11)
11a. Leaves appearing palmately compound. **Potentilla**
11b. Leaves pinnately compound. (12)
12a. Stems armed with prickles or thorns. (13)
12b. Stems unarmed. (14)
13a. Flowers white; pistils numerous on an elongated receptacle. **Rubus**
13b. Flowers variously colored; pistils numerous in a hollow receptacle. **Rosa**
14a. Leaflets entire, pinnately arranged but crowded as to appear palmate. **Potentilla**
14b. Leaflets serrate. (15)
15a. Ovary superior; stamens more than 20; leaflets 13–23; cultivated shrubs. **Sorbaria**
15b. Ovary inferior; stamens 15–20; leaflets 9–15; native shrubs or cultivated trees. **Sorbus**
16a. Leaves entire, not lobed or incised. (17)
16b. Leaves variously toothed, incised or lobed. (23)
17a. Leaves opposite; low desert shrubs; southern and southeastern Utah. **Coleogyne**
17b. Leaves alternate; plants and distribution various. (18)
18a. Plants prostrate, growing as a dense mat, usually on a rock surface. **Spiraea**
18b. Plants ascending to erect. (19)
19a. Ovary inferior, enclosed in, and adnate to, the calyx tube; fruit a pome. (20)
19b. Ovary superior, the calyx tube not adnate to ovary; fruit not a pome. (22)
20a. Leaves sessile or nearly so, alternate but fascicled at ends of branchlets; native, central to southern Utah. **Peraphyllum**
20b. Leaves short-petiolate, alternate but not fascicled; cultivated. (21)
21a. Styles 2–5, free; flowers usually in few-flowered corymbs borne on short lateral branchlets; fruit red or black. **Cotoneaster**
21b. Styles 5, free; flowers usually solitary and terminal; fruit large, yellow, densely pubescent. **Cydonia**
22a. Leaves deciduous; flowers large; petals 5; cultivated. **Exochorda**
22b. Leaves evergreen; flowers small; petals lacking; native, rarely cultivated. **Cercocarpus**
23a. Leaves distinctly palmately lobed. (24)
23b. Leaves variously toothed but not lobed. (30)
24a. Stems armed with thorns. **Crataegus**
24b. Stems unarmed. (25)

Subclass DICOTYLEDONEAE

25a. Leaves evergreen, usually less than 12 mm long. (26)
25b. Leaves deciduous, usually over 12 mm long. (28)
26a. Bractlets at base of flower giving appearance of having 10 sepals; pistils many. **Fallugia**
26b. Bractlets absent; pistils less than 10. (27)
27a. Pistils 4 or more; styles becoming elongated and plumose; leaves usually with 5 or more lobes. **Cowania**
27b. Pistils 1 or 2; styles not elongated and plumose; leaves usually with 3 lobes. **Purshia**
28a. Leaves lobulate, doubly dentate, 12–50 mm long; branchlets simply pubescent; flowers showy, in long panicles. **Holodiscus**
28b. Leaves lobed and serrate, 2–10 cm broad; flowers single or in few-flowered cymes. (29)
29a. Flowers large, solitary or in few-flowered cymes; fruit an aggregate. **Rubus**
29b. Flowers small, in umbellate cymes; fruit a cluster of papery carpels. **Physocarpus**
30a. Stems with thorns or spines. (31)
30b. Stems unarmed. (33)
31a. Leaves evergreen, finely crenate-serrulate; shrubs to 3 m tall; flowers white, in clusters; berries mostly orange. **Pyracantha**
31b. Leaves deciduous, sharply or doubly serrate. (32)
32a. Plants trees to 10 m tall; thorns stout, usually conspicuous; flowers less than 20 mm broad. **Crataegus**
32b. Plants low shrubs to 2 m tall; flowers over 20 mm broad. **Chaenomeles**
33a. Pistils 5. (34)
33b. Pistils 1. (35)
34a. Flowers in umbels, racemes, or spikes; fruit a follicle. **Spiraea**
34b. Flowers in panicles; fruit an achene. **Holodiscus**
35a. Ovary superior, free from the hypanthium. (36)
35b. Ovary inferior, adnate to the hypanthium. (39)
36a. Flowers inconspicuous; petals lacking. **Cercocarpus**
36b. Flowers showy; petals present. (37)
37a. Leaves mostly under 6 cm long; cultivated. **Exochorda**
37b. Leaves mostly over 5 cm long; cultivated or native. (38)
38a. Flowers double, rarely single, golden yellow, about 2.5 cm in diameter; branches green; cultivated. **Kerria**
38b. Flowers single or double, white, pink, or reddish, rarely over 12 mm in diameter; branches not green; cultivated or native. **Prunus**
39a. Leaves sessile or nearly so, alternate but fascicled at ends of branchlets; flowers 1 or 2–3 together; styles 2; native. **Peraphyllum**
39b. Leaves petioled, not fascicled at ends of branchlets; flowers mostly

Rosaceae

several to many in clusters; styles mostly 2–5; native or cultivated. (40)
- 40a. Leaves rounded at tip; petals strap shaped; native. **Amelanchier**
- 40b. Leaves acute; petals not strap shaped; cultivated. (41)
- 41a. Flowers borne singly, or in 3s or 4s. **Chaenomeles**
- 41b. Flowers in corymbs or racemes. (42)
- 42a. Flowers in corymbs, these 6–10 cm broad; leaves usually lobed. **Sorbus**
- 42b. Flowers in umbellike racemes; leaves not lobed, but serrate. (43)
- 43a. Leaves and branchlets pubescent, especially when young; styles fused at base. **Malus**
- 43b. Leaves and branchlets glabrous; styles free. **Pyrus**

Amelanchier Medic.   Serviceberry, Shad Bush

- 1a. Leaves and branches glabrous even when young; top of ovary also glabrous. **A. pumila** Nutt.
- 1b. Leaves and branches pubescent; top of ovary tomentose. (2)
- 2a. Styles usually 3–5; leaves finely pubescent; Utah's most common serviceberry, widely distributed and highly variable. **A. utahensis** Koehne
- 2b. Styles usually 5; leaves glabrous at maturity; usually in more moist habitats than **A. utahensis**, but intergrading with it. **A. alnifolia** Nutt.

Cercocarpus H. B. K.   Mountain Mahogany

- 1a. Leaves deciduous, flat, the margin dentate. Alder-leaf Mountain Mahogany. **C. montanus** Raf.
- 1b. Leaves evergreen, at least somewhat revolute, the margin entire or dentate. (2)
- 2a. Leaves, at least some, with the margin dentate; a hybrid. **C. montanus** Raf. x **C. ledifolius** Nutt.
- 2b. Leaves lacking teeth, usually strongly revolute. (3)
- 3a. Leaves elliptic, over 12 mm long; plants tall shrubs or small trees. Curl-leaf Mountain Mahogany. **C. ledifolius** Nutt.
- 3b. Leaves linear to narrowly oblong, usually less than 12 mm long; plants low, intricately branched shrubs. **C. intricatus** S. Wats.

Chaenomeles Lindl.   Japanese Quince

- 1a. Leaves sharply serrate, acute; branchlets smooth; flowers red to white; ornamental shrubs to 2 m tall. **C. lagenaria** Koidz.
- 1b. Leaves coarsely crenate, serrate, acutish to obtuse; branchlets scabrous; flowers mostly red; ornamental shrubs to 1 m tall. **C. japonica** Lindl.

Chamaebatiaria (T. C. Porter) Maxim.   Fern Bush

Low, densely branched aromatic shrubs, 6–20 dm tall, the herbage more or less glandular and stellate-pubescent; leaves 2-pinnate, 2–4 cm

Figs. 93-94.   93. **Moneses uniflora**, x .33.   94. **Actaea rubra**, x .22

Figs. 95-96.   95. **Ranunculus aquatilis**, x .41.   96. **Amelanchier alnifolia**, x .23

Rosaceae

long; panicle 2–10 cm long; sepals 3–5 mm long; petals about 5 mm long, white; follicles 5 mm long; mostly in western Utah. **C. millefolium** (Torr.) Maxim.

Coleogyne Torr. Blackbrush
Shrubs with rigid, often spinescent, branches; leaves entire, leathery, under 12 mm long; flowers solitary; sepals yellowish or purplish, petaloid; petals lacking; southern and southeastern Utah. **C. ramosissima** Torr.

Cotoneaster Ehrh.
1a. Leaves over 25 mm long; plant usually over 1 m tall; fruit black. Peking Cotoneaster. **C. acutifolia** Turcz.
1b. Leaves under 25 mm long; plant under 1 m tall; fruit red. Rock Cotoneaster. **C. horizontalis** Decne.

Cowania D. Don   Cliffrose
Plants shrubs to 3 m tall; leaves under 12 mm long, mostly with 5 or more lobes, leathery, persistent; flowers solitary, white to yellow; styles becoming elongated; widely distributed. **C. mexicana** D. Don

Crataegus L.   Hawthorn
1a. Leaves serrate or shallowly lobed. (2)
1b. Leaves distinctly lobed. (5)
2a. Leaves truncate to subcordate at base. Eastern Haw. **C. mollis** Scheele
2b. Leaves cuneate at base. (3)
3a. Leaves 5–8 cm long, pubescent on veins beneath; inflorescence glabrous; fruit red; native in Provo Canyon. **C. succulenta** Link
3b. Leaves 2–8 cm long, glabrous beneath; inflorescence pubescent. (4)
4a. Leaves slightly doubly serrate; fruit black; native to canyons, from central Utah northward. **C. rivularis** Nutt.
4b. Leaves sharply serrate; fruit red; cultivated ornamental. Cockspur Thorn. **C. crus-galli** L.
5a. Base of leaves truncate to subcordate; cultivated. Washington Thorn. **C. phaenopyrum** Medic.
5b. Base of leaves cuneate. (6)
6a. Leaves with 3–5 broad serrulate lobes; seeds usually 2; cultivated. English Hawthorn. **C. oxyacantha** L.
6b. Leaves with 3–7 narrow subentire lobes; seeds usually 1; cultivated. **C. monogyna** Jacq.

Cydonia Mill.   Quince
Plants shrubs or small trees; leaves entire, stipulate; flowers large, white or pink, solitary; sepals and petals 5; stamens 20; styles 5, free; fruit a many-seeded, yellow, pubescent pome to 10 cm in diameter; cultivated for fruit. **C. oblonga** Mill.

Subclass DICOTYLEDONEAE

Exochorda Lindl. Pearlbush
    Plants shrubs with elliptic leaves, these 4–6 cm long, entire or serrate; flowers white, large, in terminal racemes; sepals 5; petals 5, clawed; cultivated ornamental. **E. racemosa** Rehd.

Fallugia Endl. Apache Plume
    Low, densely branched whitish shrubs; leaves 12–20 mm long, 3- to 7-lobed, densely pubescent; flowers terminal, showy, peduncled, solitary or few; petals white; pistils many, villous; fruit of achenes; mostly in southern Utah. **F. paradoxa** (D. Don) Endl.

Fragaria L. Strawberry
1a. Native to moist slopes in mountains. **F. americana** Britt.
1b. Cultivated, with numerous varieties and hybrids. **F. chiloensis** Duchesne

Geum L. Avens
1a. Flowers yellow; native to wet meadows. **G. macrophyllum** Willd.
1b. Flowers red or purple; native to dry, rocky hillsides. **G. triflorum** Pursh

Holodiscus Maxim. Rock Spirea
1a. Leaves over 5 cm long; cultivated ornamental. **H. discolor** (Pursh) Maxim.
1b. Leaves under 5 cm long; native, rarely cultivated. (2)
2a. Leaves serrate at tip only, obovate. **H. microphyllus** Rydb.
2b. Leaves serrate to middle or below, elliptic to ovate. **H. dumosus** (Nutt.) Heller

Ivesia T. & G.
1a. Stems leafy; petals orbicular to obovate, white; stamens 20. **I. kingii** S. Wats.
1b. Stems scapose or with few leaves; petals spatulate to oblanceolate, yellow; stamens 5. (2)
2a. Calyx lobes 2–2.5 mm long, equaling spatulate petals. **I. utahensis** S. Wats.
2b. Calyx lobes 5 mm long, equaling spatulate or oblanceolate petals. **I. gordonii** (Hook.) T. & G.

Kerria DC. Goldenglow
    Plants shrubs to 2 m tall; stems green; leaves doubly serrate, mostly 2–5 cm long; flowers terminal, solitary, mostly 2.5–5 cm broad, yellow; cultivated. **K. japonica** (L.) DC.

Malus Mill. Apple
1a. Leaves on elongated shoots lobed or strongly notched; cultivated. Prairie Crab Apple. **M. ioensis** Britt.
1b. Leaves on elongated shoots not lobed or notched. (2)

Rosaceae

2a. Leaf margins obtusely crenate-serrate; cultivated. Apple. **M. sylvestris** Mill.
2b. Leaf margins sharply serrulate or serrate; cultivated ornamental. Showy Crab Apple. **M. floribunda** Sieb.

Peraphyllum Nutt. Squaw-Apple
  Intricately branched shrubs, 1–2 m tall; leaves crowded at ends of spurlike branchlets, 2–4 cm long; flowers appearing with leaves, 1–3; petals pale pink, 7–8 mm long; fruit yellowish, 8–10 mm thick, astringent. **P. ramosissimum** Nutt.

Physocarpus Maxim.   Ninebark
1a. Pistils 3–5; stamens similar. **P. capitatus** (Pursh) Kuntze
1b. Pistils 1; stamens similar or dissimilar. (2)
2a. Leaves with stellate hairs; stamens alternately long and short; shrubs to 1 m tall. **P. alternans** (M. E. Jones) J. T. Howell
2b. Leaves glabrous or nearly so; stamens about equal; shrubs to 2 m tall. (3)
3a. Styles erect; petals 4–5 mm long. **P. malvaceus** (Greene) Kuntze
3b. Styles spreading; petals less than 3 mm long. **P. monogynus** (Torr.) Coult.

Potentilla L.   Cinquefoil

1a. Plants woody shrubs; middle elevations, usually in moist sites. **P. fruticosa** L.
1b. Plants herbaceous. (2)
2a. Basal leaves palmately compound. (3)
2b. Basal leaves pinnately compound. (7)
3a. Leaflets 3. (4)
3b. Leaflets 5 or more, rarely 3. (5)
4a. Stamens 10; stems densely glandular-pubescent. **P. biennis** Greene
4b. Stamens 15–20; stems hirsute. **P. monspeliensis** L.
5a. Leaflets mostly 5, 3- to 7-toothed (incised) above middle. **P. diversifolia** Lehm.
5b. Leaflets mostly 7. (6)
6a. Leaflets incised two-thirds the distance to midrib. **P. pectinisecta** Rydb.
6b. Leaflets incised one-half the distance to midrib or less. **P. gracilis** Dougl.
7a. Plants stoloniferous, the runners to 6 dm long; leaves densely pubescent above; moist, open, often saline meadows. **P. anserina** L.
7b. Plants without stolons. (8)
8a. Leaflets dissected nearly to midrib; style attached near tip of ovary; native to rocky ridges at high elevations. (9)
8b. Leaflets dentate; style attached near base of ovary. (10)

301

Subclass DICOTYLEDONEAE

- 9a. Leaflet divisions oblong to oblanceolate. **P. ovina** Macoun
- 9b. Leaflet divisions narrowly linear. **P. wyomingensis** A. Nels.
- 10a. Leaflets usually 9–11, less than 2.5 cm long, rudimentary leaflets sometimes interspersed. **P. fissa** Nutt.
- 10b. Leaflets usually 7–9, rudimentary leaflets absent. (11)
- 11a. Flowers nearly white, in dense, short inflorescence; meadows and along streams. **P. arguta** Pursh
- 11b. Flowers bright yellow, in open cymes; hills and mountains. **P. glandulosa** Lindl.

Prunus L.

- 1a. Leaves broad-ovate to orbicular; cultivated. Apricot. **P. armeniaca** L.
- 1b. Leaves not broad-ovate to orbicular. (2)
- 2a. Leaves ovate to lanceolate. (3)
- 2b. Leaves not ovate to lanceolate. (6)
- 3a. Axillary buds 3. (4)
- 3b. Axillary buds 1. (5)
- 4a. Petiole 12–25 mm long; leaves serrulate, broadest below middle. Almond. **P. amygdalus** Batsch.
- 4b. Petiole less than 12 mm long; leaves serrate, broadest above middle. Peach. **P. persica** (L.) Batsch.
- 5a. Leaves serrate to twice serrate; flowers especially showy, often double, over 20 mm broad; fruit black. Flowering Cherry. **P. serrulata** Lindl.
- 5b. Leaves once serrulate; flowers 12 mm broad; fruit red. Pin Cherry. **P. pennsylvanica** L.
- 6a. Leaves mostly under 6 cm long. (7)
- 6b. Leaves mostly over 6 cm long. (10)
- 7a. Plants low, spreading, divaricately branched native shrubs of southwestern Utah; fruit pubescent. Desert Peach. **P. fasciculata** (Torr.) Gray
- 7b. Plants various but, if native, the fruit glabrous. (8)
- 8a. Plants shrubs; leaf tip acuminate or 3-lobed; flowers pink, double, very showy. Flowering Almond. **P. triloba** Lindl.
- 8b. Plants trees; leaf tip acute. (9)
- 9a. Leaf margin finely crenate; leaves glabrous beneath, purple in some varieties; flowers showy. Flowering Plum. **P. ceracifera** Ehrh.
- 9b. Leaf margin doubly serrate; leaves slightly pubescent beneath; cultivated for fruit. Sour Cherry **P. cerasus** L.
- 10a. Lower leaf surface glabrous except for axillary tufts; flowers in showy white racemes. (11)
- 10b. Lower leaf surface pubescent at least along veins; flowers not in racemes. (12)
- 11a. Leaf base cuneate to rounded; calyx tube glabrous inside; fruit

Rosaceae

purple; native. Choke Cherry. **P. virginiana** L. var. **melanocarpa** Sarg.
11b. Leaf base subcordate; calyx tube pubescent inside; fruit black; cultivated ornamental. May Day Cherry. **P. padus** L.
12a. Leaves pubescent over lower leaf surface; flowers greenish white, 12–25 mm broad; fruit ovoid to oblong, blue. Plum. **P. domestica** L.
12b. Leaves pubescent only along lower veins; fruit reddish. (13)
13a. Leaves obovate-oblong, smooth above; flowers white, 20–30 mm broad; fruit subglobose; bark gray brown; many commercial varieties cultivated. Plum, Prune. **P. americana** Marsh
13b. Leaves oblong-ovate, often rugose above; bark reddish, the lenticels prominent. Sweet Cherry. **P. avium** L.

Purshia DC.
Low, spreading shrubs; leaves leathery, 3-lobed or 3-toothed apically; flowers solitary, yellow; stamens 20–25 in 1 row; widespread, important browse plant. Bitterbrush. **P. tridentata** (Pursh) DC.

Pyracantha Roem. Firethorn
Thorny evergreen shrubs; leaves crenulate-serrulate; flowers white; fruit orange to red; commonly cultivated. **P. coccinea** Roem.

Pyrus L. Pear
Cultivated trees; leaves serrate, glabrous; flowers white, in umbellike racemes; stamens 20–30; ovary inferior; styles free, 2–5. **P. communis** L.

Rosa L. Rose
1a. Plants cultivated. (2)
1b. Plants native, rare in cultivation. (3)
2a. Stamens nearly as long as styles, the latter connate into a column. **R. multiflora** Thunb.
2b. Stamens about twice as long as styles, the latter free. Tea Rose. **R. odorata** Sweet
3a. Infrastipular prickles absent, stem may have scattered prickles. **R. woodsii** Lindl.
3b. Infrastipular prickles present. (4)
4a. Inflorescence 1- to 3-flowered, on laterals 3–10 cm long; foliage glandular and resin scented. **R. nutkana** Presl.
4b. Inflorescence 1- to 15-flowered, on laterals more than 10 cm long; foliage not strongly resin scented. **R. woodsii** Lindl.

Rubus L.
1a. Leaves simple, large, palmately lobed; stems unarmed; flowers single; native, middle elevations. Thimble Berry **R. parviflorus** Nutt.
1b. Leaves compound; stems prickly. (2)

Subclass DICOTYLEDONEAE

2a. Inflorescence racemose, open; fruit red. (3)
2b. Inflorescence cymose, dense; fruit black. (4)
3a. Plants cultivated, variously spiny. Red Raspberry. **R. idaeus** L.
3b. Plants native, rocky areas in mountains. Western Red Raspberry. **R. strigosis** Michx.
4a. Canes strongly armed, usually hooked; fruit purplish black. Western Black Raspberry. **R. leucodermis** Dougl.
4b. Canes weakly armed, straight; fruit black. Blackcap. **R. occidentalis** L.

Sanguisorba L.

Perennial herbs, 3–8 dm tall; leaves odd-pinnate, with 7–21 orbicular to oblong, little notched leaflets, mostly basal; lower flowers staminate, the upper perfect; petals 1; sepals 4, petallike; stamens and stigmas purplish, exserted; achene 4-ribbed; cultivated, escaping and persisting. Burnet. **S. minor** Scop.

Sibbaldia L.

Low, caespitose perennials; leaves 3-foliolate, the leaflets about 12 mm long, 3- to 5-lobed; flowers few in dense cymes, yellow; stamens 5; pistils 5–20; moist areas at high elevations. **S. procumbens** L.

Sorbaria A. Br. False Spirea

Plants shrubs with large pinnate stipulate leaves; flowers small, white, in large, showy terminal panicles; cultivated ornamental. **S. sorbifolia** A. Br.

Sorbus L. Mountain Ash

1a. Leaves simple, lobed and serrate; cultivated ornamental trees. Oakleaf Mountain Ash. **S. hybrida** L.
1b. Leaves compound. (2)
2a. Plants trees to 18 m tall; leaflets acute; cultivated. Rose-leaf Mountain Ash. **S. aucuparia** L.
2b. Plants shrubs to 4 m tall; leaflets acuminate; native, mountains at middle elevations. **S. scopulina** Greene

Spiraea L. Spirea

1a. Plants mat forming on rock surfaces, frequently on limestone; flowers small, in dense erect spikes; native. Rock Spirea. **S. caespitosa** Nutt.
1b. Plants erect, ascending, or arching shrubs; cultivated. (2)
2a. Flowers in sessile umbels of 3–6 flowers which are often double; leaves denticulate, elliptic, 25–50 mm long. **S. prunifolia** Sieb. & Zucc.
2b. Flowers in umbellike racemes; leaves rhombic-ovate, incised, serrate, 20–40 mm long. Bridal Wreath. **S. vanhoutii** Zabel

RUBIACEAE — MADDER FAMILY

Plants annual or perennial, herbaceous or shrubby; leaves opposite or appearing whorled, simple or entire; flowers perfect or imperfect, regular, 4- to 5-merous; ovary inferior; fruit a capsule, achenelike, separating into 2–4 indehiscent carpels, or fleshy.

- 1a. Plants shrubs, cultivated; flowers in dense, globose heads. **Cephalanthus**
- 1b. Plants herbaceous (rarely suffruticose in **Galium**); flowers not in globose heads. (2)
- 2a. Corolla tube about 20 mm long; ovules and seeds several in each carpel; plants usually low and caespitose. **Houstonia**
- 2b. Corolla tube less than 10 mm long; ovules and seeds solitary in each carpel; plants usually sprawling or climbing, rarely caespitose. (3)
- 3a. Corolla funnelform-salverform; leaves opposite. **Kelloggia**
- 3b. Corolla rotate; leaves apparently whorled, or opposite. (4)
- 4a. Flowers 4-merous; fruit dry. **Galium**
- 4b. Flowers 5-merous; fruit fleshy. **Rubia**

Cephalanthus L. Buttonbush

Plants shrubs, 1–6 m tall; leaves ovate to oval-lanceolate, 7–15 cm long; heads about 25 mm in diameter, long-pedunculate; flowers with bristles between them. **C. occidentalis** L.

Galium L. Bedstraw

- 1a. Flowers dioecious; fruit with long hairs, these not recurved apically; plants perennial. (Note: In addition to the species keyed below, three other taxa have been identified from Utah: **G. desereticum** Demp. & Ehren, **G. magnifolium** Demp., and **G. scabriusculum** [Ehren.] Demp. & Ehren.) (2)
- 1b. Flowers monoecious; fruit with uncinate hairs or glabrous; plants annual or perennial. (6)
- 2a. Leaves linear. **G. coloradoense** Wight
- 2b. Leaves ovate to lanceolate. (3)
- 3a. Leaves rigid and acerose. **G. stellatum** Kellogg
- 3b. Leaves thickish or thin but not rigid, sometimes apiculate but not acerose. (4)
- 4a. Inflorescence narrow, few-flowered, the flowering branches erect or ascending; leaves of midstem usually one-half to nearly equaling the length of internodes. **G. hypotrichium** Gray
- 4b. Inflorescence broad, many-flowered, the flowering branches widely spreading; leaves of midstem one-third to one-sixth the length of internodes. (5)
- 5a. Plants more or less pubescent throughout. **G. munzii** Hilend & Howell
- 5b. Plants entirely glabrous. **G. multiflorum** Kellogg

Subclass DICOTYLEDONEAE

- 6a. Plants annual; fruit with uncinate hairs. (7)
- 6b. Plants perennial; fruit glabrous or straight-hairy (uncinate in **G. triflorum**). (9)
- 7a. Leaves 6–8 per whorl; flowers in 1- to 10-flowered cymes; stem retrorsely hispid. **G. aparine** L.
- 7b. Leaves 2–4 per whorl; flowers solitary; stems glabrous or hispid, but not retrorsely so. (8)
- 8a. Flowers sessile or nearly so; leaves ovate or oblong, in whorls of 4. **G. proliferum** Gray
- 8b. Flowers pedicellate; leaves linear to linear-oblong, at least the upper ones opposite. **G. bifolium** S. Wats.
- 9a. Leaves 2-nerved; stems erect; fruit with short or long straight hairs. **G. boreale** L.
- 9b. Leaves 1-nerved; stems usually spreading; fruit glabrous or with uncinate hairs. (10)
- 10a. Fruit uncinate-hairy; leaves in whorls of 5 or 6. **G. triflorum** Michx.
- 10b. Fruit glabrous; leaves in whorls of 4. **G. trifidum** L.

Houstonia L.

Perennial, caespitose herbs to 20 cm tall; leaves mostly basal or nearly so, opposite; flowers bright pink, tubular-salverform, 10–20 mm long; southeastern Utah. **H. rubra** Cav.

Kelloggia Torr.

Perennial herbs from woody rootstocks, 1–2.5 dm tall; leaves opposite, lanceolate, 1–3.5 cm long; corolla dull pink, 5–7 mm long; fruit 4–5 mm long, densely covered with short, uncinate hairs. **K. galioides** Torr.

Rubia L.

Perennial climbing herbs; leaves lanceolate, 5–10 cm long, in whorls of 4–6; cymes terminal; flowers greenish yellow; fruit red, turning black, 2–5 mm in diameter; introduced weedy plant. Madder. **R. tinctoria** L.

RUTACEAE — RUE FAMILY

Plants shrubs or small trees; herbage glandular-punctate; leaves simple or palmately compound; flowers perfect or imperfect, regular; sepals and petals mostly 4 or 5; stamens as many, or twice as many, as petals, borne on a fleshy disk; ovary superior, 2- to 5-loculed; fruit capsular or samaroid.

- 1a. Plants large shrubs or small trees; leaves palmately 3-foliolate; fruit flat and broadly winged, samaroid. **Ptelea**
- 1b. Plants small shrubs; leaves simple, linear or narrowly spatulate, early deciduous; fruit a deeply 2-lobed capsule. **Thamnosma**

Figs. 97-98. 97. **Prunus virginiana**, x .38. 98. **Sibbaldia procumbens**, x .59

Figs. 99-100. 99. **Sorbus scopulina**, x .3. 100. **Galium aparine**, x .2

Subclass DICOTYLEDONEAE

Ptelea L. Hop-Tree
1a. Leaflets 6–12 cm long, usually glabrous below; fruit 1.8–2.5 cm broad; cultivated ornamental. **P. trifoliata** L.
1b. Leaflets 3–6 cm long, pubescent beneath; fruit 1–1.5 cm broad; native, southern Utah. **P. baldwinii** Torr.

Thamnosma Torr. & Frem. ex Frem.
Odoriferous shrubs, 3–5 dm tall; freely branching and broomlike, yellowish green, glandular-punctate; leaves few and caducous; flowers in terminal racemes; petals dark purple; fruit stipitate; southeastern Utah. **T. montana** Torr. & Frem. ex Frem.

SALICACEAE — WILLOW FAMILY

Plants trees and shrubs; leaves simple, deciduous, alternate, stipulate; plants dioecious; flowers in aments (catkins), solitary in axils of scalelike bractlets; perianth lacking; stamens 1 to many; ovary superior, solitary, 1-loculed; stigmas mostly 2; fruit a 2-valved capsule; seeds numerous, bearing apical tufts of hair.

1a. Winter buds covered by several scales; bractlets of aments laciniate (torn) or fimbriate (fringed); flowers borne on broad or cup-shaped disks; catkins pendulous; trees. **Populus**
1b. Winter buds with only 1 scale; bractlets of aments entire, merely dentate or pappuslike; flowers without disks; catkins various, but mostly upright; mostly shrubs. **Salix**

Populus L. Poplar, Cottonwood

1a. Petioles definitely flattened laterally, at least near blade. (2)
1b. Petioles not flattened laterally, round, or slightly flattened or grooved on upper surface. (5)
2a. Leaves suborbicular to ovate-orbicular, often lighter beneath; bark remaining smooth until old, whitish or cream colored; stigmas 2, filiform; typically occurring in clumps or large stands; mountains. Aspen. **P. tremuloides** Michx.
2b. Leaves not suborbicular; bark not whitish and smooth, but furrowed and darker on fairly young trunks; stigmas usually 3 or 4 and dilated. (3)
3a. Leaves with glands at base of blade, triangular-ovate with a truncate base and crenate-dentate margins with curved sinuses, abruptly acuminate at tip, very broad in relation to length. Carolina Poplar. **P. deltoides** Marsh
3b. Leaves lacking glands, the margins crenate-serrate. (4)
4a. Trees narrow, columnar; branchlets yellowish; leaves with shallowly serrate, nonciliate margins and a rounded or tapering base; staminate trees only; introduced, propagated from cuttings. Lombardy Poplar. **P. nigra** L. var. **italica** Du Roi.

Salicaceae

4b. Trees broad, spreading; branchlets orange; leaves with coarsely and irregularly serrate margins, teeth gland tipped, the margins often ciliate, the base truncate or cordate; native, banks of streams. **P. fremontii** S. Wats.

5a. Leaves of vigorous shoots whitish-tomentose beneath, often lobed and maplelike; buds tomentose. (6)

5b. Leaves of vigorous shoots glabrous or slightly pubescent beneath, not lobed; buds not tomentose. (7)

6a. Trees broad, spreading; trunks of older trees and larger branches black and deeply fissured, white or cream colored and smooth on young trees and smaller branches; introduced. White Poplar. **P. alba** L.

6b. Trees columnar; trunk and branches typically smooth and light greenish, though bark may become somewhat fissured and grayish in very large trees. Bolleana Poplar. **P. alba** var. **bolleana** Lauche

7a. Leaves lanceolate to rhomboid-lanceolate. (8)

7b. Leaves mostly ovate. (9)

8a. Leaves lanceolate to ovate-lanceolate; petiole short, not over one-third as long as blade; native, along streams. Narrowleaf Cottonwood. **P. angustifolia** James ex Long

8b. Leaves rhombic-lanceolate to rhombic-ovate, acuminate, crenate-serrate; petiole more than one-third as long, usually more than one-half as long, as blade; native; hybrid between **P. angustifolia** and **P. fremontii**. **P. x acuminata** Rydb.

9a. Trees broad, spreading; buds, branchlets and leaves pubescent; leaves longer than 12 cm, the base cordate; introduced. Balm of Gilead. **P. candicans** Ait.

9b. Trees narrow, strict; buds, branchlets and leaves glabrous; leaves less than 12 cm long, the base rounded or tapering; introduced. Chinese Poplar. **P. simonii** Carr.

Salix L. Willow

1a. All mature leaves with serrate, dentate, or crenate margins. (2)

1b. All mature leaves without serrate, dentate, or crenate margins, some or all with entire margins. (16)

2a. Plants with long, pendulous branches; cultivated. (3)

2b. Plants without long, pendulous branches; cultivated or native. (4)

3a. Leaves silky beneath, at least when young, 2–12 cm long, 0.5–2 cm wide; stamens 3–5 or more; branchlets yellow to brownish. Weeping Willow. **S. babylonica** L.

3b. Leaves glabrous, glaucous or glaucescent below, 10–15 cm long, 2.5–4 cm wide; stamens 2; branchlets olive to red-brown. Wisconsin Weeping Willow. **S. blanda** Anderss.

4a. Leaves leathery. (5)

4b. Leaves not leathery. (6)

Subclass DICOTYLEDONEAE

5a. Leaves lustrous dark green above and below, or barely paler below; overwintering buds blackish, lustrous; plants resinous-fragrant; petiole with large glands toward apex, to 1 cm long; cultivated. Laurel-leaf Willow. **S. pentandra** L.

5b. Leaves pale green above, whitened below; petiole lacking glands, to 3 cm long; stream banks, also in cultivation. Peach-leaf Willow. **S. amygdaloides** Anders.

6a. Leaves green or merely paler, not whitened beneath. (7)

6b. Leaves glaucous or whitened beneath, often pubescent. (10)

7a. Leaves with glandular-serrate margins and sometimes glands on petioles at junction with blades; glands moderately prominent; twigs reddish brown; shrubs or small trees; along mountain streams and meadows. S. caudata (Nutt.) Heller

7b. Leaves without glandular-serrate margins and glands on petioles. (8)

8a. Vigorous vegetative sprouts with leaflike, more or less persistent stipules; naturalized in moist places, widespread. Black Willow. **S. nigra** Marsh

8b. Vigorous vegetative sprouts without leaflike, persistent stipules, the stipules small and soon falling. (9)

9a. Twigs glabrous or nearly so, lustrous, very brittle at base; leaves glabrous or promptly glabrate; cultivated. Brittle Willow. **S. fragilis** L.

9b. Twigs densely pubescent in growing season, yellowish, brittle, but not excessively brittle; leaves gray-green, often more or less pubescent until half grown; southwestern Utah. Dixie Black Willow. **S. goodingii** Ball

10a. Leaves silky-pubescent beneath, at least when young. (11)

10b. Leaves glaucous beneath, but not silky-pubescent. (14)

11a. Leaves permanently pubescent on both surfaces. (12)

11b. Leaves pubescent only below. (13)

12a. Plants shrubs or small trees, to 8 m tall; petiole very short or wanting; native bars and shallows of streams or on their borders, along ditches, widespread, possibly Utah's most common native willow. Sandbar Willow. **S. exigua** Nutt.

12b. Plants trees, to 30 m tall; cultivated. White Willow. **S. alba** L.

13a. Leaves dull green above, tomentose below; branchlets and buds pubescent; leaf margin undulate; shrubs or small trees, to 8 m tall; cultivated. Golden Osier. **S. viminalis** L.

13b. Leaves when young faintly pubescent below; small shrubs; twigs, except youngest, dark red, glabrous and shining; swampy ground and banks of streams. Dusky Willow. **S. melanopsis** Nutt.

14a. Leaf apex rounded to short-acuminate; ament scales dark red or brown to blackish, white bearded, persistent, expanding long before the leaves; usually male cultivated because of showy staminate catkins. Pussy Willow. **S. discolor** Muhl.

Salicaceae

14b. Leaf apex long-acuminate to acute; petiole less than 1 cm long; cultivated. (15)
15a. Branches tortuously twisted. Corkscrew Willow. **S. matsudana** Koidz. var. **tortuosa** Rehd.
15b. Branches straight; tree umbrella shaped. **S. matsudana** Koidz. var. **umbraculifera** Rehd.
16a. Leaves, at least some, with entire margins. (17)
16b. Leaves all with entire margins. (21)
17a. Twigs usually conspicuously glaucous; leaves sparsely pubescent above, densely appressed silvery-pubescent beneath, mostly entire, those of late season rarely remotely crenulate; shrubs, to 3 m tall; along streams, pools, and in wet meadows in mountains. Blue Willow. **S. subcoerulea** Piper
17b. Twigs not conspicuously glaucous. (18)
18a. Leaves not glaucous beneath, green on both faces, subentire to shallowly glandular-serrulate, the base somewhat rounded to occasionally acute; shrubs, 2–4 m tall; seepage areas and along streams in mountains. **S. pseudocordata** (Anderss.) Rydb.
18b. Leaves glaucous beneath. (19)
19a. Leaf bases rounded to subcordate; leaves lanceolate to ligulate; twigs yellow, red on one side, to dark brown or black; stream banks and wet meadows in mountains and foothills. Yellow Willow. **S. lutea** Nutt.
19b. Leaf bases cuneate or acute. (20)
20a. Leaves obovate to oblanceolate, dark green above, often tomentose beneath; commonly shrubby, 3–4 m tall, occasionally trees to 10 m tall; branches reddish brown to almost black, twigs glabrous to densely pubescent; mountains. **S. scouleriana** Barratt ex Hook.
20b. Leaves elliptic, oblong-lanceolate to oval or obovate, sparingly hairy beneath on midrib and veins, which are often raised; shrubs or small trees, usually with a single stem and bushy top even when low (2–3 m tall), when growing with tree species to 8 m tall; riverbanks and open hillsides, mountains. Beak Willow. **S. bebbiana** Sarg.
21a. Shrubs erect, over 30 cm tall. (22)
21b. Shrubs trailing, the stems on or beneath soil, branches arising to 6 cm tall. (25)
22a. Shrubs, 2–4 (6) m tall; twigs usually conspicuously glaucous; leaves subglaucous to hairy beneath. (23)
22b. Shrubs, (2) 3–10 (20) dm tall. (24)
23a. Leaves 2–4 cm long, subglaucous beneath; moist mountains above 6,500 ft. elevation; streams, wet meadows, and bogs. **S. geyeriana** Anderss.
23b. Leaves 4–6 (8) cm long, appressed silvery-pubescent beneath;

Subclass DICOTYLEDONEAE

streams and wet meadows in mountains. Blue Willow. **S. subcoerulea** Piper

24a. Twigs glabrous, yellow when young, chestnut at maturity, slender; leaves silky-villous on both faces when young, dull green at maturity with vestiges of pubescence, puberulent or glabrate; mountains, boggy areas and slopes where snow remains late. Sierra Willow. **S. wolfii** Bebb

24b. Twigs of current season pubescent, tomentose, or pilose; leaves pubescent or pilose with long straight hairs, tardily glabrate; petiole over 4 mm long; stipules wanting; moist slopes above timberline, borders of lakes and streams. **S. pseudolapponum** Seem.

25a. Leaves early pubescent, soon glabrate except at margins, paler to subglaucous beneath, not strongly reticulate, elliptic, rarely oval or obovate, 1.5–3 cm long; alpine gravelly slopes where melting snow keeps soil moist. Arctic Willow. **S. anglorum** Cham. var. **antiplasta** Schneid.

25b. Leaves glabrous, intensely glaucous beneath, strongly reticulate, orbicular to suborbicular, elliptic-oblong or elliptic, clustered at ends of branches, 2–6 cm long, which arise from buried stems; branches glabrous to coarsely pubescent; alpine areas where snow lingers late. **S. nivalis** Hook.

SANTALACEAE — SANDALWOOD FAMILY

Perennial, partially parasitic herbs; leaves alternate, nearly or quite sessile; flowers perfect; calyx campanulate, 3- to 5-lobed; petals lacking; stamens as many as calyx lobes; ovary 1-loculed; styles 1; fruit drupaceous.

Comandra Nutt.   Bastard Toadflax

A single genus and species treated, with the characteristics of the family; flowers greenish white; widely distributed. **C. umbellata** (L.) Nutt.

SAURURACEAE — LIZARDTAIL FAMILY

Perennial herbs from creeping rootstocks; leaves mostly basal, simple; flowers in compact spikes subtended by conspicuous bracts, giving the appearance of a single flower, perfect; perianth lacking; stamens 6–8; ovary of 3–4 carpels, each 1- to 2-ovuled; styles 3–4; fruit a capsule.

Anemopsis Hook. & Arn.   Yerba Mansa

A single genus and species treated, with the characteristics of the family; mostly in southwestern Utah, but adventive elsewhere. **A. californica** Hook.

Saxifragaceae

SAXIFRAGACEAE — SAXIFRAGE FAMILY

Perennial herbs or shrubs; leaves simple, alternate, opposite or basal; flowers usually perfect and regular, usually with hypanthium and often with a disk; sepals 4–5, distinct or united; petals 4–5, separate, sometimes lacking; stamens variable in number, usually as many as petals or twice as many, sometimes numerous; pistils 1; ovary inferior to superior; fruit a follicle, capsule, or berry.

Some authors have divided this family into several families without, however, simplifying all of the resulting taxa, at least one of which remains almost as variable as the entire original family.

1a. Plants woody, prostrate or upright shrubs. (2)
1b. Plants herbaceous, not woody, at least above base. (8)
2a. Leaves alternate; fruit a berry; spines present in some. **Ribes**
2b. Leaves opposite; fruit a capsule; spines never present. (3)
3a. Flowers of 2 types in an inflorescence—small central fertile ones and enlarged marginal sterile ones; cultivated. **Hydrangea**
3b. Flowers all fertile. (4)
4a. Petals 4, rarely 5; stamens either 8 or many, never 10. (5)
4b. Petals 5; stamens 10. (6)
5a. Stamens 8; ovary inferior only at base (less than one-half). **Fendlera**
5b. Stamens 20 or more; ovary at least two-thirds inferior. **Philadelphus**
6a. Inflorescence paniculate; flowers white to bluish pink; cultivated. **Deutzia**
6b. Inflorescence cymose; flowers white; native. (7)
7a. Leaves entire, small, not over 1.5 cm long, sessile or nearly so; flowers small; petals 2.5–4 mm long. **Fendlerella**
7b. Leaves toothed, over 1.5 cm long, definitely petiolate; flowers larger; petals 8–12 mm long. **Jamesia**
8a. Stamens 10. (9)
8b. Stamens 4 or 5, with possible clusters of staminodes between. (11)
9a. Ovary 1-loculed, with 3 parietal placentae; petals 3- to 7-parted. **Lithophragma**
9b. Ovary 2-loculed, rarely more so; placentation axile. (10)
10a. Styles partially connate; petals pink to deep red; calyx campanulate, usually reddish, (5) 6–10 mm long; leaves alternate, petiolate. **Telesonix**
10b. Styles free above ovuliferous portion of ovary; petals usually white, rarely pink or red to purple, if latter colors, then either calyx not at once campanulate, red, and as long as 6 mm, or leaves sessile and opposite. **Saxifraga**
11a. Clusters of staminodes alternating with fertile stamens; flowers solitary on stems; ovary 1-loculed with 3 or 4 parietal placentae; leaves entire. **Parnassia**
11b. Clusters of staminodes lacking; flowers in clusters; ovary either

Subclass DICOTYLEDONEAE

  2-loculed or 1-loculed with 2 parietal placentae; leaves toothed or lobed. (12)
12a. Stems leafy; flowers clustered in axils of upper leaves; stamens and sepals 4; petals lacking. **Chrysosplenium**
12b. Stems scapose; leaves nearly or quite basal; flowers terminal; stamens and sepals 5; petals present, may be small or deciduous. (13)
13a. Ovary 2-loculed, with axile placentae. **Sullivantia**
13b. Ovary 1- loculed, with 2 parietal placentae. (14)
14a. Inflorescence a raceme; petals pinnatifid to 3-toothed, only rarely entire; stamens alternate or in common species, opposite to petals. **Mitella**
14b. Inflorescence a panicle (may be dense); petals entire; stamens alternate to petals. **Heuchera**

Chrysosplenium L.   Golden Saxifrage

  Perennial herbs from stoloniferous rootstocks; leaves alternate; flowers in upper leaf axils; calyx lobes 4 or 5; petals lacking; stamens 4; ovary partly inferior, 2-lobed but 1-loculed; fruit a capsule; mountains, northern Utah. **C. tetrandrum** (Lund) Fries

Deutzia Thunb.

  Plants shrubs to 2 m tall; bark shreddy; leaves opposite, rough-hairy; flowers white to bluish pink; calyx lobes 5; petals 5, 10–13 mm long; stamens 10, the filaments usually winged and toothed; ovary inferior; fruit a capsule; cultivated. **D. scabra** Thunb.

Fendlera Engelm. & Gray   Fendlerbush

  Plants shrubs to 3 m tall; leaves opposite, 5–40 mm long; flowers 1, or 2 or 3 together; calyx lobes 4; petals 4, white, 12–20 mm long; stamens 8, the filaments flattened and 2-forked at apex; ovary inferior at base, 4-loculed; fruit a capsule; southeastern Utah. **F. rupicola** Gray

Fendlerella Heller

  Plants shrubs to 1 m tall; leaves opposite; flowers perfect, in small compound cymes; calyx lobes 5; petals 5, white, 2.5–4 mm long; stamens 10, the alternating filaments long and short; ovary inferior at base, 3-loculed, fruit a capsule; eastern and southern Utah. **F. utahensis** (S. Wats.) Heller

Heuchera L.   Alumroot

1a. Stamens surpassing sepals; pistils exserted, sometimes tardily; styles slender, elongate, gradually expanded below; flowers pink or pinkish; petals much surpassing sepals. (2)
1b. Stamens not equaling sepals; pistils included. (3)
2a. Filaments distinctly widened and flattened at base; stamens inserted slightly below insertion of petals; petioles finely glandular-puberulent; widespread, mountains. **H. rubescens** Torr.

Saxifragaceae

2b. Filaments not distinctly widened and flattened at base; stamens usually inserted well below level of insertion of petals; stems commonly villous; coniferous forests, southern Utah. **H. versicolor** Greene
3a. Flowers deep pink to carmine; inflorescence relatively short and expanded, the lower branches seldom less than 2 cm long; petals much shorter than sepals; often cultivated. **H. sanguinea** Engelm.
3b. Flowers greenish or yellowish; petals often whitish; inflorescence elongate and contracted, the lower branches usually less than 2 cm long; widespread. **H. parvifolia** Nutt.

Hydrangea L.

1a. Leaves pinnately 3- to 7-lobed. **H. quercifolia** Bartram
1b. Leaves not lobed. (2)
2a. Inflorescence a pyramidal panicle, long and narrowing upward. Pee Gee Hydrangea. **H. paniculata** Sieb.
2b. Inflorescence a round- to flat-topped corymb. (3)
3a. Styles usually 3; flowers bluish or pinkish; leaves coarsely serrate, mostly tapering to petiole; common greenhouse hydrangea. **H. opuloides** Koch
3b. Styles usually 2; flowers creamy white; leaves serrate, rounded or cordate at base; cultivated. **H. arborescens** L.

Jamesia T. & G. Cliffbush

Plants shrubs to 2 m tall; leaves opposite; flowers rather large; calyx 5-lobed; petals 5, white, 8–12 mm long; stamens 10; ovary conical, about half inferior, 1-loculed finally; styles 3–5; fruit a capsule; mountains. **J. americana** T. & G.

Lithophragma Nutt. Woodland Star

1a. Leaves of stem with clusters of bulblets in all or some axils; upper part of stipules rounded; mountains. **L. bulbifera** Rydb.
1b. Bulblets seldom if ever present on stem or inflorescence; upper part of stipules usually pointed. (2)
2a. Calyx tube campanulate, the base somewhat rounded; whole calyx 3–4 mm long, adnate to ovary about one-fifth its length; mountains **L. tenella** Nutt.
2b. Calyx tube top shaped, cuneate at base; whole calyx 4–8 mm long, adnate to ovary about one-half its length or more; mountains. **L. parviflora** (Hook.) Nutt.

Mitella L. Mitrewort

1a. Stamens opposite petals; petals pinnatifid, with more than 3 filiform divisions; pedicels over 1.5 mm long; calyx lobes greenish; mountains. **M. pentandra** Hook.
1b. Stamens alternate to petals; petals entire to cleft into 3 filiform divisions; pedicels less than 1.5 mm long; calyx lobes whitish. (2)

Subclass DICOTYLEDONEAE

2a. Calyx 3–5 mm long; petals twice as long as calyx; higher mountains. **M. stauropetala** Piper
2b. Calyx 1.5–3 mm long; petals only one-half as long as calyx; high mountains. **M. stenopetala** Piper

Parnassia L.   Grass of Parnassus

1a. Petals fimbriate-pectinate on lower half; bogs, wet meadows, and stream sides, lower montane to arctic-alpine regions. **P. fimbriata** Koenig
1b. Petals not fimbriate-pectinate on lower half. (2)
2a. Leaf blades elliptic to elliptic-ovate, neither truncate nor cordate at base; bract never clasping; petals usually 5-veined, mostly 4–7 mm long; bogs, wet meadows, and stream banks, eastern Utah. **P. parviflora** DC.
2b. Leaf blades lanceolate to broadly ovate, often cordate; bract often clasping; petals 7- to 13-veined, mostly (6) 7–12 mm long; high mountains, northern Utah. **P. palustris** L.

Philadelphus L.   Mockorange

1a. Leaves entire or with 1–4 small teeth. (2)
1b. Leaves denticulate to serrate. (3)
2a. Leaves 1–1.6 cm long, entire; petals 9–11 mm long; dry canyons and hillsides, eastern Utah. **P. microphyllus** Gray
2b. Leaves 1–4 cm long, margins toothed; flowers 2.5–4 cm broad, fragrant; cultivated. **P. lemoinei** Lemoine
3a. Bark not exfoliating, often yellowish. (4)
3b. Bark exfoliating, brown. (5)
4a. Leaves glabrous, elliptic to obovate; flowers white, 4–5 cm broad, scentless; cultivated. Common Mockorange. **P. inodorus** L.
4b. Leaves pubescent on both sides or only on veins below, ovate; flowers white, 3.5–4.5 cm broad, fragrant; cultivated. **P. lewisii** Pursh
5a. Branchlets densely pubescent; leaves serrate; cultivated. **P. incanus** Koehne
5b. Branchlets glabrous or nearly so; leaves dentate. (6)
6a. Leaves pubescent below, nearly glabrous above; pedicels and calyx pubescent; flowers white, double or semidouble; styles divided to middle or below; cultivated. **P. virginalis** Rehd.
6b. Leaves glabrous above and below, except on veins below; pedicels hairy, the calyx glabrous; flowers white, fragrant; styles distinct all or part of their length; cultivated. **P. coronarius** L.

Ribes L.   Gooseberry, Currant

1a. Plants with stiff bristles or spines or both, on some or all twigs. (2)
1b. Plants lacking bristles or spines. (8)

Saxifragaceae

2a. Calyx tube above ovary short, not over 2 mm long, bowl or saucer shaped; twigs often with bristles. (3)
2b. Calyx tube above ovary over 2 mm long, campanulate to cylindric; twigs rarely with bristles. (4)
3a. Leaf blades nearly or quite glabrous, often over 3 cm wide; berry purplish black; eastern Utah. **R. lacustre** (Pers.) Poir.
3b. Leaf blades pubescent, not over 3 cm wide; berry red; eastern Utah. **R. montigenum** McClatchie
4a. Calyx tube campanulate. (5)
4b. Calyx tube cylindric. (6)
5a. Petals pink or white; calyx rarely sparsely pilose, usually glabrous; fruit glabrous, wine colored. **R. inerme** Rydb.
5b. Petals greenish; calyx pilose; fruit pubescent, rarely glabrate, greenish; cultivated. Gooseberry. **R. grossularia** L.
6a. Calyx tube glabrous, 5–8 mm long; fruit glabrous, deep purplish black; mountains, northern Utah. **R. setosum** Lindl.
6b. Calyx tube densely pubescent. (7)
7a. Calyx tube 1.5–2.5 mm long, about as broad; leaves densely pubescent, often glandular; fruit usually pubescent to glandular, yellow to red; southern Utah. **R. velutinum** Greene
7b. Calyx tube 4–8 mm long; leaves glabrous to pubescent; fruit glabrous, rarely glandular, hispid or bristly, blackish; southeastern Utah. **R. leptanthum** Gray
8a. Calyx tube above ovary saucer or bowl shaped, not over 2 mm long. (9)
8b. Calyx tube above ovary cylindric, 3 mm long or more. (13)
9a. Leaf blades without resinous glands; racemes spreading or drooping; ovary and fruit glabrous; flowers yellowish or greenish; cultivated. Common Currant. **R. sativum** Syme
9b. Leaf blades with resinous glands below; ovary glandular. (10)
10a. Ovary and usually young herbage sprinkled with sessile, yellow, crystalline glands. (11)
10b. Ovary glandless or with stipitate, nonyellow, noncrystalline glands; if herbage, as rarely, with sessile crystalline glands, then ovary always stipitate-glandular. (12)
11a. Racemes erect or spreading, mostly 4–10 cm long. **R. hudsonianum** Richards.
11b. Racemes drooping; sepals usually purplish tinged; fruit black, odorous; cultivated. Bedbug Currant. **R. nigrum** L.
12a. Calyx lobes linear to oblong-ovate, definitely longer than wide; calyx tube 1–1.5 mm long; stems stout, not trailing or prostrate; fruit glandular-bristly; often in partial shade, mountains. **R. mogollonicum** Greene
12b. Calyx lobes about as long as wide; calyx tube 0.5–1 mm long; stems

Subclass DICOTYLEDONEAE

weak, prostrate or reclining; fruit sparingly glandular-hairy; mountains. **R. coloradense** Coville

13a. Calyx bright yellow, glabrous; leaves glabrous, vernation convolute; ovary glabrous; flowers fragrant; widespread. **R. aureum** Pursh

13b. Calyx greenish, whitish, or pinkish, hairy; leaves usually hairy, vernation plicate; ovary glandular-pubescent. (14)

14a. Calyx tube above the ovary narrow, 3–4 times longer than wide; leaf blades 1–3 cm wide, sparingly if at all glandular-pubescent; fruit red, fleshy. **R. cereum** Dougl.

14b. Calyx tube above the ovary broad, about twice as long as wide; leaf blades 3–7 cm wide, glandular-pubescent to glandular-hirsute; fruit black, almost dry. **R. viscossissimum** Pursh

Saxifraga L. Saxifrage

1a. Foliage leaves all basal; flowers borne on scapes naked below the inflorescence; petals white; calyx tube short or absent, never as long as lobes. (2)

1b. Foliage leaves not all basal, the stems leafy below inflorescence; petals white, rose-purple, or yellow; calyx tube very short to long. (6)

2a. Flowers not represented by bulblets. (3)

2b. Flowers represented by clusters of bulblets or with 1 terminal flower only; petals often very unequal; high mountains, northern Utah. **S. vreelandii** (Small) Fedde ex Just

3a. Petals very unequal, 2 petals much longer than the others; plants decidedly stoloniferous; leaves cordate-orbicular, upper surface veined white, lower surface reddish; greenhouse plant. Strawberry Saxifrage. **S. sarmentosa** L.

3b. Petals equal or essentially so. (4)

4a. Filaments club shaped, wider above; leaves suborbicular to reniform, as wide as long; high mountains. **S. arguta** D. Don

4b. Filaments not club shaped, usually wider below; leaves ovate, elliptic, or narrow, longer than wide. (5)

5a. Cymules wholly or mainly aggregated into heads or into a spikelike panicle, at least at first; leaves 2–6 cm long, margins entire to crenate or dentate; widespread, moist situations, at higher elevations. **S. rhomboidea** Greene

5b. Cymules in panicles which become wide; leaves 6–20 cm long, remotely glandular-denticulate; petals with broad, clawlike bases; uncommon, bogs and stream sides, northern Utah. **S. oregana** Howell

6a. Leaves suborbicular to reniform, little if any longer than wide, 5- to 7-palmately lobed; petals white; calyx tube over 1.5 mm long. (7)

6b. Leaves narrower than suborbicular, definitely longer than wide, entire or 3-toothed or 3-lobed at apex; calyx tube long or very short. (8)

Scrophulariaceae

7a. Flowers below terminal 1 replaced by clusters of bulblets; inflorescence spikelike; stems more or less glandular-pubescent; mountains. **S. cernua** L.
7b. Flowers not replaced by bulblets; inflorescence an open cyme; stems glabrous or nearly so; among rocks in high mountains, eastern Utah. **S. debilis** Engelm.
8a. Plants with long, filiform, naked stolons, these leafy and rooting at tips; petals yellow; high mountains. **S. flagellaris** Willd. ex Sternb.
8b. Plants not stoloniferous; petals white to yellow. (9)
9a. Calyx tube long, usually becoming as long as, or longer than, lobes; stem leaves 1 to few, at least some 3-toothed or 3-lobed at apex; petals white, not over 3.5 mm long. (10)
9b. Calyx tube very short or absent; stem leaves several to many, all leaves entire; petals yellow or white, 4 mm long or longer. (11)
10a. Leaves entire or merely 3-toothed at apex; high mountains. **S. adscendens** L.
10b. Leaves, at least lower, 3-lobed or 3-cleft at apex; high mountains. **S. caespitosa** L.
11a. Petals white, often spotted with orange or purple dots; leaves spine tipped and spiny-ciliate; rocky places in high mountains, eastern Utah. **S. bronchialis** L.
11b. Petals yellow, not spotted; leaves not spiny tipped or spiny-ciliate. (12)
12a. Petals 7–13 mm long, without claws, elliptic to oblong in shape; leaves crowded but hardly rosettelike, mostly over 10 mm long; plants 6–20 cm tall; high mountains. **S. hirculus** L.
12b. Petals 5.5–7 mm long, abruptly short clawed at base, oval to suborbicular in shape; leaves rosettelike, seldom over 10 mm long; plants 3–8 cm tall; high mountains. **S. chrysantha** Gray

Sullivantia T. & G.

Perennial acaulescent herbs, 20–30 cm tall; leaves alternate; flowers in panicled cymes; calyx united, with 5 sepals shorter than the hypanthium; petals 5, white or whitish; stamens 5, alternate to petals; ovary half-inferior, 2-loculed; to be expected in southeastern Utah. **S. purpusii** (Brand) Rosendahl

Telesonix Raf.

Perennial herbs with reniform leaves; flowers in panicles, showy, reddish purple; calyx adnate to lower part of ovary; petals 5; stamens 10; high mountains. **T. jamesii** (Torr.) Raf.

SCROPHULARIACEAE — FIGWORT FAMILY

Annual or perennial herbs; stems usually terete; leaves opposite, rarely alternate or whorled, simple; flowers perfect, usually more or less irregular; calyx mostly with 5 more or less united sepals; corolla of 5

Subclass DICOTYLEDONEAE

united petals, usually 2-lipped; stamens commonly 4, or the fifth present and fertile or sterile, or 2; ovary superior, 2-loculed; styles 1; stigmas entire or 2-lobed; fruit a capsule.

- 1a. Anther-bearing stamens 5; corolla nearly regular. **Verbascum**
- 1b. Anther-bearing stamens 2 or 4 (rarely 5 in some **Penstemon** species); corolla usually irregular. (2)
- 2a. Anther-bearing stamens normally 2. (3)
- 2b. Anther-bearing stamens normally 4. (6)
- 3a. Anther cells separated by a broad connective wider than the sacs. **Gratiola**
- 3b. Anther cells contiguous at apex, sometimes confluent. (4)
- 4a. Stems leafy, the leaves opposite at least below; corolla nearly regular; stamens not conspicuously if at all exserted at anthesis. **Veronica**
- 4b. Stems scapose, the true foliage leaves all basal, cauline leaves bractlike and alternate; corolla absent or, when present, definitely irregular; stamens exserted at anthesis. (5)
- 5a. Corolla well developed, nearly regular, blue or blue-violet; filaments inconspicuously colored. **Synthyris**
- 5b. Corolla absent or, if present, definitely irregular; basal leaves merely crenate-serrate. **Besseya**
- 6a. Corolla with a spur or pouch at base. (7)
- 6b. Corolla without a spur or pouch. (9)
- 7a. Corolla with a pouch at base, on upper side of flower; plants seldom over 5–10 cm tall. **Collinsia**
- 7b. Corolla with a pouch or spur at base on lower side of flower; plants usually over 15 cm tall. (8)
- 8a. Corolla with a saccate pouch at base. **Antirrhinum**
- 8b. Corolla with a definite spur base. **Linaria**
- 9a. Stamens 5, 4 anther bearing, the fifth sterile and lacking an anther, sometimes much reduced in size; leaves opposite or whorled. (10)
- 9b. Stamens 4, all anther bearing, reduced or rudimentary stamens only very rarely present; leaves opposite, basal or alternate. (12)
- 10a. Plants annual; corolla not over 6 mm long, the middle lobe of lower lip deeply concave and enclosing the stamens; sterile stamen reduced, minute and glandlike. **Collinsia**
- 10b. Plants perennial or biennial; corolla over 6 mm long, the middle lobe of lower lip neither concave nor enclosing stamens; sterile stamen scalelike or elongate. (11)
- 11a. Sterile stamen short, scalelike, nearly as wide as long; corolla 5–10 mm long, greenish to greenish brown, not at all showy. **Scrophularia**
- 11b. Sterile stamen slender, elongated, little if any shorter than anther-bearing ones; corolla usually well over 10 mm long, white to variously colored but not greenish to greenish brown, usually showy. **Penstemon**

Scrophulariaceae

12a. Anther sacs separate and dissimilar, outer one versatile, inner pendulous by its apex or mostly smaller, sometimes rudimentary; leaves alternate, often more or less dissected or lobed. (13)
12b. Anther sacs either similar in shape and attachment or confluent, not as above; leaves usually opposite or basal, usually entire or toothed (may be alternate and dissected in **Pedicularis**). (15)
13a. Upper corolla lip much longer than the lower; plants perennial or annual in 1 species; calyx tubular, cleft above and below, lateral lobes usually toothed or cleft; flowers or bracts frequently red. **Castilleja**
13b. Upper lip of corolla little if any longer than the lower; plants annual, calyx with 4 equal or subequal lobes or cleft to base on sides or appearing as 1 sepal; flowers and bracts purplish to yellowish or white, never red. (14)
14a. Calyx of united sepals with 4 equal or nearly equal lobes; leaves entire or 3-parted. **Orthocarpus**
14b. Calyx cleft to base on sides with 2 sepals, or only upper sepal present and not united; leaves 3- to 5-parted. **Cordylanthus**
15a. Upper lip of corolla conspicuously arched upward in the middle (galeate), the corolla conspicuously 2-lipped; leaves toothed or variously lobed. (16)
15b. Upper lip of corolla not conspicuously arched (except **Mimulus**), the corolla nearly regular to somewhat 2-lipped (except for some species of **Mimulus**); leaves entire or undulate. (17)
16a. Plants annual; leaves opposite, sharply dentate or serrate; calyx 4-toothed, becoming bladderlike and veiny, completely enclosing the fruit; fruit symmetric, both cells dehiscing equally. **Rhinanthus**
16b. Plants perennial; leaves usually alternate, either crenate or variously lobed and dissected; calyx cleft on lower side, becoming distended but not bladderlike, not completely enclosing the fruit; fruit asymmetric, curved, opening chiefly or wholly on one side. **Pedicularis**
17a. Anther cells wholly confluent, appearing as a single cell; capsule 1-loculed above; corolla minute, 2–2.5 mm long; plants subscapose. **Limosella**
17b. Anthers not as above. (18)
18a. Corolla 40–50 mm long, purple or purplish, campanulate, the lower lobes overlapping in the bud. **Digitalis**
18b. Corolla less than 30 mm long or, if longer, then 2-lipped and scarlet or crimson, or bright yellow spotted red, upper lobes overlapping in the bud. **Mimulus**

Antirrhinum L.   Snapdragon

1a. Stems climbing by tendrillike peduncles. **A. filipes** Gray ex Ives
1b. Stems not climbing, commonly erect. (2)

Subclass DICOTYLEDONEAE

- 2a. Uppermost sepal not longer than others; corolla 35 mm long or more, variously colored; biennial or perennial; commonly cultivated. **A. majus** L.
- 2b. Uppermost sepal longer and larger than others; corolla 5–7 mm long, whitish with purple veins; annual; western Utah. Least Snapdragon. **A. kingii** S. Wats.

Besseya Rydb. Kitten-tails

- 1a. Corolla lacking; filaments conspicuously colored. **B. wyomingensis** (A. Nels.) Rydb.
- 1b. Corolla violet-purple; filaments not conspicuously colored; alpine regions. **B. alpina** (Gray) Rydb.

Castilleja Mutis Indian Paintbrush

- 1a. Plants annual; stems usually short and erect. **C. minor** Gray
- 1b. Plants perennial, more or less woody; stems usually elongate. (2)
- 2a. Lower lip of corolla prominent, one-half to one-third as long as upper lip (galea); galea short, one-half as long as corolla tube or less, purple; bracts purplish. **C. rubida** Piper
- 2b. Lower lip of corolla relatively small, usually less than one-third the length of galea, not over one-half; galea over one-half as long as corolla tube or even exceeding it; bracts various. (3)
- 3a. Calyx cut much deeper below than above; bracts divided into linear lobes, often not as conspicuous as calyx. (4)
- 3b. Calyx equally or subequally cut above and below; bracts divided to entire, broad and conspicuous. (5)
- 4a. Corolla 1.5–2.5 cm long; bracts yellow or rarely yellowish, red tipped; calyx yellowish. **C. flava** S. Wats.
- 4b. Corolla 3–5 cm long; bracts red or scarlet; calyx red or scarlet. **C. linariaefolia** Benth. ex DC.
- 5a. Plants mostly alpine; stems not over 20 cm tall; bracts entire or short lobed at apex, the lobes not linear, greenish yellow or somewhat streaked with red. **C. occidentalis** Torr.
- 5b. Plants not alpine; stems over 20 cm tall on normal plants, though **C. chromosa** may be as low as 10 cm tall. (6)
- 6a. Leaves, at least upper ones, deeply cleft into linear, divaricate lobes; flowering in early spring; common at lower elevations. **C. chromosa** A. Nels.
- 6b. Leaves entire or upper ones sometimes shallowly lobed near apex. (7)
- 7a. Plants viscid- or glandular-hairy, at least in the inflorescence. (8)
- 7b. Plants glabrous to puberulent, sometimes crisp-pubescent in inflorescence, not viscid or glandular. (9)
- 8a. Inflorescence yellow; bracts, calyx, and corolla never tinged with red. **C. septentrionalis** Lindl.

Scrophulariaceae

8b. Inflorescence red to scarlet, sometimes to orange or yellowish. **C. applegatei** Fern.
9a. Bracts entire or very shallowly lobed, often rose colored. **C. rhexifolia** Rydb.
9b. Bracts, at least sometimes, rather deeply cleft, never really rose colored. **C. miniata** Dougl. ex Hook.

Collinsia Nutt. Blue-eyed Mary
Annual herbs, 5–40 cm tall; leaves to 5 cm long; flowers 2–5 in a whorl, or 1; corolla 4–6 mm long; stigma 2-lobed; capsule 3–4 mm long; widely distributed. **C. parviflora** Dougl.

Cordylanthus Nutt. Birdbeak
1a. Flowers not in heads or spikes, scattered along branches. **C. parviflorus** (Ferris) Wiggans
1b. Flowers usually in headlike or short spikelike inflorescences, sometimes solitary at ends of branches. (2)
2a. Leaves and bracts entire; corolla 15–20 mm long. **C. canescens** Gray
2b. Leaves and bracts variously cleft or dissected. (3)
3a. Corolla 20–30 mm long, yellow to purple; calyx 2-parted; plants 30–50 cm tall. **C. wrightii** Gray ex Torr.
3b. Corolla 20 mm long or less; plants 30 cm tall or less. (4)
4a. Stems glandular-puberulent or glandular-villous; calyx 12–15 mm long, of 1 part; corolla purplish. **C. kingii** S. Wats.
4b. Stems cinereous-puberulent; calyx 15–25 mm long, of 2 parts; corolla yellow. **C. ramosus** Nutt. ex DC.

Digitalis L. Foxglove
Biennial herbs; leaves alternate; inflorescence racemose, many-flowered; flowers 40–50 mm long, purplish, on lower side pale with purple spots and mottlings; cultivated. **D. purpurea** L.

Gratiola L. Hedge Hyssop
Annual herbs, 7–20 cm tall; leaves opposite, 1–5 cm long; flowers solitary on axillary pedicels; corolla 8–12 mm long, the tube greenish yellow, the lobes white. **G. neglecta** Torr.

Limosella L. Madwort
Plants herbs; leaves in a basal rosette; flowers small, solitary, white to purplish, nearly regular, 2–2.5 mm long; introduced from Europe. **L. aquatica** L.

Linaria Mill. Toadflax, Butter and Eggs
1a. Flowers yellow or partly orange, usually over 15 mm long, exclusive of the spur; plants perennial. (2)
1b. Flowers whitish to violet, less than 15 mm long; plants annual or biennial. (3)

Subclass DICOTYLEDONEAE

- 2a. Stem leaves cordate-clasping, upper leaves ovate to ovate-lanceolate, some over 1 cm wide; introduced and well established in the Wasatch Mountains. **L. dalmatica** (L.) Mill.
- 2b. Stem leaves not clasping, linear or linear-lanceolate, not over 5 mm wide; occasional weed. **L. vulgaris** Mill.
- 3a. Corolla over 10 mm long, exclusive of 5–9 mm long spur. **L. texensis** Scheele
- 3b. Corolla less than 10 mm long, exclusive of 2–6 mm long spur. **L. canadensis** (L.) Du Mont de Cours

Mimulus L. Monkeyflower

- 1a. Corolla 3.3–5.5 cm long, crimson red, pink, rose, or lilac; anthers bearded. (2)
- 1b. Corolla usually less than 3.3 cm long or, if longer, then blue or yellow and with glabrous anthers, variously colored; anthers glabrous or bearded. (4)
- 2a. Plants stoloniferous; stems not over 20 cm long, procumbent or prostrate; leaf blades commonly oblong; corolla tube moderately to greatly surpassing calyx. **M. eastwoodiae** Rydb.
- 2b. Plants ascending to erect; stems over 30 cm tall. (3)
- 3a. Corolla campanulate, lobes little differentiated. **M. lewisii** Pursh
- 3b. Corolla strongly 2-lipped, upper lip arched-ascending, lower deflexed-recurved. **M. cardinalis** Dougl.
- 4a. Pedicels less than one-half as long as calyx. (5)
- 4b. Pedicels as long as, or longer than, calyx. (6)
- 5a. Corolla distinctly 2-lipped, yellow, often tinged or spotted with reddish purple; anthers glabrous. **M. parryi** Gray
- 5b. Corolla nearly regular, pink with a yellow tube, often with a bright yellow patch in throat; anthers usually hispidulous. **M. bigelovii** Gray
- 6a. Calyx lobes decidedly unequal; anthers glabrous. (7)
- 6b. Calyx lobes more or less equal. (10)
- 7a. Corolla 18 mm long or less, throat open. (8)
- 7b. Corolla 25 mm long or more, throat partly or nearly closed. (9)
- 8a. Plants annual, villous; corolla 4–9 mm long, yellow. **M. pilosus** (Benth.) S. Wats.
- 8b. Plants perennial, glabrous or nearly so; corolla 7–18 mm long, yellow. **M. glabratus** H. B. K.
- 9a. Plants perennial, 5–30 cm tall; leaves 1–3 cm long, lower short-petioled, rounded to truncate-cuneate; corolla 25–35 mm long, yellow; capsule 7–8 mm long. **M. tilingii** Regel
- 9b. Plants annual or biennial, 40–100 cm tall; leaves 1.5–9 cm long, rounded, with petioles sometimes longer than blades; corolla 30–45 mm long, yellow; capsule 10–12 mm long. **M. guttatus** Fisch.
- 10a. Mature calyx strongly inflated; corolla yellow, about 7–14 mm

Scrophulariaceae

long; leaves with definite petioles often longer than blades; plants annual. **M. floribundus** Dougl.

10b. Mature calyx not strongly inflated (may be distended by developing fruit); corolla various but not both yellow and 7–14 mm long; leaves sessile or with petioles much shorter than the blades. (11)

11a. Plants perennial, with slender stolons or underground rootstocks and bulbils, scapose to subcaulescent with short internodes; corolla 8–20 mm long. **M. primuloides** Benth.

11b. Plants annual, caulescent, commonly much branched from base, internodes well developed; corolla 5–10 mm long. (12)

12a. Stems to 5 cm tall; leaves crowded, slightly shorter to much longer than the internodes; pedicels 3–9 mm long; calyx teeth not ciliate; corolla yellow; stigma lobes unequal. **M. suksdorfii** Gray

12b. Stems to 20 cm tall; leaves not crowded, much shorter than internodes; pedicels 6–20 mm long; calyx teeth usually ciliate; corolla yellow or with the limb pink; stigma lobes equal. **M. rubellus** Gray

Orthocarpus Nutt. Owlclover

1a. Corolla white, turning purplish; inflorescence few-flowered. **O. purpureo-albus** Gray
1b. Corolla yellow; inflorescence dense. (2)
2a. Plants pubescent, hirsute, usually erect. **O. luteus** Nutt.
2b. Plants puberulent only, usually widely branched above. **O. tolmei** H. & A.

Pedicularis L. Lousewort

1a. Galea over 8 mm long, prolonged into a slender recurved beak (curved outward and upward, resembling head and trunk of an elephant). Elephantella. **P. groenlandica** Retz.
1b. Galea not over 5 mm long, beakless or with a shorter, straight or incurved beak. (2)
2a. Leaves toothed or pinnately lobed, sinuses not extending more than two-thirds the distance to midrib. (3)
2b. Leaves pinnately deeply parted or divided to midrib or nearly so, sinuses always much over two-thirds the distance to midrib. (5)
3a. Galea prolonged into an incurved beak 3–5 mm long; corolla white. **P. racemosa** Dougl. ex Hook.
3b. Galea beakless or with only small teeth near apex; corolla rarely white. (4)
4a. Leaves pinnately lobed, sinus extending one-half to two-thirds the distance to midrib; plants subcaulescent; anthers aristate-acuminate at base; usually in pinyon-juniper community. **P. centranthera** Gray
4b. Leaves doubly crenate toothed, not all lobed; usually at middle and higher elevations. **P. crenulata** Benth. ex DC.
5a. Corolla large, 25–30 mm long, the galea beakless but with 2 lateral

Subclass DICOTYLEDONEAE

teeth just below apex; flowers sordid yellow, rarely streaked with red. **P. grayi** A. Nels.
5b. Corolla smaller, less than 25 mm long, whitish or yellowish, the galea beaked or beakless and lacking teeth. (6)
6a. Galea tapering into a short beak 1-2 mm long; leaves narrow, rarely over 2 cm wide, primary divisions seldom over 1 cm long. **P. parryi** Gray
6b. Galea beakless, truncate at apex; leaves over 4 cm wide, primary divisions usually over 2 cm long. **P. paysoniana** Pennell

Penstemon Mitchell   Beardtongue

1a. Corolla 37–50 mm long. (2)
1b. Corolla up to 35 mm long. (4)
2a. Corolla campanulate to salverform, the tube slightly curved, drooping, from white to red or purple; leaves toothed or entire; inflorescence often pubescent; cultivated hybrid. **P. hartwegii** Benth. **P. x cobaea** Nutt.
2b. Corolla not campanulate. (3)
3a. Leaves entire, thick, broad, and obtuse, the stem leaves with broad bases cordate-clasping; leaves and stem very glaucous, glabrous; flowers blue, lilac, rarely white, glabrous; cultivated. **P. grandiflorus** Nutt.
3b. Leaves toothed, at least above, thick, upper clasping; leaves and stem puberulent to pubescent or glandular-pubescent; flowers purplish to white, glandular-pubescent outside, glandular within, tube abruptly inflated, limb scarcely 2-lipped; cultivated. **P. cobaea** Nutt.
4a. Flowers bright red, usually scarlet (rarely cream in **P. eatonii**). (5)
4b. Flowers white, pink, blue, lavender, or purple. (7)
5a. Calyces and pedicels definitely glandular-pubescent; anther sacs dehiscent across their joined apices and distally for less than one-half their length; southern Utah. **P. bridgesii** Gray
5b. Calyces and pedicels glabrous or pubescent, at most very obscurely glandular; anther sacs dehiscent from their free ends one-half or more toward their juncture. (6)
6a. Corolla strongly 2-lipped, the lower lip reflexed and glabrous or bearded, the upper projecting or spreading; anther sacs glabrous to long-pubescent; southern Utah. **P. barbatus** (Cav.) Roth
6b. Corolla obscurely 2-lipped, the lips subequally erect or spreading; stems glabrous to puberulent; leaves glabrous or glabrate to puberulent; north central to southern Utah. **P. eatonii** Gray
7a. Anther sacs hispid-pubescent to woolly-hairy on sides, scarcely contiguous, line of contact of sacs very short. (8)
7b. Anther sacs glabrous (in **P. hallii** sometimes with a small tuft of hairs at summit of the filament). (20)
8a. Anther sacs short-pubescent with relatively stiff hairs. (9)

Scrophulariaceae

8b. Anther sacs woolly-hairy with tangled (flexuous) hairs. (13)
9a. Anther sacs opening partially; corollas with expanded throat, deep sky blue to blue-purple, 18–23 mm long; hills and mountain slopes, Wasatch Mountains and southwestern Utah. **P. cyananthus** Hook.
9b. Anther sacs opening throughout. (10)
10a. Corolla 25–30 mm long; sepals not glandular-puberulent externally; cultivated. **P. glaber** Pursh
10b. Corolla 15–25 mm long; sepals minutely and usually obscurely glandular-puberulent externally. (11)
11a. Plants mostly densely puberulent throughout, sometimes glabrate in upper inflorescence; corolla dark bluish purple to violet-purple, 17–22 mm long; northeastern Utah. **P. fremontii** T. & G.
11b. Plants mostly glabrous, sometimes sparsely glandular-puberulent in inflorescence. (12)
12a. Corolla widening from a narrow basal tube, mostly 22–30 mm long, blue, posterior lobes united over one-half their length; sepals ovate, tip acuminate; foothills, canyons, and mountainsides. **P. subglaber** Rydb.
12b. Corolla gradually widening from a broad basal tube, 18–20 mm long, blue, posterior lobes united about one-third their length; sepals broadly ovate, acute, margin broad, white to pinkish; alpine, Uinta Mountains. **P. uintahensis** Pennell
13a. Corolla pale blue, throat scarcely exceeding relatively long basal tube; anther sacs densely woolly-hairy with long hairs; sterile filament glabrous; inflorescence lax, scarcely secund; herbage somewhat glaucous; rocky or gravelly sagebrush slopes, southeastern quarter of Utah. **P. comarrhenus** Gray
13b. Corolla deep blue, rose-red, or lilac-purple, throat obviously exceeding basal tube in length; anther sacs less densely woolly-hairy with long hairs or densely hairy with tufted hairs; sterile filaments more or less bearded or glabrous; inflorescence secund or condensed; herbage green, glabrous, glandular-pubescent, sometimes glaucous. (14)
14a. Anther sacs woolly-pubescent with slender hairs, these shorter than, or about equaling, width of anther sac; flowers deep blue. (15)
14b. Anther sacs woolly with twisted hairs, these mostly much exceeding width of sac. (17)
15a. Leaves ovate to lanceolate; stem leaves lanceolate, obtuse to acute, largest 3–7 cm long; sepals with obscure, narrow, scarious margin, acute; corolla 15–20 mm long; east central Utah. **P. cyanocaulis** Payson
15b. Leaves lanceolate; stem leaves narrowly lanceolate, mostly acuminate, largest 6–10 cm long; sepals with broad, scarious margin, acuminate to attenuate; corolla 20–30 mm long. (16)
16a. Sepals 8–9 mm long, the margin finely denticulate distally and with

Subclass DICOTYLEDONEAE

an accuminate tip, this nearly or quite equaling the sepal body; corolla 30 mm long, posterior lobes projecting; leaves lanceolate, 9–11 mm wide; central Wasatch and Uinta mountains. **P. scariosus** Pennell

16b. Sepals 4–6 mm long, the margin coarsely denticulate distally with an acuminate tip, this much shorter than sepal body; corolla 20 mm long, posterior lobes apparently more spreading; leaves linear-lanceolate, 7–9 mm wide; northern Utah. **P. garrettii** Pennell

17a. Corolla deep blue; plants upright, glabrous to puberulent; anther sacs less densely woolly-hairy; leaves lanceolate to linear; inflorescence secund; eastern Utah. **P. strictus** Benth.

17b. Corolla rose-red or lilac-purple; plants low, often decumbent, considerably woody, glabrous or glandular-pubescent; leaves leathery, sometimes more or less densely canescent below; inflorescence subracemose, glandular-pubescent. (18)

18a. Corolla lilac-purple, 30–35 mm long, 9–10 mm wide, pressed; plants shrubby, forming loose clumps 10–20 cm tall; leaves narrowly elliptic to ovate, acute, sharply serrate, light green, glandular-pubescent; north central Utah. **P. montanus** Greene

18b. Corolla rose. (19)

19a. Plants decumbent or creeping, 15–30 cm tall; leaves 15–40 mm long, 8–16 mm wide; corolla rose-red to purplish, 22–30 mm long; cultivated. **P. newberryi** Gray

19b. Plants depressed mostly less than 10 cm tall; leaves 8–20 mm long, 6–12 mm wide; corolla deep rose, 27–35 mm long; cultivated. **P. rupicola** (Piper) Howell

20a. Anther sacs opening partially. (21)

20b. Anther sacs opening throughout. (28)

21a. Anthers horseshoe shaped, opening at top only. (22)

21b. Anthers not horseshoe shaped, opening from tips. (25)

22a. Corolla glandular-pubescent; calyx and summit of pedicels bearing club-shaped glands; anthers 1.1–1.2 mm long; to be expected in northwestern Utah. **P. kingii** S. Wats.

22b. Corolla glabrous; calyx and summit of pedicels without stalked club-shaped glands; anthers 1.2–2.1 mm long. (23)

23a. Sepals obovate or broadly oblong, truncate at tip except for small abrupt point (mucro); calyx 1.8–3.2 mm long; stems glabrous; summit of pedicels glabrous; plants blue-glaucous; flowers lavender or violet; north central Utah. **P. sepalulus** A. Nels.

23b. Sepals lanceolate to ovate, acuminate or sometimes acute; calyx 3.5–5.5 mm long; stems puberulent; summit of pedicels glandular; plants not blue-glaucous. (24)

24a. Corolla 15–18 mm long, the lobes deep blue, throat blue-violet, tube violet; anthers retuse at apex, sometimes merely truncate or rounded, 1.1–1.3 mm long; stem leaves mostly oblanceolate; north central and southwestern Utah. **P. leonardii** Rydb.

Scrophulariaceae

24b. Corolla 22–25 (30) mm long, the lobes, throat, and tube violet or lavender; anthers obtuse or rounded at apex, 1.6–2.1 mm long; stem leaves mostly elliptic; northern Utah. **P. platyphyllus** Rydb.

25a. Corolla glandular-puberulent externally; sterile filament glabrous; plants 20–25 cm tall, densely cinerous-puberulent; mountains, Sevier Co. **P. wardii** Gray

25b. Corolla glabrous externally; sterile filament bearded or glabrous; plants 20–80 cm tall, glabrous or puberulent. (26)

26a. Plants puberulent, glabrate above, glaucous; corolla 15–20 mm long, blue; sepals 3–5 mm long, the margin narrowly or not at all scarious, slightly denticulate, narrowly ovate; stem leaves slightly clasping from narrowed base; San Pitch Mountains, central Utah. **P. tidestromii** Pennell

26b. Plants glabrous to pruinose-puberulent, sometimes glaucescent; corolla 20–35 mm long. (27)

27a. Corolla bright blue-purple, 25–35 mm long; sterile filament glabrous, rarely bearded distally; lower leaves entire, lanceolate to oblanceolate or spatulate, stem leaves linear-lanceolate, merely sessile; dry plains and hillsides. **P. speciosus** Dougl.

27b. Corolla blue, 20–30 mm long; sterile filament bearded; lower leaves ovate, stem leaves clasping by rounded base; southwestern Utah. **P. laevis** Pennell

28a. Leaves, at least some, more or less regularly toothed, though sometimes merely denticulate. (29)

28b. Leaves all entire or only remotely undulate toothed. (39)

29a. Corolla less than 18 mm long. (30)

29b. Corolla mostly over 18 mm long or, if less, then leaves mostly toothed or strigillose, or lower lip of corolla projecting-spreading much beyond upper. (32)

30a. Corolla 8–10 mm long; sterile filament white bearded; known only from type collection at Tunnel Springs, about 10 mi. east of Garrison, Millard Co. **P. concinnus** Keck

30b. Corolla usually more than 10 mm long and not white bearded. (31)

31a. Plants 10–20 cm tall, pruinose-puberulent; leaves subentire to sharply serrate, broadly ovate, obtuse or acute, 10–25 mm long, all but uppermost petiolate; corolla violet or deep pink, drying purple with faint guidelines, 15–17 mm long, glabrous, prominently bearded across lower lip; Beaver Dam Mountains, southwestern Utah. **P. petiolatus** Brandegee

31b. Plants 20–60 cm tall, glabrous or glandular-puberulent; leaves of sterile shoots 10–50 mm long, narrowed to short, slender petiole, those of fertile shoots sessile or clasping, usually coarsely dentate-serrate; corolla 8–16 mm long, fragrant, externally glandular, white or ochroleucous with fine violet-purple guidelines on all sides; northwestern Utah. **P. deustus** Dougl. ex Lindl.

Subclass DICOTYLEDONEAE

32a. Upper leaves connate-perfoliate. (33)
32b. Upper leaves not connate-perfoliate, though sometimes cordate-clasping. (34)
33a. Corolla abruptly inflated from short tube about equaling calyx, strongly 2-lipped, whitish suffused with pink or lilac, 22–35 mm long, fragrant; leaves gray-glaucous, irregularly spinulose-dentate, sometimes subentire to entire; stems 50–140 cm tall; southern and western Utah. **P. palmeri** Gray
33b. Corolla abruptly inflated from tube at least twice as long as calyx, lavender-purple with blue lobes, strongly 2-lipped, 25–33 mm long; leaves green or glaucescent, glabrous throughout, coarsely serrate; stems 80–120 cm tall; cultivated. **P. spectabilis** Thurb.
34a. Staminode included, glabrous; corolla expanded, especially below from tube about twice as long as calyx, 22–30 mm long, rose-pink; leaves gray-glaucous; cultivated. **P. floridus** Brandegee
34b. Staminode exserted, bearded or glabrous; corolla more or less expanded, not rose-pink; leaves not glaucous; native or cultivated. (35)
35a. Corolla obviously glandular within, at least in fresh material. (36)
35b. Corolla not glandular within. (38)
36a. Inflorescence of divergent loose clusters, glandular-pubescent; corolla ampliate and 2-lipped, pale lavender to whitish; staminode slightly exserted, prominently bearded; cultivated. **P. ovatus** Dougl.
36b. Inflorescence densely flowered, glandular pubescent; staminode slightly to well exserted, glabrous to bearded. (37)
37a. Corolla with 2 longitudinal grooves on underside of tube, light wine-purple to black-purple, sometimes dirty white to yellowish; widespread at high elevations. **P. whippleanus** Gray
37b. Corolla not 2-grooved below, round in cross section, glandular-pubescent outside, blue-lavender; southern Utah. **P. jamesii** Benth.
38a. Capsule and ovary glabrous; corolla 30–35 mm long, light to deep lavender, glandular-pubescent without, sparsely bearded across base of lower lobes; staminode exserted, densely bearded on all sides with short orange-yellow hairs throughout its length; plants 12–18 cm tall; Uintah Co. **P. grahamii** Keck
38b. Capsule and ovary glandular-puberulent; corolla 16–25 mm long, purple, exceedingly pilose at base of lower lip; staminode slightly exserted, strongly bearded with long, golden yellow, septate hairs throughout its length; plants 10–20 cm tall; southwestern Wyoming and adjacent Utah. **P. cleburnei** M. E. Jones
39a. Leaves all narrowly linear or narrowly oblanceolate, up to 5 mm wide and mostly under 20 mm long, sometimes to 40 mm long in some species. (40)
39b. Leaves various, but at least some other than linear and regularly over 5 mm wide and 20 mm long. (47)

Scrophulariaceae

40a. Staminode glabrous, much shorter than corolla tube; corolla 15–25 mm long, white, pink to dull rose on tube and reverse of lobes, limb scarcely 2-lipped; plants 20–50 cm tall, diffusely branched; southern Utah. **P. ambiguus** Torr.
40b. Staminode bearded, as long as, or longer than, corolla tube. (41)
41a. Plants ascending to erect from noncreeping caudex; corolla 2-ridged within, upper lip spreading, 16–24 mm long, tubular-campanulate, bright blue with purplish tube; southern Utah. **P. linarioides** Gray
41b. Plants caespitose, the caudex more or less creeping, forming mats; corolla nearly tubular or somewhat expanded, the throat 2-ridged within, forming a palate, upper lip erect. (42)
42a. Leaves puberulent to canescent (often sparingly so in some **P. caespitosus**); calyx lobes not prominently scarious margined. (43)
42b. Leaves glabrate at least apically (except in some **P. crandallii**); calyx lobes usually prominently scarious margined (except in **P. acaulis**). (45)
43a. Leaves greenish, linear to oblanceolate to spatulate; inflorescence viscid; corolla 3–4 mm wide at throat; widespread. **P. caespitosus** Nutt. ex Gray
43b. Leaves cinereous-whitened. (44)
44a. Leaves oblanceolate to spatulate-oblong or obovate, mucronate; inflorescence obscurely viscid; corolla 4–5 mm wide at throat; southern Utah. **P. thompsoniae** (Gray) Rydb.
44b. Leaves linear-oblanceolate to narrowly spatulate, obtuse; inflorescence glandular-pubescent; corolla 3.5–7 mm broad at throat; Beaver and Millard cos., southwestern Utah. **P. nanus** Keck
45a. Plants essentially acaulescent; leaves scabro-glutinous, not long tipped; flowers borne in a basal rosette of foliage, deep blue, 14–16 mm long, strongly golden bearded within throat; Daggett Co. **P. acaulis** Williams
45b. Plants with stems elongate; leaves long tipped, not scabro-glutinous. (46)
46a. Corolla with guidelines within throat; leaves 12–30 mm long, linear to oblanceolate or spatulate; San Juan Co. **P. crandallii** A. Nels.
46b. Corolla without guidelines; leaves mostly 10 mm long, linear; central Utah. **P. abietinus** Pennell
47a. Plants pubescent to some degree, either throughout or at least in the inflorescence. (48)
47b. Plants glabrous throughout (sometimes scabrous-pubescent in **P. angustifolius**). (61)
48a. Corolla glabrous on outside; inflorescence glabrous or pubescent. (49)
48b. Corolla distinctly glandular-pubescent on outside; inflorescence glandular-pubescent or glabrous. (52)
49a. Corolla mostly more than 16 mm long, pale violet, violet-blue,

Subclass DICOTYLEDONEAE

white, or rarely pink, usually marked with purple-red guidelines, 2-lipped; sterile stamen glabrous; southern Utah. **P. virgatus** Gray

49b. Corolla mostly less than 16 mm long, dark, not markedly 2-lipped; sterile stamen bearded. (50)

50a. Calyx 2.5–3.5 mm long; flowers somewhat congested, but individual pedicels evident; basal rosette absent; widespread. **P. watsonii** Gray

50b. Calyx 3–6 mm long; flowers in dense fascicles with individual pedicels obscured; basal rosette present. (51)

51a. Corolla 6–10 mm long; flowers usually declined; scarious margins of sepals not conspicuous; northern Utah. **P. procerus** Dougl. ex R. Grah.

51b. Corolla 10–16 mm long; flowers usually horizontal; scarious margins of sepals more or less conspicuous; corolla dark violet-purple; central to northern Utah. **P. rydbergii** A. Nels.

52a. Leaves of lower stem wholly glabrous. (53)

52b. Leaves of lower stem pubescent either throughout or at least on veins and lower margins near base. (56)

53a. Leaves of lower stem dull or dark green or glaucous; inflorescence glandular-pubescent; corolla dull white to dark purple or blue; staminode bearded to glabrous. (54)

53b. Leaves of lower stem leathery, glaucous, glabrous or basal, rough or with hairy margins; inflorescence glabrous; corolla carmine to rose-lavender; staminode glabrous to papillate-bearded. (55)

54a. Corolla 17–28 mm long, dull white to dark purple; staminode well exserted, tufted-bearded near apex or occasionally glabrous; higher elevations, throughout Utah. **P. whippleanus** Gray

54b. Corolla 25–30 mm long, blue; staminode included, slightly bearded to glabrous; mountains, southwestern Utah. **P. leiophyllus** Pennell

55a. Corolla 18–24 mm long, tubular, carmine, viscid within; anther sacs broader than long, explanate (flat-opening); staminode usually glabrous; southern Utah. **P. utahensis** Eastw.

55b. Corolla 14–20 (22) mm long, slightly ampliate, reddish purple to rose-lavender, viscidulous to glabrous within; anther sacs longer than broad, flat-opening; staminode usually bearded; central to southern Utah. **P. confusus** M. E. Jones

56a. Corolla somewhat flattened beneath, with 2 strongly marked longitudinal ridges; Summit Co. **P. radicosus** A. Nels.

56b. Corolla tube more or less rounded, without marked grooves and ridges. (57)

57a. Corolla 8–14 mm long; calyx 3–5 mm long; leaves lanceolate to oblanceolate-spatulate; plants to 30 cm tall; central to northern Utah. **P. humilis** Nutt. ex Gray

57b. Corolla 14–24 mm long. (58)

58a. Plants cinereous-puberulent throughout with reflexed hairs; corolla

Scrophulariaceae

15–18 mm long, blue; sterile filament scarcely exserted, moderately bearded; northeastern and central Utah. **P. dolius** M. E. Jones
58b. Plants not cinereous-puberulent throughout. (59)
59a. Anther sacs widely spreading and flat-opening; sterile filament usually exserted; plants 10–45 cm tall; corolla to 32 mm long, orchid to purplish, prominently white bearded within; southern Utah. **P. jamesii** Benth.
59b. Anther sacs widely spreading but not flat-opening; sterile filament mostly included. (60)
60a. Plants 5–10 cm tall, not glandular above, few-flowered; leaves 2–2.5 cm long, 4–5 mm wide, slightly clasping above, smaller; corolla 20 mm long, blue; sterile stamen about equaling posterior stamens, glabrous; southern Utah. **P. parvus** Pennell
60b. Plants 10–30 cm tall; leaves 4–8 cm long, 1–2.5 mm wide, cordate-clasping above in taller specimens; corolla 15–20 mm long, blue to purple; sterile stamen included or barely exserted, sparsely bearded about one-half its length; east central Utah. **P. moffatii** Eastw.
61a. Staminode not strongly widened toward apex, glabrous or bearded; leaves dark green, at least strongly glaucous, and thin. (62)
61b. Staminode strongly widened toward apex, mostly with prominent dense bearding as well; leaves firm-thickened or fleshy-thickened, glaucous. (64)
62a. Corolla 17–30 mm long, immediately enlarged above calyx with scarcely any tube proper; plants 10–25 cm tall; alpine regions, southeastern Utah. **P. hallii** Gray
62b. Corolla 6–16 mm long. (63)
63a. Corolla 11–16 mm long; calyx 2–3.5 mm long; widespread. **P. watsonii** Gray
63b. Corolla 6–10 mm long; calyx 3–6 mm long; northern Utah. **P. procerus** Dougl. ex R. Grah.
64a. Corolla 35 mm or more long; cultivated. **P. grandiflorus** Nutt.
64b. Corolla less than 35 mm long; native. (65)
65a. Basal leaves broadly oblanceolate to obovate-spatulate or orbicular, usually broader than those of upper stem. (66)
65b. Basal leaves linear to narrowly or somewhat broadly oblanceolate, usually narrower than those of upper stem. (68)
66a. Inflorescence more or less secund; peduncles elongate; flowers not congested; plants 20–30 cm tall; corolla dark blue or light purple to bluish purple, 18–24 mm long; sterile filament not or only slightly exserted, much widened toward apex, densely to sparsely bearded; southeastern Utah. **P. lentus** Pennell
66b. Inflorescence not secund; peduncles short; flowers in congested fascicles. (67)
67a. Sterile filament linear, about 1 mm wide, slightly enlarged at apex,

Subclass DICOTYLEDONEAE

densely bearded with hairs much longer than width of filament; anthers flat-opening; eastern and southern Utah. **P. pachyphyllus** Gray
67b. Sterile filament 2–3 mm wide, cuneately widened and densely bearded at apex; Uintah Co. **P. osterhoutii** Pennell
68a. Corolla 10–15 (17) mm long, lobes of upper lip larger than those of lower or subequal; sterile filament densely golden bearded (sometimes purple) near apex and sparsely so in 2 lateral lines; Uintah Co. **P. arenicola** A. Nels.
68b. Corolla 15–20 mm long, lobes of upper lip markedly smaller than those of lower; sterile filament strongly widened upward, bearded; southwestern Utah. **P. angustifolius** Nutt. ex Pursh

Rhinanthus L. Rattleweed

Annual caulescent herbs, 20–70 cm tall; leaves opposite, sessile; inflorescence leafy bracted, forming a spikelike raceme; 4-toothed calyx bladderlike; corolla yellow, 2-lipped; stamens 4, didynamous; parallel anther sacs equal; capsule loculicidal. **R. crista-galli** L.

Scrophularia L. Figwort

Caulescent herbs, 50–200 cm tall; stems 4-angled; leaves opposite, petiolate; corolla 6–10 mm long, greenish purple or brownish; capsule ovoid-conical; widespread, mountains. **S. lanceolata** Pursh

Synthyris Benth. ex A. DC.

Perennial herbs; leaves mostly basal, the blades reniform, orbicular, or ovate; inflorescence a narrow terminal raceme; calyx 4-parted; corolla 4-parted, nearly regular; stamens 2; capsule loculicidal; higher mountains. **S. pinnatifida** S. Wats.

Verbascum L. Mullein

Mostly biennial herbs; stems 30–200 cm tall; major leaves in a basal rosette; stem leaves 10–40 cm long, lanate-woolly; inflorescence dense, spikelike; calyx about as long as capsule; corolla 1.5–2.5 cm wide, yellow; capsule 6–10 mm long; waste places. **V. thapsus** L.

Veronica L. Speedwell

1a. Flowers in axillary racemes; main stem never terminating in an inflorescence; leaves opposite throughout; aquatic or semiaquatic. (2)
1b. Flowers in terminal racemes, or axillary and solitary, never in axillary racemes; leaves opposite or alternate; wet places but seldom actually aquatic. (4)
2a. Leaves with short petioles, the blades ovate to lance-oblong, entire, serrate, or crenate; corolla blue or white; capsule broader than long, emarginate. **V. americana** Schwein ex DC.
2b. Leaves sessile, lanceolate to linear, entire to serrulate or denticulate. (3)
3a. Leaves broadly lanceolate; pedicels rarely over 7 mm long. **V. anagallis-aquatica** L.

Solanaceae

3b. Leaves linear-lanceolate to linear; pedicels filiform, 10 mm long or more. **V. scutellata** L.
4a. Plants perennial; only the upper leaf axils flower bearing; inflorescence appearing to form a dense raceme. (5)
4b. Plants annual; most leaf axils flower bearing; inflorescence appearing to be of axillary flowers only. (6)
5a. Leaves, at least the lower ones, petiolate; corolla white or purplish, small; capsule retuse, puberulent. **V. serpyllifolia** L.
5b. Leaves all sessile; corolla blue or violet. **V. wormskjoldii** Roem. & Schult.
6a. Pedicels definitely longer than sepals, often longer than subtending leaves; corolla over 4 mm wide. **V. persica** Poir.
6b. Pedicels shorter than sepals, much shorter than leaves; corolla 2–3 mm wide. (7)
7a. Leaves ovate or oval, petiolate, or upper nearly sessile; hairs obscurely if at all glandular. **V. arvensis** L.
7b. Leaves oblanceolate, spatulate or oblong, sessile, or lower may be petiolate; hairs usually gland tipped. **V. peregrina** L.

Solanaceae — Potato Family

Plants herbaceous or woody; leaves alternate, rarely opposite, mostly simple; flowers perfect, usually regular and usually cymose; calyx usually of 5 more or less united sepals; corolla of 5 united petals, rotate, campanulate, funnelform or salverform; stamens as many as corolla lobes and alternate with them, or fewer; ovary superior, 2-loculed (except in cultivated tomato); placenta axile; fruit a berry or capsule.

1a. Corollas funnelform to salverform; fruit a capsule (berry in **Lycium**). (2)
1b. Corollas rotate, rotate-campanulate, or open campanulate. (7)
2a. Stamens in 2 sets of 2, the fifth one by itself and often smaller, abortive or lacking; cultivated. (3)
2b. Stamens not in 2 sets of 2, all 5 fertile and approximately equal in size. (4)
3a. Perfect stamens 5. **Petunia**
3b. Perfect stamens 4. **Salpiglossis**
4a. Plants spiny shrubs or vines; fruit a berry. **Lycium**
4b. Plants herbs; fruit a capsule. (5)
5a. Capsule circumscissile near apex; corollas about 2 cm long. **Hyocyamus**
5b. Capsule opening by longitudinal valves; corollas 2–20 cm long. (6)
6a. Capsule completely included in the calyx, not spiny, less than 15 mm long; flowers in terminal racemes or panicles. **Nicotiana**
6b. Capsules not included in the calyx, spiny, to 40 mm long or more; flowers solitary in the upper forks of the stem. **Datura**

Figs. 101-2.   101. **Ribes aureum**, x .63.   102. **Collinsia parviflora**, x .38

Figs. 103-4.   103. **Mimulus guttatus**, x .3.   104. **Veronica wormskjoldii**, x .42

Solanaceae

7a. Plants more or less spiny shrubs or vines. **Lycium**
7b. Plants herbs, if somewhat woody then not spiny, if spiny then herbaceous. (8)
8a. Anthers connivent (closely appressed but not united) around the style. (9)
8b. Anthers not connivent around the style. (10)
9a. Anthers opening by a terminal pore or slit; cultivated or not cultivated. **Solanum**
9b. Anthers opening by a longitudinal slit from base to apex; cultivated. **Lycopersicon**
10a. Calyx thick, not expanding in fruit; fruit frequently pungent and more or less hollow within. **Capsicum**
10b. Calyx papery, expanding at maturity and covering the fruit or nearly so; fruit not pungent or hollow. (11)
11a. Calyx in fruit not angled, the lobes spreading, exposing the top of the fruit. **Chamaesarcha**
11b. Calyx in fruit angled, the lobes connivent apically, covering the fruit. **Physalis**

Capsicum L. Pepper

Herbaceous annuals, 2–6 dm tall or more; leaves simple, entire; flowers white or greenish white, solitary or in 2s or 3s, 10–20 mm broad; stamens 5, not closely connivent, mostly bluish; fruit a podlike, indehiscent berry of many sizes and shapes; cultivated. **C. frutescens** L.

Chamaesarcha Gray

Herbaceous perennials, 10–20 cm tall; leaves oblong-lanceolate to linear; calyx 3–5 mm long; corolla 8–13 mm wide, greenish white or tinged with purple; southern Utah. **C. coronopus** (Dunal) Gray

Datura L. Jimsonweed
1a. Corollas 15–20 cm long; calyx 7–12 cm long; leaves and stems canescent-puberulent; capsules nodding, globose, opening irregularly; widely distributed, southern Utah, cultivated and escaping elsewhere in Utah; poisonous. **D. meteloides** Dunal ex DC.
1b. Corollas 5–10 cm long; calyx 3–6 cm long; leaves and stems glabrous or sparsely puberulent; capsules erect, ovoid, opening by 4 valves; uncommon, southern Utah; poisonous. **D. stramonium** L.

Hyocyamus L. Henbane

Annual or biennial herbs; leaves 6–20 cm long, irregularly lobed, cleft or pinnatifid; calyx 2–2.5 cm long; corolla about 1.5–2 cm long, greenish yellow to white with pinkish or purplish veins; capsule 10–14 mm long; naturalized from Europe, occasional in waste places; poisonous. **H. niger** L.

Lycium L. Matrimony Vine
1a. Stems slender, climbing or sprawling; cultivated and escaping

Subclass DICOTYLEDONEAE

around homesites and cemeteries, especially in northern Utah. Tea Vine, Matrimony Vine. **L. halimifolium** Mill.
1b. Stems stout, erect; native, infrequent or not at all in cultivation, southern and southwestern Utah. (2)
2a. Corolla about 20 mm long; calyx lobes at least 2 mm long, the lobes about equal in length; southernmost tier of counties. Pale Wolfberry, Tomatilla. **L. pallidum** Miers
2b. Corolla 8–15 mm long; calyx lobes usually less than 2 mm long, the lobes more or less unequal in length; various distribution. (3)
3a. Leaves 10–25 mm long; corolla lobes shortly but densely white-ciliate; Washington, Kane, and Garfield cos. Torrey Wolfberry. **L. torreyi** Gray
3b. Leaves 4–12 mm long; corolla lobes smooth or ciliate, but not densely white-ciliate; Washington Co. northward to Millard Co. Anderson Wolfberry. **L. andersonii** Gray

Lycopersicon Mill. Tomato

Annual herbs; stems 1–2 m long or more; leaves alternate, odd-pinnate, 15–45 cm long; flowers 3–7, nodding, 7–20 mm broad or more; fruit a red, pink, or yellow berry; cultivated, occasionally escaping but not persisting. **L. esculenta** Mill.

Nicotiana L. Tobacco

1a. Corolla pink to red; cultivated tobacco. **N. tabacum** L.
1b. Corolla whitish or greenish yellow; native. (2)
2a. Leaves mostly cordate- or auriculate-clasping, sessile or with short, broad petioles; corolla externally rather copiously pubescent; southern Utah. **N. trigonophylla** Dunal ex DC.
2b. Leaves petiolate, not cordate- or auriculate-clasping at base; corolla externally glabrous or sparsely pubescent; widely distributed. **N. attenuata** Torr. ex S. Wats.

Petunia Juss. Petunia

Annual herbs; stems 2–6 dm long; leaves alternate, simple, entire, upper may be opposite; flowers variously colored, 5–8 cm long, funnelform; fruit a capsule; cultivated; probable hybrid between **P. axillaris** BSP and **P. violacea** Lindl. **P. hybrida** Vilm.

Physalis L. Groundcherry

1a. Pubescence at least partly of forked or stellate hairs. **P. fendleri** Gray
1b. Pubescence of simple hairs. (2)
2a. Stems and leaves inconspicuously pubescent or glabrate, the leaves minute, not or inconspicuously viscid. (3)
2b. Stems and leaves conspicuously pubescent, at least some hairs spreading, viscid. (4)

Solanaceae

3a. Leaf blades broadly ovate, rounded-deltoid or suborbicular; herbage persistently puberulent; flowering stems diffusely branched. **P. crassifolia** Benth.
3b. Leaf blades lanceolate or oblong-lanceolate; herbage sparsely pubescent or glabrate; flowering stems little branched; Utah's most common species. **P. longifolia** Nutt.
4a. Leaf blades usually at least 5 cm long, acute or acuminate; pubescence of 2 types—short glandular hairs and long flat ones; corolla 15 mm broad or more; fruiting calyx 2.5–4 cm long. **P. heterophylla** Nees
4b. Leaf blades usually less than 5 cm long, obtuse to acute; pubescence glandular-pilose, with few or no long flat hairs; corolla usually less than 15 mm broad; fruiting calyx less than 2.5 cm long. **P. hederaefolia** Gray

Salpiglossis Ruiz & Pav.

Annual herbs, 3–8 dm tall; leaves alternate, simple, entire or nearly so; flowers 5–7 cm long and broad, variously colored; fruit a capsule; cultivated ornamental. **S. sinuata** Ruiz & Pav.

Solanum L. Nightshade
1a. Plants climbing, somewhat woody below; leaves hastately lobed or parted near base; flowers usually bright violet to blue-purple; fruit ripening red; cultivated and escaping. European Bittersweet. **S. dulcamara** L.
1b. Plants not climbing or woody below; leaves variously shaped; flowers variously colored but not usually violet to blue-purple. (2)
2a. Plants cultivated; tubers large. Potato. **S. tuberosum** L.
2b. Plants seldom or not cultivated; tubers small or lacking. (3)
3a. Fruit closely covered by enlarged calyx, this densely armed with long, straight, very sharp spines; plants annual. Buffalo Bur. **S. rostratum** Dunal
3b. Fruit not closely covered by enlarged calyx; plants without spines or, if present, plants perennial. (4)
4a. Herbage and calyx spiny; corolla 20–30 mm in diameter, purple or violet. White Horsenettle. **S. elaeagnifolium** Cav.
4b. Herbage and calyx not spiny; corolla usually less than 20 mm broad. (5)
5a. Plants perennial from globose tubers; leaves pinnately compound; corollas 12–18 mm wide. Wild Potato. **S. jamesii** Torr.
5b. Plants annual, lacking tubers; leaves entire, sinuate-dentate to pinnatifid; corollas 6–12 mm wide. (6)
6a. Leaf blades sinuate-dentate to pinnatifid; berries green at maturity. **S. triflorum** Nutt.
6b. Leaf blades entire to sinuate-dentate, never pinnatifid; berries black or yellow at maturity. (7)

Subclass DICOTYLEDONEAE

7a. Stems and leaves glabrate, puberulent or strigose; berries black when ripe. Black Nightshade. **S. nigrum** L.
7b. Stems and leaves viscid-villous; berries yellow when ripe. **S. sarachoides** Sendt. ex Mart.

TAMARICACEAE — TAMARIX FAMILY

Large shrubs or small trees; branches covered when young with small, appressed, imbricated scalelike leaves; flowers in slender terminal spikes or grouped in paniculate clusters, regular and perfect; sepals 4 or 5; petals distinct, 4 or 5; stamens as many, or twice as many, as the petals; ovary superior, 1-loculed with basal placenta; styles 3–5; fruit a many-seeded capsule with 3–5 valves; seeds with a tuft of hairs.

Tamarix L. Tamarix

1a. Leaves deciduous, dull yellow-green, ovate-lanceolate; introduced from Europe and widely naturalized. **T. pentandra** Pall.
1b. Leaves evergreen, inconspicuous; branchlets jointed; rare in cultivation, southwestern Utah. **T. aphylla** (L.) Karst.

TROPAEOLACEAE — TROPAEOLUM FAMILY

Annual or perennial herbs; leaves alternate, digitately angled or peltate, sometimes lobed or dissected; sepals 5; petals 5, or sometimes less, often cut or fringed, the upper ones unlike the others and usually smaller and inserted in the opening of a spur; stamens 8, unequal; styles 1, apical; stigmas 3; ovary 3-lobed and 3-loculed.

Tropaeolum L.

A single genus and species treated, with the characteristics of the family. Garden Nasturtium **T. majus** L.

ULMACEAE — ELM FAMILY

Plants trees or shrubs, with pith finely chambered at nodes; leaves alternate, simple, unequal at base, deciduous, somewhat palmately veined; flowers perfect or imperfect, axillary, solitary or in small clusters; perianth 4- to 6-parted; stamens 4–5; ovary superior, 1-loculed; styles none; stigmas 2; fruit a drupe with thin flesh and hard-pitted seed or a samara.

1a. Fruit a samara; flowers on last year's branches, perfect, before leaves appear (except in **U. parvifolia**). **Ulmus**
1b. Fruit a drupe; flowers on the young growth, after leaves appear. (2)
2a. Calyx campanulate, 4- to 5-lobed; style not central; leaves with 7 or more pairs of parallel veins, the lowest pair not prominent; winter buds somewhat spreading. **Zelkova**
2b. Calyx of distinct sepals, 5- to 6-parted; style central; leaves with

Ulmaceae

3 nerves at base, pairs of veins usually less than 6; winter buds appressed. **Celtis**

Celtis L. Hackberry
1a. Leaves serrate; cultivated trees. **C. occidentalis** L.
1b. Leaves entire or rarely with a few teeth. (2)
2a. Leaves ovate to oblong-lanceolate, long-acuminate, usually broadly cuneate at base; fruit stalk longer than petiole; cultivated. **C. laevigata** Willd.
2b. Leaves ovate to ovate-oblong, acute or short-acuminate, rounded or subcordate at base; fruit stalk about as long as petiole; native. **C. reticulata** Torr.

Ulmus L. Elm
1a. Leaves once serrate (obscurely twice serrate in U. pumila). (2)
1b. Leaves twice serrate. (3)
2a. Trees flowering in late summer or early autumn; bark with orange lenticels; small trees to 17 m tall, not weedy, generally desirable. Chinese Elm. **U. parvifolia** Jacq.
2b. Trees flowering in spring; bark with lenticels inconspicuous; common weedy trees, generally undesirable. Siberian Elm. **U. pumila** L.
3a. Samara ciliate on margins; flowers pendulous. (4)
3b. Samara glabrous on margins; flowers upright. (6)
4a. Samara glabrous over seed; branchlets without corky wings; flowers fascicled; tall, ascending, umbrella-shaped trees to 35 m tall. American Elm. **U. americana** L.
4b. Samara tomentose on sides; branchlets with corky wings; flowers racemose. (5)
5a. Leaves not ciliate, 5–10 cm long, with 16–20 pairs of veins; buds pubescent; samaras 9–15 mm broad; trees to 30 m tall. Rock Elm. **U. thomasii** Sarg.
5b. Leaves minutely ciliate, 3–6 cm long, with 8–12 pairs of veins; buds glabrous; samaras 3–5 mm broad; trees to 20 m tall. Winged Elm. **U. alata** Michx.
6a. Branchlets glabrous; trees to 35 m tall. Dutch Elm. **U. x hollandica** Mill.
6b. Branchlets pubescent. (7)
7a. Buds covered with rusty hairs, obtuse; branchlets scabrous; leaves ciliate; fruit pubescent in the middle; trees to 20 m tall. Slippery Elm. **U. fulva** Michx.
7b. Buds pale-pubescent or glabrous; branchlets glabrous; leaves not ciliate; fruit glabrous. (8)
8a. Leaves 7.5–15 cm long, with 16–20 pairs of veins; bark smooth; samaras 1–2 cm broad, glabrous throughout, seed in the center. Scots Elm. **U. glabra** Huds.
8b. Leaves 5–7.5 cm long, with 8–18 pairs of veins; bark fissured; sa-

Subclass DICOTYLEDONEAE

maras 1–1.5 cm broad, glabrous throughout, seed toward end of samara. English Elm. **U. procera** Salisb.

Zelkova Spach

Trees to 30 m tall; leaves ovate to oblong-ovate, rounded or slightly cordate at base, 2.5–8 cm long; flowers polygamous; calyx campanulate; stamens 4–5; fruit a 1-seeded drupe; native in Eurasia. **Z. serata** Makino

UMBELLIFERAE — CARROT FAMILY

Acaulescent or caulescent, annual, biennial, or perennial herbs; leaves alternate, rarely opposite, or basal, compound or rarely simple, usually much incised or divided, with usually sheathing petioles; flowers small, regular in compound umbels, sometimes capitate; rays sometimes subtended by bracts forming an involucre; umbellets usually subtended by bractlets forming an involucel; calyx teeth small or obsolete; petals 5; stamens 5, inserted on an epigynous disk; ovary inferior, 2-loculed, with 1 ovule in each cell; styles 2, sometimes swollen at base and forming a stylopodium; fruit a schizocarp.

1a. Inflorescence capitate, with obsolete rays and pedicels, not a distinct umbel. **Cymopterus**
1b. Inflorescence a distinct umbel with evident rays and usually pedicels, more or less spreading, never capitate. (2)
2a. Ovary and fruit armed with bristles, prickles, tubercles, papillae, or callous teeth. (3)
2b. Ovary and fruit not armed, sometimes pubescent. (5)
3a. Ovary and fruit bristly-hispid, the bristles never uncinate or barbed or armed with callous teeth. **Osmorhiza**
3b. Ovary and fruit armed with uncinate or barbed bristles or prickles or merely tuberculate or papillate. (4)
4a. Petals cuneate to obovate; fruit flattened laterally, armed with uncinate bristles; seed face deeply sulcate. **Caucalis**
4b. Petals obcordate, unequally cleft; fruit flattened dorsally, with some or all of bristles glochidiate; seed face shallowly concave to nearly plane. **Daucus**
5a. Fruit and ovary terete or flattened laterally; ribs of fruit not prominently winged. (6)
5b. Fruit flattened dorsally; some or all ribs of fruit winged. (26)
6a. Plants annual, slender, mostly low and diffuse, caulescent, rarely pubescent; leaves mostly small, the leaflets usually linear to filiform; flowers white or rosy; stylopodium depressed to conic. **Apium**
6b. Plants biennial or perennial, acaulescent or caulescent, glabrous or pubescent; leaves mostly larger, the leaflets broader; flowers white, yellow, or purple; stylopodium lacking or present. (7)
7a. Plants low, acaulescent or short-caulescent, subscapose or with slender, naked, unbranched peduncles from a cluster of basal leaves. (8)

Umbelliferae

7b. Plants tall, caulescent, with several to many stem leaves. (14)
8a. Leaves entire or reduced to fistulous, septate phyllodes. **Orogenia**
8b. Leaves variously compound. (9)
9a. Stylopodium conic or low conic. (10)
9b. Stylopodium wanting, at least when fruit is mature. (11)
10a. Leaves ternately-pinnately compound, the leaflets linear to ovate; involucel inconspicuous or wanting; pedicels slender, spreading-ascending. **Ligusticum**
10b. Leaves 1- to 2-pinnate, the leaflets lanceolate to orbicular; involucel conspicuous; pedicels obsolete or short and flattened or winged. **Podistera**
11a. Plants from tuberous roots; fruit with a corky projection extending down the middle of the inner faces their entire length. **Orogenia**
11b. Plants from taproots; fruit without a corky projection. (12)
12a. Carpophore (slender axis between the fruit halves) wanting; ribs corky winged, the wings broadly linear to subovate in cross section; seed slightly flattened in cross section; strengthening cells at base of wings. **Oreoxis**
12b. Carpophore present; ribs not winged, sometimes prominent; seed subterete in cross section; strengthening cells absent or inconspicuous. (13)
13a. Plants scaberulous, at least in the inflorescence; fruit scaberulous to glabrate. **Musineon**
13b. Plants glabrous to puberulent, not scaberulous; fruit glabrous. **Aletes**
14a. Leaves ternately, pinnately, or ternately-pinnately compound, the leaflets distinct, mostly large, lanceolate to suborbicular or obovate, variously serrate, dentate or lobed, occasionally incised. (15)
14b. Leaves ternately, pinnately, or ternately-pinnately more than once compound, the leaflets usually somewhat confluent, small, filiform to ovate, pinnately incised to parted. (22)
15a. Flowers bright yellow; stylopodium lacking. **Zizia**
15b. Flowers white or greenish yellow, rarely purple; stylopodium present, depressed to conic. (16)
16a. Involucel of numerous, usually conspicuous, foliaceous bractlets (lacking in **Cicuta**); ribs prominent, corky, subequal or unequal, or the entire pericarp thick and corky; mostly aquatic or semiaquatic. (17)
16b. Involucel lacking, rarely conspicuous, rarely foliaceous; ribs not conspicuously prominent, not corky, subequal, filiform or obscure. (20)
17a. Leaves conspicuously heteromorphic; plants stoloniferous; ribs filiform, obscure in thick, corky pericarp. **Berula**
17b. Leaves homomorphic; plants not stoloniferous; ribs evident, corky, subequal or unequal. (18)

Subclass DICOTYLEDONEAE

18a. Leaflets lanceolate to suborbicular; umbels subsessile in the axils and terminal on the branches, the terminal mostly short-pedunculate. **Apium**

18b. Leaflets linear-lanceolate to ovate; umbels terminal on the branches, mostly long-pedunculate. (19)

19a. Involucre of conspicuous subfoliaceous bracts; rays few; ribs subequal in cross section. **Sium**

19b. Involucre wanting or of few inconspicuous bracts; rays usually numerous; ribs unequal in cross section. **Cicuta**

20a. Plants from tuberous or fusiform fascicled roots; involucel of usually scarious or colored bractlets; calyx teeth conspicuous. **Perideridia**

20b. Plants from slender or thick fascicled roots, taproots or rootstocks; involucel wanting or foliaceous, never scarious or colored; calyx teeth obsolete or minute. (21)

21a. Fruit linear-oblong to linear-fusiform, usually several times longer than broad; plants from slender or thick fascicled roots. **Osmorhiza**

21b. Fruit orbicular to oblong, usually 2–3 times longer than broad; plants from fibrous root crowns surmounting taproots. **Ligusticum**

22a. Involucel of numerous, usually foliaceous bractlets; ribs prominent, corky, or entire pericarp thick and corky; aquatic or semiaquatic. (23)

22b. Involucel present or absent; ribs obscure to somewhat prominent, never corky; mostly of dry land. (24)

23a. Leaves conspicuously heteromorphic; plants stoloniferous; ribs filiform, obscure in thick, corky pericarp. **Berula**

23b. Leaves not conspicuously heteromorphic; plants not stoloniferous; ribs evident, corky. **Sium**

24a. Involucre of numerous, entire or divided, more or less conspicuous bracts; introduced weed with spotted stems. **Conium**

24b. Involucre lacking or inconspicuous. (25)

25a. Plants from fibrous root crowns; pedicels subequal; native. **Ligusticum**

25b. Plants without fibrous root crowns; pedicels unequal; introduced weeds or cultivated herbs. **Carum**

26a. Fruit with both lateral and dorsal wings developed or the dorsal ribs prominent. (27)

26b. Fruit usually strongly flattened dorsally, the dorsal wings absent, the dorsal ribs filiform or obsolete, the lateral wings more or less prominent. (33)

27a. Plants acaulescent or short-caulescent, low, usually slender; stems subscapose. (28)

27b. Plants caulescent, mostly tall, sometimes stout; stems leafy. (30)

28a. Calyx teeth prominent, linear-lanceolate, acuminate, often unequal. **Pteryxia**

Umbelliferae

28b. Calyx teeth obsolete or evident and ovoid to deltoid. (29)
29a. Plants subcaulescent to caulescent; peduncles conspicuously hirtellous-pubescent at base of umbel. **Pseudocymopterus**
29b. Plants acaulescent to subcaulescent; peduncles glabrous to rarely subscaberulous. **Cymopterus**
30a. Stylopodium absent. (31)
30b. Stylopodium present. (32)
31a. Calyx teeth prominent, linear-lanceolate, attenuate, often unequal. **Pteryxia**
31b. Calyx teeth obsolete or evident and ovate. **Pseudocymopterus**
32a. Plants annual, with a strong anise odor; involucel usually wanting; cultivated dill. **Anethum**
32b. Plants perennial, anise odor lacking; involucel usually present; native. **Angelica**
33a. Leaves, at least the cauline, more than once compound with filiform to linear segments. (34)
33b. Leaves pinnate, ternate, or ternately-pinnately compound, with broad, entire, serrate, crenate, or lobed leaflets, or the leaves reduced to hollow, acute, septate phyllodes. (36)
34a. Fruit with a corky projection down middle of inner face of each fruit half. **Orogenia**
34b. Fruit without a corky projection. (35)
35a. Peduncles conspicuously hirtellous-pubescent at base of umbel. **Pseudocymopterus**
35b. Peduncles glabrous or pubescent throughout. **Lomatium**
36a. Stylopodium lacking. **Lomatium**
36b. Stylopodium present. (37)
37a. Plants slender, from fascicled tubers; leaves simply pinnate or ternate, sometimes reduced to phyllodes, frequently glaucous; marshes and low ground. **Oxypolis**
37b. Plants stout, from taproots or fascicled roots; leaves pinnately or ternately compound, not reduced. (38)
38a. Outer petals of inflorescence radiant and often 2-cleft; plants tomentose-pubescent. **Heracleum**
38b. Outer petals of inflorescence neither radiant nor 2-cleft; plants not tomentose, sometimes glabrous. (39)
39a. Leaves pinnate; involucel usually lacking; flowers yellow or red; introduced weed or cultivated. **Pastinaca**
39b. Leaves mostly 2-pinnate or ternate-pinnate; involucel usually present; flowers white, pink, or purplish. **Angelica**

Aletes Coult. & Rose

Caespitose acaulescent perennials, 6–25 cm tall; leaves 2–8 cm long, oblong in outline, 1- to 2-pinnate; rays 4–8; flowers yellow; calyx teeth

Subclass DICOTYLEDONEAE

conspicuous; fruit 3-8 mm long; canyons and mesas, southeastern Utah. **A. macdougalli** Coult. & Rose

Anethum L.   Dill

Caulescent herbs, 4-17 dm tall; leaves oblong to obovate in outline, more than once pinnately compound; rays numerous; flowers yellow; calyx teeth obsolete; fruit ovoid, about 4 mm long; cultivated. **A. graveolens** L.

Angelica L.

1a. Ovaries glabrous; northern Utah. **A. arguta** Nutt. ex T. & G.
1b. Ovaries pubescent or roughened. (2)
2a. Petals pubescent; mountains, western Utah. **A. kingii** (S. Wats.) Coult. & Rose
2b. Petals glabrous. (3)
3a. Involucel present; high mountains, frequently above timberline. **A. roseana** Henderson
3b. Involucel lacking. (4)
4a. Fruit orbicular, 3-6 mm long, glabrous; leaves oblong to oval, pinnate to incompletely 2-pinnate; widely distributed, mountains. **A. pinnata** S. Wats.
4b. Fruit oblong-oval, 5-7 mm long, 3-5 mm broad, pubescent; leaves deltoid, ternately-pinnately divided; northern to central Utah. **A. wheeleri** S. Wats.

Apium L.   Celery

Perennial herbs, 5-15 dm tall; leaves oblong to obovate in general outline; rays rather few; flowers white or greenish; calyx teeth minute but evident; fruit about 1.5 mm long; adventive weed or cultivated for its succulent petioles (var. **dulce** Pers.). **A. graveolens** L.

Berula Hoffm. ex Besser   Water Parsnip

Caulescent, stoloniferous, aquatic herbs, 20-80 cm tall; leaves 10-30 cm long, narrowly oblong in outline; rays few; flowers white; calyx teeth subulate, minute; fruit 2-3 mm long; moist sites. **B. erecta** (Huds.) Coville

Carum L.   Caraway

Biennial or perennial herbs, 3-10 dm tall; leaves oblong to oval in outline, 10-25 cm long; peduncles 4-12 cm long; rays 7-14; flowers white; calyx teeth obsolete; fruit 3-4 mm long. **C. carvi** L.

Caucalis L.   False Carrot

Slender perennial herbs, 8-40 cm tall; leaves oblong to ovate in outline, blades 2-5 cm broad; peduncles 2-10 cm long; involucre of several pinnately compound bracts; rays 1-9, ascending; flowers white; fruit oblong, 3-7 mm long, armed with rows of hooked bristles. **C. microcarpa** H. & A.

Umbelliferae

Cicuta L. Water Hemlock
Caulescent perennials from tuberous bases, 60–120 cm tall; leaves 1- to 3-pinnate or ternate-pinnate; rays numerous; flowers white or greenish; calyx teeth evident; fruit 2–4 mm long; marshes and stream banks. **C. douglasii** (DC.) Coult. & Rose

Conium L. Poison Hemlock
Biennial herbs with spotted stems, 5–30 dm tall; leaves broadly ovate in outline; rays numerous, 15–25 mm long; flowers white; calyx teeth obsolete; fruit 2–2.5 mm long; frequently in moist places. **C. maculatum** L.

Cymopterus Raf.
1a. Rays obsolete, the umbels thus discoid; bractlets paleaceous. **C. globosus** S. Wats.
1b. Rays present, 0.2–9 cm long, the umbels thus subcompact to spreading, not discoid; bractlets not paleaceous. (2)
2a. Bracts scarious, united; bractlets conspicuous, scarious, usually prominently nerved and sometimes united. (3)
2b. Bracts usually wanting, never scarious; bractlets conspicuous or inconspicuous, occasionally scarious margined. (5)
3a. Bractlets purple or greenish white, conspicuously many-nerved; pedicels less than 1 mm long or obsolete. **C. multinervatus** (Coult. & Rose) Tidestr.
3b. Bractlets white or whitish, few-nerved; pedicels 3–12 mm long. (4)
4a. Umbels somewhat spreading, the rays 10–50 mm long; fruit ovoid-oblong, the wings usually narrower than body. **C. bulbosus** A. Nels.
4b. Umbels densely globose, the rays 4–10 mm long; fruit ovoid, the wings 2 to 3 times as wide as body. **C. purpurascens** (Gray) M. E. Jones
5a. Leaves neither fleshy nor coriaceous, green to gray-green, not glaucescent. (6)
5b. Leaves somewhat fleshy or coriaceous (membranous in **C. duchesnensis**), pallid and glaucescent. (9)
6a. Plants caespitose; bractlets not foliaceous; fruiting pedicels evident, 2–13 mm long; fruit wings membranous. **C. bipinnatus** S. Wats.
6b. Plants not caespitose; bractlets conspicuously foliaceous; fruiting pedicels obsolete or less than 1 mm long; fruit wings corky-spongy. (7)
7a. Pseudoscape present; leaflets usually longer than broad. (8)
7b. Pseudoscape absent; leaflets usually broader than long. **C. newberryi** (S. Wats.) M. E. Jones
8a. Peduncles shorter than or equaling the leaves; bracts usually wanting; flowers usually white; central umbellet pedicellate. **C. acaulis** (Pursh) Raf.

Subclass DICOTYLEDONEAE

8b. Peduncles exceeding the leaves; bracts present; flowers usually yellow; central umbellet sterile and sessile. **C. fendleri** Gray
9a. Pseudoscape conspicuous, usually fleshy, up to 16 cm long. (10)
9b. Pseudoscape inconspicuous or lacking, never fleshy, up to 2 cm long. (13)
10a. Umbels subcompact; bractlets conspicuous. (11)
10b. Umbels spreading; bractlets inconspicuous or, if conspicuous, involucre present. (12)
11a. Peduncles shorter than or equaling the leaves; bracts usually wanting; flowers white; central umbellet pedicellate. **C. acaulis** (Pursh) Raf.
11b. Peduncles exceeding the leaves; bracts present; flowers yellow; central umbellet sterile or sessile. **C. fendleri** Gray
12a. Mature pseudoscape 1–7 cm long; dried leaves finely wrinkled, appearing granulate-pubescent; flowers white; lateral wings about equaling body; western Utah. **C. ibapensis** M. E. Jones
12b. Mature pseudoscape 5–16 cm long; dried leaves smooth; flowers yellow, rarely white; lateral wings broader than body; central Utah. **C. longipes** S. Wats.
13a. Leaves pinnate, ternate, or digitate, the leaflets lobed. (14)
13b. Leaves ternate-pinnate or ternate-2-pinnate, the leaflets dentate. (18)
14a. Leaves orbicular-reniform to cordate-oblong, ternate or digitate. **C. basalticus** M. E. Jones
14b. Leaves ovate-oblong to broadly elliptic, pinnate (sometimes ternate in **C. duchesnensis**). (15)
15a. Pedicels obsolete; fruit wings corky, constricted at base, narrower than body. **C. newberryi** (S. Wats.) M. E. Jones
15b. Pedicels evident; fruit wings thin or spongy, not constricted, broader at base. (16)
16a. Umbels subcompact, the rays 2–10 mm long; flowers white. **C. coulteri** (M. E. Jones) Mathias
16b. Umbels somewhat spreading, the rays 5–35 mm long; flowers yellow or purple. (17)
17a. Plants more or less scaberulous; leaflets coriaceous, shallowly dentate, the teeth 1–2 mm long; fertile rays 3–4, 5–20 mm long; pedicels 3–6 mm long. **C. rosei** M. E. Jones
17b. Plants glabrous; leaflets membranous, deeply dentate, the teeth 4–8 mm long; fertile rays 6–10, 15–35 mm long; pedicels 2–4 mm long. **C. duchesnensis** M. E. Jones
18a. Plants 3–8 cm high, scabrous-puberulent throughout; flowers white; fruit 4–5 mm long, about 3 mm broad; Cedar Breaks, Iron Co. **C. minimus** Mathias
18b. Plants 10–50 cm high, glabrous or sparsely scaberulous on the

Umbelliferae

leaves and peduncles; flowers yellow or purple; fruit 6–12 mm long, 5–14 mm broad. (19)

19a. Flowers yellow or greenish yellow; fruit ovoid-oblong or oblong, 6–12 mm long, 5–10 mm broad; wings slightly or not at all inflated at base, equaling or a little broader than body; seed face concave. **C. purpureus** S. Wats.

19b. Flowers purple; fruit broadly ovoid to ellipsoid, 10–12 mm long, 10–14 mm broad; wings conspicuously inflated at base, several times the width of body; seed face deeply sulcate. **C. jonesii** Coult. & Rose

Daucus L. Carrot

Biennial herbs, 1.5–12 dm tall; stems solitary; leaves oblong in outline; involucres 3–30 mm long; involucel entire, or rarely pinnate; flowers whitish; fruit 3–4 mm long, ovoid; weed of waste places; var. sativa DC. is the cultivated carrot. **D. carota** L.

Heracleum L. Cow Parsnip

Stout, pubescent biennial or perennial herbs, 1–3 m tall; leaves ternately compound, orbicular to reniform in outline; rays 15–30; petals white; calyx teeth minute or obsolete; fruit 8–12 mm long; widely distributed, river and stream bottoms. **H. lanatum** Michx.

Ligusticum L. Lovage

1a. Plants slender; leaflets linear, 1–3 cm broad; mountains. **L. filicinum** S. Wats.

1b. Plants stout; leaflets ovate, oblong, or lanceolate, 5–40 mm broad; mountains. **L. porteri** Coult. & Rose

Lomatium Raf. Desert Parsley

1a. Peduncles conspicuously swollen and inflated at apex; western Utah. **L. nudicaule** (Pursh) Coult. & Rose

1b. Peduncles not conspicuously inflated at apex. (2)

2a. Plants mostly low from globose or somewhat elongate or irregular tubers; leaves mostly small. (3)

2b. Plants usually stouter, from more or less thickened, elongate taproots, sometimes with a deep-seated tuber. (4)

3a. Involucel absent or inconspicuous; northern Utah. **L. ambiguum** (Nutt.) Coult. & Rose

3b. Involucel of conspicuous bractlets; northern Utah. **L. leptocarpum** (T. & G.) Coult. & Rose

4a. Leaves compound, dissected into numerous small divisions. (5)

4b. Leaves with mostly few large divisions, ternately or pinnately divided, the divisions mostly remote. (10)

5a. Ovaries and young, sometimes mature, fruit variously pubescent or roughened. (6)

Subclass DICOTYLEDONEAE

5b. Ovaries and fruit glabrous. (7)
6a. Bractlets with a conspicuously scarious margin, never tomentose or villous; western Utah. **L. nevadense** (S. Wats.) Coult. & Rose
6b. Bractlets not conspicuously scarious margined, more or less tomentose or villous; widely distributed. **L. macdougalii** Coult. & Rose
7a. Fruit 12–16 mm long, 6–10 mm broad, the wings very narrow and corky-thickened; widely distributed. **L. dissectum** (Nutt.) Math. & Const.
7b. Fruit 5–13 mm long, 3–7 mm broad, the wings thin and membranous. (8)
8a. Plants more or less pubescent; flowers yellow; northern Utah. **L. juniperinum** (M. E. Jones) Coult. & Rose
8b. Plants glabrous or occasionally scaberulous, never pubescent; flowers yellow or purple. (9)
9a. Ultimate leaf divisions 1–2 mm broad; bractlets equaling flowers; fruit 4–5 mm broad; northern Utah. **L. donnellii** Coult. & Rose
9b. Ultimate leaf divisions 0.1–0.25 mm broad; bractlets shorter than flowers; fruit 5–8 mm broad; northern and western Utah. **L. grayi** Coult. & Rose
10a. Plants mostly caulescent, tall; leaves ternately-pinnately or quinately-pinnately divided; widely distributed (includes **L. simplex** [Nutt.] Macbr.). **L. triternatum** (Pursh) Coult. & Rose
10b. Plants mostly caulescent or short-caulescent; leaves 1- to 2-pinnate, rarely 3-pinnate. (11)
11a. Foliage variously pubescent. (12)
11b. Foliage glabrous. (13)
12a. Leaves pinnate; plants 1 dm or more high; Bryce Canyon and Panguitch Plateau. **L. minimum** Mathias
12b. Leaves mostly 2- to 3-pinnate; plants 1 dm or more high; southwestern Utah. **L. scabrum** (Coult. & Rose) Mathias
13a. Leaves 2- to 3-pinnate. (14)
13b. Leaves pinnate, rarely 2-pinnate. (15)
14a. Peduncles equaling or somewhat exceeding leaves; pedicels 10–17 mm long; wings equaling or somewhat broader than body; southeastern Utah. **L. parryi** (S. Wats.) Macbr.
14b. Peduncles usually greatly exceeding leaves; pedicels 3–7 mm long; wings about half as broad as body; widely distributed. **L. nuttallii** (Gray) Macbr.
15a. Plants less than 1 dm high; leaf blades less than 2.5 cm long; Bryce Canyon and Panguitch Plateau. **L. minimum** Mathias
15b. Plants more than 1 dm high; leaf blades more than 2.5 cm long. (16)
16a. Umbel 3- to 6-rayed; pedicels 4–10 mm long; widely distributed. **L. nuttallii** (Gray) Macbr.
16b. Umbels 4- to 11-rayed; pedicels 1–4 mm long; southeastern Utah. **L. latilobum** (Rydb.) Mathias

Umbelliferae

Musineon Raf.
1a. Stems dichotomously branched; dry hills and plains. **M. divaricatum** (Pursh) Nutt. ex DC.
1b. Stems not dichotomously branched or acaulescent. **M. lineare** (Rydb.) Mathias

Oreoxis Raf.
1a. Bractlets linear, entire, green; high mountains. **O. alpina** (Gray) Coult. & Rose
1b. Bractlets obovate, toothed at apex, usually purplish; high mountains. **O. bakeri** Coult. & Rose

Orogenia S. Wats. Indian Potato
Plants 5-15 cm tall, from a globose tuber 5-12 mm in diameter; leaves ovate in outline, blades 2-8 cm long, 1- to 2-ternate, leaflets linear to lanceolate, 1.5-7 cm long, 1-7 mm broad, the first leaf occasionally simple; involucel of 1 or more linear bractlets or lacking; fertile rays 1-4, 2-25 mm long; flowers white; calyx teeth obsolete; fruit oblong-oval, 3-4 mm long, 2-2.5 mm broad; common in oak brush zone, northern Utah. **O. linearifolia** S. Wats.

Osmorhiza Raf. Sweet Cicely
1a. Fruit glabrous or sparsely bristly toward base, obtuse at base, not caudate; rays ascending or spreading-ascending; petals yellow; hillsides and valleys. **O. occidentalis** (Nutt.) Torr.
1b. Fruit bristly-hispid, caudate at base with conspicuous tails; rays spreading-ascending to divaricate and reflexed; petals greenish white or pinkish. (2)
2a. Rays and pedicels spreading-ascending; fruit linear-oblong, cylindric, beaked; woods. **O. chilensis** H. & A.
2b. Rays and pedicels divaricate; fruit clavate, obtuse; woods. **O. depauperata** Phil.

Oxypolis Raf.
Caulescent perennials from fascicled tubers, 6-10 dm tall; leaves oblong in general outline, pinnate or ternate; rays 5-14; flowers white or purple; calyx teeth conspicuous; fruit 3-5 mm long; subalpine stream banks, southeastern Utah. **O. fendleri** (Gray) Heller

Pastinaca L. Parsnip
Stout biennial or perennial herbs, 3-10 dm tall; leaves pinnately compound, oblong to ovate in outline; rays 15-25; petals yellow; calyx teeth minute or obsolete; fruit 5-6 mm long; widely naturalized weed (var. **sylvestris** DC.) or cultivated as a root crop. **P. sativa** L.

Perideridia Reichenb. False Caraway
1a. Basal leaves 1- to 2-pinnate or 1- to 2-ternate, the petioles and rachis not dilated. **P. gairdneri** (H. & A.) Mathias

Subclass DICOTYLEDONEAE

1b. Basal leaves ternate-pinnate or pinnately compound, the petioles and rachis dilated. **P. bolanderi** (Gray) Nels. & Macbr.

Petroselinium Hoffm. Parsley

Biennial herbs, 3–13 dm high; leaves deltoid in outline; peduncles 3–8 cm long; rays 10–20; flowers yellow or greenish yellow; calyx teeth obsolete; fruit 2–4 mm long; cultivated. **P. crispum** (Mill.) Mansf.

Podistera S. Wats.

Caulescent perennial herbs, 7–30 cm tall; leaves oblong in general outline, the leaflets ovate, 10–15 mm long, 10–15 mm broad, deeply 2- to 3-lobed; peduncles 1–3 dm long; rays 5–8; flowers greenish yellow; calyx teeth conspicuous; fruit 3–4 mm long. **P. eastwoodiae** (Coult. & Rose) Math. & Const.

Pseudocymopterus Coult. & Rose

Subcaulescent to caulescent perennial herbs; stems 5–85 cm tall; leaves 1- to 3-pinnate, ovate-oblong to broadly ovate in outline; rays few; flowers yellow or purple; calyx teeth evident; fruit 3–7 mm long; widely distributed. **P. montanus** (Gray) Coult. & Rose

Pteryxia Nutt. ex Coult. & Rose Desert Parsley

Caespitose, acaulescent or caulescent perennial herbs, 10–60 cm tall; leaves petiolate, 2-pinnate to pinnately or ternately-pinnately compound, ovate-oblong to broadly ovate in outline; rays few to several; flowers yellow, whitish or purple; calyx teeth prominent; fruit 7–11 mm long and broad. **P. terbinthina** (Hook.) Coult. & Rose

Sium L. Water Parsnip

Caulescent perennial herbs from fascicled roots, aquatic or semi-aquatic, 50–100 cm tall; leaves oblong or ovate in outline; flowers white; calyx teeth minute to obsolete; fruit 2–3 mm long. **S. suave** Walt.

Zizia Koch Alexanders

Caulescent perennial herbs, 30–60 cm tall; basal leaves 4–8 cm long, simple, cordate, oval or broadly ovate, crenate-dentate; cauline leaves ternately divided; involucre wanting; involucel of a few bractlets; rays 12–16, spreading-ascending, 1–3 cm long; flowers yellow; calyx teeth prominent; fruit 2–4 mm long; moist meadows. **Z. aptera** (Gray) Fern.

URTICACEAE — NETTLE FAMILY

Annual or perennial herbs; leaves alternate or opposite; flowers in axillary cymes, inconspicuous, greenish, monoecious, dioecious, polygamous, mostly 4-merous; petals none; stamens as many as sepals and opposite them; styles 1; ovary 1-loculed; fruit an achene.

1a. Leaves opposite, toothed; plants with stinging hairs. **Urtica**
1b. Leaves alternate, entire; plants without stinging hairs. **Parietaria**

Valerianaceae

Parietaria L. Pellitory
1a. Leaves rounded at base, ovate-oblong to oblong; involucral bracts oblong, obtuse; plants more or less villous, branching from base. **P. obtusa** Rydb.
1b. Leaves cuneate at base, lanceolate; involucral bracts linear, obtuse; plants more or less puberulent or villous, simple or branching from base. **P. pennsylvanica** Muhl.

Urtica L. Nettle
1a. Stipules oblong or broadly lanceolate, obtuse or acute; leaves lanceolate to cordate, more or less densely pubescent beneath, sparingly so above. **U. breweri** S. Wats.
1b. Stipules linear to narrowly lanceolate; leaves lanceolate, acute or acuminate, cordate or rounded at base, strigose to glabrate on both faces. **U. gracilis** Ait.

VALERIANACEAE — VALERIAN FAMILY

Annual or perennial, frequently odoriferous herbs; leaves basal or opposite, simple or pinnately divided, flowers more or less irregular, perfect or polygamous, in corymbose, paniculate, or capitate cymes; ovary inferior, 1- to 3-loculed, 1-ovuled; corolla 5-lobed, usually more or less zygomorphic, often gibbous or spurred at base.

1a. Stamens 1; corolla with a long spur, usually magenta or red, rarely white, more than 1 cm long; perennial; adventive from Europe. **Centranthus**
1b. Stamens 3; corolla gibbous or spurred, white, pink, or bluish, less than 1 cm long. (2)
2a. Calyx limb composed of plumose bristles in fruit; perennial mostly with rhizomes or taproots; native. **Valeriana**
2b. Calyx limb absent; annual. (3)
3a. Flowers in cymose clusters forming a more or less flat-topped inflorescence; stem dichotomously branched; annual; adventive from Europe. **Valerianella**
3b. Flowers densely clustered in capitate or interrupted, spikelike inflorescences; native. **Plectritis**

Centranthus DC. Centranth
Glaucous perennial herbs to 1 m tall; leaves simple, entire, or sometimes toothed at base; calyx inrolled in flower, becoming a plumose pappus in fruit; corolla tube usually with a long spur, limb somewhat zygomorphic; fruit somewhat compressed, crowned by persistent plumose calyx; cultivated. Red Valerian. **C. ruber** DC.

Plectritis DC.
Annual herbs; leaves simple, opposite; flowers in dense, headlike or interrupted clusters; calyx absent; corolla 2-lipped, spurred at base

Subclass DICOTYLEDONEAE

white to pale pink; stamens 3; fruit with a thick wing bounded by a cylindric margin, hairy; dry hillsides. Corn Salad. **P. macrocera** T. & G.

Valeriana L. Valerian

1a. Leaves mostly ligulate-spatulate, gradually decurrent to clasping leaf bases, with stem leaves often pinnatifid but more or less decurrent; plants from vertical, usually forked taproots; hillsides and meadows in mountains. **V. edulis** Nutt. ex T. & G.
1b. Leaves mostly pinnatifid, generally petiolate; plants from rhizomes or stolons. (2)
2a. Leaves mostly oblong in outline; corolla with a short tube and widely flaring limb (rotate or nearly so), 2–3.5 mm long; plants relatively robust and leafy, 30–90 cm tall; fruit sparsely to densely pubescent, rarely glabrous; eastern and central Utah. **V. occidentalis** Heller
2b. Leaves usually ovate to spatulate in outline; corolla funnelform to salverform, 3–19 mm long, the tube obviously gibbous; stamens exserted, longer than corolla lobes. (3)
3a. Leaves ascending-ciliate, terminal lobe linear to elliptic or oblanceolate, lateral lobes of basal leaves in 8–12 pairs; cultivated. Common Valerian. **V. officinalis** L.
3b. Leaves glabrous, or with spreading pubescence; plants relatively slender, 10–60 cm tall; cauline leaves essentially sessile; central Utah, type locality Mt. Nebo, Juab Co. **V. capitata** Pall. ex Link

Valerianella Mill.

Annual herbs; basal leaves entire; stem leaves sessile and often dentate; flowers in terminal clusters; calyx limb short or obsolete; corolla bluish, white, or pink, nearly regular; stamens 3; ovary 3-loculed, with 1 cell fertile; naturalized from Europe. Lamb's Lettuce. **V. locusta** (L.) Betcke

VERBENACEAE — VERBENA FAMILY

Annual or perennial herbs; leaves opposite, simple; flowers perfect, irregular; calyx mostly 4- to 5-lobed; corolla of 4–5 united petals; stamens 4, didynamous; ovary superior, entire or 2- to 4-loculed; fruit separating at maturity into 4 nutlets.

Verbena L.

1a. Spikes panicled at apices of stems and branches; fruiting calyces 1.7–3 mm long; corolla limb 2–4.5 mm broad, blue; stem erect; perennial. **V. hastata** L.
1b. Spikes 1 or 3 at tips of stems or branches, often pencillike, with appressed-ascending short floral bracts several to a branch, with divergent bracts much exceeding calyx; stem prostrate. **V. bracteata** Lag. & Rodr.

Figs. 105-6.   105. **Angelica pinnata,** x .2.   106. **Osmorhiza chilensis,** x .22

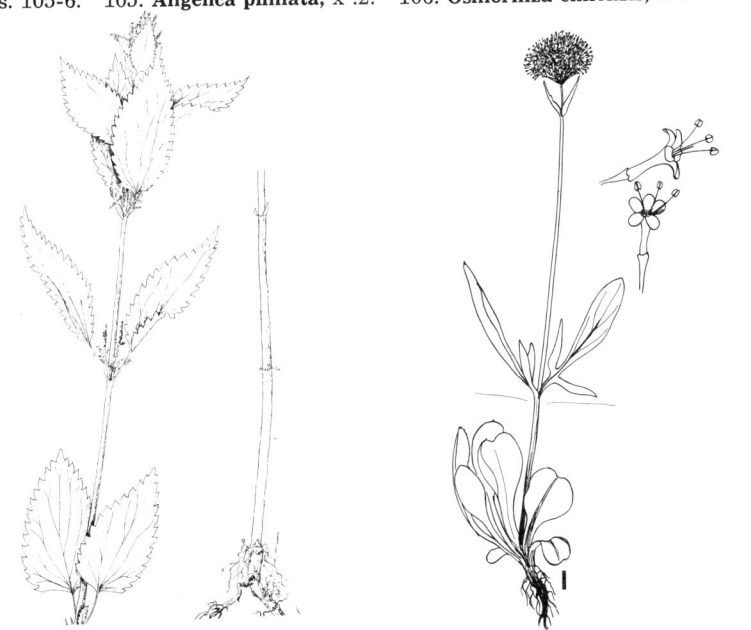

Figs. 107-8.   107. **Urtica gracilis,** x .21.   108. **Valeriana occidentalis**, x .26

Subclass DICOTYLEDONEAE

VIOLACEAE — VIOLET FAMILY

Annual or perennial herbs; leaves basal or cauline and alternate; flowers irregular, 5-merous, the lower petal gibbous or spurred at base; stamens 5; anthers connivent over pistil; ovary superior, 1-loculed; styles and stigmas 1; fruit a capsule.

Viola L. Violet

1a. Leaves deeply divided; flowers 2-colored; dry sagebrush slopes. Western Pansy Violet. **V. beckwithii** T. & G.
1b. Leaves crenate, dentate, or lobed but not divided. (2)
2a. Plants lacking a conspicuous stem. (3)
2b. Plants with a stem. (5)
3a. Plants not stoloniferous; leaves broadly cordate-ovate, crenate-serrate; moist woods or wet meadows. Bog Violet. **V. nephrophylla** Greene
3b. Plants stoloniferous. (4)
4a. Flowers white to pale violet; native to marshy sites along streams. Marsh Violet. **V. palustris** L.
4b. Flowers deep violet-purple; cultivated fragrant ornamentals. Sweet Violet. **V. odorata** L.
5a. Plants annuals or short-lived perennials; stipules nearly as large as leaf blade, pinnately parted toward base; flowers variable; cultivated ornamentals, Pansy. **V. tricolor** L.
5b. Plants perennials; stipules small. (6)
6a. Flowers yellow. (7)
6b. Flowers purple or white, not yellow. (9)
7a. Leaf blades mostly less than 25 mm long, coarsely few-toothed; veins purplish and conspicuous below; dry hillsides. Pine Violet. **V. purpurea** Kell.
7b. Leaf blades mostly over 25 mm long, nearly entire. (8)
8a. Leaf blades about 25 mm long, usually pubescent; sagebrush foothills. Yellow Violet. **V. vallicola** A. Nels.
8b. Leaf blades 50–75 mm long, nearly glabrous; moist, wooded sites. Yellow Violet. **V. praemorsa** Dougl.
9a. Petals white on inner face, purplish on outer; moist, mostly shaded wooded areas. White Violet. **V. canadensis** L.
9b. Petals violet, laterals bearded; moist, shaded hills and valleys. Blue Violet. **V. adunca** J. E. Smith

ZYGOPHYLLACEAE — CALTROP FAMILY

Plants herbs or shrubs; leaves usually opposite, pinnately compound or digitately 2- to 7-foliolate, the leaflets entire; flowers perfect, regular or nearly so; sepals usually 5, distinct or united at base; petals usually

Subclass MONOCOTYLEDONEAE

5; stamens as many as petals or 2–3 times as many; ovary 2- to 12-loculed; fruit an angled capsule, or splitting into several nutlets.

1a. Flowers purple; stipules spiny; leaflets palmately 1- to 7-foliolate. **Fagonia**
1b. Flowers yellow; stipules not spiny; leaves pinnately compound. (2)
2a. Plants woody shrubs; fruit densely villous, globose. **Larrea**
2b. Plants prostrate herbs; fruit spiny, splitting into 5 or fewer spiny nutlets. **Tribulus**

Fagonia L.
Plants perennial, woody below; leaves opposite, 1- to 7-foliolate, the leaflets more or less spinose tipped; stipules spinulose tipped; flowers solitary, purple; sepals 5, caducous; petals 5, caducous; stamens 10; ovary 5-loculed; fruit ovoid, deeply 5-angled, separating into 5 nutlets; southern Utah. F. californica Benth.

Larrea Cav. Creosote Bush
Evergreen shrubs, 1–2 m tall or more, resinous; leaves a single pair of leaflets, sessile; flowers solitary; sepals 5, caducous; petals 5, yellow, 6–8 mm long; stamens 10; ovary 5-loculed; fruit globose, separating into 5 nutlets; southwestern Utah. L. tridentata (DC.) Coville

Tribulus L. Puncture Vine
Prostrate annual herbs, the branches 30–50 cm long or more; leaves opposite, pinnately compound; flowers solitary; sepals 5; petals 5, yellow or white, 2–4 mm long; stamens 10; ovary 5-loculed; fruit depressed, 5-angled, spinose, separating at maturity into 5 bony nutlets, each with 2 stout spines and resembling a goat head; widely distributed adventive weed. T. terrestris L.

CLASS ANGIOSPERMAE
SUBCLASS MONOCOTYLEDONEAE
Key to the Families

1a. Plants small, free-floating aquatics, without true stems. LEMNACEAE, p. 399.
1b. Plants with stems and leaves, terrestrial or aquatic, but not free floating. (2)
2a. Perianth lacking or reduced and inconspicuous, its parts often bristles or scales, not petallike in color and texture. (3)
2b. Perianth well developed, at least the inner segments petaloid in color and texture. (14)
3a. Flowers sessile in axils of chaffy or husklike scales; leaves with sheathing bases. (4)

Subclass MONOCOTYLEDONEAE

3b. Flowers not in axils of chaffy bracts, sessile or pedicellate; leaves with or without sheathing bases. (5)

4a. Leaf sheaths split lengthwise on side opposite blade; leaves 2-ranked; stems mostly hollow and terete; anthers versatile; flowers subtended by 2 bracts. GRAMINEAE, p. 370.

4b. Leaf sheaths not split; leaves 3-ranked; stems mostly solid and triangular in cross section; anthers basifixed; flowers subtended by 1 bract. CYPERACEAE, p. 365.

5a. Plants floating or submerged, usually not raised above surface of water. (6)

5b. Plants terrestrial or in shallow water, usually both leaves and flowers emergent. (10)

6a. Flowers in spikes or heads. (7)

6b. Flowers axillary and solitary or very few in clusters. (8)

7a. Flowers imperfect, in globose heads, the lower pistillate, the upper staminate. SPARGANIACEAE, p. 412.

7b. Flowers perfect, in peduncled or axillary spikes; sepals 4. POTAMOGETONACEAE, p. 411.

8a. Leaves alternate. RUPPIACEAE, p. 412.

8b. Leaves opposite. (9)

9a. Leaves 3–10 cm long or more; carpels 2 or more. ZANNICHELLIACEAE, p. 414.

9b. Leaves mostly less than 3 cm long; carpels 1. NAJADACEAE, p. 407.

10a. Flowers on a spadix (a fleshy spikelike structure) surrounded by or subtended by a spathe (an enlarged bract). ARACEAE, p. 364.

10b. Flowers not on a spadix. (11)

11a. Inflorescence a dense, elongate spike. (12)

11b. Inflorescence in subglobose heads, racemes or otherwise, but not in dense spikes. (13)

12a. Inflorescence a double spike, staminate above and pistillate below; plants usually over 10 dm tall. TYPHACEAE, p. 414.

12b. Inflorescence a single spike, the sexes intermingled or the flowers perfect; plants usually under 6 dm tall. JUNCAGINACEAE, p. 398.

13a. Flowers imperfect, the lower heads pistillate; perianth inconspicuous, of chaffy scales. SPARGANIACEAE, p. 412.

13b. Flowers perfect, usually not in heads; perianth usually 6-parted, in 2 whorls. JUNCACEAE, p. 396.

14a. Pistils several to many, 1-loculed and 1-ovuled, maturing into a cluster or whorl of achenes. ALISMATACEAE, p. 360.

14b. Pistils 1 to a flower, 3- to 12-loculed, maturing into a capsule or a berry. (15)

Agavaceae

15a. Ovary superior. (16)
15b. Ovary inferior. (17)
16a. Sepals green, rarely reddish; petals colored; flowers in umbels. COMMELINACEAE, p. 364.
16b. Sepals and petals colored alike or, if unlike, the flowers not in umbels. LILIACEAE, p. 399.
17a. Plants aquatic with mostly submersed, whorled leaves. HYDROCHARITACEAE, p. 393.
17b. Plants terrestrial, the leaves neither submersed nor whorled. (18)
18a. Plants quite woody; leaves long, sword shaped. AGAVACEAE, p. 359.
18b. Plants herbaceous; leaves various. (19)
19a. Inflorescence a scapose umbel, sometimes reduced to a single flower, subtended by more or less membranous bracts; stamens 6. AMARYLLIDACEAE, p. 360.
19b. Inflorescence various but not an umbel, if subtended by a bract, this not membranous; stamens 3 or less. (20)
20a. Fertile stamens 3; petaloid staminodes lacking. IRIDACEAE, p. 393.
20b. Fertile stamens 1 or 2, the others often developing into petaloid staminodes. (21)
21a. Ovary twisted; stamens adnate to style forming a gynandrium. ORCHIDACEAE, p. 407.
21b. Ovary not twisted; stamens not adnate to style. CANNACEAE, p. 364.

AGAVACEAE — AGAVE FAMILY

Rhizomatous perennials; leaves rigid, fleshy, mostly with spiny edges, and spine tipped, arranged in a basal rosette; perianth more or less funnelform, with a short tube, the 6 segments narrow and nearly equal; stamens inserted at throat or in tube; anthers versatile; fruit a loculicidal capsule.

Agave L. Century Plant

1a. Leaves smooth edged, spiny tipped, 3-angled, flat above, marked lengthwise with white lines, 2.5 cm broad, 15 cm long, many in a globelike cluster; cultivated. **A. victoria-reginae** Moore
1b. Leaves spiny margined. (2)
2a. Leaves 12–20 cm long, 20–25 mm wide, margin more or less wavy, with white deltoid teeth about 2 mm long, terminal spine 20–35 mm long; flower stalk 1.5–2.5 m tall; flowers in clusters of 2, 4, or rarely 6, yellow; native in southern Utah. **A. utahensis** Engelm.
2b. Leaves up to 2 m long, to 25 cm wide, indented between the spines, spines usually dark, longer than 2 mm, some plants with white margins; rarely producing flowers; cultivated, usually indoors. **A. americana** L.

Subclass MONOCOTYLEDONEAE

ALISMATACEAE — WATER PLANTAIN FAMILY

Aquatic or marsh herbs; leaves with broad blades and sheathing bases; flowers regular, perfect or unisexual, in racemes or panicles; calyx of 3 sepals; petals 3, separate; stamens 6 or more; pistils several to many in a ring or dense head; ovaries superior, 1-celled, usually 1-ovuled; fruit a compressed achene.

1a. Leaves ovate to linear, never sagittate; flowers perfect; achenes arranged in a ring or on receptacle. **Alisma**
1b. Leaves sagittate; flowers monoecious, dioecious, or sometimes polygamous; achenes densely packed over surface of receptacle. **Sagittaria**

Alisma L. Water Plantain

1a. Leaf blades 2–10 cm wide; petals 3–6 mm long. **A. plantago-aquatica** L.
1b. Leaf blades 0.1–1.5 cm wide; petals 1–3 mm long. **A. geyeri** Torr. ex Nicolett

Sagittaria L. Arrowhead

1a. Terminal leaf lobes much longer than basal ones, often twice as long or longer; achene beaks minute, erect; fruiting heads seldom over 1.5 cm broad. **S. cuneata** Sheld.
1b. Terminal leaf lobes subequal to or slightly longer than lateral ones; achene beaks long, horizontal; fruiting heads usually over 1.5 cm broad. **S. latifolia** Willd.

AMARYLLIDACEAE — AMARYLLIS FAMILY

Perennial mostly acaulescent herbs from corms, rhizomes, bulbs, or fibrous roots; leaves alternate, entire; flowers perfect, the perianth of 6 parts; stamens usually 6; stigma usually 3-lobed; ovary inferior, 3-celled; fruit a capsule, mostly loculicidal; cultivated.

1a. Flowers lacking a corona, no scales or teeth between filaments. (2)
1b. Flowers with a corona formed by expanded filaments or by teeth, scales, or a tube. (13)
2a. Ovules few, 2–6 in each cell; fruit a berry or capsule. (3)
2b. Ovules numerous; fruit a capsule. (6)
3a. Leaves persistent; perianth 4–6 cm long. **Clivia**
3b. Leaves not persistent. (4)
4a. Perianth 2 cm long; fruit a berry. **Haemanthus**
4b. Perianth 3 cm long or more; fruit a capsule. (5)
5a. Tube of perianth equaling or exceeding segments. **Crinum**
5b. Tube of perianth much shorter than segments. **Lycoris**

Amaryllidaceae

6a. Perianth tube absent or very short; stamens epigynous or inserted near base of segments. (7)
6b. Perianth tube present; stamens inserted on tube. (10)
7a. Scape solid; segments of perianth unequal. **Galanthus**
7b. Scape hollow. (8)
8a. Perianth regular; flowers 1 to few. **Leucojum**
8b. Perianth declined or irregular. (9)
9a. Flowers 1, 2-lipped, the segments narrow. **Sprekelia**
9b. Flowers in umbels, sometimes 1, the segments broad. **Amaryllis**
10a. Flowers several together; scape solid. **Crinum**
10b. Flowers 1 or 2. (11)
11a. Filaments very short; flowers fragrant, night blooming. **Cooperia**
11b. Filaments well developed; flowers not fragrant. (12)
12a. Perianth erect; stamens of 2 lengths. **Zephyranthes**
12b. Perianth not erect, oblique to declined; stamens of 4 lengths. **Habranthus**
13a. Corona a cup formed by expanded filaments. **Hymenocallis**
13b. Corona not formed by expanded filaments. (14)
14a. Perianth with a corona which is ring shaped, or tubular and separate from filaments. **Narcissus**
14b. Perianth with a corona which consists of separate, often inconspicuous scales between filaments. (15)
15a. Flowers 1, 2-lipped. **Sprekelia**
15b. Flowers borne in an umbel, sometimes 1. (16)
16a. Ovules numerous in each locule; perianth 10–12 cm long. **Amaryllis**
16b. Ovules 2–3 in each locule; perianth 3.5–8 cm long. **Lycoris**

Amaryllis L.

1a. Stigma definitely 3-branched. (2)
1b. Stigma capitate, obscurely 3-lobed or 3-notched. **A. reginae** L.
2a. Valves of spathelike involucre 4 cm long or less; peduncles slightly compressed, 30 cm long, about as long as leaves, the tube cylindric. **A. striata** Lam.
2b. Valves of spathelike involucre 5–7.6 cm long; peduncles round, 60–90 cm long, the tube funnel shaped; leaves 45–60 cm long. (3)
3a. Perianth segments white, striped mauve. **A. vittata** L'Her
3b. Perianth segments deep uniform red. **A. x johnsonii** Hort.

Clivia Lindl. Kaffirlily

Plants perennials with fleshy roots, the base slender-bulbous; leaves many, persistent, 2-ranked; flowers 10–20 or more, scarlet to orange, yellow inside; perianth 5–7.5 cm long; ovary globose, inferior, 5–6 ovules per cell; fruit a red berry with large, globose, translucent seeds; common greenhouse plants. **C. miniata** Regel

Subclass MONOCOTYLEDONEAE

Cooperia Herb.   Rain- or Prairielily
1a. Neck of bulb less than 4 cm long; perianth tube 7.5–12 cm long. **C. drummondii** Herb.
1b. Neck of bulb over 5 cm long; perianth tube 3.5–5 cm long. **C. pedunculata** Herb.

Crinum L.   Crinumlily
1a. Perianth funnelform, the upper part of curved tube much broadened, passing gradually into limb; segments usually as much as 2.5 cm broad; stamens and style close together, commonly declined; flowers curving outward or even drooping. (2)
1b. Perianth salverform, the tube long, slender, standing nearly or quite erect, not markedly expanded at summit; segments usually linear or narrowly lanceolate; stamens spreading, not contiguous to style; flowers erect or strongly ascending, tube straight. **C. asiaticum** L.
2a. Leaves glaucous, gradually tapering; flowers white or pink, not porcelain smooth. **C. bulbispermum** Milne-Redhead & Schweickerdt
2b. Leaves green, sword shaped; flowers white or pink, porcelain smooth. **C. x powellii** Hort.

Galanthus L.   Snowdrop

Small, hardy, bulbous plants with 2–3 leaves about 6 mm wide and 2.5 cm long; flowers declined or nodding, borne singly on a solid scape, usually less than 30 cm tall; outer segments white, 1–2.5 cm long, inner segments half as long, white with green at sinuses; flowering in early spring; cultivated. **G. nivalis** L.

Habranthus Herb.   Rainlily
1a. Flowers yellow to copper inside, striped purple outside; perianth 2.5 cm long. **H. texanus** Herb.
1b. Flowers never yellow, either pink or red; perianth 3 cm or more long. (2)
2a. Corolla pink, deep claret red in throat; scape 30 cm long; anthers linear. **H. brachyandrus** (Baker) Sealy
2b. Corolla light pink, throat white or greenish; scape 10–15 cm long, stout; anthers crescent shaped. **H. robustus** Herb. ex Sweet

Haemanthus L.   Bloodlily
1a. Leaves thick and fleshy, usually only 2, deciduous, reappearing after flowers; flower head 10 cm in diameter, consisting of very many blood red flowers on a red-spotted scape. **H. coccineus** L.
1b. Leaves various, 2 or more, evergreen or nearly so, if deciduous then appearing with flowers. (2)
2a. Leaves thick, evergreen, the margins ciliate; flowers white, about 100 in a small umbel. **H. albiflos** Jacq.
2b. Leaves thin, glabrous; flowers bright red, about 125 in a head to 30 cm wide. **H. katherinae** Baker

Amaryllidaceae

Hymenocallis Salisb. Spiderlily
1a. Free filaments 4–5 cm long; perianth segments linear; leaves sword shaped. **H. littoralis** Salisb.
1b. Free filaments 1.5 cm long or less; perianth segments lanceolate; leaves strap shaped. **H. calathina** Nichols.

Leucojum L. Snowflake
1a. Pedicel about as long as ovary, much shorter than spathelike involucre; flowers usually 1; early spring. **L. vernum** L.
1b. Pedicel slender, equaling or exceeding spathelike involucre; flowers usually more than 1; middle to late spring. **L. aestivum** L.

Lycoris Herb. Amaryllis
Bulbous perennials with summer-deciduous leaves, new leaves emerging before flowers in late summer; scape solid; flowers rose-lilac or pink, fragrant, 8 cm long; stamens about as long as perianth segments; style exserted. **L. squamigera** Maxim.

Narcissus L. Daffodil
1a. Leaves nearly round in cross section, narrowly channeled on face; corona much shorter than perianth segments; flowers yellow, 2–6 per stalk, fragrant. Jonquil. **N. jonquilla** L.
1b. Leaves flat or nearly so; corona various. (2)
2a. Corona equaling or exceeding perianth segments, these spreading. Daffodil. **N. pseudo-narcissus** L.
2b. Corona less than half as long as perianth segments; flowers fragrant. (3)
3a. Corona a red-edged, shallow cup; flowers usually 1 per scape. Poets' Narcissus. **N. poeticus** L.
3b. Corona not red edged; flowers 4 or more per scape. Polyanthus Narcissus. **N. tazetta** L.

Sprekelia Heister Jacobeanlily
Bulbous perennials, the bulb long necked; leaves developing with flowers; flowers borne singly, the 3 lower segments rolled together around the stamens and pistil, the 3 upper segments erect or nearly so; ovary 3-celled with many ovules. **S. formosissima** Herb.

Zephyranthes Herb. Zephyrlily
1a. Flowers white to rose. (2)
1b. Flowers light yellow. **Z. ajax** Spreng.
2a. Stigma deeply 3-parted. (3)
2b. Stigma obscurely 3-lobed or 3-notched. **Z. candida** Herb.
3a. Flowers white or tinged purplish. **Z. atamasco** Herb.
3b. Flowers rose-red. **Z. grandiflora** Lindl.

Subclass MONOCOTYLEDONEAE

ARACEAE — ARUM FAMILY

Plants perennials from a creeping rhizome; leaves parallel veined; inflorescence a densely flowered spadix, subtended by a spathe; flowers variously disposed, perfect; perianth of 6 segments; stamens 6; ovary superior; fruit berrylike.

Acorus L. Sweetflag

A single genus and species treated, with the characteristics of the family; leaves 6–20 dm long, 8–20 mm wide; scape 3-angled, bearing 1 apparently lateral, green spadix; spathe leaflike and appearing as an extension of scape; represented in Utah by introduced plants in Salt Lake Co., to be expected elsewhere in the state. **A. calamus L.**

CANNACEAE — CANNA FAMILY

Perennial herbs with leafy stems; leaves alternate in 2 ranks; inflorescence terminal; flowers clustered, perfect, irregular; sepals 3, green; petals 3, pale red, united below; stamens represented by staminodes and comprising showy part of flower, only 1 anther bearing; ovary inferior, 3-celled, many-ovuled; fruit a capsule.

Canna L.

A single genus and species treated, with the characteristics of the family; flowers showy, bright red, yellow, speckled or variegated; cultivated in greenhouses or in summer plantings outside. **C. indica L.**

COMMELINACEAE — SPIDERWORT FAMILY

Perennial herbs; leaves alternate, entire; flowers perfect, usually almost regular; sepals 3, green, rarely reddish; petals 3, separate or united below, colored; stamens usually 6; filaments often hairy; styles 1; stigma simple or 2- to 3-lobed; fruit a capsule.

1a. Petals united into a tube; usually a greenhouse plant. **Zebrina**
1b. Petals free or nearly so; cultivated or native. **Tradescantia**

Tradescantia L. Spiderwort

1a. Sepals glabrous to villous, not glandular; common cultivated spiderwort. **T. virginiana L.**
1b. Sepals usually glandular-puberulent; native in southern Utah. **T. occidentalis** (Britt.) Smyth

Zebrina Schnizl. Wandering Jew

Decumbent or pendulous herbs; leaves striped white and green above, purple beneath. **Z. pendula Schnizl.**

Cyperaceae

CYPERACEAE — SEDGE FAMILY

Plants grasslike; stems (culms) usually solid and triangular; leaves 3-ranked, with closed sheaths; flowers perfect or imperfect, each subtended by 1 or sometimes 2 scales; inflorescence spikelike; perianth present or absent; stamens 1–3; ovary superior, 1-celled, 1-ovuled; stigmas 2 or 3; fruit an achene.

1a. Florets all unisexual; fruit naked or enclosed. (2)
1b. Florets perfect or perfect and staminate; fruit naked. (3)
2a. Fruit enclosed in a perigynium (a saclike structure) except for apex; wide distribution. **Carex**
2b. Fruit without a closed perigynium, the enclosing scale split to the base; restricted to high alpine meadows. **Kobresia**
3a. Scales of the spikelet 2-ranked; perianth bristles none. **Cyperus**
3b. Scales of the spikelet spirally imbricate. (4)
4a. Perianth bristles many, long-silky; inflorescence appearing like a tuft of cotton. **Eriophorum**
4b. Perianth none or few and short, not at all cottonlike. (5)
5a. Spikes 1 to a culm. (6)
5b. Spikes 2 or more to a culm. (7)
6a. Involucral leaves lacking; style base persistent as a tubercle at achene apex. **Eleocharis**
6b. Involucral leaves present, 1 or more; style base deciduous, the tubercle lacking. **Scirpus**
7a. Perianth bristles lacking; style bases swollen. **Fimbristylis**
7b. Perianth bristles present; style bases not swollen. **Scirpus**

Carex L. Sedge

A large, cosmopolitan genus of tremendous complexity. The species treated herein are representative of the more than ninety reported for Utah.

1a. Spikes 1 to an inflorescence. (2)
1b. Spikes more than 1 to an inflorescence. (3)
2a. Spikes staminate above, pistillate below; open woodlands throughout Utah. **C. geyeri** Boott
2b. Spikes entirely staminate or pistillate; alpine tundra. **C. pseudoscirpoides** Rydb.
3a. Stigmas 2. (4)
3b. Stigmas 3. (19)
4a. Lateral spikes sessile, short. (5)
4b. Lateral spikes pedunculate or elongate. (14)
5a. Plants not caespitose; rhizomes long creeping. (6)
5b. Plants caespitose; rhizomes seldom present or, if so, not long creeping. (8)

Subclass MONOCOTYLEDONEAE

6a. Spikes aggregated into globose or ovoid heads; meadows and moist sites in spruce-fir forests. **C. vernacula** Bailey
6b. Spikes not aggregated, at least the lower distinct. (7)
7a. Leaves involute; culms obtusely triangular; dry meadows at middle and higher elevations. **C. douglasii** Boott ex Hook.
7b. Leaves flat; culms sharply triangular; alkaline bottomlands, widely distributed. **C. praegracilis** Boott
8a. Spikes staminate above, pistillate below. (9)
8b. Spikes not as above; upper spike pistillate above or staminate below, or staminate; lateral spikes pistillate above, staminate below or pistillate. (11)
9a. Perigynium beak obliquely cleft dorsally; sagebrush grass at middle elevations. **C. vallicola** Dewey
9b. Perigynium beak tridentate. (10)
10a. Spikes densely capitate; damp sites throughout north central Utah. **C. hoodii** Boott ex Hook.
10b. Spikes lax in the inflorescence, lower ones somewhat separate; varied habitats, northern Utah. **C. occidentalis** Bailey
11a. Perigynia at most thin margined, lower part of body spongy-thickened; wet meadows and bogs, widely distributed. **C. interior** Bailey
11b. Perigynia wing margined, not spongy-thickened at base. (12)
12a. Bracts not conspicuously exceeding inflorescence. (13)
12b. Bracts conspicuously exceeding inflorescence; meadows and stream sides, widely distributed. **C. arthrostachya** Olney
13a. Scales shorter and narrower than perigynia, largely exposing perigynia, meadows and moist hillsides at middle elevations. **C. festivella** Mack.
13b. Scales about as long as perigynia, and nearly the same width, nearly concealing the perigynia; open, rocky slopes or grasslands at higher elevations. **C. phaeocephala** Piper
14a. Perigynia beakless or short beaked, not lustrous. (15)
14b. Perigynia beak 0.5 mm long or more, blackish and lustrous. (18)
15a. Lowest bract long-sheathing; perigynia golden yellow at maturity; bogs, meadows, and stream sides, common throughout Utah. **C. aurea** Nutt.
15b. Lowest bract sheathless or short-sheathing; perigynia not golden yellow at maturity. (16)
16a. Perigynia with conspicuous raised nerves. (17)
16b. Perigynia nerveless except for marginal ribs; widespread, wet sites. **C. aquatilis** Wahl.
17a. Plants caespitose; rhizomes slender; perigynia ovate, 1.5–3 mm long; beaks entire; marshes, wet meadows, stream sides, common in northern Utah. **C. kelloggii** Boott ex S. Wats.

Cyperaceae

17b. Plants loosely caespitose; rhizomes stout and well developed; perigynia obovate, 2.7–3.5 mm long; beaks bidentate; common, widely distributed. **C. nebraskensis** Dewey
18a. Rhizomes short-creeping; culms ascending; swamps and wet meadows, aspen-fir zone and above. **C. vesicaria** L.
18b. Rhizomes short or long; culms erect; meadows and marshes, common throughout Utah. **C. rostrata** Stokes
19a. Perigynia pubescent or puberulent. (20)
19b. Perigynia glabrous. (21)
20a. Pistillate flowers 1–20; fruit closely enveloped; open woodlands, widespread and common. **C. rossii** Boott ex Hook.
20b. Pistillate flowers 15–200; fruit loosely enveloped; meadows and stream banks, river canyons to aspen-fir zone. **C. lanuginosa** Michx.
21a. Perigynia not or only slightly inflated, nerveless or moderately nerved; beak not or slightly bidentate; styles deciduous; plants usually not coarse and tall. (22)
21b. Perigynia strongly inflated, smooth, shining and strongly ribbed; beak bidentate; styles persistent; plants usually stout and tall. (23)
22a. Lower bract long-sheathing; wet meadows and alpine tundra, higher mountains. **C. capillaris** L.
22b. Lower bract sheathless (or sheath less than 2 mm long); mountain meadows and stream sides, spruce-fir to alpine regions. **C. nova** Bailey
23a. Rhizomes short-creeping; culms ascending; wet meadows, aspen-fir zone and above. **C. vesicaria** L.
23b. Rhizomes short or long; culms erect; meadows and marshes, common throughout Utah. **C. rostrata** Stokes

Cyperus L. Nutgrass

1a. Spikelets persistent on spike; scales deciduous. (2)
1b. Spikelets deciduous above basal pair of scales, breaking into 1-fruited joints; scales persistent. **C. strigosus** L.
2a. Rachis not winged or slightly so. (3)
2b. Rachis wings conspicuously developed. (4)
3a. Stamens 1, rarely 2; plants annual. **C. aristatus** Rottb.
3b. Stamens 3, rarely 2; plants perennial. **C. schweinitzii** Torr.
4a. Plants robust annuals; fruit distinctly mucronate apically. **C. erythrorhizos** Muhl.
4b. Plants perennials; fruit obtuse apically. **C. esculentus** L.

Eleocharis R. Br. Spikerush

1a. Styles bifid; achenes lenticular. (2)
1b. Styles trifid; achenes plump or triangular. (3)
2a. Spikelets with 1 basal empty scale, nearly encircling base. **E. calva** Torr.

Subclass MONOCOTYLEDONEAE

2b. Spikelets with 2 or 3 basal empty scales. **E. machrostachya** Britt.
3a. Fruit with several longitudinal ridges and many fine horizontal lines. **E. acicularis** (L.) R. & S.
3b. Fruit not longitudinally ribbed. (4)
4a. Spikelets 2- to 7-flowered. **E. pauciflora** (Lightf.) Link
4b. Spikelets 10- to many-flowered. (5)
5a. Style bases confluent with achenes, not tuberclelike. **E. rostellata** Torr.
5b. Style bases tuberclelike. **E. bolanderi** Gray

Eriophorum L. Cottongrass

1a. Spikes solitary; involucre lacking; bristles white to reddish brown. **E. chamissonis** C. A. Meyer
1b. Spikes several, some distinctly pedicelled; involucre of 1–4 leaflike bracts; bristles white. (2)
2a. Leaves less than 2 mm wide; usually only 1 bract present; bristles seldom over 2.5 cm long. **E. gracile** Koch
2b. Leaves over 2 mm wide; usually 2 or more bracts present; bristles often over 2.5 cm long. **E. angustifolium** Roth

Fimbristylis Vahl.

Plants perennials with rhizomes; culms 20–70 cm tall; leaves basal, blades flat, 1–2.5 cm wide; inflorescence a simple or compound umbel; spikelets 8–15 mm long, many-flowered; fruit lenticular, about 1.5 mm long; moist meadows and swamps. **F. thermalis** S. Wats.

Kobresia Willd.

Plants perennials; culms 3–45 cm tall, densely tufted; leaves filiform, revolute; inflorescence densely flowered, 15–30 mm long; perigynium 3–3.5 mm long; alpine meadows. **K. bellardii** (All.) Deyland ex Loisel.

Scirpus L. Bulrush (Contributed by L. B. Barnett)
1a. Involucral bracts 2 or more, leaflike; culms leafy. (2)
1b. Involucral bracts solitary, often appearing as part of culm; culms naked, the leaves at base. (4)
2a. Spikelets 3–12, 1–4 cm long, in capitate or umbellate clusters; moist, alkaline soils, widely distributed. **S. paludosus** A. Nels.
2b. Spikelets numerous, 3–6 mm long, in compound umbels or umbellate heads. (3)
3a. Styles bifid; achenes lenticular; moist sites near springs and streams, common along Wasatch Front. **S. microcarpus** Presl.
3b. Styles trifid; achenes triangular; moist sites along streams or in seeps, uncommon. **S. pallidus** (Britt.) Fern.
4a. Spikelets 1, terminal; bract seldom exceeding spikelet; rare, Uinta Mountains. **S. caespitosus** L.

Figs. 109-10.   109. **Sagittaria cuneata**, x .2.   110. **Carex aquatlis**, x .19

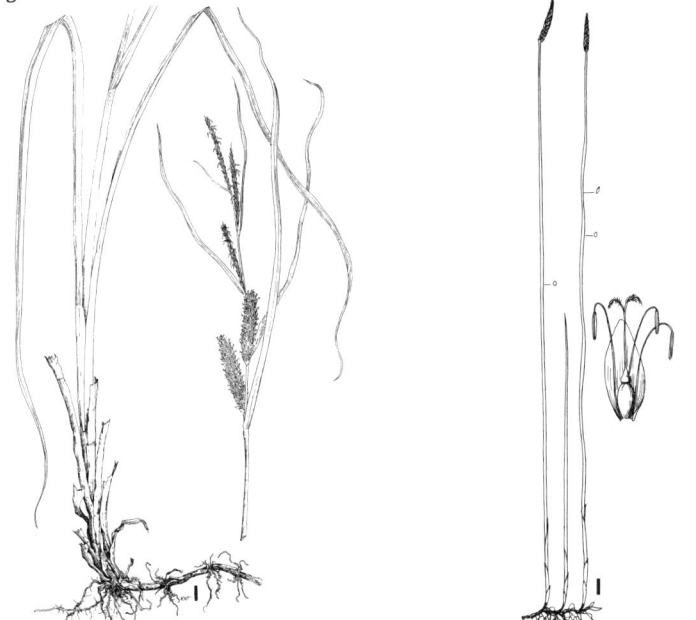

Figs. 111-12.   111. **Carex rostrata**, x .19.   112. **Eleocharis machrostachya**, x .21

Subclass MONOCOTYLEDONEAE

- 4b. Spikelets many; bract longer than spikelet. (5)
- 5a. Culms triangular or subterete, leafy at base; spikelets 1–12, capitate. (6)
- 5b. Culms stout and terete; leaves reduced to basal sheaths with short blades; spikelets in umbels. (8)
- 6a. Scales not awned; bristles one-half the length of achenes; culms subterete; rare, western Utah. **S. nevadensis** S. Wats.
- 6b. Scales awned; bristles as long as, or longer than, achenes; culms triangular, at least near inflorescence. (7)
- 7a. Culms triangular or subterete below and triangular near inflorescence; involucral bracts 3–10 cm long; widespread, wet soils throughout Utah. **S. americanus** Pers.
- 7b. Culms concavely triangular; leaf blades flat; involucral bracts 1–3 cm long; marshes and mud flats near lakes and ponds. **S. olneyi** Gray
- 8a. Roots swollen and spongy; spikelets subcylindric; scales much longer than achenes; bristles shorter than achenes; lakes and sloughs, usually an emergent. **S. acutus** Muhl.
- 8b. Roots fibrous; spikelets ovoid; scales barely covering achenes; bristles longer than achenes; moist soils of swamps and seeps. **S. validus** Vahl.

GRAMINEAE — GRASS FAMILY

(Contributed by Seville Flowers)

Herbaceous annuals or perennials; stems (culms) usually hollow in internodes; leaves 2-ranked, divided into a basal sheath and a terminal blade, ligule often present at juncture of blade and sheath; inflorescence a spike, raceme, or panicle, rarely otherwise; flowers perfect or imperfect, in florets consisting of a flower and 2 bracts (lemma and palea), the florets arranged in clusters (spikelets) on an axis (rachilla) which are subtended by 1 or 2 bracts (glumes); perianth reduced, usually consisting of 2 fleshy lodicules; stamens usually 3; ovary superior, 1-loculed, 1-ovuled, usually with 2 plumose stigmas; fruit a caryopsis. A large and complex group. The present work treats only about half of the species known to occur in Utah.

Key to Subfamilies

- 1a. Spikelets bearing 1 to many florets; staminate, neuter, or rudimentary florets, if present, always above perfect one (except in **Phalaris**); when mature the spikelets flattened laterally. FESTUCOIDEAE, p. 371.
- 1b. Spikelets bearing 1 perfect floret, a few with a staminate, neuter, or rudimentary floret below perfect one; when mature, the spikelets flattened front to back. PANICOIDEAE, p. 374.

Gramineae

SUBFAMILY FESTUCOIDEAE

1a. Spikelets with 2 slender, hairy rudimentary lemmas below the single perfect floret; lemmas hard and shiny, pale yellowish; tall grass growing in wet soil. **Phalaris**, p. 388.
1b. Spikelets without rudimentary lemmas below, imperfect florets, if any, above perfect ones. (2)
2a. Spikelets sessile on a continuous, jointed rachis or on branches of a raceme. (3)
2b. Spikelets on pedicels in open or contracted panicles. (5)
3a. Spikelets densely crowded on one side of branches of a racemose or paniculate inflorescence, erect or spreading. KEY III, p. 373.
3b. Spikelets alternate on opposite sides of a zigzag rachis, forming a slender or dense spike. (4)
4a. Plants monoecious; spikelets with tufts of long white hairs at base, 3 at a joint, the middle one with 1 pistillate floret, the lateral ones with 2 staminate florets; short desert grass with curled leaves. **Hilaria**, p. 386.
4b. Plants perfect; spikelets without tufts of hairs at base, 1 to several at joints, all perfect; glumes persistent; taller grasses of various regions. KEY V, p. 374.
5a. Spikelets 1-flowered. KEY I, p. 371.
5b. Spikelets 2- to many-flowered. (6)
6a. Glumes as long as lowest floret, often as long as entire spikelet; lemmas awned from back below tip (except in **Koeleria**). KEY II, p. 372.
6b. Glumes shorter than first floret; lemmas without awns or awned from tip or from between 2-toothed tip in **Bromus** and **Tridens**. KEY IV, p. 373.

KEY I. Spikelets 1-flowered.

1a. Lemmas hard and closely investing the seed when mature; nerves obscure; callus oblique and bearded. (2)
1b. Lemmas thin or firm but not hard, not or only loosely investing seed when mature; nerves evident. (4)
2a. Awn 3-divided, with the lateral divisions often shorter, indistinctly merging with lemma; grasses less than 5 dm tall, in dense bunches on dry plains and hillsides. **Aristida**, p. 377.
2b. Awn single, with a distinct line of junction with lemma. (3)
3a. Awn remaining on lemma, twisted many times, much longer than slender seed; tall grasses, not or only moderately bunched. **Stipa**, p. 392.
3b. Awn deciduous, straight or only slightly twisted, rather short, 2–3 times longer than plump, dark seed. **Oryzopsis**, p. 388.

Subclass MONOCOTYLEDONEAE

4a. Glumes shorter than lemmas. (5)
4b. Glumes longer than lemmas. (6)
5a. Lemmas awned or with short, sharp tips, 3-nerved; pericarp not slipping from seed when moistened. **Muhlenbergia**, p. 387.
5b. Lemmas without awns, 1-nerved; pericarp slipping from seed when moistened; seed falling from lemma and palea. **Sporobolus**, p. 392.
6a. Panicle very dense, spikelike, cylindric, ovoid or globose. (7)
6b. Panicle open or contracted but not forming dense heads. (9)
7a. Panicle thick and soft-hairy; glumes with long, soft awns and falling with a floret. **Polypogon**, p. 391.
7b. Panicle cylindric to globose, often very long and slender, not or only slightly hairy. (8)
8a. Glumes with short, sharp points, stiffly hairy on keels; lemmas awnless. **Phleum**, p. 388.
8b. Glumes blunt, soft-hairy on keels; lemmas with short awns. **Alopecurus**, p. 376.
9a. Florets on short but distinct pedicels; rachilla prolonged behind palea; tall drooping grass of damp or wet shady places. **Cinna**, p. 379.
9b. Florets sessile; wet or dry places. (10)
10a. Florets naked at base or with short hairs; palea lacking in some species. **Agrostis**, p. 376.
10b. Florets with conspicuous long hairs at base; palea present, the rachilla prolonged behind it. **Calamagrostis**, p. 378.

KEY II. Glumes as long as lowest floret; lemmas awned from back below tip.

1a. Spikelets awnless or upper lemmas short pointed; glumes shorter than lowest floret; panicle dense and narrow, spikelike. **Koeleria**, p. 387.
1b. Spikelets awned; glumes exceeding lowest floret. (2)
2a. Florets 2, upper one perfect, lower one staminate; awn exserted, twisted and bent. **Arrhenatherum**, p. 377.
2b. Florets 2 or more, all perfect. (3)
3a. Spikelets less than 8 mm long. (4)
3b. Spikelets 15 mm long or more; glumes longer than florets. (5)
4a. Lemmas rounded at back. **Deschampsia**, p. 379.
4b. Lemmas keeled at back. **Trisetum**, p. 393.
5a. Lemmas awned from below middle of back; awns not flattened; spikelets nodding. **Avena**, p. 377.
5b. Lemmas awned from between the 2-toothed apex; awns flat, twisted; spikelets on stiff pedicels, not nodding. **Danthonia**, p. 379.

Gramineae

KEY III. Spikelets densely crowded on one side of branches of a racemose or paniculate inflorescence.

1a. Spikelet with a modified awned floret above perfect one; spikelike branches flat, spreading or reflexed. **Bouteloua**, p. 377.
1b. Spikelets with perfect florets only, falling entire; spikelike branches erect. (2)
2a. Glumes unequal, narrow and pointed; perennials. **Spartina**, p. 392.
2b. Glumes equal, very broad, boat shaped; spikelets as broad as long, rounded and flat; annuals. **Beckmannia**, p. 377.

KEY IV. Glumes shorter than first floret; lemmas without awns or awned from tip.

1a. Panicle large and plumelike; florets obscured by long silky hairs; tall reeds with broad leaves. **Phragmites**, p. 390.
1b. Panicle not plumelike; florets not obscured by hairs; small annuals or perennials, if large, not as above. (2)
2a. Inflorescence a small head hidden among the short, spine-tipped leaves; low, spreading annuals. **Munroa**, p. 388.
2b. Inflorescence an open or contracted panicle; leaves not spiny. (3)
3a. Lemmas 3-nerved. (4)
3b. Lemmas 5- to many-nerved. (6)
4a. Nerves of lemmas with long silky hairs; small desert grasses. **Tridens**, p. 392.
4b. Nerves of lemmas glabrous. (5)
5a. Lemmas broad, obtuse, hyaline at apex; water or very wet soil. **Catabrosa**, p. 379.
5b. Lemmas acute or acuminate, not hyaline at apex; dry or moist soils. **Eragrostis**, p. 384.
6a. Plants dioecious; lemmas smooth and firm; nerves indistinct; saline soil. **Distichlis**, p. 381.
6b. Plants with perfect flowers (except some **Poa** species). (7)
7a. Spikelets strongly flattened, crowded into dense flat or 1-sided heads. (8)
7b. Spikelets not strongly flattened, not in flat or 1-sided heads. (9)
8a. Plants low, spreading annuals; spikelets in 2 opposite rows forming a flat or 1-sided racemelike panicle. **Schlerochloa**, p. 391.
8b. Plants tall; spikelets in dense 1-sided clusters at ends of stiff, spreading or erect branches. **Dactylis**, p. 379.
9a. Lemmas keeled on back; spikelets less than 10 mm long, mostly much smaller, awnless; leaf margins more or less parallel; apex boat shaped. **Poa**, p. 390.
9b. Lemmas rounded on back, slightly keeled at apex in some species. (10)

Subclass MONOCOTYLEDONEAE

10a. Glumes papery, rather loose; stems bulbous at base. **Melica**, p. 387.
10b. Glumes not papery; stems not bulbous at base. (11)
11a. Lemmas obtuse; nerves not converging near apex. (12)
11b. Lemmas acute or awned; nerves converging toward apex. (13)
12a. Nerves of lemmas very prominent on back; tall nodding grasses of wet soil. **Glyceria**, p. 386.
12b. Nerves of lemmas faint; plants low to medium; wet or damp, saline soil. **Puccinellia**, p. 391.
13a. Lemmas entire, pointed or awn tipped. **Festuca**, p. 384.
13b. Lemmas cleft at apex, 2-toothed, awned from between teeth. **Bromus**, p. 377.

KEY V. Plants perfect; spikelets without tufts of hairs at base; glumes persistent.

1a. Spikelets mostly 1 at each joint of rachis, sometimes 2 at a joint near base only. (2)
1b. Spikelets 2-3 per joint of rachis or, if single, the glumes tapering from base of attachment to slender awnlike tips and nerves not evident. (6)
2a. Spikelets placed edgewise on rachis; glumes narrow, the upper one toward rachis lacking. **Lolium**, p. 387.
2b. Spikelets placed broadside to rachis; glumes broad. (3)
3a. Plants annual; escaped cultivation or introduced weed. (4)
3b. Plants perennial. **Agropyron**, p. 375.
4a. Spikes very narrowly cylindric, about 3 times as thick as peduncles; awns erect and appressed. **Aegilops**, p. 375.
4b. Spikes more broadly cylindric or more or less compressed; awns divergent. (5)
5a. Lemmas very broad and coarse, toothed or awned; seeds plump. **Triticum**, p. 393.
5b. Lemmas narrow, fringed with long, stiff, spreading hairs. **Secale**, p. 391.
6a. Spikelets 1-flowered, 3 at a joint, the 2 lateral ones on short pedicels and reduced to awns. **Hordeum**, p. 386.
6b. Spikelets 2- to 6-flowered, usually 2-3 at a joint, all alike. (7)
7a. Rachis not disarticulating when dry or at maturity; glumes and lemmas with or without awns, the awns not forked. **Elymus**, p. 381.
7b. Rachis readily disarticulating when dry; glumes and lemmas long awned. **Sitanion**, p. 391.

SUBFAMILY PANICOIDEAE

1a. Spikelets solitary; glumes membranous, the sterile lemma like glumes in texture; fertile lemma and palea hardened, or at least firmer than glumes. (2)

Gramineae

1b. Spikelets in pairs, usually one sessile and fertile, the other one pedicellate and staminate, neuter or lacking with only the pedicel present; sessile spikelet with lowermost florets staminate or neuter and the upper perfect; glumes 2, firm or indurated, awnless; lemmas thin or hyaline, with or without awns. **Andropogon**, p. 376.
2a. Spikelets subtended or surrounded by 1 to many bristles or spines. (3)
2b. Spikelets not subtended or surrounded by bristles or spines. (4)
3a. Bristles slender, distinct, persistent; spikelets deciduous; inflorescence a bristly spikelike panicle. **Setaria**, p. 391.
3b. Bristles stout and united, forming a spiny subglobose bur, falling with spikelets inclosed; burs sessile on a slender axis. **Cenchrus**, p. 379.
4a. Inflorescence a slender digitate raceme, the branches flat and with sessile spikelets on one side only. **Digitaria**, p. 381.
4b. Inflorescence an open raceme or panicle, not digitate. (5)
5a. Spikelets large, with awns or spiny tips, on dense 1-sided, spikelike branches of an open raceme; large coarse grass. **Echinochloa**, p. 381.
5b. Spikelets small and terminal, in open, diffuse panicles. **Panicum**, p. 388.

Aegilops L. Goatgrass
Plants annuals with erect stems and flat leaves; introduced, in dry or moist soils around farms and waste places. **A. cylindrica** Host.

Agropyron Gaertn. Wheatgrass
1a. Plants with creeping rootstocks. (2)
1b. Plants without creeping rootstocks. (4)
2a. Glumes rigid, gradually tapering to sharp or shortly awned tips; native. Bluestem. **A. smithii** Rydb.
2b. Glumes not rigid, acute or abruptly shortly awned. (3)
3a. Lemmas glabrous; frequent along waysides, common lawn weed. Quackgrass. **A. repens** (L.) Beauv.
3b. Lemmas hairy; frequent on plains and in mountains, important forage grass. Thickspike Wheatgrass. **A. dasystachyum** (Hook.) Scribn.
4a. Spikes broad; spikelets strongly compressed, densely crowded, more or less spreading and overlapping sidewise; introduced in dry desert areas to prevent erosion. Crested Wheatgrass. **A. desertorum** (Fisch.) Schult.
4b. Spikes narrow; spikelets neither strongly compressed nor noticeably spreading or overlapping sidewise. (5)
5a. Spikelets without awns or very shortly awn tipped. (6)
5b. Spikelets with conspicuous awns. (7)

Subclass MONOCOTYLEDONEAE

6a. Glumes narrow, shorter than slender spikelets; rachilla finely scabrous; frequent, lowlands and mountains. Beardless Wheatgrass. **A. inerme** (Scribn. & Smith) Rydb.
6b. Glumes broad, nearly as long as broader spikelets; rachilla hairy; usually in dense bunches at lower elevations or with few stems at high elevations. Slender Wheatgrass. **A. trachycaulum** (Link) Malte
7a. Awn straight or nearly so when dry; spike quite narrow; spikelets strongly overlapping vertically; fields, waysides, and mountains. Bearded Wheatgrass. **A. subsecundum** (Link) A. S. Hitchc.
7b. Awn divergent outward when dry; spike not noticeably narrow; spikelets rather distinct, barely overlapping; plains to mountains. Bluebunch Wate Wheatgrass. **A. spicatum** (Pursh) Scribn. & Smith

Agrostis L.   Bentgrass

1a. Panicles diffuse, erect to widely spreading; branches long. (2)
1b. Panicles erect, narrow, spikelike and dense; branches very short; plants tall, bright green; wet or damp soil, mostly in mountains, also along ditch banks. Spike Redtop. **A. exarata** Trin.
2a. Panicles 4–18 cm long, widely spreading; spikelets numerous; plants about 12–15 dm tall; common in fields, becoming very tall in wet places, a common hay grass. Redtop. **A. alba** L.
2b. Panicles less than 8 cm long, erect when young; spikelets few. (3)
3a. Palea present, well developed with the rachilla prolonged into a bristle behind it; frequent in wet soil in mountains. Thurber Redtop. **A. thurberiana** A. S. Hitchc.
3b. Palea lacking or very short, the rachilla not prolonged. (4)
4a. Pedicels short; panicle branches more or less permanently erect and bearing spikelets nearly to base; open places in woods or meadows. Ross Redtop. **A. rossae** Vasey
4b. Pedicels long; panicle branches widely spreading to drooping in flower and bearing spikelets only above middle; purplish, feathery tops especially characteristic; common in mountains. Ticklegrass. **A. hiemalis** (Walt.) BSP

Alopecurus L.   Foxtail

Loosely tufted, low, erect or ascending perennials; panicle dense and spikelike; common in wet places around lakes, ponds, and boggy areas. Marsh Foxtail. **A. geniculatus** L.

Andropogon L.   Bluestem

1a. Racemes single on each peduncle; rootstocks lacking; mostly in dry soil, Colorado River Basin and southern Utah. Little Bluestem. **A. scoparius** Michx.
1b. Racemes 2–4 on each peduncle. (2)
2a. Rootstocks lacking; joints of sterile pedicel and rachis shortly villous; awn of sessile spikelet well developed; dry soils, southern Utah. Big Bluestem. **A. gerardii** Vitman

Gramineae

2b. Rootstocks present; joints of sterile pedicel and rachis conspicuously villous; awns reduced or lacking; sandy soil of Colorado River Basin. Sand Bluestem. **A. hallii** Hack.

Aristida L. Threeawn

Slender, densely tufted perennials, glabrous throughout; stems numerous, 1–4 dm tall; common bunchgrass of dry plains and hillsides. Red Threeawn. **A. longiseta** Steud.

Arrhenatherum Beauv. Oatgrass

Tall perennials with open panicles; infrequent around farms and along waysides, resembles small oat species. Tall Oatgrass. **A. elatius** (L.) Presl.

Avena L. Oats

1a. Lemmas hairy; awns long and bent or twisted; annuals, 3–10 dm tall or more, stout; common in cultivated land and along waysides. Wild Oats. **A. fatua** L.
1b. Lemmas not hairy; awns lacking or short and straight; panicles more nodding; cultivated, common around farms. Oats. **A. sativa** L.

Beckmannia Host. Sloughgrass

Coarse annuals, 3–10 dm tall; stems stout, prominently jointed, glabrous; wet places on ditch banks and around ponds and swamps. **B. syzigachne** (Steud.) Fern.

Bouteloua Lag. Grama

1a. Branches (spikes) 1–2, spreading and curved, 3–10 cm long, not falling entire at maturity; spikelets numerous, cluster flat; common in dry soil on desert plains and hillsides, important range grass, mostly in southern Utah. Blue Grama. **B. gracilis** (H. B. K.) Lag.
1b. Branches 30–50, small and pendulous, straight, 1–2 cm long, falling entire at maturity from main axis; spikelets few, 5–8, appressed to ascending; dry soils on plains and hillsides, mainly eastern and southern Utah. Side-oats Grama. **B. curtipendula** (Michx.) Torr.

Bromus L. Bromegrass

1a. Plants annuals; introduced weeds. (2)
1b. Plants perennials; mostly native species. (9)
2a. Spikelets and lemmas narrow, long-acuminate; awn usually longer than body. (3)
2b. Spikelets and lemmas broad, abruptly narrowed, blunt. (6)
3a. Panicle nodding; pedicels very slender and weak; glumes up to 12 mm long; panicle often becoming reddish with age; extremely common weed everywhere, matures early in June and constitutes an acute fire hazard on plains and foothills. Cheatgrass. **B. tectorum** L.

Subclass MONOCOTYLEDONEAE

3b. Panicle erect or slightly nodding; pedicels stout and stiff; glumes mostly over 12 mm long. (4)
4a. Panicle erect and contracted into a dense, compact head; dry and often saline soil, mainly western and southern Utah. Foxtail Chess. **B. rubens** L.
4b. Panicle erect to slightly spreading, open. (5)
5a. Lemmas, including awn, less than 6 cm long; first glume about 8 mm long; similar to cheatgrass but more robust. **B. sterilis** L.
5b. Lemmas, including awn, 6–8 cm long; first glume about 15 mm long; more vigorous than **B. sterilis**, the awn very stout and spinulose. Ripgut Grass. **B. rigidus** Roth
6a. Lemmas awnless, or very short awned; spikelets very broad and papery in texture at maturity; large nodding spikelets rattle when moved; common on hillsides and waysides. Rattlesnake Brome. **B. brizaeformis** Fisch. & Mey.
6b. Lemmas awned. (7)
7a. Panicle open, the branches spreading, similar to **B. racemosus**. Hairy Chess. **B. commutatus** Schrad.
7b. Panicle contracted, the branches erect or ascending. (8)
8a. Lemmas hairy; common in fields. Soft Chess. **B. mollis** L.
8b. Lemmas glabrous or minutely scabrous, not hairy; common in fields and waste places. **B. racemosus** L.
9a. Spikelets without awns, or the awns very short; plants with creeping rootstocks; oblong, erect appearance of spikelets especially characteristic; common in meadows and canyons. Smooth Brome. **B. inermis** Leyss.
9b. Spikelets awned; plants not producing creeping rootstocks. (10)
10a. Panicle nodding, with conspicuous appressed silky hairs; tall, spreading and partly drooping panicles with soft hairy spikelets characteristic; frequent in mountains. Fringed Brome. **B. ciliatus** L.
10b. Panicle erect, slightly spreading or slightly nodding, if hairy, not conspicuously so. (11)
11a. Spikelets rounded, not noticeably flattened; panicles spreading and finally nodding; lemmas hairy, not keeled; mountains and lower canyons. Nodding Brome. **B. anomalus** Rupr.
11b. Spikelets flattened conspicuously; panicles erect, finally slightly spreading; lemmas keeled at apex. (12)
12a. Lemmas glabrous or minutely scabrous; spikelets relatively narrow. **B. polyanthus** Scribn.
12b. Lemmas hairy at apex and on margins; spikelets relatively broad; possibly the commonest perennial brome, an important forage grass. California Brome. **B. marginatus** Nees

Calamagrostis Adans. Reedgrass
1a. Panicle open, often spreading in lower branches; wet or damp soil,

Gramineae

often in shallow water, common in lowlands. Bluejoint. **C. canadensis** (Michx.) Beauv.
1b. Panicle contracted, narrow and often spikelike. (2)
2a. Awns straight; panicles narrow but scarcely spikelike; damp or wet soil. Northern Reedgrass. **C. inexpansa** Gray
2b. Awns bent and protruding from side; panicles narrow and spikelike; frequent in moist or wet soils. Pinegrass. **C. rubescens** Buckl.

Catabrosa Beauv. Brookgrass

Aquatic perennials with creeping, decumbent stems rooting at the nodes; leaves 3–12 cm long, 2–6 mm wide; panicle open and spreading; common in brooklets and ponds, mainly in mountains. **C. aquatica** (L.) Beauv.

Cenchrus L. Sandbur

Plants annuals or occasionally perennials; stems mostly decumbent and ascending; leaf blades flat; sandy soil in lowlands and waste ground; the sharp spines inflict painful wounds. **C. pauciflorus** Benth.

Cinna L. Woodreed

Tall, slender, erect perennials; leaves numerous and broad; frequent in wet or moist soil, mainly in mountains. **C. latifolia** (Trev.) Griseb.

Dactylis L. Orchard Grass

Erect perennials, 6–12 dm tall; leaf sheaths with prominent white midveins; blades to 25 cm long, 3–6 mm wide; panicle to 20 cm long; stiff common grass along ditches, often cultivated for hay. **D. glomerata** L.

Danthonia DC. Oatgrass
1a. Leaf sheaths and sometimes blades with long, widely spreading soft hairs; spikelets large and solitary on end of stem; single spikelet usually characteristic; usually in open mountain valleys. One-spike Oatgrass. **D. unispicata** Munro
1b. Leaf sheaths and blades very short-hairy or without hairs; spikelets few, in dense and contracted panicles; frequent on hillsides and in mountains. Timber Oatgrass. **D. intermedia** Vasey

Deschampsia Beauv. Hairgrass

1a. Plants annual with sparse foliage; leaf blades filiform, less than 1.4 mm broad; glumes 4–8 mm long; awns geniculate; near marshes, often in saline areas, in Cache and Salt Lake cos. Annual Hairgrass. **D. danthonioides** (Trin.) Munro ex Benth.
1b. Plants perennial with dense foliage; leaf blades sometimes greater than 1.5 mm wide; glumes sometimes less than 5 mm long; awns not geniculate. (2)

Figs. 113-14.   113. **Eriophorum angustifolium,** x .21.   114. **Avena fatua,** x .21

Figs. 115-16.   115. **Beckmannia syzigachne,** x .2.   116. **Catabrosa aquatica,** x .21

Gramineae

2a. Panicle narrow; glumes equaling or exceeding upper floret, lemmas awned just below midlength; plants densely tufted; invades under heavy grazing, in moist, sandy or gravelly banks and slopes, across northern part of Utah. Slender Hairgrass. **D. elongata** (Hook.) Munro ex Benth.
2b. Panicle typically open; glumes not reaching upper floret, lemmas awned from near base; plants not densely tufted; open areas to partial shade, transition zone to alpine areas, widespread. Tufted Hairgrass. **D. caespitosa** (L.) Beauv.

Digitaria Heister   Crabgrass

1a. Sheaths glabrous; lemma brown; spikelets 2 mm long; raceme branches few, scarcely whorled; common lawn weed, smaller than the next species. Smooth Crabgrass. **D. ischaemum** (Schreb.) Muhl.
1b. Sheaths pilose or villous; lemma pale; spikelets 2.5–3.5 mm long; raceme branches 6–10, whorled; common lawn and garden weed. Crabgrass. **D. sanguinalis** (L.) Scop.

Distichlis Raf.   Saltgrass

Plants from underground rootstocks, sending up aerial stems in straight rows, 2–5 cm tall, with coarse, overlapping scales at base; abundant in saline soils. **D. stricta** (Torr.) Rydb.

Echinochloa Beauv.   Cockspur

Coarse annuals with erect, ascending or decumbent stems, 4–12 dm long; blades large, to 6 dm long; common in gardens and around farms. Barnyard Grass. **E. crusgallii** (L.) Beauv.

Elymus L.   Wildrye

1a. Plants with stout or creeping rhizomes. (2)
1b. Plants lacking rhizomes. (9)
2a. Glumes long-acuminate, 1.5–2.8 cm long; culms 80–200 cm tall; introduced from Eurasia, cultivated and persisting. Giant Wildrye. **E. giganteus** Vahl.
2b. Glumes less than 1.5 cm long. (3)
3a. Lemmas 3- to 5-nerved, never 7-nerved; introduced from Eurasia, cultivated and persisting. **E. sabulosus** Bieb.
3b. Lemmas obscurely nerved, or 1- to 3-nerved (7-nerved in some **E. triticoides**). (4)
4a. Spikelets 2–3 cm long; lemmas coarsely pubescent; low valleys in drifting sand, known from Juab and Millard cos. Sand Ryegrass. **E. arenicola** Scribn. & Smith
4b. Spikelets less than 2 cm long. (5)
5a. Lemmas in lower part, or nearly over whole surface except apex, covered with subappressed soft hair; introduced in experimental plantings, Sanpete Co. **E. dasystachys** Trin. ex Ledeb.

Figs. 117-18.   117. **Cinna latifolia**, x .19.   118. **Dactylis glomerata**, x .2

Figs. 119-20.   119. **Danthonia intermedia**, x .21.   120. **Deschampsia caespitosa**, x .2

Gramineae

- 5b. Lemmas either glabrous or scabrous. (6)
- 6a. Spike dense, with spikelets 3–5 at each node; common in valleys, especially along watercourses, widespread. Gray Ryegrass. **E. cinereus** Scribn. & Merr.
- 6b. Spike slender, with spikelets 1 or 2 at each node. (7)
- 7a. Lemmas about 7 mm long, with rigid awn 4–6 mm long; riverbanks, alkaline flats, drifting sand, and rocky slopes, from Wayne Co. northward to Davis and Duchesne cos. **E. simplex** Scribn. & Merr.
- 7b. Lemmas 7–10 mm long, glabrous or scabrous, with awns less than 4 mm long. (8)
- 8a. Lemmas long-acute, with awns about 3 mm long, the nerves distinct near the tip; spikelets 3- to 5-flowered; known from Weber, Davis, Duchesne, Sanpete, and Kane cos.; often confused with **Agropyron smithii**. Beardless Ryegrass. **E. triticoides** Buckl.
- 8b. Lemmas long-acute, with awns less than 2 mm long, the nerves obscure; rocky slopes, sagebrush hills, and saline soils, Grand, Garfield, Sanpete, Sevier, and Tooele cos. Salina Wildrye. **E. salinus** M. E. Jones
- 9a. Rachis tardily disarticulating; glumes and lemmas awned; mountains or wet meadows, Sanpete Co. north through Cache Co. Macoun's Wildrye. **E. macounii** Vasey
- 9b. Rachis persistent, not disarticulating. (10)
- 10a. Glumes subulate or subsetaceous, nerveless to obscurely nerved. (11)
- 10b. Glumes lanceolate or narrower, broadened above base, strongly 3- to several-nerved. (14)
- 11a. Lemmas 7–8 mm long, covered with short, stiff hairs or long, soft ones, the awns 15–20 mm long; introduced from Eurasia, in experimental planting, Utah Co. **E. junceus** Fisch.
- 11b. Lemmas 8–10 mm long, glaucous or pubescent, with awns less than 15 mm long. (12)
- 12a. Spikes dense; spikelets 3–5 at a node; culms 75–200 cm tall. Widespread. Gray Ryegrass. **E. cinereus** Scribn. & Merr.
- 12b. Spikes less dense; spikelets 1–2 per node. (13)
- 13a. Glumes 4–10 mm long; lemmas awnless or awn tipped, the awns less than 2 mm long; south central Utah. Salina Wildrye. **E. salinus** M. E. Jones
- 13b. Glumes 8–12 mm long; lemmas awn tipped, the awns 2–5 mm long; open slopes, canyons, and rocky hillsides, Millard Co. eastward to southeastern Utah (includes **E. ambiguus** authors, not Vasey and Scribn.) Bull Grass. **E. salmonis** C. L. Hitchc.
- 14a. Glumes relatively thin, not indurate at base, flat and several-nerved. (15)
- 14b. Glumes firm, indurate at base. (16)
- 15a. Lemmas glabrous or scabrous, with awns 10–20 mm long; most common along streams in mountains, also along roads and on dry hill-

Subclass MONOCOTYLEDONEAE

sides and in meadows, Sevier and Iron cos. north to Box Elder, Rich, and Uintah cos. Blue Wildrye. **E. glaucus** Buckl.
15b. Lemmas short awned (less than 4 mm); spikelets imbricate; Davis and Utah cos. Pacific Rye Grass. **E. virescens** Piper
16a. Awns of lemmas divergently curved when dry; base of glumes not terete; stream banks, fence rows, and open wooded areas, widely distributed, probably in every county. Canada Wildrye. **E. canadensis** L.
16b. Awns of lemmas straight or lacking; base of glumes terete, bowed out from rachilla; woods and open ground, northwestern quarter of Utah. **E. virginicus** L. var. **submuticus** Hook.

Eragrostis Beauv. Lovegrass

1a. Plants creeping by decumbent stolons which root at the nodes; spikelets few and small; damp or wet places, often around farms. Creeping Lovegrass. **E. hypnoides** (Lam.) BSP
1b. Plants often decumbent at base but not forming stolons; panicle ample; spikelets larger; weed of farms and gardens. Stinkgrass. **E. cilianensis** (All.) Link

Festuca L. Fescue

1a. Plants spring annuals; stamens usually 1, sometimes 3; flowers usually self-pollinated. (2)
1b. Plants perennials; culms simple; stamens 3. (7)
2a. Spikelets usually more than 5-flowered, often 8-flowered; margins of lemmas inrolled, not scarious. (3)
2b. Spikelets 5- or fewer-flowered; lemmas usually with scarious margins. (4)
3a. Backs of lemmas glabrous or scabrous, not pubescent or hirtellous, the awns commonly 3–5 (7) mm long; open sterile ground, sometimes the dominant species in the spring flora of the desert, Salt Lake and Daggett cos. south in eastern Utah, from Juab Co. south to the western tier of counties. **F. octoflora** Walt. var. **octoflora**
3b. Backs of lemmas pubescent or hirtellous, sometimes strongly scabrous only, the awns mostly 2–4 mm long; central eastern Utah and Washington and Utah cos. **F. octoflora** Walt. var. **hirtella** Piper
4a. Awn 5–8 mm long; spikelets reflexing at maturity; in disturbed areas or rocky slopes, southwestern Utah. **F. reflexa** Buckl.
4b. Awn generally more than 8 mm long. (5)
5a. Panicle narrow, the branches appressed; lemmas not pubescent on back but ciliate toward apex; apparently rare, open sterile or disturbed ground, below 5,500 ft., southwestern Utah. **F. megalura** Nutt.
5b. Panicle spreading; spikelets often spreading. (6)
6a. Spikelets glabrous; apparently rather uncommon, as only 1 collec-

Gramineae

tion known—from open ground in Zion National Park, southwestern Utah. **F. pacifica** Piper

6b. Spikelets pubescent, the pubescence on glumes or lemmas or both; reported for Utah but not in state's herbaria, to be looked for in southwestern Utah. **F. grayi** (Abrams) Piper

7a. Leaf blades flat or loosely involute on ends. (8)

7b. Leaf blades involute and narrow, less than 3 mm wide. (11)

8a. Awn 5–20 mm long; moist soil, up to 7,500 ft., central Utah. **F. subulata** Trin.

8b. Awn less than 2 mm long or absent. (9)

9a. Spikelets oblong to linear, mostly 6- to 10-flowered; florets generally more than 10 mm long; lemmas rarely with short awn (1 mm or less); naturalized from Eurasia, found at 4,000–8,000 ft. in northeast quarter of Utah and from Beaver Co. in the west through to San Juan Co. **F. elatior** L.

9b. Spikelets ovate or oval, generally 5- or fewer-flowered; florets less than 10 mm long. (10)

10a. Panicles narrow, the branches short and not spreading; leaf blades rather firm and erect; dry mountains and hills, 6,000–10,500 ft. (includes **F. kingii** Cassidy) **Hesperochloa kingii** (S. Wats.) Rydb.

10b. Panicles open, the branches spreading; leaf blades lax and spreading; open woods, 6,000–9,000 ft., reportedly in southeastern Utah. **F. sororia** Piper

11a. Ligules 2–6 mm long; awn wanting; dry slopes and rocky hillsides, 7,500–10,500 ft., central and southeastern Utah. **F. thurberi** Vasey

11b. Ligules less than 1 mm long. (12)

12a. Panicle branches densely ciliate at angles; leaf blades about 1 mm wide, flat or folded; "rocky slopes, rare, Utah" (A. S. Hitchcock). **F. dasyclada** Hack.

12b. Panicle branches not ciliate at angles. (13)

13a. Culms often geniculate at base, rather coarse, sometimes reddish or purplish at base; leaf blades smooth; awn about same length as lemma; cooler areas (mountains?), in meadows, hills, bogs, and marshes, Utah and Uintah cos. (and between?). **F. rubra** L.

13b. Culms erect, neither geniculate, decumbent, nor reddish or purplish at base; culms and leaves filiform. (14)

14a. Leaf blades less than half as long as culms, which are generally less than 30 cm tall; panicles narrow, almost spikelike, usually less than 10 cm long; spikelets mostly 4- or 5-flowered. (15)

14b. Leaf blades more than half as long as culms; panicles somewhat open, not spikelike. (17)

15a. Leaves glaucous, elongate; a cultivated plant, probably not yet naturalized, distribution discontinuous. Blue Fescue. **F. ovina** L.

15b. Leaves neither glaucous nor elongate; not cultivated in Utah. (16)

16a. Culms mostly 5–20 cm tall; leaf blades smooth, short and rather

Subclass MONOCOTYLEDONEAE

lax; alpine areas, Iron, Garfield, and Grand cos. in the south, Juab Co. northeast to Summit and Uintah cos. in the north. Alpine Fescue. **F. ovina** L. var. **brachyphylla** (Schult.) Piper

16b. Culms mostly 20–30 (40) cm tall, densely tufted; leaves very scabrous to glabrous, slender and involute; the most common fescue in Utah, open woods and stony slopes, Iron Co. east to San Juan Co., Juab and Sanpete cos. north to Weber Co. and northeast to Daggett, Uintah, and Grand cos. Sheep Fescue. **F. ovina** L. var. **ovina**

17a. Awns 2–4 mm long; open woods and rocky slopes, 7,000–10,000 ft., Box Elder and San Juan cos., Sevier Co. northeast by north to Duchesne and Summit cos. **F. idahoensis** Elmer

17b. Awns less than 1 mm long or absent; open pine woods, reported from Utah and to be expected in southern Utah. **F. arizonica** Vasey

Glyceria R. Br. Mannagrass

1a. Plants slender; leaf blades 2–6 mm wide; spikelets 2–4 mm long, short and plump; lower glume up to 1 mm long; lemma ovate, to 2 mm long; common on ditch banks and ponds at all elevations. Fowl Mannagrass. **G. striata** (Lam.) A. S. Hitchc.

1b. Plants large and stout; leaf blades 6–12 mm wide; spikelets 4–6 mm long; lower glume 1.5 mm long; lemma narrowly oval, 2–3 mm long; swamps and streams. American Mannagrass. **G. grandis** S. Wats.

Hilaria H. B. K.

Plants perennials with stems densely clustered or decumbent, often with scaly creeping rhizomes, 1–2 dm tall; common on deserts, dry foothills and mesas; important forage grass. Galleta. **H. jamesii** (Torr.) Benth.

Hordeum L. Barley

1a. Plants annual; spikes erect, strict; awns stout. (2)
1b. Plants perennial; spikes erect to slightly nodding; awns very slender. (3)
2a. Glumes widest at base and suddenly tapering, subulate; spikes short, about 2 cm wide, including awn tips, 1.5–2 times longer than wide; spikelets widely divergent; common on saline plains and around wet places. Mediterranean Barley. **H. gussonianum** Parl.
2b. Glumes widened above narrower base, fringed with stiff spreading hairs; spikes elongate, 1–1.5 cm wide, including awn tips, 2–3 times longer than wide; spikelets erect and appressed; a very common weed. Mouse Barley. **H. murinum** L.
3a. Awns about 1 cm long; spikes very narrow, usually erect; frequent in damp soil in lowlands. Meadow Barley. **H. nodosum** L.
3b. Awns 2–5 cm long; spikes plumelike, much broader, usually curved to slightly nodding; common in wet or dry soils, lowlands and waste places. Foxtail Barley. **H. jubatum** L.

Gramineae

Koeleria Pers. Junegrass
Caespitose, slender perennials, 3–6 dm tall; panicles narrow and contracted, shiny, spinelike; mountains, on hillsides, open woods and meadows, often among rocks. Junegrass. **K. cristata** (L.) Pers.

Lolium L. Ryegrass
1a. Lemmas awnless or with very short awns; common lawn grass, coarse and undesirable. Perennial Ryegrass. **L. perenne** L.
1b. Lemmas awned, sometimes lower spikelets awnless; taller and more robust than L. perenne; common on ditch banks and around farms. Italian Ryegrass. **L. multiflorum** Lam.

Melica L. Melicgrass
Plants perennials, often swollen at base, 3–6 dm tall; mainly in open valleys and slopes. Oniongrass. **M. bulbosa** Geyer

Muhlenbergia Schreb. Muhly Grass
1a. Plants with creeping rootstocks covered with scales. (2)
1b. Plants without creeping rootstocks; stems decumbent and often rooting at nodes. (7)
2a. Panicle open and diffusely branched; delicate grass along river banks and seeps, especially in saline soil. Scratchgrass. **M. asperifolia** (Nees & Mey.) Parodi
2b. Panicle erect, spikelike or somewhat open, not diffuse. (3)
3a. Callus bearing numerous long white hairs as long as lemmas; plants 5–10 dm tall; common in meadows, mountains, and on slopes. Foxtail Muhly. **M. andina** (Nutt.) A. S. Hitchc.
3b. Callus nearly naked or, if hairy, hairs less than one-half the length of lemmas. (4)
4a. Panicle rather open, branches long and very slender; leaves divergent at a wide angle, pungent; common in dry, sandy soil on deserts, sand dunes, and rocky canyons, mostly in eastern and southern Utah. Pungent Muhly. **M. pungens** Thurb.
4b. Panicle contracted and spikelike; pedicels very slender and short. (5)
5a. Leaf blades to 7 cm long, 2–5 mm wide, flat; plants 6–10 dm tall; largest of Utah species; rather common in wet places in mountains. Marsh Muhly. **M. racemosa** (Michx.) BSP
5b. Leaf blades short, less than 3 cm long and 2 mm wide, inrolled; plants usually under 3 dm tall. (6)
6a. Lemmas very long awned; glumes about three-fourths the length of lemma body; leaves finely hairy; dry, rocky soil of canyons and hillsides in southern Utah. Shortleaf Muhly. **M. curtifolia** Scribn.
6b. Lemmas blunt or merely mucronate; glumes very short; leaves glabrous or very finely scabrous on upper surface; dry or moist places in lowlands and mountains, often in saline soil. Mat Muhly. **M. squarrosa** (Trin.) Rydb.

Subclass MONOCOTYLEDONEAE

7a. Lemmas mucronate, glabrous; glumes bluntly acute; most common species, in mountains, dry hillsides, woods, and drier borders of wet places. Slender Muhly. **M. filiformis** (Thurb.) Rydb.
7b. Lemmas long awned and long white-hairy; glumes of 2 types—upper glume 3-toothed, lower acute; dry soil in mountains, hillsides, and mesas. Mountain Muhly. **M. montana** (Nutt.) A. S. Hitchc.

Munroa Torr. False Buffalograss

Low, depressed and spreading annuals; common on dry deserts, benches, and hillsides. **M. squarrosa** (Nutt.) Torr.

Oryzopsis Michx. Indian Ricegrass

Mostly tufted, slender perennials with flat or inrolled leaves; panicles narrow or open and spreading; important forage grass; common in dry sandy soils, especially sand dunes. **O. hymenoides** (R. & S.) Riker

Panicum L. Panicgrass

1a. Plants annual; leaf sheaths dilated, long-hairy; panicle diffuse; common garden weed. Witchgrass. **P. capillare L.**
1b. Plants perennial; leaf sheaths narrower, not noticeably dilated. (2)
2a. Rootstocks or stolons present; leaf blades long and narrow. (3)
2b. Rootstocks or stolons absent; leaf blades much shorter and wider, lanceolate to linear; spikelets obovate. (4)
3a. Panicles diffuse; spikelets pointed; damp or wet soils, mostly in meadows. Switchgrass. **P. virgatum L.**
3b. Panicles erect and contracted; spikelets obovate, obtuse; sandy or gravelly soil on banks of streams, often drooping down from steep banks. Vine Mesquite. **P. obtusum H. B. K.**
4a. Spikelets about 3.3 mm long; blades scabrous to sparingly hairy on upper side; stems rather stout; damp soil along streams. Scribner's Panicgrass. **P. scribnerianum** Nash
4b. Spikelets 1.5–1.8 mm long; blades, when young, hairy on upper side, especially toward base; stems more slender; damp or wet soil along streams and ditches, mainly in southeastern Utah. **P. huachucae** Ashe

Phalaris L. Canarygrass

Coarse perennials, 6–15 cm tall; leaves 8–25 cm long; panicles dense or somewhat open; common on ditch banks around farms and in swampy places. Reed Canarygrass. **P. arundinacea** L.

Phleum L. Timothy

1a. Panicle long and cylindric, 6–10 times as long as thick; plants erect; cultivated hay grass. Timothy. **P. pratense L.**
1b. Panicle shorter and thicker, ovoid to oblong-cylindric, 1–5 times as long as thick; plants erect or decumbent; common in alpine meadows. Mountain Timothy. **P. alpinum L.**

Figs. 121-22.　121. **Hordeum jubatum**, x .2.　122. **Lolium multiflorum**, x .2

Figs. 123-24.　123. **Phalaris arundinacea**, x .2.　124. **Phleum alpinum**, x .22

Subclass MONOCOTYLEDONEAE

Phragmites Trin.   Common Reed
Largest native grass in Utah, to 4 m tall; damp or very wet soils around lakes, ponds, swamps, and seeps. **P. communis** Trin.

Poa L.   Bluegrass

1a. Aerial stems bulbous at base; upper floret of each spikelet modified with ovary as a conspicuous purple bulblet; seeds not formed; introduced weed in fields and foothills. Bulbous Bluegrass. **P. bulbosa** L.
1b. Aerial stems not bulbous at base; upper floret normal. (2)
2a. Plants annual, pale green, to 2 dm tall; lemmas floccose; common as a lawn weed. Annual Bluegrass. **P. annua** L.
2b. Plants perennial. (3)
3a. Stems arising from creeping rootstocks, these often short. (4)
3b. Stems not arising from creeping rootstocks. (7)
4a. Stems flattened, elliptic in cross section; panicle small, more or less erect; common in fields, foothills, and mountains. Canadian Bluegrass. **P. compressa** L.
4b. Stems round or only slightly compressed; panicle open. (5)
5a. Lemmas floccose on nerves at back; spikelets numerous; leaf tip prominently keeled; common in lawns and damp places. Kentucky Bluegrass. **P. pratensis** L.
5b. Lemmas glabrous to scabrous on nerves or finely hairy on back, not floccose. (6)
6a. Panicle branches reflexed below; frequent in shaded woods and open slopes in mountains. Shortleaf Bluegrass. **P. curta** Rydb.
6b. Panicle branches erect to spreading, but not reflexed; wet or damp soil in mountains. Wheeler Bluegrass. **P. nervosa** (Hook.) Vasey
7a. Lemmas long-hairy on nerves or floccose at base; spikelets flattened. (8)
7b. Lemmas short-hairy on back between nerves as well as on them, sometimes with a tuft of longer hairs at base. (11)
8a. Panicle branches reflexed below; similar to **P. curta** except that lemmas are shorter and floccose at base and creeping rootstocks are lacking; frequent in mountains, in woods or open slopes. Nodding Bluegrass. **P. reflexa** Vasey & Scribn.
8b. Panicle branches not reflexed, erect to widely spreading. (9)
9a. Panicles large and spreading; spikelets about 4 mm long; plants pale green; damp or wet places in mountains. Fowl Bluegrass. **P. palustris** L.
9b. Panicles small, more or less contracted; spikelets about 8 mm long; dry soil, deserts and slopes. (10)
10a. Ligule 6–10 mm long, tapering, acute; plants erect and densely tufted. Longtongue Muttongrass. **P. longiligula** Scribn. & Will.

Gramineae

10b. Ligule shorter, truncate or obtuse; plants densely tufted. Muttongrass. **P. fendleriana** Vasey
11a. Panicle small, spreading; spikelets flattened; plants small, pale green; high elevations in mountains. Alpine Bluegrass. **P. alpina** L.
11b. Panicle tall, erect or contracted, spikelike; spikelets plump, not markedly flattened; plants mostly tall. (12)
12a. Lemmas finely hairy on back with tuft of longer hairs at base; stems very slender; dry plains and foothills. Sandberg Bluegrass. **P. secunda** Presl.
12b. Lemmas finely scabrous to nearly smooth, without longer hairs at base; stems stout and leafy. (13)
13a. Ligule 4–6 mm long, acute or truncate-acute; leaf sheaths scabrous; frequent on dry plains, often in saline soil. Nevada Bluegrass. **P. nevadensis** Vasey
13b. Ligule shorter, 2–3 mm long, truncate, rounded or bluntly acute; leaf sheaths smooth; larger than **P. nevadensis**; moist and dry soil, fields and ravines. Big Bluegrass. **P. ampla** Merr.

Polypogon Desf. Rabbitfoot

Densely tufted annuals, 1–6 dm tall; leaves broad and flat; panicles erect and contracted, soft-hairy; common in damp or wet saline soils in lowlands. **P. monspeliensis** (L.) Desf.

Puccinellia Parl. Alkaligrass

Low, smooth and pale perennials with short rootstocks, 3–6 dm tall or more; leaf sheaths somewhat swollen; moist soils in salt marshes and sloughs. **P. airoides** (Nutt.) Walt. & Coult.

Schlerochloa Beauv. Hardgrass

Small annuals, decumbent, forming tufts to 3 dm in diameter and 3–15 cm tall; common weed of barren waste ground around farms, often in saline soils. **S. dura** (L.) Beauv.

Secale L. Rye

Plants annuals or biennials; stems simple or tufted, 8–15 dm tall or more in favorable sites; around farms and waysides, escaped from cultivation. **S. cereale** L.

Setaria Beauv. Bristlegrass

1a. Second glume about two-thirds as long as spikelet; fertile lemmas strongly transversely wrinkled; bristles 5–16 below each spikelet, these bristles yellowish at maturity. **S. lutescens** (Wieg.) F. T. Hubb.
1b. Second glume three-fourths to as long as spikelet; fertile lemmas very finely wrinkled; bristles 1–3 directly below each spikelet, these bristles green at maturity. Green Bristlegrass. **S. viridis** (L.) Beauv.

Sitanion Raf. Squirreltail

Densely tufted perennials, 2–15 dm tall; general resemblance to

Subclass MONOCOTYLEDONEAE

Hordeum but coarser; common on dry plains and foothills. **S. hystrix** (Nutt.) J. G. Smith

Spartina Schreb. Cordgrass
Plants perennials with simple, coarse, rigid, erect stems from creeping rootstocks; common in saline and sandy soils around marshes. **S. gracilis** Trin.

Sporobolus R. Br. Dropseed
1a. Panicles open and diffusely branched; about as long as wide, branches stiff; throat of leaf sheath not hairy; abundant in dry or moist saline soil. Alkali Sacaton. **S. airoides** Torr.
1b. Panicles contracted or, if open, then narrowly pyramidal in outline, not diffuse, branches not stiff; throat of leaf sheath with a tuft of dense, straight white hairs. (2)
2a. Panicles contracted and slenderly spikelike, more or less cylindric, bearing spikelets to the base of branches; dry, sandy soil along streams, in sand dunes and on hillsides. Spike Dropseed. **S. contractus** A. S. Hitchc.
2b. Panicles more or less open and spreading, at least in upper part, lower part often included in leaf sheath, rarely entirely so; bases of branches naked. (3)
3a. Branches of panicle flexuous; spikelets on ends of slender peduncles, not crowded; dry soils. Mesa Dropseed. **S. flexuosus** (Thurb.) Rydb.
3b. Branches of panicle straight; spikelets crowded along main branches, entirely enclosed in leaf sheath; dry, sandy plains and hillsides. Sand Dropseed. **S. cryptandrus** (Torr.) Gray

Stipa L. Needlegrass
1a. Awns less than 5 cm long. (2)
1b. Awns 5 cm long or more, often very long. (3)
2a. Lemmas with hairs at apex; leaves inrolled when fresh; canyons and mountain slopes. Lettermann's Needlegrass. **S. lettermannii** Vasey
2b. Lemmas without hairs at apex; leaves flat when fresh; common, dry soils in canyons and on mountainsides. Columbia Needlegrass. **S. columbiana** Macoun
3a. Awns not hairy on upper part; plants stout, erect and tufted, 3–9 dm tall; common, dry plains and hillsides, especially in sandy soil. Needle-and-Thread Grass. **S. comata** Trin. & Rupr.
3b. Awns plumose with rather dense hairs; plants erect, smooth and tufted, 3–10 dm tall; common in dry, rocky canyons and mesas, southern Utah. New Mexican Feathergrass. **S. neomexicana** (Thurb.) Scribn.

Tridens Roem. & Schult.
1a. Plants low and spreading by stolons; panicles subtended by fas-

cicle of leaves which exceeds them; dry deserts and hillsides, mainly in southern Utah. Fluffgrass. **T. pulchella** (H. B. K.) A. S. Hitchc.

1b. Plants erect, without stolons; panicles on long stems not subtended by fascicle of leaves; dry desert plains, mostly southern Utah. Hairy Tridens. **T. pilosa** (Buckl.) A. S. Hitchc.

Trisetum Pers.

Tufted perennials with flat narrow leaves and densely contracted, spikelike panicles; common in meadows, steep, rocky slopes of high mountains. **T. spicatum** (L.) Richt.

Triticum L. Wheat

1a. Glumes prominently keeled only at apex; lemmas with or without awns; stems 5–10 dm tall, hollow; common around farms and roadsides. Wheat. **T. aestivum** L.
1b. Glumes with a prominent winglike keel extending to base; lemmas with very long awns; stems usually with a pith; common around fields and farms, escaped. Durum Wheat. **T. durum** Desf.

HYDROCHARITACEAE — FROGBIT FAMILY

Usually dioecious perennial, submerged or floating, aquatic herbs; leaves opposite or whorled; flowers regular, sessile or on scapelike peduncles in a spathe; calyx of 3 sepals; petals 3 or 0; staminate flowers with 3–21 stamens; ovary inferior.

Elodea Michx. Waterweed

A single genus and species treated, with the characteristics of the family; mountain ponds and streams. **E. canadensis** Michx.

IRIDACEAE — IRIS FAMILY

Perennials from short rhizomes; leaves narrow, equitant; flowers perfect, regular or irregular, subtended by spathaceous bracts; perianth of 6 segments, in 2 series of 3 each; ovary inferior, 3-loculed; style 3-cleft; fruit a capsule.

1a. Plants lacking an aerial scape, blooming in autumn or early spring; flowers sessile on the corm which is several cm below ground level; cultivated. **Crocus**
1b. Plants with an aerial scape; flowers on elongate stems. (2)
2a. Flowers irregular or the perianth oblique and curved, secund; stamens usually more or less disposed on one side. **Gladiolus**
2b. Flowers regular, not curved or secund; stamens regularly disposed. (3)
3a. Style branches large and petaloid; flowers large, usually over 5 cm wide; sepals and petals unlike. **Iris**

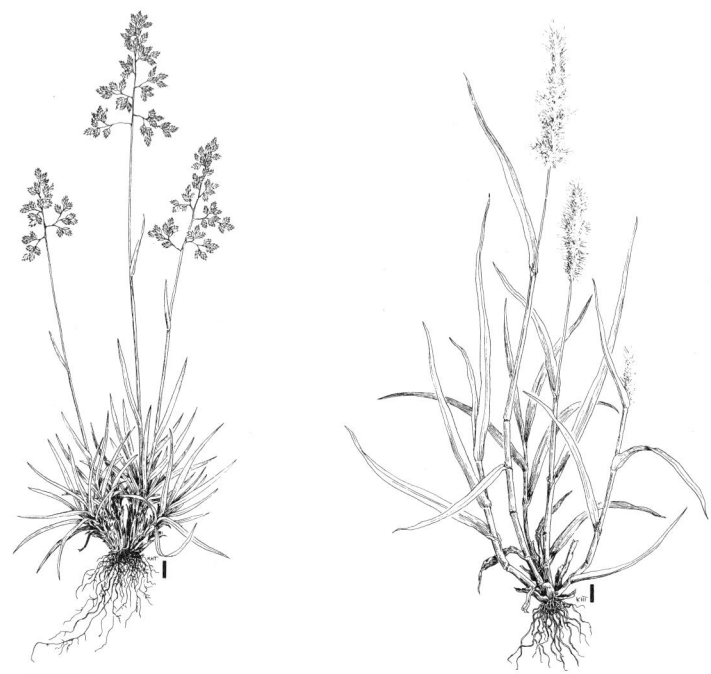

Figs. 125-26. 125. **Poa alpina**, x .23. 126. **Polypogon monspeliensis**, x .23

Figs. 127-28. 127. **Secale cereale**, x .2. 128. **Trisetum spicatum**, x .21

Iridaceae

3b. Style branches not petaloid; flowers less than 3 cm wide; sepals and petals alike. **Sisyrinchium**

Crocus L.

1a. Plants blooming in autumn. **C. sativus** L.
1b. Plants blooming in early spring. (2)
2a. Flowers yellow, at least inside. (3)
2b. Flowers white, lilac, or purple. (4)
3a. Style branches pale yellow. **C. maesiacus** Ker.
3b. Style branches orange-red. **C. susianus** Ker.
4a. Perianth segments 2-colored without, the outer ones purplish veined, the inner ones rose-purple. **C. imperati** Terr.
4b. Perianth segments of 1 color. (5)
5a. Perianth throat yellow or orange within. **C. biflorus** Mill.
5b. Perianth throat not yellow or orange within. **C. vernus** Wulfen

Gladiolus L.

A genus much confused by hybridization. The origins of many cultivated forms are lost in a tangle of unkept records. However, several names are applied; chief among them is the one applied to the yellow-flowered species. **G. primulinus** Baker.

Iris L. (Contributed by Roberta Wilson)
1a. Rootstock bulblike; usually flowering in the summer. Spanish Iris (Dutch Iris here also). **I. xiphium** L.
1b. Rootstock rhizomatous; usually flowering in the spring. (2)
2a. Flower segments bearded. (3)
2b. Flower segments not bearded. (6)
3a. Petals (standards) bearded; sepals (falls) bearded. **I. hoogiana** Dykes
3b. Petals not bearded; sepals bearded. (4)
4a. Flower spike shorter than leaves, which are less than 3 dm tall. **I. pumila** L.
4b. Flower spike equaling or longer than leaves, which are more than 3 dm tall. (5)
5a. Flower spike equaling leaves; flowers yellow and red-brown. **I. variegata** L.
5b. Flower spike exceeding leaves; flowers violet to white. **I. pallida** Lam.
6a. Stalk hollow; flower segments unequal. **I. siberica** L.
6b. Stalk solid; flower segments almost equal. (7)
7a. Rhizome slender; common in cultivation. **I. spuria** L.
7b. Rhizome thick, stout; native, uncommon in cultivation. **I. missouriensis** Nutt.

Sisyrinchium L.   Blue-eyed Grass

A confusing genus, much in need of revision.

Subclass MONOCOTYLEDONEAE

1a. Perianth 1.5–2 cm long, reddish purple; filaments united at base only. **S. douglasii** A. Dietr.
1b. Perianth less than 1.5 cm long; filaments united almost to the top. (2)
2a. Outer bract of spathe conspicuously longer than inner ones. (3)
2b. Outer bract of spathe equaling or slightly exceeding inner ones. (4)
3a. Perianth segments more or less retuse, deep violet, 10–12 mm long; capsule 4–6 mm long. **S. montanum** Greene
3b. Perianth segments not retuse, white or purple tinged, 6–10 mm long; capsule 6–7 mm long. **S. segetum** Bicknell
4a. Stems with several peduncles from leafy nodes. **S. radicatum** Bicknell
4b. Stems simple and leafless. (5)
5a. Perianth 12–14 mm long, purple; segments rounded apically; capsule subglobose, glabrate. **S. occidentale** Bicknell
5b. Perianth less than 10 mm long, bluish purple; segments abruptly acuminate; capsule puberulent. **S. halophilum** Greene

JUNCACEAE — RUSH FAMILY

Annual or perennial grasslike herbs; leaves alternate or basal, sheathing; flowers small, perfect, regular; perianth of 6 separate, similar, scalelike segments; stamens 3–6; ovary 1- or 3-loculed; stigmas 3; fruit a capsule with 3 to several seeds.

1a. Capsule 1- to 3-loculed with many seeds; leaf sheaths open; plants glabrous. **Juncus**
1b. Capsule 1-loculed with 3 seeds; leaf sheaths closed; plants glabrous or hairy. **Luzula**

Juncus L. Rush

1a. Bract which subtends the inflorescence erect, terete, and appearing like a continuation of the peduncle; inflorescence appearing lateral. (2)
1b. Bract which subtends the inflorescence not erect or, if so, not terete; inflorescence appearing terminal. (7)
2a. Perianth segments with hyaline margins and a brown stripe down each side of midrib. **J. balticus** Willd.
2b. Perianth segments green, often pale with age but not with hyaline margins or a brown stripe. (3)
3a. Bract of inflorescence exceeding or nearly equaling stem. **J. filiformis** L.
3b. Bract of inflorescence much shorter than stem. (4)
4a. Flowers numerous in a more or less compound panicle; stems stout. **J. cooperi** Engelm.
4b. Flowers 1–5, all but one pedicellate; stems slender. (5)

Juncaceae

5a. Uppermost leaf sheaths merely bristle tipped, the blade rudimentary. **J. drummondii** C. A. Meyer
5b. Uppermost leaf sheaths, at least some, bearing well-developed blades. (6)
6a. Capsules acute apically; perianth segments 5–7 mm long. **J. parryi** Engelm.
6b. Capsules retuse at apex; perianth segments 4–5 mm long. **J. hallii** Engelm.
7a. Leaves transversely flattened, the flat leaf surface facing toward and away from stem, not at all septate. (8)
7b. Leaves terete or ensiform, with sharp edges toward and away from stem, hollow or septate. (16)
8a. Plants small delicate annuals. (9)
8b. Plants perennials. (11)
9a. Flowers solitary at ends of peduncles which are 2–3 cm long. **J. uncialis** Greene
9b. Flowers numerous, subsessile, in a diffuse panicle. (10)
10a. Perianth 4–6 mm long, the segments rather appressed to the oblong capsules which are 3–4.5 mm long. **J. bufonius** L.
10b. Perianth 3–4 mm long, the segments usually spreading from the subglobose capsules which are 2–3 mm long. **J. sphaerocarpus** Nees
11a. Flowers each in axil of a bract and with 2 bractlets at base of perianth, variously displayed but not in heads. (12)
11b. Flowers each in axil of a bract but lacking bractlets, in 1 or more heads. (14)
12a. Perianth segments deep purplish brown, with broad green midrib, obtuse; plants from elongate rhizomes; introduced, in salt marshes, Salt Lake Co. **J. gerardii** Lois
12b. Perianth segments pale greenish or yellowish green, acute or acuminate; plants densely tufted, elongate rhizomes lacking; native, of various distribution. (13)
13a. Outer perianth segments scarious to apex, 3.5–4 mm long, exceeding oblong capsule. **J. confusus** Coville
13b. Outer perianth segments not scarious at aristate apex, 3–4.5 mm long, the oval capsule three-fourths as long. **J. tenuis** Willd.
14a. Heads usually hemispheric; perianth parts deep brown with lighter middle; seeds long tailed. **J. regelii** Buch.
14b. Heads mostly not hemispheric; perianth parts pale brown, largely scarious; seeds not tailed. (15)
15a. Sheaths markedly differentiated from blades; perianth smooth and shining. **J. longistylis** Torr.
15b. Sheaths not clearly differentiated from blades; perianth dull, minutely roughened. **J. orthophyllus** Coville
16a. Leaf blade usually channeled along upper side; septae usually

Subclass MONOCOTYLEDONEAE

 imperfect, not externally evident; inflorescence in 1–4 heads; high elevations. **J. albescens** (Lange) Fern.

16b. Leaf blade not channeled along upper side; septae usually perfect and externally evident; inflorescence usually of several to many heads; not of high elevations. (17)

17a. Leaves ensiform (sharp edges toward and away from the stem). (18)
17b. Leaves terete. (22)
18a. Heads 5- to 12-flowered, usually numerous. (19)
18b. Heads 15- to 25-flowered, 1 or few. (20)
19a. Plants stout; heads greenish or light brown. **J. xiphoides** E. Meyer
19b. Plants slender; heads deep brown. **J. saximontanus** A. Nels.
20a. Heads light brown. **J. tracyi** Rydb.
20b. Heads very dark brown. (21)
21a. Stamens 6; ligules usually auriculate. **J. saximontanus** A. Nels.
21b. Stamens 3; ligules not auriculate. **J. ensifolius** Wikstr.
22a. Capsules obtuse, acute, or mucronate. (23)
22b. Capsules subulate pointed. (24)
23a. Capsule much exceeding perianth. **J. tweedyi** Rydb.
23b. Capsule shorter than perianth. **J. mertensianus** Bong.
24a. Capsule much exceeding perianth; heads 6–9 mm in diameter; stems slender, 15–60 cm tall. **J. nodosus** L.
24b. Capsule about equaling perianth; heads 10–16 mm in diameter, mostly clustered; stems 30–90 cm tall. **J. torreyi** Coville

Luzula DC. Woodrush

1a. Flowers 1–3 together on slender pedicels in open panicles. **L. wahlenbergii** Rupr.
1b. Flowers many together, congested in globose or oblong spikes, sessile or nearly so. (2)
2a. Inflorescence erect; spikes borne on peduncles. **L. multiflora** (Retz.) Lejeune
2b. Inflorescence nodding; spikes sessile or nearly so. **L. spicata** (L.) DC.

JUNCAGINACEAE — ARROWGRASS FAMILY

 Perennial herbs growing in marshes; leaves basal, linear, sheathing at base; flowers small in terminal spikes, perfect or imperfect; perianth segments 3 or 6; greenish or reddish stamens usually 6; carpels 3 or 6, superior; style short or absent; fruit a cluster of 3–6 1-seeded carpels.

Triglochin L. Arrowgrass
1a. Carpels and stigmas 3; high mountains. **T. palustris** L.
1b. Carpels and stigmas 6; low and middle elevations. **T. maritima** L.

Liliaceae

LEMNACEAE — DUCKWEED FAMILY

Small floating or submerged aquatics consisting of leafless, flat, disklike bodies bearing 1 or more roots or rootless; flowers from a saclike spathe, rare, consisting of a single stamen or pistil, often 2 staminate and 1 pistillate flower per spathe.

1a. Plants without roots. **Wolfia**
1b. Plants with roots. (2)
2a. Roots 1 on each segment; segments 1–3 mm long, underside green or streaked with brown. **Lemna**
2b. Roots several from each segment; segments 3–6 mm long, underside usually reddish. **Spirodela**

Lemna L. Duckweed

1a. Joints of plants long and narrow stalked; plants forming chainlike colonies. **L. trisulca** L.
1b. Joints of plants rounded, not stalked; plants 1 or a few together. **L. minor** L.

Spirodela Schleid.

A single species; common in marshes and ponds. **S. polyrrhiza** (L.) Schleid.

Wolfia Horkel

A single species. **W. punctata** Griseb.

LILIACEAE — LILY FAMILY

Herbaceous, rarely somewhat woody, perennials from bulbs, corms, rhizomes, woody caudices or rarely with a perennial stem; flowers regular or nearly so, perfect, of 6 separate or united segments; stamens 6; ovary superior, 3-loculed; styles 1 or 3; stigmas 3-lobed or 3; fruit a capsule or berry.

1a. Plants from a woody caudex or else treelike; leaves numerous, in large rosettes at apex of caudex or branches, sword shaped, often spine tipped. **Yucca**
1b. Plants neither from a woody caudex nor treelike; leaves various, but not sword shaped in large rosettes as above. (2)
2a. Perianth segments unlike, the 3 outer much narrower and sepaloid, the 3 inner broader and petaloid. **Calochortus**
2b. Perianth segments alike or nearly so. (3)
3a. Inflorescence umbellate, or subumbellate and nearly sessile. (4)
3b. Inflorescence racemose or scapose, or corymbose, not umbellate. (7)
4a. Perianth tube 3–8 cm long; lobes white; umbel subsessile. **Leucocrinum**

Subclass MONOCOTYLEDONEAE

4b. Perianth to 2 cm long; lobes variously colored but mostly other than white. (5)
5a. Perianth segments distinct, or united only at base; plants with onion or garlic odor. **Allium**
5b. Perianth segments united below middle; plants lacking distinctive onion odor. (6)
6a. Stamens with filaments united into a tube, this with toothlike lobes between anthers. **Androstephium**
6b. Stamens with filaments separate, no toothlike lobe between anthers. **Brodiaea**
7a. Stems much branched, from thick, tuberous roots; leaves scalelike, subtending filiform leaflike branchlets (phyllodia). **Asparagus**
7b. Stems simple or sparingly branched; underground parts various, usually not of thick, tuberous roots, but occasionally so; leaves neither scalelike nor subtending phyllodia. (8)
8a. Plants 1–2 m tall; leaves 20–30 cm long or more; flowers in large panicles, these usually over 20 cm long. **Veratrum**
8b. Plants usually less than 1 m tall; leaves often less than 20 cm long, if inflorescence paniculate, then smaller. (9)
9a. Plants with scapes rather than leafy stems; all leaves basal or nearly so, if some cauline leaves present, these much smaller than basal ones. (10)
9b. Plants with leafy stems; basal leaves not conspicuously larger than stem leaves. (18)
10a. Perianth segments distinct or nearly so. (11)
10b. Perianth segments distinctly united, at least below. (15)
11a. Plants not bulbous, short-rhizomatous, with thickened roots. **Eremocrinum**
11b. Plants from bulbs or corms. (12)
12a. Anthers basifixed; flowers 1 or few. (13)
12b. Anthers dorsifixed (versatile); flowers several to many. (14)
13a. Flowers nodding or pendulous. **Erythronium**
13b. Flowers erect. **Tulipa**
14a. Filaments conspicuously dorsally flattened; flowers whitish. **Ornithogalum**
14b. Filaments not flattened, or only basally; flowers bluish or purplish. **Camassia**
15a. Plants not bulbous, producing a rhizome, or with thickened, often tuberous underground parts. (16)
15b. Plants from bulbs or corms. (17)
16a. Flowers 7–12 cm long, orange-red. **Hemerocallis**
16b. Flowers about 6 mm long, white. **Convallaria**
17a. Perianth urn shaped, not expanded at throat; lobes very short. **Muscari**

Liliaceae

17b. Perianth bell shaped, open at throat; lobes long. **Hyacinthus**
18a. Plants not bulbous, producing a rhizome, or with thickened, often tuberous underground parts. (19)
18b. Plants from bulbs or corms. (21)
19a. Flowers terminal; perianth segments not recurved. (20)
19b. Flowers axillary; perianth segments recurved or spreading. **Streptopus**
20a. Flowers in racemes or panicles, not nodding; perianth less than 7 mm long; stems simple. **Smilacina**
20b. Flowers 1 or few in umbellike inflorescence, nodding; perianth 8 mm long or more; stems branching. **Disporum**
21a. Styles 2–3, distinct to base; perianth segments seldom over 10 mm long; inflorescence a many-flowered raceme or panicle. **Zygadenus**
21b. Styles 1, more or less lobed at top; perianth segments usually more than 10 mm long; inflorescence 1- or few-flowered. (22)
22a. Anthers dorsifixed (versatile); flowers over 4 cm long. **Lilium**
22b. Anthers basifixed; flowers usually less than 4 cm long (occasionally longer in **Tulipa**). (23)
23a. Outer perianth segments essentially like inner ones, not sepallike. (24)
23b. Outer perianth segments smaller, sepallike. **Calochortus**
24a. Flowers white; segments with purple veins; dwarf alpine plants. **Lloydia**
24b. Flowers usually other than white (white occasionally in **Tulipa**). (25)
25a. Flowers nodding or pendulous. (26)
25b. Flowers erect. **Tulipa**
26a. Leaves 2; flowers yellow, over 22 mm long. **Erythronium**
26b. Leaves 3 or more; flowers purple, brown, yellow, or orange, if yellow then less than 22 mm long. **Fritillaria**

Allium L.   Onion

1a. Flowers mostly transformed into bulblets. **A. rubrum** Osterh.
1b. Flowers not transformed into bulblets. (2)
2a. Leaves hollow, terete. (3)
2b. Leaves not hollow, plane, flat or keeled. (4)
3a. Scapes inflated; cultivated onion. **A. cepa** L.
3b. Scapes not inflated; native or cultivated. Chives. **A. schoenoprasum** L.
4a. Umbels more or less nodding; stamens and styles exserted. **A. cernuum** Roth
4b. Umbels not nodding; stamens included. (5)
5a. Bulbs borne on short, stout rhizomes; leaves over 4 mm wide; ovary not crested at apex. **A. brevistylum** S. Wats.

Subclass MONOCOTYLEDONEAE

5b. Bulbs lacking rhizomes; leaves less than 4 mm wide; ovary crested in most species or not crested. (6)
6a. Outer bulb coat traversed with coarse, interwoven fibers. (7)
6b. Outer bulb coat not coarsely fibrous. (9)
7a. Perianth 10 mm long or more, segments long-attenuate apically; peduncles often 2–3 from same loose sheaths. **A. macropetalum** Rydb.
7b. Perianth 10 mm long or less, segments not long-attenuate apically; peduncles 1 from close sheaths. (8)
8a. Leaves usually 2 to a stem; flowers usually white; scapes usually less than 20 cm tall. **A. textile** Nels. & Macbr.
8b. Leaves usually 3 or more to a stem; flowers usually pink; scapes usually more than 25 cm tall. **A. geyeri** S. Wats.
9a. Perianth segments serrulate. **A. acuminatum** Hook.
9b. Perianth segments entire. (10)
10a. Perianth segments gibbous at base. (11)
10b. Perianth segments not gibbous at base. **A. diehlii** M. E. Jones
11a. Capsules crested apically, usually with 6 crests. (12)
11b. Capsules not crested. **A. brandegei** S. Wats.
12a. Leaves 1 to a stem. (13)
12b. Leaves 2 or more to a stem. (14)
13a. Perianth segments twice as long as stamens, 10 mm long or more. **A. cristatum** S. Wats.
13b. Perianth segments only slightly exceeding stamens. **A. nevadense** S. Wats.
14a. Staminal filaments more or less dilated; flowers light pink. **A. biceptrum** S. Wats.
14b. Staminal filaments filiform; flowers deep pink to white. **A. campanulatum** S. Wats.

Androstephium Torr.   Funnellily

Scapose herbs, 10–30 cm tall; leaves several, equaling or exceeding scapes; umbel 3- to 12-flowered; pedicels 10–20 mm long; perianth 15–20 mm long, variously colored; bud usually whitish or pinkish, segments longer than tube; widely distributed. **A. breviflorum** S. Wats.

Asparagus L.

1a. Plants erect, dioecious, grown for food; escaping and persisting. **A. officinalis** L.
1b. Plants climbing or drooping, not dioecious, grown for ornament; a common greenhouse plant. **A. sprengeri** Regel

Brodiaea Sm.   Wild Hyacinth

1a. Anthers basifixed; southern Utah. **B. pulchella** (Salisb.) Greene
1b. Anthers versatile; northern Utah. **B. douglasii** S. Wats.

Liliaceae

Calochortus Pursh   Sego Lily
1a. Petals bright orange; southern Utah. **C. kennedyi** T. C. Porter
1b. Petals cream colored, purplish, lilac, white, blue, or golden yellow. (2)
2a. Anthers acute; southeastern Utah. **C. gunnisonii** S. Wats.
2b. Anthers obtuse. (3)
3a. Petals abruptly acuminate; glands oblong; doubtfully reported for Utah. **C. macrocarpus** Dougl.
3b. Petals rounded or merely acute; glands not broader than long. (4)
4a. Stems flexuous and decumbent, often branched; flowers usually pale blue; southern Utah. **C. flexuosus** S. Wats.
4b. Stems erect or nearly so, usually simple; flowers white, rarely lilac, or golden yellow in var. **aureus** (S. Wats.) Ownbey; widely distributed; native (state flower). **C. nuttallii** T. & G.

Camassia Lindl.   Camas
Bulbous, scapose plants, 2.5–6 dm tall; leaves 2–5 dm long, 5–20 cm wide; flowers regular or irregular, blue-violet; segments 12–30 mm long; capsules ovoid, 8–25 mm long; northern Utah. **C. quamash** Greene

Convallaria L.   Lily-of-the-Valley
Rhizomatous, scapose herbs, 1.5–3 dm tall; leaves 2, 1–2 dm long, 3.5–7.5 cm wide or more; flowers white, about 6 mm long; fruit a red berry; cultivated. **C. majalis** L.

Disporum Salisb.   Fairybells
Rhizomatous, caulescent herbs, 3–6 dm tall; leaves 3–10 cm long, ovate to oblong-lanceolate; flowers 1 or few; perianth 8–15 mm long, white or yellowish; fruit globose, 8–10 mm in diameter. **D. trachycarpum** (S. Wats.) Benth. & Hook.

Eremocrinum M. E. Jones
Tuberous-rooted, scapose herbs, 1.5–3 dm tall; leaves 2 or more; flowers few in a raceme, 8–12 mm long, whitish with green veins; fruit a capsule, 5–7 mm long; sandy situations in southern Utah. **E. albomarginatum** M. E. Jones

Erythronium L.   Dogtooth Violet
Cormous, scapose herbs, 1.5–3 cm tall; leaves 2, 10–25 cm long; flowers 1 to few, rarely several; segments 20–40 mm long, golden yellow, nodding, usually reflexed; style long and slender; stigma short lobed; widely distributed in mountains. **E. grandiflorum** Pursh

Fritillaria L.   Fritillary
1a. Flowers beneath a terminal whorl of leaves; cultivated ornamental. Crown Imperial. **F. imperialis** L.
1b. Flowers terminal on stem; native, mountains. (2)

Subclass MONOCOTYLEDONEAE

2a. Flowers yellow or orange, not mottled; style shortly 3-lobed. Yellowbells. **F. pudica** (Pursh) Spreng.
2b. Flowers brown or purplish, mottled; style divided into long lobes. Leopard Lily. **F. atropurpurea** Nutt.

Hemerocallis L. Daylily

Tuberous-rooted, scapose herbs, 6-10 dm tall; leaves 60 cm long or more, about 2.5 cm broad; flowers 6–10, bright orange-red, 7–12 cm long; cultivated, escaping and persisting. Orange Daylily. **H. fulva** L.

Hyacinthus L. Hyacinth

Bulbous, scapose herbs, 1.5–4 dm tall; leaves several, thick; flowers in racemes, of many colors, fragrant, about 2.5 cm long, the lobes widely spreading or reflexed; early spring ornamental. Common Hyacinth. **H. orientalis** L.

Leucocrinum Nutt. Starlily

Plants herbs from short, vertical rhizomes and fleshy roots; leaves several, 2–6 mm wide, linear; flowers 4–8 in umbellike, sessile clusters, the tube 3–8 cm long, lobes 20–25 mm long; ovaries and pedicels below ground; high elevations down the central highlands. **L. montanum** Nutt.

Lilium L. Lily

1a. Flowers erect, open bowl shaped, the segments widely spreading. (2)
1b. Flowers horizontal to nodding, the segments usually not widely spreading. (4)
2a. Perianth segments 2.5–5 cm long; style not longer than ovary. Starlily. **L. concolor** Salisb.
2b. Perianth segments 5–10 cm long; style more than twice as long as ovary. (3)
3a. Pedicels erect, not curved; perianth segments 5–25 mm wide. Candlestick Lily. **L. dauricum** Ker.
3b. Pedicels curved, spreading; perianth segments 25–30 mm wide. **L. bulbiferum** L.
4a. Perianth segments curved only near apex. (5)
4b. Perianth segments strongly curved at middle or below. (7)
5a. Perianth tube slender, little widened to middle. Easter Lily. **L. longiflorum** Thunb.
5b. Perianth tube widening from base upward. (6)
6a. Segments 5–7 cm long; flowers white. Madonna Lily. **L. candidum** L.
6b. Segments 10–18 cm long; flowers white. Regal Lily. **L. regale** Wils.
7a. Flowers white, tinged with pink, spotted red. **L. speciosum** Thunb.
7b. Flowers orange-red, spotted black. Tiger Lily. **L. tigrinum** Ker.

Lloydia Salisb. Alplily

Plants with bulbs arising from rhizomes, 5–15 cm tall; leaves 1–2 mm wide, several; flowers 1–4, erect, about 1 cm wide, whitish with

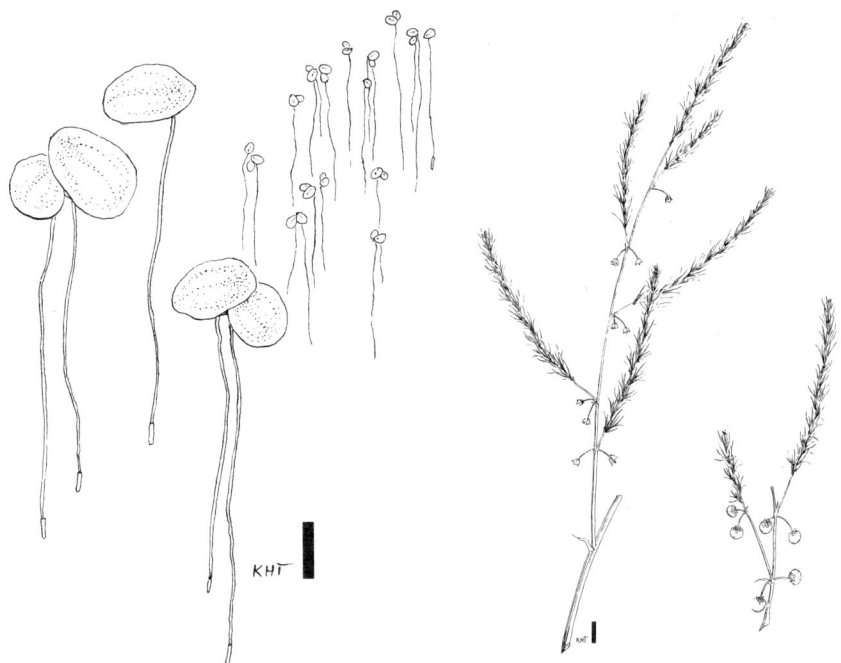

Figs. 129-30. 129. **Lemna minor**, x .79. 130. **Asparagus officinalis**, x .28

Figs. 131-32. 131. **Erythronium grandiflorum**, x .2. 132. **Lloydia serotina**, x .44

Subclass MONOCOTYLEDONEAE

purple veins; capsules obovoid, about 8 mm long; high elevations, mountains. **L. serotina** (L.) Sweet

Muscari Mill. Grape Hyacinth

Bulbous, scapose herbs, 1–3 dm tall; leaves linear, to 5 mm broad; flowers in a raceme, subglobose, to 5 mm long; teeth spreading, very short; cultivated ornamental. **M. botryoides** Mill.

Ornithogalum L.

Bulbous, scapose plants, 2–4 dm tall; leaves linear; flowers corymbose, about 25 mm broad, white within, green keeled without; staminal filaments expanded; common greenhouse plant. Star-of-Bethlehem. **O. umbellatum** L.

Smilacina Desf. False Solomonsseal

1a. Flowers in racemes; perianth segments 5–7 mm long. **S. stellata** (L.) Desf.

1b. Flowers in panicles; perianth segments about 2 mm long. **S. racemosa** (L.) Desf.

Streptopus Michx. Twistedstalk

Rhizomatous, caulescent herbs, 3–10 dm tall; leaves cordate-clasping; flowers 1–2, in axillary inflorescences, 8–12 mm long, greenish white, the segments widely spreading; fruit a globose berry, 8–15 mm long; canyons and slopes. **S. amplexifolius** (L.) DC. & Lam.

Tulipa L. Tulip

1a. Perianth segments 7–10 cm long, red with a dark blotch at base; early flowering. **T. fosteriana** Hoog.

1b. Perianth segments 4–6 cm long, variously colored, usually lacking a dark blotch; early to late flowering. (2)

2a. Stems pubescent and somewhat scabrous. Duc Van Thol Tulip. **T. suaveolens** Roth

2b. Stems glabrous and smooth. Common Tulip. **T. gesneriana** L.

Veratrum L. False Hellebore

Plants herbs, from thick rhizomes, 50–200 cm tall; leaves 20–30 cm long or more, 8–20 cm broad; inflorescence paniculate, to 50 cm long; perianth segments 6–15 mm long, greenish white; moist meadows at middle and higher elevations. **V. californicum** Dur.

Yucca L.

1a. Leaves lacking free marginal fibers; caulescent, 2–10 m tall; restricted to southwestern Utah. Joshua Tree. **Y. brevifolia** Engelm.

1b. Leaves with marginal fibers; acaulescent or nearly so, usually less than 2 m tall, exclusive of inflorescence. (2)

2a. Leaves thin and bright green; common in cultivation. **Y. smalliana** Fern.

Orchidaceae

2b. Leaves thick and glaucous; native, uncommon in cultivation. (3)
3a. Leaves 3 cm wide or more; perianth segments over 6 cm long; fruit fleshy, often over 15 cm long, pendulous. Datil Yucca. **Y. baccata** Torr.
3b. Leaves seldom over 3 cm wide; perianth segments less than 6 cm long; fruit dry, dehiscent, erect. (4)
4a. Leaves 2–10 mm wide, not rigid, often very flexible; capsules 3.5–5 cm long. **Y. angustissima** Engelm. ex Trel.
4b. Leaves over 10 mm broad or, if less, then rigid; capsules (4) 5–6.5 cm long. (5)
5a. Leaves narrowly linear, yellow-green; style white; capsule usually not constricted. **Y. standleyi** McKelvey
5b. Leaves linear to spatulate-lanceolate, glaucous or finally green; capsule often constricted; common native yucca. **Y. harrimaniae** Trel.

Zygadenus Michx. Death Camas

1a. Perianth segments 9–11 mm long, rarely shorter; ovary partly inferior; glands of perianth obcordate; common in mountains at high elevations, at low elevations confined to moist hanging gardens in southern Utah. **Z. elegans** Pursh
1b. Perianth segments 4–8 mm long; ovary superior; glands of perianth ovate or semicircular. (2)
2a. Perianth segments usually less than 4 mm long, acute or acuminate; common death camas of foothills. **Z. paniculatus** (Nutt.) S. Wats.
2b. Perianth segments 5–8 mm long, obtuse or rounded apically; uncommon in meadows and moist situations. **Z. venenosus** S. Wats.

NAJADACEAE — WATERNYMPH FAMILY

Annual, usually submerged herbs; leaves subopposite or whorled, with a sheathing base; flowers imperfect, minute, axillary, solitary, the staminate with 1 stamen and a 2-lipped perianth, the pistillate naked, with a single 1-ovuled carpel; fruit indehiscent.

Najas L.
1a. Leaves coarsely toothed; internodes and back of leaves often spiny. **N. marina** L.
1b. Leaves minutely toothed or almost entire; internodes and back of leaves unarmed. **N. guadalupensis** Morong.

ORCHIDACEAE — ORCHID FAMILY

Perennial herbs with perfect, irregular flowers; sepals 3, the lower sometimes united; petals 3 with the central forming a prominent lip; stamens 1 or 2, adnate to style and forming a column (gynandrium); ovary inferior, 3-carpelled; fruit a many-seeded capsule.

Subclass MONOCOTYLEDONEAE

1a. Plants reddish or yellowish, lacking green leaves. **Corallorhiza**
1b. Plants with green leaves. (2)
2a. Lip a large inflated sac often 2.5 cm long. **Cypripedium**
2b. Lip not saclike or, if so, less than 12 mm long. (3)
3a. Flowers with a distinct spur projecting down from base of lip. **Habenaria**
3b. Flowers without a spur. (4)
4a. Leaves 1 or 2, ovate to suborbicular, never lanceolate. (5)
4b. Leaves 3 or more, rarely 2, variously shaped. (6)
5a. Plants scapose, with single leaf and flower stalk arising from a corm. **Calypso**
5b. Plants not scapose, with 2 leaves opposite near the middle of the stem. **Listera**
6a. Flowers 15–25 mm long; leaves prominently several-ribbed; floral bracts foliaceous. **Epipactis**
6b. Flowers less than 15 mm long; leaves not prominently ribbed; floral bracts not foliaceous. (7)
7a. Leaves in a basal rosette, evergreen, commonly reticulately veined. **Goodyera**
7b. Leaves not in a basal rosette, not evergreen or reticulately veined. **Spiranthes**

Calypso Salisb.   Fairy Slipper

Plants herbs from a corm; leaf 1; scape 1-flowered or rarely 2-flowered; sepals and petals purplish, rarely white, 12–20 mm long or more, the lip larger than the rest of flower; capsule erect; a very showy flower of mountains, usually in spruce-fir or lodgepole pine woods. **C. bulbosa** (L.) Oakes

Corallorhiza (Hall) Chat.   Coralroot

1a. Lip not lobed; woods in mountains. **C. striata** Lindl.
1b. Lip with 2 large or small teeth near base. (2)
2a. Lateral sepals 1-nerved; lip 3–5 mm long; plants usually less than 1.5 dm tall. **C. trifida** Chat.
2b. Lateral sepals 3-nerved; lip 5–9.5 mm long; plants usually more than 1.5 dm tall. **C. maculata** Raf.

Cypripedium L.   Lady's Slipper

1a. Leaves 3 or more, alternate along stem; flowers yellowish; rare, bogs and meadows at middle elevations. Yellow Lady's Slipper. **C. calceolus** L.
1b. Leaves 2, opposite or nearly so; flowers purplish; coniferous forests at middle and higher elevations. Clustered Lady's Slipper. **C. fasciculatum** Kell. ex S. Wats.

Figs. 133-34.   133. **Smilacina stellata**, x .22.   134. **Streptopus amplexifolius**, x .23

Figs. 135-36.   135. **Zygadenus elegans**, x .2.   136. **Calypso bulbosa**, x .28

Subclass MONOCOTYLEDONEAE

Epipactis L. C. Rich. Helleborine

Leafy herbs from creeping rootstocks; flowers racemose, greenish or purplish; sepals and petals distinct and similar; woods and moist areas at middle and lower elevations, widely distributed. **E. gigantea** Dougl. ex Hook.

Goodyera R. Br. Rattlesnake Plantain

Scapose herbs from creeping rootstocks; leaves 2 or more, mostly basal; flowers white tinged with green, in a spicate raceme. **G. oblongifolia** Raf.

Habenaria Willd. Bog Orchid

1a. Leaves 1 or 2 (3), basal or essentially so; stem without bracts or provided with a solitary bract about middle; damp or wet soil in open or dense coniferous forests, bogs, swamps, and along stream banks at high elevations. **H. obtusata** (Banks ex Pursh) Richards.
1b. Leaves several, cauline or subbasal; stem leafy or conspicuously bracteate. (2)
2a. Lip unequally tridentate at apex, with middle tooth minute; moist or wet soil in a variety of habitats at middle and higher elevations. **H. viridis** (L.) R. Br.
2b. Lip entire at apex. (3)
3a. Leaves always clustered at or near base of stem, usually withering before or during anthesis; stem provided with numerous scalelike bracts; lip somewhat truncate at base or angled on each side at base; sepals 1-nerved; dry or moist soils in many habitats at middle and lower elevations. **H. unalascensis** (Spreng.) S. Wats.
3b. Leaves scattered on stem or sometimes clustered at or near base of stem, usually persisting after anthesis; stem provided with a few foliaceous bracts; lip not truncate or angled at base; sepals 3-nerved. (4)
4a. Lip rhombic-lanceolate, prominently and rather abruptly dilated at base; flowers usually white, rarely greenish; moist situations at middle and higher elevations. **H. dilatata** (Pursh) Hook.
4b. Lip linear to broadly lanceolate, not prominently dilated at base; flowers usually greenish, sometimes marked with purple. (5)
5a. Lip lanceolate to elliptic; raceme densely flowered and short, seldom over 10 cm long; moist or wet soil in a variety of habitats, usually at high elevations. **H. hyperborea** (L.) R. Br. ex Ait.
5b. Lip linear to oblong-linear or lanceolate; raceme usually laxly flowered and elongated, frequently over 10 cm long. (6)
6a. Spur slender, usually exceeding lip; leaves variable, scattered along stem or clustered near base of stem, elliptic to linear-lanceolate. **H. sparsiflora** S. Wats.
6b. Spur saccate at base, clavate, shorter than lip; leaves cauline, typically elliptic; not definitely known from Utah. **H. saccata** Greene

Potamogetonaceae

Listera R. Br.   Twayblade
1a. Lip cuneate to obovate, not auriculate, broadest at apex. **L. convallarioides** (Sw.) Nutt.
1b. Lip oblong, auriculate, broadest at base. **L. borealis** Morong.

Spiranthes L. C. Rich.   Ladies' Tresses
Erect caulescent herbs; leaves alternate; flowers spirally arranged around rachis; moist meadows at middle and higher elevations. **S. romanzoffiana** Cham. & Schlecht.

POTAMOGETONACEAE — PONDWEED FAMILY

Perennial herbs of fresh water, rhizomatous; stems jointed, leafy, rooting at nodes; leaves 2-ranked, alternate or opposite, sheathing at base; flowers perfect, in pedunculate, axillary spikes; perianth of 4 valvate segments; stamens 4; carpels 4, 1-celled and 1-ovuled; fruit drupaceous.

Potamogeton L.   Pondweed
1a. Stipules adnate to leaf base, forming a sheath enfolding the stem; leaf appearing as though arising from top of sheath. (2)
1b. Stipules axillary and free from leaf base, either free from one another or united to form a cylinder around the stems, or margins enfolding the stem, not united. (5)
2a. Leaves all linear-filiform, sometimes setaceous, less than 1 mm broad; free part of stipules not united to form a ligule. (3)
2b. Leaves more than 1 mm broad; free part of stipules united to form a ligule. (4)
3a. Leaves with obtuse or rounded apex; fruit lacking a beak, 2–2.6 mm long, style lacking. **P. filiformis** Pers.
3b. Leaves with a long tapering apex; fruit with a beak, 2.6–4.2 mm long; style present. **P. pectinatus** L.
4a. Leaf margins entire; lobes of ligule entire; leaf blades not auricled at junction with sheath; nutlets usually lacking a keel. **P. latifolius** (Robbins) Morong.
4b. Leaf margins minutely serrulate; lobes of ligule often dissected; leaf blades often auricled at junction with sheath; nutlets prominently 3-keeled. **P. robbinsii** Oakes
5a. Leaves all submersed and essentially alike, the petiole short or lacking. (6)
5b. Leaves of 2 kinds, submersed and floating, the floating leaves with broad blades and long petioles. (12)
6a. Leaves lanceolate, oblong or ovate. (7)
6b. Leaves linear-filiform or setaceous. (9)
7a. Leaves tapering at base, sessile or petioled. **P. illinoensis** Morong.
7b. Leaves cordate at base, sessile, or partially clasping stem. (8)

Subclass MONOCOTYLEDONEAE

8a. Stems straight, usually not white; stipules not conspicuous; leaf base completely clasping stem or nearly so. **P. richardsonii** (Bennett) Rydb.

8b. Stems zigzag, usually white; stipules conspicuous; leaf base clasping not more than one-half of stem. **P. praelongus** Wulf. ex Roem.

9a. Leaves broadly linear and grasslike, 2–5 mm broad. **P. zosteriformis** Fern.

9b. Leaves filiform or setaceous, less than 2 mm broad. (10)

10a. Stipules strongly fibrous, becoming whitish and chartaceous; leaves usually revolute; peduncles thickened apically. **P. strictifolius** Ar. Benn.

10b. Stipules not fibrous, scarious or subherbaceous, usually greenish or brownish; peduncles not thickened apically. (11)

11a. Spikes capitate; keel of fruit conspicuously dentate, winged. **P. foliosus** Raf.

11b. Spikes short-cylindric; keel of fruit not dentate. **P. pusillus** L.

12a. Submersed leaves sessile or petioled, lanceolate, oblong, or ovate, at least some over 5 mm wide. (13)

12b. Submersed leaves linear or filiform, usually less than 5 mm wide or, if wider, then very unequal in size and shape. (14)

13a. Floating leaves often absent, when present, delicate and thin, no sharp distinction between blade and petiole. **P. alpinus** Balbis

13b. Floating leaves coriaceous, with sharp distinction between blade and petiole. **P. nodosus** Poir. ex Lam.

14a. Submersed leaves linear, usually bladeless, 0.8–2 mm wide, the blade, when present, linear-lanceolate and long-petiolate; floating leaves broad, many-veined, base of blade subcordate. **P. nutans** L.

14b. Submersed leaves linear to lanceolate or oblanceolate, 1–15 mm wide, sessile. **P. gramineus** L.

RUPPIACEAE — DITCHGRASS FAMILY

Slender, widely branched aquatic herbs; leaves submersed, opposite or alternate, sheathing at base; flowers perfect, small, in terminal spikes; perianth lacking; stamens 2; carpels 4 or more; stigma broad and flat; ovules 1.

Ruppia L. Ditchgrass

A single genus and species treated, with the characteristics of the family. **R. maritima** L.

SPARGANIACEAE — BURREED FAMILY

Herbaceous perennials; leaves alternate, sheathing at base; flowers monoecious, in globose heads, the uppermost heads staminate; perianth

Figs 137-38. 137. **Cypripedium calceolus,** x .22. 138. **Habenaria dilatata,** x .22

Figs. 139-40. 139. **Spiranthes romanzoffiana,** x .24. 140. **Potamogeton nutans,** x .21

Subclass MONOCOTYLEDONEAE

of 3-6 linear-subulate scales; stamens usually 5; ovary 1- or 2-celled; fruit nutlike.

Sparganium L. Burreed

1a. Stigmas mostly 2-lobed; achenes broadly obovoid, obpyramidal; plants usually over 1 m tall. **S. eurycarpum** Engelm. ex Gray
1b. Stigmas simple, rarely 2-lobed; achenes fusiform; plants less than 1 m tall. (2)
2a. Stipe and beak of achene short, less than 1 mm long; fruiting heads 8-12 mm broad; staminate heads usually solitary. **S. minimum** (Hartm.) Fries
2b. Stipe and beak of achene each 2 mm long or more; fruiting heads 15-25 mm broad; staminate heads usually more than 1. (3)
3a. Leaves 1-5 mm broad, rounded on back, usually floating; fruiting heads 10-20 mm broad. **S. angustifolium** Michx.
3b. Leaves 5-12 mm broad, flat or rounded above to keeled below, usually erect and emergent, rarely floating; fruiting heads 20-25 mm broad. **S. emersum** Rehmann

TYPHACEAE — CATTAIL FAMILY

Rhizomatous perennials; leaves distichous, linear, sheathing at base; flowers monoecious, in spikes, the staminate above; pistillate flowers pedicellate, interspersed with bracts and sterile flowers; ovary on a stipe bearing long hairs; staminate flowers soon falling; stamens on branched filaments.

Typha L. Cattail

1a. Leaf sheaths usually closed at throat, the margins free but parallel, auricled above; leaves dark green to glaucous, 5-6 mm wide; possibly not present in Utah, the specimens so assigned likely belong to the next species. **T. angustifolia** L.
1b. Leaf sheaths usually open at throat, the margins free, tapering to blade, rarely auricled; leaves yellow-green or light green, 6-20 mm wide. (2)
2a. Leaves 6-12 mm wide, convex on back; pistillate spike buff colored, 6-10 times as long as thick; stigmas linear. **T. domingensis** Pers.
2b. Leaves 8-20 mm wide, usually flat on back; pistillate spike dark brown, usually about 6 times as long as thick. **T. latifolia** L.

ZANNICHELLIACEAE — HORNED PONDWEED FAMILY

Perennial, aquatic, monoecious or dioecious herbs from a creeping rhizome; stems slender, branched; leaves alternate or opposite or crowded at nodes, sheathing at base, sheaths usually ligulate; flowers

Figs. 141-42. 141. **Ruppia maritima**, x .22. 142. **Sparganium emersum**, x .22

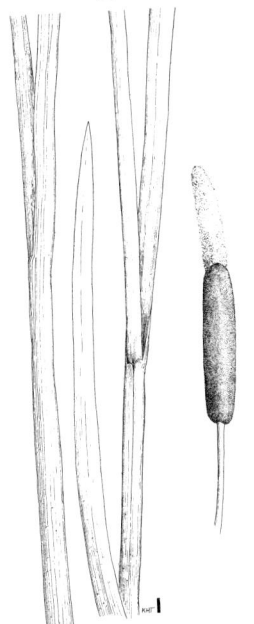

Fig. 143. **Typha latifolia**, x .2

Subclass MONOCOTYLEDONEAE

minute, axillary, solitary or cymose; perianth 3-scaled or lacking; stamens 3, 2, or 1; carpels separate, 1–9.

Zannichellia L.   Horned Pondweed

A single genus and species treated, with the characteristics of the family. **Z. palustris** L.

Illustrated Glossary

ABORTIVE. Not developing; rudimentary; barren.
ACAULESCENT. Apparently without a stem, the leaves basal.
ACCRESCENT. Enlarging after flowering, usually sepals.
ACCESSORY FRUIT. Fruit with large and succulent receptacle, as in the strawberry.
ACEROSE. Needle shaped.
ACHENE. Small, dry, 1-loculed, 1-seeded, indehiscent fruit, the seed attached at a single point.
ACTINOMORPHIC. Regular; radially symmetric.
ACULEATE. Armed with prickles, as the stem of a rose.
ACUMEN. Apex.
ACUMINATE. Gradually tapering to a point, the sides somewhat concave.
ACUTE. Tapering to the apex with the sides straight.
ADNATE. Union of 2 unlike parts, as with stamens attached to petals.
ADVENTITIOUS. Organ developing in an unusual position.
AGGREGATE FRUIT. Fruit having all the carpels of the flower attached to a common receptacle, the carpels separate, as in the raspberry.
ALLUVIAL. Pertaining to or composed of alluvium as sand, gravel, or similar detrital material deposited by running water.
ALTERNATE. Borne between, not in front of, as with stamens and petals; or borne singly, one per node, as with leaves.
AMENT. Catkin; a spike of unisexual flowers, as in the poplar.
AMPLIATE. Enlarged.
ANDROGYNOUS. Staminate and pistillate flowers in the same inflorescence, the staminate above, as in **Carex**.
ANNUAL. Plants growing from seed and producing flowers and seeds and dying the same year.
ANNULAR. In the form of a ring.
ANNULUS. In ferns, the organ which partially invests the sporangium and at maturity bursts it.
ANTERIOR. The side away from the axis.
ANTHER. The pollen-bearing part of the stamen.
ANTHERIFEROUS. Anther-bearing.
ANTHESIS. Time period when the flower is open.
APETALOUS. Without petals.
APICULATE. Ending in an abrupt slender tip.
APPRESSED. Lying close and flat against some other part.
APPROXIMATE. Close together but not united.
AQUATIC. Living in water.
ARACHNOID. Beset with cobwebby or entangled hairs.
ARCUATE. Arching or curved like a bow.
AREOLE. A small space on or beneath the surface—often used in leaves

Illustrated Glossary

for the area between small veins; in cacti, the structure which bears flowers, or spines, or glochids, or all three; also spelled areola.
ARGILLACEOUS. Clayey; growing in clay; clay colored.
ARIL. An appendage growing at or above the hilum of a seed.
ARISTATE. With an awn or stiff bristle.
ARTICULATING. With a joint or node separating at maturity.
ASCENDING. Rising obliquely upward, often curving.
ATTENUATE. Gradually narrowing to a tip or base.
AURICLE. An ear-shaped appendage.
AWL-SHAPED. Narrowly triangular; short, sharp pointed, like an awl.

ATTENUATE

AWN. A bristlelike appendage, as on the tips of glumes and lemmas of many grasses.
AXIL. Upper angle formed by a leaf or branch with the stem.
AXILE. Belonging to, or situated in, the axis, as with a placenta situated in the axis of a pistil.

BANNER. Upper petal of a papilionaceous flower, as in the sweet pea.
BARBED. Bearing sharp, rigid, reflexed points, like the barb of a fishhook.
BARBELLATE. With short, usually stiff hairs.
BARK. Outer layer of a woody stem, usually including all tissues external to the vascular cambium.
BASAL. Related to, or situated at, the base.
BASIFIXED. Attached by the base, as an anther attached by its base to the filament.
BEARDED. Bearing long hairs.
BERRY. A pulpy indehiscent fruit, usually with several to many seeds, as the tomato.
BI-. Prefix signifying two, twice, or doubly.
BIDENTATE. Having 2 teeth.
BIENNIAL. Plants living 2 years and generally flowering only during the second.
BIFID. 2-cleft.
BIFURCATE. Twice forked.
BILABIATE. With 2 lips.

BILABIATE

BILOCULAR. With 2 locules, as an anther or an ovary.
BIPINNATE. Doubly or 2-pinnate.
BIPINNATIFID. Twice pinnately cleft.
BLADDERY. Thin and inflated.
BLADE. The expanded part of a leaf or petal.
BRACT. A reduced leaf subtending a flower, usually associated with inflorescences.
BRACTEATE. Provided with bracts, as at the base of a pedicel in an inflorescence.
BRACTEOLATE. Provided with bracteoles, as at the base of a flower near the apex of a pedicel.

Illustrated Glossary

BRACTEOLE. A small bract, especially on a floral axis; also called bractlet.
BRISTLY. Bearing stiff hairs.
BUD. A rudimentary stem bearing immature leaves or flowers or both.
BULB. An underground leaf bud with thickened scales or coats, as the onion.
BULBLET. A small bulb, usually axillary, as in members of the Liliaceae.
BUR. A fruit, fruit part, or associated structure with spines or appendages.
CADUCOUS. Falling off very early or prematurely.
CAESPITOSE. Growing in tufts.
CALCAREOUS. Growing on limestone or in soil impregnated with lime.
CALLUS. The thickened extension at the base of the lemma in some grasses.
CALYX. Outer circle of flowering parts, usually green; collective term for sepals.
CALYX LOBE. In a gamosepalous calyx, the free parts.
CAMPANULATE. Bell shaped.
CANESCENT. Covered with grayish white or hoary fine hairs.
CAPILLARY. Hairlike; exceedingly slender.
CAPITATE. Head shaped, or grouped into a ball or head.
CAPSULE. A dry fruit of more than 1 carpel which opens to release the seeds.
CARPOPHORE. That part of the receptacle which is prolonged between the carpels.
CARPEL. A simple pistil, or 1 of the modified leaves forming a compound pistil.
CARUNCLE. A wart or protuberance near the hilum of a seed.
CARUNCULAR. Of or pertaining to a caruncle.
CARYOPSIS. The grain or fruit of grasses.
CATKIN. A scaly, deciduous spike of unisexual flowers; ament.
CAUDATE. Bearing a tail or a slender taillike appendage.
CAUDEX. The woody base of an otherwise herbaceous perennial.
CAULESCENT. With an obvious leafy stem.
CAULINE. Belonging to the stem.
CELL. A cavity of an anther containing the pollen, or of an ovary containing ovules.
CERACEOUS. Waxy in appearance or color.
CERNUOUS. Nodding, pendulous, applied to such flowers as **Erythronium**.
CHAFF. Thin dry scales.
CHARTACEOUS. With the texture of writing paper.
CILIATE. Fringed with hairs on the margin.
CINEREOUS. Ash colored; light gray.
CIRCINATE. Coiled from the top downward with the apex as a center, as in the developing leaves of ferns.

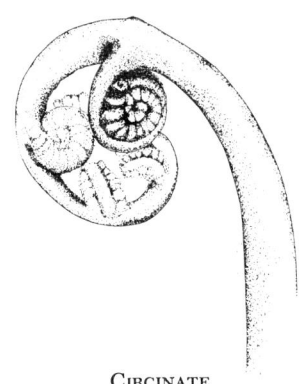

CIRCINATE

Illustrated Glossary

CIRCUMSCISSILE. Dehiscing by a transverse line around the fruit or anther, the top falling as a lid.
CLASPING. Leaf partly or wholly surrounding the stem; amplexicaul.
CLAVATE. Club shaped, gradually thickened toward the apex.
CLAW. Narrowed base of the petals of some flowers.
CLEFT. Split nearly to the middle.
CLEISTOGAMOUS. Small flowers self-fertilizing without opening, as in the violet.
COALESCENT. Said of organs of one kind that have grown together.
COHERENT. United with other organs of the same kind.
COLLAR. The area on the outside of a grass leaf at the junction of the blade and the sheath.

CIRCUMSCISSILE

COLUMN. Body formed by the union of stamens and pistils, as in orchids, mallows, and milkweeds.
COMA. A tuft of hairs, as in a seed.
COMMISSURE. The face by which two carpels cohere, as in Umbelliferae.
COMOSE. Having a tuft of hairs.
COMPLETE. Having all the parts belonging to it, as a flower with sepals, petals, stamens, and pistils.
COMPLICATE. Folded together.
COMPOUND. Having 2 or more similar parts in one organ.
COMPRESSED. Flattened.
CONCAVE. Hollowed out.
CONDUPLICATE. Folded together lengthwise, as the leaves of many grasses.

COMA

CONFLUENT. Blending of one part into another.
CONGESTED. Crowded together.
CONIC. Cone shaped.
CONNATE. The union of like structures.
CONNECTIVE. Portion of the filament connecting the 2 cells of an anther.
CONNIVENT. Converging but not united.
CONTORTED. Twisted or bent.
CONTRACTED. Narrowed in a particular place.
CONVEX. Rounded on the surface.
CONVOLUTE. Rolled up longitudinally.
CORDATE. Heart shaped.
CORIACEOUS. Leathery.
CORM. A short, bulblike, underground stem, with only papery scale leaves.
COROLLA. Inner whorl of floral parts; collective name for petals.
CORONA. A crown; the whorl of structures between petals and stamens, cuplike in **Narcissus,** or consisting of separate parts in **Asclepias.**
CORYMB. A flat-topped or convex racemose flower cluster, the lower or outer pedicels longer, their flowers opening first.
CORYMBOSE. Arranged in corymbs.

Illustrated Glossary

COTYLEDON. The primary leaf or leaves of the embryo.
CRENATE. Having margins with rounded teeth.
CRENULATE. Crenate, but the teeth themselves small.
CREOSOTE. A colorless, oily, aromatic liquid, as in reference to **Larrea tridentata,** an odoriferous shrub.
CRISPED. Curled; wavy.
CRISTATE. Crested or with a terminal tuft.
CROWN. The persistent base of a herbaceous perennial; the top of a tree; a corona.
CRUCIFORM. Cross shaped.
CULM. The type of hollow or pithy slender stem found in grasses and sedges.
CUNEATE. Wedge shaped; triangular, with the narrow part at point of attachment.
CUPULATE. Furnished with or subtended by a small cup.
CUSPIDATE. Tipped with a cusp or a sharp, short, rigid point.
CYATHIUM. The inflorescence of **Euphorbia,** consisting of a 3-loculed ovary and a cuplike involucre bearing male flowers with solitary stamens.
CYME. A flat-topped or convex paniculate flower cluster, with central flowers opening first.
CYMULE. A small cyme.

DECIDUOUS. Falling off; not evergreen.
DECLINED. Curved downward.
DECOMPOUND. More than once divided or compounded.
DECUMBENT. Resting on ground, but with tip of stem ascending.
DECURRENT. Extending down the stem below the insertion, as with leaves or stipules.
DECUSSATE. Opposite pairs, usually applied to leaves, alternating at right angles with those above or below.
DEFLEXED. Turned abruptly downward.
DEHISCENT. Opening spontaneously when ripe to discharge the contents, as an anther or fruit.
DELTOID. Equilaterally triangular; shaped like the Greek letter delta.
DENTATE. Having the margin cut with sharp teeth which are not directed forward.
DENTICULATE. Minutely toothed.
DEPRESSED. Low and flattened from above.
DIADELPHOUS. Stamens united by their filaments into 2 sets.
DICHOTOMOUS. Repeatedly forked in pairs.
DIDYMOUS. Twin; found in pairs.
DIDYNAMOUS. Having 4 stamens disposed in pairs, 2 long and 2 short, as in many Labiatae.
DIGITATE. Fingered; shaped as an open hand; palmate.
DILATED. Flattened and broadened, as an expanded filament.
DIMORPHIC. Having 2 forms.

Illustrated Glossary

DIOECIOUS. Having staminate and pistillate flowers on different plants.
DISCOID. Disklike.
DISK. A fleshy development of the receptacle about the base of an ovary; in Compositae, the tubular flowers (disk florets) of the head as distinct from the ray; also spelled disc.
DISSECTED. Deeply divided into numerous fine segments.
DISTAL. Opposite the point of attachment; apical; away from the axis.
DISTICHOUS. In 2 vertical rows or ranks.
DISTINCT. Separate; not united with parts in the same whorl.
DIVARICATE. Widely divergent.
DIVERGENT. Extending away from each other by degrees.
DIVIDED. Separated to the base.
DORSAL. Pertaining to the back; the surface turned away from the axis.
DORSIFIXED. Attached to the back.
DRUPE. Fleshy fruit in which the inner layer of the ovary wall becomes hard, as in the peach.
EBRACTEATE. Without bracts.
ELLIPSOID. An elliptic solid.
ELLIPTIC. In the form of a flattened circle, more than twice as long as broad, widest in the center and the 2 ends equal.

ELLIPTIC

EMARGINATE. With a small notch at the apex.
EMERSED. Raised above the water.
ENATION. An outgrowth on the surface of an organ.
ENDOCARP. The inner layer of the pericarp.
ENFOLD. To fold inward or toward one another.
ENSIFORM. Sword shaped, as the leaves of an iris.
ENTIRE. Undivided; the margin continuous, not incised or toothed.
EPIGYNOUS. Attached by the base to the top of the ovary, or apparently so.
EQUITANT. Overlapping in 2 ranks.
ERECT. Upright in relation to the ground, or sometimes perpendicular to the surface of attachment.

EMARGINATE

EROSE. Irregularly toothed as if gnawed.
EVANESCENT. Soon disappearing; lasting only a short time.
EXFOLIATE. To come off in scales or flakes, as the bark of some trees.
EXOCARP. The outer layer of the pericarp.
EXPLANATE. Spread out flat.
EXSERTED. Protruding, as stamens projecting beyond the corolla; not included.
FALCATE. Sickle shaped.
FARINACEOUS. Having a mealy texture or surface.
FASCICLE. A close cluster or bundle of flowers, leaves, stems, or roots.

FASCICULATE. Connected or drawn into a bundle.
FASTIGIATE. Clustered, parallel, erect branches.
FERTILE. Said of pollen-bearing stamens and seed-bearing fruits.
FILAMENT. A thread, especially the stalk of an anther.
FILIFORM. Threadlike.
FIMBRIATE. Fringed.
FISTULOSE. Hollow, often rather enlarged.
FLESHY. Thick and juicy; succulent.
FLEXUOSE. Zigzag.
FLOCCOSE. Bearing tufts of soft woolly hair.
FLORET. Lemma and palea with the small included flower of a grass; also the small flower of the Compositae.
FOLIACEOUS. Leaflike.
FOLIOLATE. Having leaflets.
FOLIOSE. Closely clothed with leaves; leaflike.
FOLLICLE. A dry, dehiscent fruit, consisting of a single carpel which opens only on the ventral suture, as in the milkweed.
FORNIX. A little scale.
FOVEATE. Pitted.

FOLLICLE

FREE. Not joined to other organs.
FROND. Leaf of a fern.
FRUIT. The ripened pistil with all of its accessory parts.
FRUTESCENT. Shrubby or bushy in the sense of being woody.
FUNNELFORM. Gradually widening upwards, like a funnel.
FUSIFORM. Spindle shaped, thickest near the middle and tapering toward each end.

GALEA. The upper lip in certain 2-lipped corollas.
GALEATE. Hollow and vaulted, as in many lipped corollas.
GAMOPETALOUS. Corolla with petals united; also sympetalous.
GAMOSEPALOUS. Calyx with sepals united.
GEMINATE. Arranged in pairs.
GENICULATE. Abruptly bent.
GIBBOUS. Swollen on one side; ventricose.
GLABRESCENT. Becoming glabrous.
GLABROUS. Without hairs.
GLAND. A depression, protuberance, or appendage which secretes a usually sticky fluid.
GLAUCESCENT. Slightly glaucous; becoming glaucous.
GLAUCOUS. Covered or whitened with a bloom, as a cabbage leaf.
GLOBOSE. Spherical or nearly so.
GLOCHID. A barbed hair or bristle.
GLOCHIDIATE. Pubescent with barbed prickles or heads.
GLOMERATE. Densely compacted in clusters.
GLOMERULATE. Arranged in small, compact clusters.
GLOMERULE. A compact capitate cyme.

Illustrated Glossary

GLUMES. The pair of bracts at the base of a grass spikelet.
GLUTINOUS. With a gluey exudation.
GRADUATE. Marked with small regular distances.
GRANULAR. Covered with very small grains or granules; minutely mealy.
GYNAECANDROUS. Having staminate and pistillate flowers in the same spikelet, the latter above the former.
GYNANDRIUM. A column bearing stamens and pistils.
GYNANDROUS. Stamens adnate to the pistil.
GYNOECIUM. Collectively, the pistils of a flower.
HASTATE. Arrowhead shaped but with the basal lobes turned outward.
HEAD. A dense globular cluster of sessile or subsessile flowers arising essentially from the same point on the peduncle.
HERB. A plant without a persistent woody stem, at least above ground.
HERBACEOUS. Pertaining to an herb; opposed to woody; having the texture or odor of a foliage leaf; dying to the ground each year.
HERBAGE. Collectively, the green parts of a plant.
HETEROMORPHIC. Of more than 1 kind or form.
HILUM. A scar on a seed, as on a bean, marking the point of attachment of the ovule.
HIRSUTE. Rough with coarse, stiff hairs.
HIRSUTULOUS. Minutely hirsute; hirtellous.
HIRTELLOUS. Minutely hirsute.
HISPID. Rough, with stiff or bristly hairs.
HISPIDULOUS. Minutely hispid.
HOARY. Covered with white hairs.
HOLOSERICEOUS. Covered with fine and silky hairs.
HOMOMORPHOUS. All alike, as the nutlets in some Boraginaceae; having perfect flowers of only 1 type.
HOMOSPOROUS. With spores all alike in size and shape.
HOMOSTYLOUS. With anthers and styles the same length.
HOST. A plant which nourishes a parasite.
HYALINE. Colorless, translucent, or transparent.
HYPANTHIUM. A cup-shaped enlargement of the receptacle on which the calyx, corolla, and often the stamens are inserted; in perigyny, the "calyx tube."
HYPOCOTYL. The axis of a plant embryo or seedling below the cotyledon.
HYPOGEOUS. Growing or living below the surface of the ground.
HYPOGYNOUS. Borne on the receptacle below or free of the pistil, as with sepals, petals, and stamens.

IMBRICATE. Overlapping as shingles on a roof.
IMMERSED. Growing under water.
INCISED. Sharply lobed.
INCLUDED. Not protruding beyond the surrounding organ or envelope.

Illustrated Glossary

INDEHISCENT. Not splitting open a maturity.
INDIGENOUS. Native to the area.
INDURATE. Hardened.
INDUSIUM. In ferns, the epidermal outgrowth that covers or invests the sorus.
INFERIOR. Lower or beneath, as with the ovary when it is below the other parts of a flower.
INFLATED. Blown up; bladdery.
INFLORESCENCE. The flower cluster of a plant.
INNATE. Borne on the apex of the support; in an anther the antithesis of adnate.

INFERIOR

INROLLED. Involute; curled or curved inward.
INSERTED. Attached to or growing upon.
INTERNODE. The portion of stem between 2 consecutive nodes.
INTERRUPTED. Not continuous.
INVOLUCEL. A secondary involucre, as the bracts subtending the secondary umbels in the Umbelliferae.
INVOLUCRE. A whorl of bracts subtending a flower cluster, as in the heads of Compositae.
INVOLUTE. With the edges rolled inward toward the upper side; not revolute.
IRREGULAR. Showing a lack of uniformity; bilaterally symmetric, as a zygomorphic flower.

KEEL. A prominent dorsal ridge, analogous to the keel of a boat; the two lower united petals of a papilionaceous corolla.

LABIATE. Lipped; a member of the Labiatae.
LACERATE. Appearing irregularly torn or cleft.
LACINIATE. Cut into narrow lobes or segments.
LAMINA. The blade or expanded part of a leaf, petal, etc.
LANATE. Woolly; densely clothed with long entangled hairs.
LANCEOLATE. Lance shaped; much longer than broad, tapering from below the middle to the apex and to the base.
LATERAL. At or on the side.
LAX. Loose, with component parts separated.

LANCEOLATE

LEAFLET. A segment of a compound leaf.
LEGUME. A superior 1-loculed fruit of a simple pistil usually dehiscent into 2 valves, having the seeds attached along the ventral suture; a leguminous plant.
LEMMA. In grasses, the lower of the 2 bracts immediately enclosing the floret.
LENTICELS. Corky spots on young bark, arising in relation to epidermal stomates.
LENTICULAR. Lens shaped.

Illustrated Glossary

LIGNEOUS. Woody; of or resembling wood.
LIGULATE. Strap shaped; furnished with a ligule.
LIGULE. The strap-shaped part of a ray corolla in Compositae; the thin, collarlike appendage on the inside of the blade at the junction with the sheath in grasses.
> LIMB. The expanded flat part of an organ, especially the expanded part of a gamopetalous corolla.
> LINEAR. Resembling a line; long and narrow, of uniform width, as the leaf blades of grasses.
> LOBE. A division or segment of an organ, usually rounded or obtuse.
> LOBULATE. Made up of lobules.
> LOBULE. A small lobe.
> LOCALLY ABUNDANT. Abundant where found, but not occurring uniformly.
> LOCULE. The "cell" or cavity of an organ, applied to pistils and stamens.
> LOCULICIDAL. Dehiscent longitudinally through the middle of the back of a pericarp, between the partitions into the cavity.

LINEAR LODICULES. The 2 or 3 minute hyaline scales at the base of the stamens in grasses, representing the perianth.
LOMENT. A legume which is constricted between the seeds.
LYRATE. Lyre shaped; pinnatifid, with the terminal lobe large and rounded, the lower lobes small.
MACROPHYLL. A large leaf, as the complex, usually few- to many-veined leaves of a Pteropsida.
MACROSPORANGIUM. The organ in which macrospores are produced.
MACROSPORE. The larger of the 2 kinds of spores in Selaginellaceae.
MACROSPOROPHYLL. The modified leaf that bears the macrosporangium.
MACULATE. Blotched or spotted.
MALPIGHIAN HAIRS. Straight appressed hairs attached by the middle and tapering to the free tips.
MARCESCENT. Withering without falling off, as the persistent leaves at the bases of some plants.
MEGASPORE. A synonym of macrospore.
MEMBRANOUS. Of the nature of a membrane; thin, soft, and pliable.
-MEROUS. A suffix denoting parts or numbers, as 3-merous.
MESOCARP. The middle layer or coat of a fruit.
MICROPHYLL. A small leaf, as the simple, usually 1-veined leaves of Lycopsida and Sphenopsida.
MICROSPORE. The smaller of the 2 kinds of spores, as in Selaginellaceae.
MIDRIB. The central rib of a leaf or other organ.
MONADELPHOUS. Stamens united by their filaments into a tube surrounding the gynoecium, as in malvaceous flowers.
MONOCARPIC. Composed of only 1 carpel.

Illustrated Glossary

MONOECIOUS. Having staminate and pistillate flowers on the same plant, but not perfect ones.
MUCILAGINOUS. Moist and viscid; slimy; composed of mucilage.
MUCRO. A small and short abrupt tip of an organ.
MUCRONATE. Possessing a short and straight point (mucro), as some leaves.
MULTIFID. Cleft into many narrow lobes or segments.
MURICATE. Rough with short and firm sharp excrescences.
NECTARY. An organ which secretes nectar.
NERVE. A simple vein or slender rib of a leaf or bract.

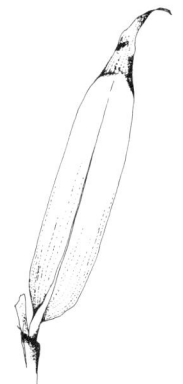

NEUTER. Lacking functional stamens or gynoecium.
NODE. The joint of a stem; the point of insertion of a leaf or leaves.
NODULOSE. Having minute nodules; finely knobby.
NUT. A hard, indehiscent, usually 1-seeded fruit, produced from a compound ovary.
NUTLET. A small nut.

OB-. A Latin prefix usually signifying inversion.
OBCONIC. Conical, but attached at the narrower end.
OBLANCEOLATE. Inversely lanceolate.
OBLONG. Much longer than broad, with nearly parallel sides.

OBLONG

OBOVATE. Shaped like the longitudinal section of an egg, but with the broadest part toward the tip.
OBSOLETE. Rudimentary or not evident, as an organ that is almost entirely suppressed; vestigial.
OBTUSE. Blunt or rounded at the end.
OCHROLECOUS. Yellowish white.
OCREA. A sheath around the stem derived from the stipules; chiefly in the Polygonaceae.
ODD-PINNATE. Having a terminal leaflet instead of a tendril or pair of leaflets.
ODORIFEROUS. Having an odor.
OPPOSITE. Set against, as leaves when 2 at a node; 1 part in front of another, as a stamen in front of a petal.
ORBICULAR. Approximately circular in outline.
OVAL. Broadly elliptic.
OVARY. The part of the pistil that contains the ovules.
OVATE. With the outline of an egg in longitudinal section, the broader end downward.

OVAL

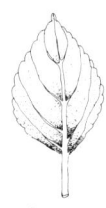
OVATE

427

Illustrated Glossary

OVULE. The megasporangium and associated integuments of a seed plant; the body in an ovary which becomes a seed.
OVULIFEROUS. Bearing ovules.
PALEA. One of the chafflike scales on the receptacle of many Compositae; the inner bract of a grass floret, often partly surrounded by the lemma.
PALEACEOUS. Covered with or resembling chaffy scales.
PALMATE. Hand shaped with the fingers spread; in a leaf, having the lobes or divisions radiating from a common point.
PANDURATE. Fiddle shaped.
PANICLE. A compound racemose inflorescence.
PAPILIONACEOUS. Applied to the butterflylike corolla of the pea, with banner, wings, and keel.
PAPILLAE. Soft superficial glands or protuberances.
PAPILLATE. Having papillae.
PAPPUS. The modified calyx limb on Compositae, consisting of a crown of bristles or scales at the summit of the achene.
PARIETAL. Attached to the wall of the ovary, instead of to the axis.
PARTED. Deeply cleft nearly to the base.
PECTINATE. With narrow closely set divisions like the teeth of a comb.
PEDATE. Palmate, with the lateral lobes 2-cleft.
PEDICEL. The stalk of a single flower in a flower cluster, or of a spikelet in grasses.
PEDICELLATE. Having a pedicel, as a flower.
PEDUNCLE. The stalk of a flower or of a cluster of flowers.
PEDUNCULATE. Having a peduncle, as a flower cluster.
PELTATE. Shield shaped; a flat body having a stalk attached to the lower surface instead of at the base or margin.
PENDULOUS. Hanging downward; pendent.
PEPO. An indehiscent, fleshy, 1-celled or falsely 3-celled, many-seeded berry, usually with a hard rind, as in the cucumber.
PERENNIAL. Lasting 3 or more years.
PERFECT. A flower having both stamens and pistils.
PERFOLIATE. With the leaf entirely surrounding the stem.
PERIANTH. The floral envelopes; collectively, the calyx and corolla, especially when they are alike.
PERICARP. The ripened walls of the ovary, referring to a fruit.
PERIGYNIUM. The scalelike organ surrounding the pistil in **Carex.**
PERIGYNOUS. Borne around the ovary in contrast to beneath it, as the stamens and corolla are inserted on the floral tube.
PERNICIOUS. Highly injurious or destructive in character; deadly.
PERSISTENT. Remaining attached, as a calyx on the fruit.
PETAL. One of the leaves of a corolla, usually colored.
PETALOID. Resembling a petal.
PETIOLE. A leaf stalk.
PETIOLULE. The petiole of a leaflet of a compound leaf.

Illustrated Glossary

PHYLLODE. A leaflike petiole with no blade, as in some **Astragalus** species.
PILOSE. Bearing soft and straight spreading hairs.
PILOSULOSE. Bearing very small, soft, straight, spreading hairs.
PINNA. A leaflet or primary division of a pinnate leaf.
PINNATE. A compound leaf, having leaflets arranged on each side of a common petiole; featherlike.
PINNATIFID. Pinnately cleft into narrow lobes not reaching to the midrib.
PINNATISECT. Pinnately divided to the midrib.
PINNULE. The secondary pinna, as in 3-compound leaves of ferns.
PISTIL. The ovule-bearing organ of a flower, consisting of stigma and ovary, usually with a style between; gynoecium.
PISTILLATE. Provided with pistils and without stamens; female.
PLACENTA. The ovule-bearing surface in the ovary.
PLACENTATION. The arrangement or orientation of the placenta.
PLICATE. Plaited; folded as a fan.
PLUMOSE. Feathery; having fine hairs on each side as a plume.
POD. Any dry dehiscent fruit, especially a legume.
POLLINIA. The pollen masses of the orchids and milkweed.
POLYCHROME. Many colored.
POLYGAMOUS. Bearing unisexual and bisexual flowers on the same plant.
POME. A fleshy fruit made up mostly of a modified floral tube.
PORICIDAL. Opening by means of pores, as in the capsules of poppies.
PORRECT. Like a parrot beak.
POSTERIOR. On the side toward the axis; the upper side of the flower.
PRICKLE. Sharp outgrowth of the bark or epidermis.
PROCUMBENT. Trailing on the ground, but not rooting.
PROSTRATE. Lying flat upon the ground.
PROXIMAL. Nearest the axis or base, as contrasted with distal.
PRUINOSE. Having a waxy powdery surface; covered with whitish dust or bloom.
PSEUDOSCAPE. A false scape, as in a tulip where not all leaves are basal.
PUBERULENT. Minutely pubescent.
PUBESCENT. Covered with short soft hairs; downy.
PULVINATE. Cushion shaped.
PUNCTATE. Dotted with punctures or with translucent pitted glands or with colored dots.
PUNGENT. Ending in a rigid, sharp point or prickle; acrid.
PUSTULATE. Bearing irregular blisterlike swellings or pustules, mostly at the bases of hairs.

RACEME. A simple, elongated, indeterminate inflorescence with each flower subequally pedicelled.
RACEMOSE. Having racemes; racemelike.
RACHILLA. A small rachis, specifically the axis of a grass spikelet.
RACHIS. The axis of a spike or raceme, or of a compound leaf.
RADIATE. Spreading from a common center; bearing rays.

Illustrated Glossary

RADICLE. That portion of the embryo below the cotyledons.
RAPHE. A ridge along the side of a seed adjacent to the hilum.
RAY. A primary branch of an umbel; in Compositae, the ligule of a ray flower bears the flowers in the head.
RECEPTACLE. That portion of the floral axis upon which the flower parts are borne; in Compositae, that which bears the flowers in the head.
RECURVED. Curved backward or downward.
REFLEXED. Bent downward.
REGULAR. Said of a flower having radial symmetry, with the parts in each series alike.
REMOTE. Distantly spaced.
RENIFORM. Kidney shaped.
REPAND. With an undulating margin, less strongly wavy than sinuate.
REPENT. Prostrate and rooting.
REPLICATE. Folded backward.
RETICULATE. With a network; net veined.
RETRORSE. Bent backward or downward.
RETUSE. Notched shallowly at a rounded apex.
REVOLUTE. Rolled backward from both margins toward the underside.
RHIZOMATOUS. Having rhizomes.
RHIZOME. An underground stem or rootstock, with scales at the nodes and producing leafy shoots on the upper side and roots on the lower side.
RHOMBIC. Somewhat diamond shaped.
RHOMBOIDAL. Approaching a rhombic outline; quadrangular, with the lateral angles obtuse; diamond shaped.
ROSETTE. A crowded cluster of radiating leaves appearing to rise from the ground.
ROSULATE. With a collection of clustered leaves; a rosette.
ROTATE. Wheel shaped; said of a sympetalous corolla with obsolete tube and with a flat and circular limb.
ROTUND. Rounded in outline.
RUDIMENTARY. Imperfectly developed; vestigial.
RUGA. A wrinkle or fold.
RUGOSE. Wrinkled.
RUGULOSE. Somewhat wrinkled.
RUNCINATE. Sharply pinnatifid or incised, the lobes pointing downward.

SAC. The cavity of an anther.
SACCATE. Bag shaped.
SAGITTATE. Arrowhead shaped, with the basal lobes turned downward.
SALVERFORM. A corolla with slender tube abruptly expanding into a flat limb.

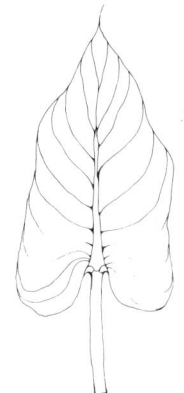

SAGITTATE

Illustrated Glossary

SAMARA. An indehiscent, winged fruit.
SCABROUS. Rough to the touch, owing to the structure of the epidermis or to the presence of short stiff hairs.
SCALE. Any thin, scarious bract; usually a vestigial leaf.
SCAPE. A leafless peduncle rising from the ground in acaulescent plants.
SCARIOUS. Thin, dry, and membraneous, not green.
SCURFY. Clothed with small branlike scales.
SECUND. Arranged on one side only; unilateral.
SEED. The ripened ovule.
SEGMENT. A division or part of a leaf or other organ.
SELENIFEROUS. Bearing selenium.
SELENOPHILE. A plant which takes up selenium from the soil.
SEPAL. A leaf or segment of the calyx.
SEPALOID. Sepallike.
SEPTICIDAL. Dehiscence of a capsule through the septa and between the locules.
SEPTUM. A partition between cavities.
SERICEOUS. Silky; clothed with appressed fine and straight hairs.
SERRATE. Saw-toothed, the sharp teeth pointed forward.
SESSILE. Attached directly by the base, not stalked, as a leaf without a petiole.
SHEATH. The tubular basal part of the leaf that encloses the stem, as in the grasses and sedges.
SILICLE. A short silique, not much longer than wide.
SILIQUE. A many-seeded capsule of the Cruciferae, with 2 valves splitting from the bottom and leaving the placentae with the false partition (replum) between them.
SIMPLE. Unbranched, as a stem or hair; not compound, as a leaf; single, as a pistil of one carpel.
SINUATE. With a strongly wavy margin.
SINUOUS. Sinuate, with a deeply wavy margin.
SINUS. The cleft or recess between 2 lobes of an expanded organ such as a leaf.
SONORAN HABITAT. Warm to hot desert habitat, as at St. George, Washington Co.
SORUS. The fertile portion of a fern frond; the place where sporangia are borne.
SPADIX. A floral spike with a fleshy or succulent axis usually enclosed in a spathe.
SPATHACEOUS. Spathe bearing; like a spathe.
SPATHE AND SPADIX SPATHE. A bract or pair of bracts enclosing a flower cluster.
SPATULATE. Like a spatula, which is a knife rounded above and gradually tapering to the base.
SPICATE. Having the form of or arranged in a spike.
SPIKE. An elongated rachis of sessile flowers or spikelets.

Illustrated Glossary

SPIKELET. A secondary spike; the ultimate flower cluster in grasses, consisting of 2 glumes and 1 or more florets, also in sedges.
SPINE. A sharp-pointed, stiff, woody body, arising from below the epidermis; commonly the counterpart of a stipule.
SPORANGIUM. A spore case or sac.
SPORE. The reproductive body of lower plants.
SPOROCARP. The receptacle containing sporangia or spores.
SPOROPHYLL. A spore-bearing leaf.
SPREADING. Diverging almost to the horizontal; nearly prostrate.
SPUR. A slender, saclike, nectariferous process from petal or sepal.
SQUARROSE. Rough or scurfy with spreading and outstanding processes, at the tips of bracts.
STAMEN. The male organ of the flower which bears pollen.
STAMINATE. Having stamens but not pistils; said of a flower or plant that is male, hence not seed-bearing.
STAMINODE. A sterile stamen, or what corresponds to a stamen.
STELLATE. Star shaped.
STERILE. Infertile or barren, as a stamen lacking an anther.
STIGMA. The receptive part of the pistil on which the pollen germinates.
STIPE. The stalk beneath an ovary which is inserted upon the receptacle.
STIPEL. The stipule of a leaflet.
STIPULATE. Bearing stipules.
STIPULE. One of the pair of usually foliaceous appendages found at the base of the petiole in many plants.
STOLON. A modified stem bending over and rooting at the tip, or creeping and rooting at the nodes; a horizontal stem that gives rise to a new plant at its tip.
STOLONIFEROUS. Having stolons.
STOMATE. A breathing pore or aperture in the epidermis.
STRIATE. Marked with fine longitudinal lines or furrows.
STRICT. Very straight and upright, not at all lax or spreading.
STRIGOSE. Clothed with sharp and stiff appressed straight hairs.
STROBILUS. Conelike aggregation of sporophylls.
STYLE. The contracted portion of the pistil between the ovary and the stigma.
STYLOPODIUM. An enlargement or disklike expansion at base of the style as in Umbelliferae.
SUB-. A prefix usually signifying somewhat, slightly, rather, or almost.
SUBTEND. To be below and close to, as the leaf subtends the shoot borne in its axis.
SUBULATE. Awl shaped.
SUCCULENT. Juicy; fleshy and soft.
SUFFRUTESCENT. Obscurely shrubby; very slightly woody, but not necessarily low.
SUFFRUTICOSE. Woody but very low; diminutively shrubby.
SULCATE. Longitudinally grooved, furrowed, or channeled.

Illustrated Glossary

SUPERIOR. Growing above, as an ovary that is free from the other floral organs.
SUTURE. The line of dehiscence of fruits or anthers; the line of a natural union or division between coherent parts.
SYMMETRIC. Said of a flower having the same number of parts in each circle.
SYMPETALOUS. With petals united in a 1-piece corolla; gamopetalous.
TALUS. The accumulation of detritus, usually boulders and smaller particles, at the foot of a cliff or along a steep slope.
TAXON. Any taxonomic unit, as a genus, species, subspecies, variety, etc.
TENDRIL. A slender, coiling or twining organ by which a climbing plant grasps its support.
TEPALS. A division of the perianth, sepal, or petal, as in Polygonaceae.
TERETE. Cylindric; round in cross section.
TERNATE. In 3s, as a leaf consisting of 3 leaflets.
TESSELATE. Checkered.
THORN. A sharp-pointed stiff woody body derived from a modified branch.
THROAT. The orifice of a gamopetalous corolla; the expanded portion between limb and tube proper.
THYRSE. A compact, ovate panicle; strictly, a panicle with main axis indeterminate, but other axes cymose.
TOMENTOSE. With tomentum; covered with a rather short, densely matted, soft white wool.
TOMENTULOSE. Slightly tomentose.
TOMENTUM. A cobwebby covering of hairs.
TOOTH. Any small marginal lobe.
TOROSE. With swellings at intervals.
TORTUOUS. Bent or twisted in different directions.
TORULOSE. Constricted between the seeds.
TRICHOTOMOUS. 3-forked.
TRIDENTATE. 3-toothed.

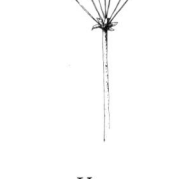

UMBEL

TRIFOLIOLATE. With 3 leaflets.
TRIQUETROUS. 3-edged.
TRITERNATE. Thrice or 3-ternate.
TRUNCATE. As if cut off squarely at the end.
TUBER. A thickened, solid, and short underground stem with many buds.
TUBERCLE. A small tuberlike prominence or nodule; the persistent base of the style in some Cyperaceae.
TUBEROUS. Producing or resembling a tuber.
TURBINATE. Top shaped.
TURGID. Swollen; inflated.

UMBEL. A flat or convex flower cluster in which the pedicels arise from a common point, like the rays of an umbrella.
UNCINATE. Hooked at the tip.

Illustrated Glossary

UNDULATE. Wavy; repand; with less pronounced waves than sinuate.
UNILATERAL. 1-sided, or turned to 1 side of an axis; secund.
URCEOLATE. Pitcherlike; hollow and contracted at the mouth like an urn or pitcher.
UTRICLE. A small, bladdery, 1-seeded fruit.

VALVE

VALVATE. Opening by valves, as in most dehiscent fruits and some anthers.
VALVE. One of the segments into which a dehiscent capsule or legume separates.
VEIN. A vascular bundle of a leaf or other flat organ.
VELUTINOUS. Velvety.
VENATION. The arrangement of the veins of a leaf; nervation.
VENTRAL. Relating to the inward surface of an organ, in relation to the axis.
VENTRICOSE. Swollen or inflated on 1 side, as in some corollas.
VERNATION. The arrangement of foliage leaves within the bud.
VERRUCOSE. Warty.
VERSATILE. An anther attached near the middle.
VERTICIL. A whorl, or circular arrangement of similar parts about the same point on an axis.
VESTIGIAL. Reduced to a vestige or trace of a part or organ at one time more perfectly developed.
VILLOUS. Bearing long, soft, and unmatted hairs; shaggy.
VISCID. Sticky; glutinous.
WHORL. A ring of similar organs radiating from a node; a verticil.
WING. A thin, usually dry extension bordering an organ; a lateral petal of a papilionaceous flower.
WOOLLY. Having long, soft, entangled hairs; lanate.
ZYGOMORPHIC. Bilaterally symmetric; that which can be bisected only in one plane into similar halves.

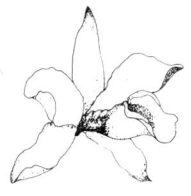

ZYGOMORPHIC

Selected References

Anderson, L. C. 1964. Taxonomic notes on the *Chrysothamnus viscidiflorus* complex (Astereae, Compositae). Madroño 17:222–227.

Bailey, L. H. 1949. Manual of cultivated plants. New York: The Macmillan Company.

Barneby, R. C. 1964. Atlas of North American *Astragalus*. Memoirs New York Botanical Garden 13:1–1088.

Beetle, A. A. 1960. A study of sagebrush. The section Tridentatae of *Artemisia*. Univ. Wyo. Agr. Exp. Sta. Bul. 368. 88 p.

Beetle, D. E. 1944. A monograph of the North American species of *Fritillaria*. Madroño 7:133–159.

Blake, S. F. 1961. Geographical guide to the floras of the world. Part II. Western Europe. U.S. Dept. Agr. Misc. Pub. 797. 742 p.

Blake, S. F., and A. C. Atwood. 1942. Geographical guide to the floras of the world. U.S. Dept. Agr. Misc. Pub. 401. 336 p.

Christensen, E. M. 1967. Bibliography of Utah botany and wildland conservation. Brigham Young Univ. Sci. Bul. Bio. Ser. 9(1):1–136.

Clausen, R. T. 1938. A monograph of the Ophioglossaceae. Mem. Torrey Bot. Club 19(2):1–177.

Constance, L. 1941. The genus *Nemophila* Nutt. Univ. Calif. Pub. Bot. 19(10):341–398.

Constance, L., and R. H. Shan. 1948. The genus *Osmorhiza* (Umbelliferae). Univ. Calif. Pub. Bot. 23(3):111–156.

Cronquist, A. 1947. Revision of the North American species of *Erigeron* north of Mexico. Brittonia 6:121–302.

Cronquist, A., et al. 1972. Intermountain flora. Vol. I. New York: Hafner Publishing Company, Inc.

Cutler, H. C. 1939. Monograph of the North American species of the genus *Ephedra*. Ann. Missouri Bot. Gard. 26:373–427.

Davidson, J. F. 1950. The genus *Polemonium* (Tournefort) L. Univ. Calif. Pub. Bot. 23(5):209–282.

Davis, R. J. 1952. Flora of Idaho. Provo: Brigham Young Univ. Press.

Detling, L. E. 1939. A revision of the North American species of *Descurainia*. Amer. Midl. Nat. 22:481–520.

Galway, D. H. 1945. The North American species of *Smilacina*. Amer. Midl. Nat. 33:644–666.

Garrett, A. O. 1936. Spring flora of the Wasatch region. 5th Ed. Salt Lake City: Stevens and Wallis, Inc.

Hall, H. M., and F. E. Clements. 1923. The phylogenetic method in taxonomy. The North American species of *Artemisia*, *Chrysothamnus*, and *Atriplex*. Carnegie Inst. Wash. Pub. 326. 355 p.

Harrington, H. D. 1954. Manual of the plants of Colorado. Denver: Sage Books.

Selected References

Hitchcock, A. S., and A. Chase. 1950. Manual of the grasses of the United States. U. S. Dept. Agr. Misc. Pub. 200. 1051 p.

Hitchcock, C. L. 1941. A revision of the Drabas of western North America. Univ. Wash. Pub. Biol. 11:1–132.

Hitchcock, C. L. 1952. A revision of the North American species of *Lathyrus*. Univ. Wash. Pub. Biol. 15:1–104.

Hitchcock, C. L., and B. Maguire. 1947. A revision of the North American species of *Silene*. Univ. Wash. Pub. Biol. 13:1–73.

Hitchcock, C. L., et al. 1969. Vascular plants of the Pacific Northwest. 5 Vols. Seattle: Univ. of Washington Press.

Holmgren, A. H. 1948. Handbook of the vascular plants of the northern Wasatch. San Francisco: Lithotype Process Company.

Jones, M. E. 1923. Revision of North American species of *Astragalus*. Salt Lake City: Publ. by author.

Kearney, T. H. 1935. The North American species of *Sphaeralcea*; subgenus Eusphaeralcea. Univ. Calif. Pub. Bot. 19(1):1–128.

Kearney, T. H., and R. H. Peebles. 1960. Arizona flora. Berkeley: Univ. of California Press.

Lewis, M. E. 1958. *Carex*—Its distribution and importance in Utah. Brigham Young Univ. Sci. Bul. Biol. Ser. 1(2):1–43.

Maguire, B. 1946. A monograph of the genus *Arnica* (Senecioneae, Compositae). Brittonia 4:386–510.

Maguire, B. 1951. Studies in the Caryophyllaceae—V. *Arenaria* in North America north of Mexico. A conspectus. Amer. Midl. Nat. 46:493–511.

Munz, P. A. 1965. Onagraceae. N. Amer. Fl. II. 1–225.

Sargent, C. S. 1926. Manual of the trees of North America (exclusive of Mexico). 2nd Ed. Boston: Houghton Mifflin Company.

Solbrig, O. T. 1960. Cytotaxonomic and evolutionary studies in the North American species of *Gutierrezia* (Compositae). Contrib. Gray Herb. 188:1–63.

Tidestrom, I. 1925. Flora of Utah and Nevada. Contrib. U.S. Natl. Herb. 25:1–625.

Index

Abies, 18
 balsamea, 18
 concolor, 18
 grandis, 18
 holophylla, 18
 lasiocarpa, 18, 20
 nordmanniana, 18
Abronia, 232
 fragrans, 232
 nana, 232
 pumila, 232
 salsa, 232
 villosa, 232
Abutilon, 226
 pictum, 226
 theophrasti, 226
Acacia, 187
 greggii, 187, 189
 Rose, 218
Acamptopappus, 98
 sphaerocephalus, 98
Acer, 36
 campestris, 36
 circinatum, 37
 diabolicum, 37
 ginnala, 37
 glabrum, 36, 37
 grandidentatum, 37
 negundo, 36
 palmatum, 36
 platanoides, 37
  schwedleri, 37
 pseudoplatanus, 37
 rubrum, 37
 saccharinum, 36
 saccharum, 37
Achillea, 98
 millefolium, 98, 100
Aconitum, 287
 columbianum, 287
Acorus, 364
 calamus, 364
Actaea, 287
 rubra, 287, 298
Adiantum, 7
 capillus-veneris, 7
 pedatum, 7, 8
Adonis, 287
 annua, 287
Aegilops, 375
 cylindrica, 375
Aesculus, 169
 glabra, 169
 hippocastanum, 169
 octandra, 169
Agastache, 179
 pallidiflora, 180
 urticifolia, 180
Agave, 359

 americana, 359
 utahensis, 359
 victoria-reginae, 359
Agoseris, 98
 arizonica, 98
 aurantiaca, 98, 100
 glauca, 98
 grandiflora, 98
 heterophylla, 98
Agropyron, 375
 dasystachyum, 375
 desertorum, 375
 inerme, 376
 repens, 375
 smithii, 375
 spicatum, 376
 subsecundum, 376
 trachycaulum, 376
Agrostis, 376
 alba, 376
 exarata, 376
 hiemalis, 376
 rossae, 376
 thurberiana, 376
Aguilegia, 288
 chrysantha, 288
 coerulea, 288
 elegantula, 288
 flavescens, 288
 formosa, 288
 micrantha, 288
 scopulorum, 288
Air-Plant, 139
Ajuga, 180
 reptans, 180
Albizia, 188
 julibrissin, 188, 189
Alder, 44
 Thin-leaved, 44
Aletes, 345
 macdougalli, 346
Alexanders, 352
Alfalfa, 214
Alisma, 360
 geyeri, 360
 plantago-aquatica, 360
Allenrolfea, 80
 occidentalis, 80
Allionia, 232
 choisyi, 232
 incarnata, 232
Allium, 401
 acuminatum, 402
 biceptrum, 402
 brandegei, 402
 brevistylum, 401
 campanulatum, 402
 cepa, 401
 cernum, 401

Index

cristatum, 402
diehlii, 402
geyeri, 402
macropetalum, 402
nevadense, 402
rubrum, 401
schoenoprasum, 401
textile, 402
Almond, 302
  Flowering, 302
Alnus, 44
  tenuifolia, 44
Alopecurus, 376
  geniculatus, 376
Althaea, 226
  rosea, 226
Alumroot, 314
Alyssum, 143
  alyssoides, 143
  saxatile, 143
  Sweet, 152
Amaranth, 38
Amaranthus, 38
  albus, 38
  graecizans, 38
  hybridus, 38
  powellii, 38
  retroflexus, 38, 58
Amaryllis, 361, 363
  x johnsonii, 361
  reginae, 361
  striata, 361
  vittata, 361
Ambrosia, 98
  acanthicarpa, 99
  artemisiifolia, 99
  coronopifolia, 99
  dumosa, 99
  eriocentra, 99
Amelanchier, 297
  alnifolia, 297, 298
  pumila, 297
  utahensis, 297
Amorpha, 188
  fruticosa, 188, 189
Amphipappus, 99
  fremontii, 99
Amsinckia, 47
  intermedia, 47
  retrorsa, 47
  tesselata, 47
Amsonia, 39
  brevifolia, 39
  eastwoodiana, 39
  jonesii, 39
  tomentosa, 39
Anaphalis, 99
  margaritacea, 99, 100
Anchusa, 47
  officinalis, 48
Andropogon, 376
  gerardii, 376
  hallii, 377
  scoparius, 376

Androsace, 282
  carinata, 282
  filiformis, 282
  occidentalis, 282
  septentrionalis, 282, 285
Androstephium, 402
  breviflorum, 402
Anemone, 288
  Desert, 288
  globosa, 288
  japonica, 288
  tuberosa, 288
Anemopsis, 312
  californica, 312
Anethum, 346
  graveolens, 346
Angelica, 346
  arguta, 346
  kingii, 346
  pinnata, 346, 355
  roseana, 346
  wheeleri, 346
Antennaria, 99
  anaphaloides, 101
  concinna, 101
  dimorpha, 99
  luzuloides, 101
  media, 101
  microphylla, 101
  neglecta, 101
  parvifolia, 99, 101
  rosea, 101, 107
  rosulata, 99
  umbrinella, 101
Anthemis, 101
  cotula, 101
  tinctoria, 101
Antirrhinum, 321
  filipes, 321
  kingii, 322
  majus, 322
Apache Plume, 300
Apium, 346
  graveolens, 346
  dulce, 346
Apocynum, 40
  androsaemifolium, 40, 58
  cannabinum, 40
  medium, 40
  sibiricum, 40
Apple, 300, 301
  Prairie Crab, 300
  Showy Crab, 301
  Squaw, 301
Applebush, 84
Apricot, 302
Arabis, 143
  demissa, 144
  divaricarpa, 144
  drummondii, 143
  fendleri, 144
  glabra, 143
  gunnisoniana, 144
  hirsuta, 143

Index

holboellii
  pinetorum, 144
  retrofracta, 144
lemonii, 144
lignifera, 144
lyallii, 143
microphylla, 143
nuttallii, 143
pendulina, 144
perennans, 144
pulchra, 144
selbyi, 143
sparsiflora, 144
Araucaria, 14
  columnaris, 14
  excelsa, 14
Arborvitae, 16
  American, 16
  Oriental, 16
Arceuthobium, 223
  americanum, 223
  campylopodium, 223
  douglasii, 223
  vaginatum, 223
Arctium, 101
  minus, 101
Arctomecon, 245
  humilis, 245
Arctostaphylos, 159
  nevadensis, 159
  patula, 159
  pungens, 160
  uva-ursi, 159, 166
Arenaria, 72
  aculeata, 72
  congesta, 72, 73
  eastwoodiae, 73
  fendleri, 73
  filiorum, 73
  hookeri, 73
  kingii, 73
  lanuginosa, 72
  lateriflora, 72
  macrodenia, 73
  nuttallii, 73
  obtusiloba, 73
  rubella, 73
Argemone, 245
  corymbosa, 245
  munita, 245
Aristida, 377
  longiseta, 377
Arnica, 101
  chamissonis, 101
  cordifolia, 102
  diversifolia, 102
  fulgens, 102
  longifolia, 102
  mollis, 102, 107
  parryi, 102
  rydbergii, 102
  sororia, 102
Arrhenatherum, 377
  elatius, 377

Arrowhead, 360
Arrowleaf, 122
Artemisia, 102
  absinthium, 102
  biennis, 102
  bigelovii, 103
  cana, 103
  dracunculus, 102
  filifolia, 103
  frigida, 102
  ludoviciana, 102
  nova, 103
  spinescens, 103
  tridentata, 103
  tripartita, 103
Asclepias, 40
  asperula, 40
  cryptoceras, 41
  curassavica, 41, 42
  cutleri, 41
  erosa, 41
  fascicularis, 41
  hallii, 42
  incarnata, 42
  involucrata, 41
  labriformis, 41
  latifolia, 41
  macrosperma, 41
  rusbyi, 40
  ruthiae, 41
  speciosa, 42
  subverticillata, 41
  tuberosa, 41
Ash, 235
  Black, 235
  Blue, 235
  European, 235
  German, 235
  Modesto, 235
  Mountain, 304
  Oak-leaf Mountain, 304
  Red, 235
  Rose-leaf Mountain, 304
  Single-leaf, 235
  White, 235
Asparagus, 402
  officinalis, 402, 405
  sprengeri, 402
Aspen, 308
Asperugo, 48
  procumbens, 48
Asplenium, 7
  adiantium-nigrum, 9
  septentrionale, 7
  trichomanes, 9, 12
  viride, 9
Astephanus, 42
  utahensis, 42
Aster, 103, 123
  arenosus, 103
  brachyactis, 103
  chilensis, 104
  eatonii, 104
  engelmannii, 104

439

Index

falcatus, 104
foliaceus, 105
frondosus, 103
glaucodes, 104
Golden, 121
hesperius, 104
hirtifolius, 103
integrifolius, 104
junciformis, 104
laevis, 104
occidentalis, 105
pansus, 104
paucicapitatus, 104
perelegans, 104
subspicatus, 105
Astragalus, 188
adanus, 106
adsurgens
  robustior, 191
agrestis, 196
alpinus, 195
amphioxys
  amphioxys, 192
  vespertinus, 192
ampullarius, 196
argophyllus
  argophyllus, 202
  martinii, 202
  panguicensis, 202
asclepiadoides, 192
beckwethii
  beckwethii, 203
  purpureus, 203
bisulcatus
  bisulcatus, 196
  haydenianus, 195, 196
bodinii, 194
brandegei, 198
bryantii, 206
calycosus, 191
canadensis
  brevidens, 191
  canadensis, 191
castaneiformis
  consobrinus, 192
ceramicus, 190
chamaeleuce, 192
chloodes, 190
cibarius, 206, 216
coltonii
  coltonii, 197, 205
  moabensis, 198, 205
convallarius
  convallarius, 194
  finitimus, 194
cronquistii, 198
cymboides, 192
desereticus, 202
desperatus
  conspectus, 201, 203
  desperatus, 199
detritalis, 191
diversifolius, 193
drummondii, 194
duchesnensis, 197
eastwoodiae, 196, 204, 205, 206
episcopus, 198, 205
eremiticus, 204
eurekensis, 202
flavus
  argillosus, 191
  candicans, 191
  flavus, 191
flexuosus
  diehlii, 195
  flexuosus, 195
fucatus, 195
geyeri, 190
gilviflorus, 190
hamiltonii, 197, 205
harrisonii, 197
humistratus
  humivagans, 191
iodanthus, 199
jejunus, 194
kentrophyta
  coloradoensis, 188
  elatus, 188
  implexus, 188
lancearius, 198
lentiginosus, 205
  araneosus, 200
  chartaceous, 200
  diphysus, 200
  palans, 200, 206
  platyphyllidius, 200
  salinus, 200
  scorpionis, 200
  stramineus, 200
  ursinus, 200
  vitreus, 200
limnocharis, 194
loanus, 192
lonchocarpus, 197, 205
lutosus, 203
malacoides, 204
marianus, 202
megacarpus, 201, 203
minthorniae
  gracilior, 206
miser
  oblongifolius, 193, 195
  tenuifolius, 193, 195
missouriensis
  amphibolus, 192
moencoppensis, 195
mollissimus
  thompsonae, 200
monumentalis, 199
musiniensis, 192, 201
nelsonianus, 193
newberryi
  newberryi, 202
nidularius, 197, 205
nuttallianus
  imperfectus, 188
  micranthiformis, 188
oophorus

Index

caulescens, 203
lonchocalyx, 203
pardalinus, 190, 199
pattersonii, 204, 206
perianus, 194
pinonis, 198
platytropis, 194
praelongus
　ellisiae, 204, 206
　lonchopus, 204
　praelongus, 204, 206
preussii
　laxiflorus, 205, 206
　preussii, 205
pubentissimus, 188, 190, 199
purshii
　glareosus, 201
　purshii, 201
racemosus
　treleasei, 196
rafaelensis, 193
sabulonum, 190
sabulosus, 204, 206
saurinus, 193
scopulorum, 196
serpens, 199
sesquiflorus, 191
spatulatus, 191
straturensis, 198
striatiflorus, 194
subcinereus, 195
tenellus, 195
tephrodes
　brachylobus, 202
tetrapterus, 196, 203
toanus, 193
utahensis, 201
wardii, 199
wetherillii, 199
wingatanus, 193, 195
woodruffii, 190, 196
zionis, 202
Athyrium, 9
　americanum, 9
　felix-foemina, 9
Atrichoseris, 105
　platyphylla, 105
Atriplex, 80
　argentea, 80
　bonnevillensis, 80
　canescens, 80
　confertifolia, 81
　corrugata, 81
　cuneata, 81
　elegans, 82
　falcata, 81
　gardneri, 81
　garrettii, 81
　graciliflora, 82
　hortensis, 82
　hymenelytra, 80
　lentiformis, 80
　Mat, 81
　navajoensis, 81

obovata, 81
patula, 82
powellii, 82
rosea, 82
semibaccata, 80
torreyi, 80
tridentata, 81, 85
truncata, 82
welshii, 81
wolfii, 82
Avena, 377
　fatua, 377, 380
　sativa, 377
Avens, 300
Azolla, 11
　caroliniana, 11

Babysbreath, 74
Baccharis, 105
　emoryi, 105
　glutinosa, 105
　sergiloides, 105
　viminea, 105
Bachelor's Buttons, 108
Bahia, 105
　dissecta, 105
　nudicaulis, 105
　oblongifolia, 105
　ourolepis, 105
Baileya, 105
　multiradiata, 105
　pleniradiata, 105
Balloon Flower, 67
Balm of Gilead, 309
Balsamorrhiza, 106
　hirsuta, 106
　hookeri, 106
　macrophylla, 106
　sagittata, 106, 107
Balsamroot, 106
　Arrow-leaf, 106
　Hooker, 106
　Large-leaf, 106
　Smaller, 106
Baptisia, 207
　leucantha, 207
Barbarea, 145
　orthoceras, 145, 146
Barberry, 43
　Common, 43
　Japanese, 43
　Juliana, 43
　Sargent, 43
Barley, 386
　Foxtail, 386
　Meadow, 386
　Mediterranean, 386
　Mouse, 386
Basil, 181
Basket-of-Gold, 143
Bay, Bull, 224
　Sweet, 224
Bean, 215
　Castor, 162

441

Index

Beardtongue, 326
Beautybush, 69
Beckmannia, 377
  syzigachne, 377, 380
Bedstraw, 305
Beech, 162
  American, 162
  European, 162
Beeplant, 68
  Rocky Mountain, 68
  Yellow, 68
Beggar's Ticks, 106
Begonia, 42
  semperflorens, 42
  tuberhybrida, 42
  Tuberous, 42
Bellflower, 66
  Peach-leaf, 67
  Tussock, 66
Bells, Canterbury, 67
  Fairy, 403
  Whispering, 170
  Yellow, 404
Berberis, 43
  julianae, 43
  sargentiana, 43
  thunbergii, 43
  vulgaris, 43
Berry, Bane, 287
  Bear, 159
  Blue, 160
  Buffalo, 158
  China, 230
  Coral, 70
  Goose, 316, 317
  Hack, 341
  Rasp. See Raspberry
  Service, 297
  Silver, 158
  Snow. See Snowberry
  Straw, 300
  Thimble, 303
  Wolf. See Wolfberry
Berula, 346
  erecta, 346
Besseya, 322
  alpina, 322
  wyomingensis, 322
Betony, 183
Betula, 44
  glandulosa, 44
  occidentalis, 44
Bidens, 106
  cernua, 106, 107
  frondosus, 106
Big-Tree, 23
Bindweed, 136
Birch, 44
  River, 44
Birdbeak, 323
Bitterbrush, 303
Bittercress, 145
Bittersweet, European, 339
Bitterroot, 279

Blackbrush, 299
Blackcap, 304
Blackeyed Susan, 128
Bladderpod, 150
Bladderwort, 220
Blanketflower, 118
Blazing Star, 221
Bleeding Heart, 164
Bloodweed, 63
Bluebell, 60
Bluebuttons, 158
Blue-eyed Mary, 323
Bluegrass, Alpine, 391
  Annual, 390
  Big, 391
  Bulbous, 390
  Canadian, 390
  Fowl, 390
  Kentucky, 390
  Nevada, 391
  Nodding, 390
  Sandberg, 391
  Shortleaf, 390
  Wheeler, 390
Bluejoint, 379
Blue Sailors, 111
Bluestem, 375, 376
  Big, 376
  Little, 376
  Sand, 377
Boisduvalia, 237
  glabella, 237
Borage, 48
Borago, 48
  officinalis, 48
Botrychium, 5
  boreale, 6
  lanceolatum, 6
  lunaria, 6, 8
  matricariaefolium, 6
  simplex, 5
Bouncing Bet, 75
Bouteloua, 377
  curtipendula, 377
  gracilis, 377
Bracken, 11
Brassica, 145
  campestris, 145, 146
  juncea, 145
  nigra, 145
Brickellbush, 106
Brickellia, 106
  californica, 106
  grandiflora, 106
  longifolia, 106
  microphylla, 106
  oblongifolia, 106
  scabra, 106
  watsonii, 106
Bridal Wreath, 304
Brodiaea, 402
  douglasii, 402
  pulchella, 402
Brome, California, 378

Index

Fringed, 378
Nodding, 378
Rattlesnake, 378
Smooth, 378
Bromus, 377
  anomalus, 378
  brizaeformis, 378
  ciliatus, 378
  commutatus, 378
  inermis, 378
  marginatus, 378
  mollis, 378
  polyanthus, 378
  racemosus, 378
  rigidus, 378
  rubens, 378
  sterilis, 378
  tectorum, 377
Broomrape, 243
Buddleja, 222
  davidii, 223
  utahensis, 223
Bugloss, 47
  Vipers, 57
Burdock, 101
Bursage, 98
Bugseed, 83
Buttercup, 290
  Bur, 290
  Hillside, 292
Buckthorn, 294
Butterflyweed, 239
Buckwheat, Wild, 254
Bugleweed, 180
Butternut, 177
Buckeye, Ohio, 169
  Yellow, 169
Buffalo Bur, 339
Burnet, 304
Burreed, 414
Butter and Eggs, 323
Buttonbush, 305

Cabbage, Wild, 145
Cactus, Ball, 64
  Fishhook, 64
  Hedgehog, 64
Calamagrostis, 378
  canadensis, 379
  inexpansa, 379
  rubescens, 379
Callirrhoe, 226
  involucrata, 226
Callitriche, 66
  heterophylla, 66
  palustris, 66
Calochortus, 403
  flexuosus, 403
  gunnisonii, 403
  kennedyi, 403
  macrocarpus, 403
  nuttallii, 403
    aureus, 403
Caltha, 288

  leptosepala, 288
Calypso, 408
  bulbosa, 408, 409
Camas, 403
  Death, 406
Camassia, 403
  quamash, 403
Camelina, 145
  microcarpa, 145, 151
Camomile, 101
  Yellow, 101
Campanula, 66
  carpatica, 66
  medium, 67
  parryi, 66
  persicifolia, 67
  rotundifolia, 66
Campion, 74
Campsis, 45
  radicans, 45
Candytuft, 150
  Edging, 150
Canna, 364
  indica, 364
Cannabis, 230
  sativa, 230
Cantaloupe, 156
Capsella, 145
  bursa-pastoris, 145, 151
Capsicum, 337
  frutescens, 337
Caragana, 207
  arborescens, 207, 216
Caraway, 346
  False, 351
Cardamine, 145
  breweri, 145
  cordifolia, 145
  pennsylvanica, 145
  unijuga, 145
Cardaria, 145
  draba, 145
  pubescens, 145
Carduus, 106
  nutans, 106
Carex, 365
  aquatilis, 366, 369
  arthrostachya, 366
  aurea, 366
  capillaris, 367
  douglasii, 366
  festivella, 366
  geyeri, 365
  hoodii, 366
  interior, 366
  kelloggii, 366
  lanuginosa, 367
  nebraskensis, 367
  nova, 367
  occidentalis, 366
  phaeocephala, 366
  praegracilis, 366
  pseudoscirpoides, 365
  rossii, 367

Index

rostrata, 367, 369
vallicola, 366
vernacula, 366
vesicaria, 367
Carnation, 74
Carrot, 349
  False, 346
Carthamnus, 108
  tinctorius, 108
Carum, 346
  carvi, 346
Carya, 177
  illinoensis, 177
  ovata, 177
Castanea, 162
  dentata, 162
Castilleja, 322
  applegatei, 323
  chromosa, 322
  flava, 322
  linariaefolia, 322
  miniata, 323
  minor, 322
  occidentalis, 322
  rhexifolia, 323
  rubida, 322
  septentrionalis, 322
Catabrosa, 379
  aquatica, 379, 380
Catalpa, 45
  bignonioides, 45
  bungei, 45
  speciosa, 45
  Western, 45
Catchfly, 75
Catchweed, 48
Catnip, 181
Catsclaw, 187
Catseye, 49
Cattail, 414
Caucalis, 346
  microcarpa, 346
Caulanthus, 145
  crassicaulis, 147
Ceanothus, 293
  fendleri, 293
  greggii, 293
  martinii, 293
  velutinus, 293
Cedar, 18
  Atlas, 18
  Deodar, 18
  Incense, 16
  Port Orford, 14
Cedar of Lebanon, 18
Cedrus, 18
  atlantica, 18
  deodara, 18
  libani, 18
Celery, 346
Celosia argentea, 38
Celtis, 341
  laevigata, 341
  occidentalis, 341
  reticulata, 341
Cenchrus, 379
  pauciflorus, 379
Centaurea, 108
  cyanus, 108
  repens, 108
Centaurium, 164
  calycosum, 165
  exaltatum, 165
  nuttallii, 165
Centaury, 164
Centranth, 353
Centranthus, 353
  ruber, 353
Century Plant, 359
Cephalanthus, 305
  occidentalis, 305
Cephalotaxus, 14
  drupacea, 14
Cerastium, 73
  brachypodum, 74
  nutans, 74
  tomentosum, 74
  vulgatum, 74
Ceratophyllum, 78
  demersum, 78
Cercis, 207
  canadensis, 207
  occidentalis, 207
Cercocarpus, 297
  intricatus, 297
  ledifolius, 297
  montanus, 297
  montanus x ledifolius, 297
Chaenactis, 108
  alpina, 108
  carphoclinia, 108
  douglasii, 108
  fremontii, 108
  macrantha, 108
  stevioides, 108
Chaenomeles, 297
  japonica, 297
  lagenaria, 297
Chaffbush, 99
Chamaebatiaria, 297
  millefolium, 299
Chamaechaenactis, 108
  scaposa, 108
Chamaecyparis, 14
  lawsoniana, 14
Chamaesarcha, 337
  coronopus, 337
Cheilanthes, 9
  covillei, 9
  eatonii, 9
  feei, 9
  gracillima, 9
  siliquosa, 9
Chenopodium, 82
  album, 83, 85
  ambrosioides, 82
  atrovirens, 83
  berlandieri, 83

Index

botrys, 82
capitatum, 82
fremontii, 83
gigantospermum, 83
glaucum, 83
humile, 83
incanum, 83
leptophyllum, 82
paganum, 83
pratericola, 83
rubrum, 83
Cherry, Choke, 303
  Cornelian, 138
  Flowering, 302
  Ground, 338
  May Day, 303
  Pin, 302
  Sour, 302
  Sweet, 303
Chess, Foxtail, 378
  Hairy, 378
  Soft, 378
Chestnut, 162
  American, 162
  Horse, 169
Chickory, 111
Chickweed, 73, 77
Chilopsis, 45
  linearis, 45
Chimaphila, 285
  umbellata, 285
Chinchweed, 126
Chionanthus, 234
  virginicus, 234
Chives, 401
Chlorocrambe, 147
  hastatus, 147
Chorispora, 147
  tenella, 147
Chorizanthe, 254
  brevicornu, 254
  rigida, 254
  thurberi, 254
Chrysanthemum, 108
  balsamita, 109
  leucanthemum, 109
Chrysosplenium, 314
  tetrandrum, 314
Chrysothamnus, 109
  albidus, 109
  depressus, 109
  greenei, 109
  linifolius, 109
  nauseosus, 110
    albicaulis, 111
    consimilis, 111
    gnaphaloides, 111
    graveolens, 111
    junceus, 111
    leiospermus, 111
    salicifolius, 110
    turbinatus, 110
  paniculatus, 109
  parryi, 110

attenuatus, 110
howardii, 110
nevadensis, 110
parryi, 110
pulchellus, 109
vaseyi, 109
viscidiflorus, 109
  elegans, 110
  lanceolatus, 110
  puberulus, 110
  stenophyllus, 110
Cicely, Sweet, 351
Cichorium, 111
  intybus, 111
Cicuta, 347
  douglasii, 347
Cinna, 379
  latifolia, 379, 382
Cinquefoil, 301
Circaea, 238
  pacifica, 238
Cirsium, 111
  arizonicum, 113
  arvense, 111, 129
  bipinnatum, 113
  calcareum, 112, 113
  canescens, 112
  canovirens, 112
  clavatum, 112
  eatonii, 112
  neomexicanum, 112
  nidulum, 113
  parryi, 112
  pulchellum, 113
  rothrockii, 112
  rydbergii, 111
  scopulorum, 112
  undulatum, 112
  utahense, 112
  vulgare, 111
Citrullus, 156
  vulgaris, 156
Cladrastris, 207
  lutea, 207
Clammyweed, 68
Clarkia, 238
  rhomboidea, 238
Claytonia, 279
  lanceolata, 279, 280
  megarrhiza, 279
Clematis, 288
  columbiana, 289
  hirsutissima, 289
  jackmanii, 288
  ligusticifolia, 289
  orientalis, 289
  pseudoalpina, 289
  Purple, 288
Cleome, 68
  lutea, 68
  serrulata
    angusta, 68
    serrulata, 68

Index

Cleomella, 68
  palmerana, 68
  plocosperma, 68
Cliffbrake, 10
Cliffbush, 315
Clivia, 361
  miniata, 361
Clover, 219
  Castle Valley, 81
  Owl, 325
  Prairie, 215
  Red, 219
  Sweet, 214
  White Dutch, 219
  White Sweet, 214
  Yellow Sweet, 214
Cocklebur, 136
Cockscomb, 38
Cockspur, 381
Coffee-Tree, Kentucky, 207
Coldenia, 48
  canescens, 48
  hispidissima, 48
  nuttallii, 48
Coleogyne, 299
  ramosissima, 299
Coleus, 180
  blumei, 180
Collinsia, 323
  parviflora, 323, 336
Collomia, 248
  grandiflora, 248
  linearis, 248, 252
  tenella, 248
Columbine, 288
  Crimson, 288
  Golden, 288
  Yellow, 288
Colutea, 207
  arborescens, 207
Comandra, 312
  umbellata, 312
Combseed, 62
Cone Flower, 128
  Tall, 128
  Western, 128
Conium, 347
  maculatum, 347
Conringia, 147
  orientalis, 147
Convallaria, 403
  majalis, 403
Convolvulus, 136
  arvensis, 137
  sepium, 136
Conyza, 114
  canadensis, 114, 115
Cooperia, 362
  drummondii, 362
  pedunculata, 362
Copperweed, 126
Corallorhiza, 408
  maculata, 408
  striata, 408
  trifida, 408
Coralroot, 408
Cordylanthus, 323
  canescens, 323
  kingii, 323
  parviflorus, 323
  ramosus, 323
  wrightii, 323
Corispermum, 83
  hyssopifolium, 83
Corn Salad, 354
Cornus, 138
  mas, 138
  stolonifera, 138, 146
Corydalis, 164
  aurea, 164
  caseana, 164
  Golden, 164
  ochroleuca, 164
Corylus, 44
  avellana, 44
Coryphantha, 64
  vivipara, 64
Costmary, 109
Cotoneaster, 299
  acutifolia, 299
  horizontalis, 299
  Peking, 299
  Rock, 299
Cotton, 226
  Upland, 226
Cottonwood, 308
  Narrowleaf, 309
Cowania, 299
  mexicana, 299
Cranesbill, 168
Crassula, 139
  argentea, 139
Crataegus, 299
  crus-galli, 299
  mollis, 299
  monogyna, 299
  oxyacantha, 299
  phaenopyrum, 299
  rivularis, 299
  succulenta, 299
Cream Cups, 246
Creosote Bush, 357
Crepis, 113
  acuminata, 113
  atrabarba, 113, 114
  intermedia, 114
  modocensis, 113
  nana, 113
  occidentalis, 114
  runcinata
    glauca, 113
    hispidulosa, 113
    runcinata, 113
Cress, 153
Cressa, 137
  truxillensis, 137
Crinum, 362
  asiaticum, 362

Index

bulbispermum, 362
  x powellii, 362
Crocus, 395
  biflorus, 395
  imperati, 395
  maesiacus, 395
  sativus, 395
  susianus, 395
  vernus, 395
Croton, 161
  longipes, 161
  texensis, 161·
Crownbeard, 135
Crown Imperial, 403
Cryptantha, 49
  abata, 56
  affinis, 49
  ambigua, 50, 51
  augustifolia, 50
  bakeri, 53, 56
  barbigera, 51
  barnebyi, 52, 54
  breviflora, 55
  caespitosa, 56
  capitata, 52
  circumscissa, 51
  compacta, 55
  confertiflora, 52
  crassisepala
    elachantha, 51
  decipiens, 51
  dumetorum, 50
  elata, 56
  fendleri, 50
  flaccida, 49
  flava, 52, 58
  flavoculata, 53
  fulvocanescens
    echinoides, 53
    fulvocanescens, 53
  gracilis, 50
  grahamii, 57
  humilis, 54, 55, 56, 57
    commixta, 55
    nana, 55
    ovina, 55
    shantzii, 55
  inaequata, 49, 50
  jamesii
    disticha, 54
    multicaulis, 54
    pustulosa, 54
    setosa, 54
  johnstonii, 52
  jonesiana, 52
  kelseyana, 51
  longiflora, 53
  mensana, 55
  micrantha, 50, 51
  nevadensis, 51
  ochroleuca, 56
  osterhoutii, 56
  paradoxa, 53
  pattersonii, 50, 51

  pterocarya
    cycloptera, 49
    pterocarya, 49
  racemosa, 49, 50
  recurvata, 51
  rollinsii, 53, 54
  rugulosa, 54
  scoparia, 51
  semiglabra, 52
  sericea, 56
  setosissima, 54
  stricta, 54
  tenuis, 53
  torreyana, 49
  utahensis, 49, 51
  virginensis, 56
  watsoni, 50
  wetherillii, 53
Cryptogramma, 9
  crispa, 9, 12
  stelleri, 9
Cucumber, 156
  Squirting, 157
  Wild, 157
Cucumber Tree, 224
Cucumis, 156
  melo, 156
  sativus, 156
Cucurbita, 156
  foetidissima, 157
  maxima, 157
  moschata, 157
  pepo, 157
Cudweed, 118
Cupressus, 15
  arizonica, 15
Currant, 316
  Bedbug, 317
  Common, 317
Cuscuta, 137
  approximata, 137
  campestris, 137
  cephalanthii, 137
  cuspidata, 137
  denticulata, 137
  indecora, 137
  occidentalis, 137
  salina, 138
Cushaw, 157
Cycas, 16
  revoluta, 16
Cydonia, 299
  oblonga, 299
Cymopterus, 347
  acaulis, 347, 348
  basalticus, 348
  bipinnatus, 347
  bulbosus, 347
  coulteri, 348
  duchesnensis, 348
  fendleri, 348
  globosus, 347
  ibapensis, 348
  jonesii, 349

Index

longipes, 348
minimus, 348
multinervatus, 347
newberryi, 347, 348
purpurascens, 347
purpureus, 349
rosei, 348
Cynoglossum, 57
officinale, 57
Cyperus, 367
aristatus, 367
erythrorhizos, 367
esculentus, 367
schweinitzii, 367
strigosus, 367
Cypress, 15
Arizona, 15
Bald, 23
False, 14
Summer, 84
Cypripedium, 408
calceolus, 408, 413
fasciculatum, 408
Cystopteris, 9
bulbifera, 10
fragilis, 9, 12

Dactylis, 379
glomerata, 379, 382
Daffodil, 363
Daisy, Oxeye, 109
Woolly, 117
Dalea, 207
fremontii, 207
lanata, 207
polyadenia, 207
thompsonae, 207
Dandelion, 132
Mountain, 98
Danthonia, 379
intermedia, 379, 382
unispicata, 379
Datura, 337
meteloides, 337
stramonium, 337
Daucus, 349
carota, 349
sativa, 349
Deerbrush, 293
Delphinium, 289
ajacis, 289
amabile, 289
nelsonii, 289
occidentale, 289
scaposum, 289
Deschampsia, 379
caespitosa, 381, 382
danthonioides, 379
elongata, 381
Descurainia, 147
californica, 147
pinnata, 147
richardsonii, 147
sophia, 147

Deutzia, 314
scabra, 314
Dianthus, 74
barbatus, 74
caryophyllus, 74
plumarias, 74
Dicentra, 164
spectabilis, 164
uniflora, 164
Dicoria, 114
brandegei, 114
canescens, 114
Digitalus, 323
purpurea, 323
Digitaria, 381
ischaemum, 381
sanguinalis, 381
Dill, 346
Diplotaxis, 147
muralis, 147
Dipsacus, 158
sylvestris, 158
Disporum, 403
trachycarpum, 403
Distichlis, 381
stricta, 381
Dithyrea, 147
wislizenii, 147
Dock, 276
Dodder, 137
Dodecatheon, 282
alpinum, 283
dentatum, 283
pauciflorum, 283
Dogbane, 40
Dogweed, 114
Dogwood, 138
Red Osier, 138
Dollar Plant, 152
Downingia, 67
laeta, 67
Draba, 147
apiculata, 149
aurea, 148
brachystylis, 148
crassa, 148
crassifolia, 149
cuneifolia, 148
densifolia, 149
fladnizensis, 148
lanceolata, 148
maguirei, 149
nemorosa, 148, 151
nivalis, 148
oligosperma, 149
rectifructa, 148
reptans, 148
sobolifera, 149
spectabilis, 148
stenoloba, 148
subalpina, 148
ventosa, 149
zionensis, 148

Index

Dracocephalum, 180
  nuttallii, 180
  virginianum, 180
Dragonhead, 181
  False, 180
Dropseed, 392
  Mesa, 392
  Sand, 392
  Spike, 392
Dryopteris, 9
  felix-mas, 10
Duckweed, 399
Dunebroom, 215
Dutchman's Breeches, 164
Dyssodia, 114
  thurberi, 114

Ecballium, 157
  elaterium, 157
Echinocactus, 64
  acanthodes, 64
  johnsonii, 64
  whipplei, 64
    parviflorus, 64
    spinosior, 64
Echinocereus, 64
  engelmannii, 64
  fendleri, 64
  triglochidiatus, 64
Echinochloa, 381
  crusgallii, 381
Echinocystis, 157
  lobata, 157
Echinopsilon, 83
  hyssopifolium, 83
Echium, 57
  vulgare, 57
Elaeagnus, 158
  angustifolia, 158
  commutata, 158
Elder, 70
  American, 70
  Blackbeard, 70
  Box, 36
  European, 70
  Marsh, 122
  Red, 70
Elderberry, 70
Eleocharis, 367
  acicularis, 368
  bolanderi, 368
  calva, 367
  machrostachya, 368, 369
  pauciflora, 368
  rostellata, 368
Elephantella, 325
Elm, 341
  American, 341
  Chinese, 341
  Dutch, 341
  English, 342
  Scots, 341
  Siberian, 341
  Slippery, 341
  Winged, 341
Elodea, 393
  canadensis, 393
Elymus, 381
  ambiguus, 383
  arenicola, 381
  canadensis, 384
  cinereus, 383
  dasystachys, 381
  giganteus, 381
  glaucus, 384
  junceus, 383
  macounii, 383
  sabulosus, 381
  salinus, 383
  salmonis, 383
  simplex, 383
  triticoides, 383
  virescens, 384
  virginicus
    submuticus, 384
Emmenanthe, 170
  penduliflora, 170
Encelia, 114
  farinosa, 114
  frutescens, 114
Enceliopsis, 114
  argophylla, 115
  nudicaulis, 114
  nutans, 114
Ephedra, 16
  cutleri, 17
  fasciculata, 17
  nevadensis, 17
  torreyana, 17
  viridis, 17
Epilobium, 238
  adenocaulon, 238
  alpinum, 239
  angustifolium, 238
  brevistylum, 238
  clavatum, 239
  halleanum, 238
  hornemanii, 238
  latifolium, 238
  palustre, 238
  paniculatum, 238
  saximontanum, 238
Epipactis, 410
  gigantea, 410
Equisetum, 4
  arvense, 4, 8
  hyemale, 4
  kansanum, 4
  laevigatum, 4
  prealtum, 4
  variegatum, 4
Eragrostis, 384
  cilianensis, 384
  hypnoides, 384
Eremocrinum, 403
  albomarginatum, 403

Index

Eretrichium, 57
  nanum
    elongatum, 57, 58
Eriastrum, 248
  diffusum, 248
  filifolium, 248
  sapphirinum, 248
Erigeron, 115
  arenarioides, 116
  argentatus, 116
  bellidiastrum, 115
  caespitosus, 117
  compositus, 116
  concinnus, 117
  controversus, 117
  coulteri, 115
  divergens, 115
  eatonii, 166
  engelmannii, 116
  flagellaris, 115
  glabellus, 115, 116
  leiomeris, 116
  lonchophyllus, 115
  macranthus, 116
  peregrinus, 115
  pumilus, 117
  simplex, 116
  speciosus, 116
  ursinus, 117
  vetensis, 117
Eriogonum, 254
  alatum, 269
  ammophilum, 256
  aretioides, 266
  batemanii, 263
  bicolor, 260, 264
  brachypodum, 272
  brevicaule
    brevicaule, 261
    cottamii, 261
    laxifolium, 261, 265
    wasatchense, 263
  caespitosum, 269
  cernuum
    cernuum, 273
    viminale, 273
  chrysocephalum, 261
  clavellatum, 260
  contortum, 261, 264
  corymbosum, 258
    albogilvum, 258
    corymbosum, 257
    davidsei, 258
    erectum, 258
    glutinosum, 258
    orbiculatum, 258
    velutinum, 258
  cronquistii, 263
  davidsonii, 274
  deflexum
    deflexum, 272
    nevadense, 272
  desertorum, 265
  devaricatum, 275
  x duchesnense, 258, 259
  ephedroides, 261
  eremicum, 263
  fasciculatum
    polifolium, 255
  flexum, 271
  gordonii, 270
  grayi, 265
  heermannii
    subracemosum, 260
    sulcatum, 260
  heracleoides, 267
  hookeri, 272
  howellianum, 270
  humivagans, 262
  hylophilum, 259
  inflatum
    fusiforme, 271
    inflatum, 270
  insigne, 272
  intermontanum, 262
  jamesii
    flavescens, 269
    rupicola, 269
  kearneyi, 256
  lancifolium, 259
  leptocladon
    leptocladon, 256
    papiliunculum, 256
    ramosissimum, 256
  leptophyllum, 259
  loganum, 261, 264
  maculatum, 274
  microthecum
    foliosum, 257, 259
    laxiflorum, 257
    mortonianum, 257
  nanum, 265
  nidularium, 275
  nummulare, 256, 263
  nutans, 273
  ostlundii, 263
  ovalifolium
    multiscapum, 267
    nivale, 267
    ovalifolium, 267
  palmerianum, 275
  panguicense
    alpestre, 265
    panguicense, 265
  parryi, 272
  pharnaceoides, 274
  plumatella, 260
  polycladon, 275
  puberulum, 275
  pusillum, 273
  racemosum, 264
  revealinaum, 257
  salsuginosum, 271
  saurinum, 259
  scabrellum, 272
  shockleyi
    longilobum, 266
    shockleyi, 266

Index

smithii, 256
spathuiforme, 263
spathulatum, 261
subalpinum, 269
subreniforme, 274
thomasii, 273
thompsonae
  albiflorum, 263
  thompsonae, 262
trichopes, 270
triste, 269
tumulosum, 266
umbellatum
  aureum, 268
  dichrocephalum, 268
  majus, 269
  porteri, 268
  subaridum, 268
  umbellatum, 268
villiflorum, 266
viridulum, 262
wetherillii, 273
zionis, 264
Eriodictylon, 170
  angustifolium, 170
Eriophorum, 368
  angustifolium, 368, 380
  chamissonis, 368
  gracile, 368
Eriophyllum, 117
  lanatum, 117
  lanosum, 117
  wallacei, 117
Erodium, 167
  cicutarium, 168
Erysimum, 149
  asperum, 149
  cheiranthoides, 149
  inconspicuum, 149
  repandum, 149
Erythronium, 403
  grandiflorum, 403, 405
Eschscholtzia, 245
  californica, 245
  glyptosperma, 245
  mexicana, 245
  minutiflora, 245
Eucrypta, 170
  micrantha, 170
Eupatorium, 117
  maculatum, 117
Euphorbia, 161
  albomarginata, 161
  cyparissias, 161
  dentata, 161
  esula, 161
  fendleri, 161
  glyptosperma, 161
  marginata, 161
  parryi, 161
  pulcherrima, 161
  robusta, 161
Eurotia, 84
  lanata, 84

Exochorda, 300
  racemosa, 300

Fagus, 162
  grandifolia, 162
  sylvatica, 162
Fairy Slipper, 408
Fagonia, 357
  californica, 357
Fallugia, 300
  paradoxa, 300
Falsemermaid, 221
Fame Flower, 281
Fendlera, 314
  rupicola, 314
Fendlerbush, 314
Fendlerella, 314
  utahensis, 314
Fern, Bladder, 9
  Cloak, 10
  Gold, 10
  Grape, 5
  Holly, 10
  Lady, 9
  Lance-leaved Grape, 6
  Lip, 9
  Little Grape, 5
  Maidenhair, 7
  Male, 10
  Matricary Grape, 6
  Mosquito, 11
  Northern Grape, 6
  Wood, 10
Fern Bush, 297
Fescue, 384
  Alpine, 386
  Blue, 385
  Sheep, 386
Festuca, 384
  arizonica, 386
  dasyclada, 385
  elatior, 385
  grayi, 385
  idahoensis, 386
  kingii, 385
  megalura, 384
  octoflora
    hirtella, 384
    octoflora, 384
  ovina, 385
    brachyphylla, 386
    ovina, 386
  pacifica, 385
  reflexa, 384
  rubra, 385
  sororia, 385
  subulata, 385
  thurberi, 385
Fiddleneck, 47
Figwort, 334
Filago, 117
  californica, 117
Fimbristylis, 368
  thermalis, 368

Index

Fir, 18
  Alpine, 18
  Balsam, 18
  Douglas, 21
  Giant, 18
  Needle, 18
  Nordman, 18
  White, 18
Firethorn, 303
Fireweed, 238
Flaveria, 117
  campestris, 118
Flax, 221
  False, 145
Fleabane, 115
Floerkea, 221
  proserpinacoides, 221
Flower-of-an-Hour, 226
Forestiera, 234
  neomexicana, 235
Forget-me-not, 62
Forsellesia, 78
  meionandra, 78
  nevadensis, 78
Forsythia, 235
  intermedia, 235
  ovata, 235
  suspensa, 235
  viridissima, 235
Four O'Clock, 232
Foxglove, 323
Foxtail, 376
  Marsh, 376
Fragaria, 300
  americana, 300
  chiloensis, 300
Fraxinus, 235
  americana, 235
  anomala, 235
  excelsior, 235
  holotricha, 235
  nigra, 235
  pennsylvanica, 235
    lanceolata, 235
  quadrangulata, 235
  velatina, 235
Fringe-Tree, 234
Fritillaria, 403
  atropurpurea, 404
  imperialis, 403
  pudica, 404
Fritillary, 403
Fuchsia, 239
  hybrida, 239
Funastrum, 42
  heterophyllum, 42
Fumaria, 164
  officinalis, 164
Fumitory, 164

Gaillardia, 118
  aristata, 118
  arizonica, 118
  flava, 118
  gracilis, 118
  parryi, 118
  pinnatifida, 118
  spathulata, 118
Galanthus, 362
  nivalis, 362
Galium, 305
  aparine, 306, 307
  bifolium, 306
  boreale, 306
  coloradoense, 305
  desereticum, 305
  hypotrichium, 305
  magnifolium, 305
  multiflorum, 305
  munzii, 305
  proliferum, 306
  scabriusculum, 305
  stellatum, 305
  trifidum, 306
  triflorum, 306
Galleta, 386
Gaultheria, 160
  humifusa, 160
Gaura, 239
  parviflora, 239
Gayophytum, 239
  nuttallii, 239
  racemosum, 239
  ramosissimum, 239
Gentian, 165
  Green, 167
Gentiana, 165
  affinis, 165
  barbellaia, 165
  calycosa, 165, 166
  forwoodii, 165
  fremontii, 165
  heterosepala, 165
  parryi, 165
  plebeia, 167
  romanzovii, 165
  strictiflora, 167
  tenella, 165
  tortuosa, 167
Geraea, 118
  canescens, 118
Geranium, 168
  atropurpureum, 168
  carolinianum, 168
  fremontii, 168
  Fremont's, 168
  marginale, 168
  parryi, 168
  Parry's, 168
  pusillum, 168
  richardsonii, 168
  Richardson's, 168
Germander, 184
Geum, 300
  macrophyllum, 300
  triflorum, 300
Gilia, 248
  aggregata, 249

Index

arizonica, 249
calcarea, 249
congesta, 248
depressa, 248
filiformis, 249
gunnisonii, 248
hutchinsifolia, 249
latifolia, 249
laxiflora, 249
leptomeria, 249
longiflora, 249
mcvickerae, 249
polycladon, 248
pumila, 248
scopulorum, 249
sinuata, 249
spicata, 248
stenothyrsa, 249
subnuda, 249
tenerrima, 249
Ginkgo, 17
  biloba, 17
Gladiolus, 395
  primulinus, 395
Glaux, 283
  maritima, 283
Gleditsia, 207
  triacanthos, 207
Globeflower, 292
Glyceria, 386
  grandis, 386
  striata, 386
Glycyrrhiza, 207
  lepidota, 207
Glyptopleura, 118
  marginata, 118
  setulosa, 118
Gnaphalium, 118
  chilense, 119
  macounii, 119
  palustre, 118
  uliginosum, 119
  wrightii, 119
Goatsbeard, 135
Goldenglow, 300
Goldenhead, 98
Goldenrod, 131
Goldenweed, 119
Goodyera, 410
  oblongifolia, 410
Goosefoot, 82
Gossypium, 226
  hirsutum, 226
Gourd, 156, 157
  Bottle, 157
Grama, 377
  Blue, 377
  Side-oats, 377
Grass, Alkali, 391
  American Manna, 386
  Annual Hair, 379
  Arrow, 398
  Barnyard, 381
  Bent, 376

  Blue. See Bluegrass
  Blue-eyed, 395
  Bristle, 391
  Brome, 377
  Brook, 379
  Bull, 383
  Canary, 388
  Cheat, 377
  Columbia Needle, 392
  Cord, 392
  Cotton, 368
  Crab, 381
  Creeping Love, 384
  Ditch, 412
  False Buffalo, 388
  Fluff, 393
  Fowl Manna, 386
  Goat, 375
  Green Bristle, 391
  Hair, 379
  Hard, 391
  Indian Rice, 388
  June, 387
  Lettermann's Needle, 392
  Longtongue Mutton, 390
  Love, 384
  Manna, 386
  Melic, 387
  Muhly, 387
    See also Muhly
  Mutton, 391
  Needle, 392
  Needle-and-Thread, 392
  New Mexican Feather, 392
  Northern Reed, 379
  Nut, 367
  Oat. See Oatgrass
  Onion, 387
  Orchard, 379
  Panic, 388
  Pine, 379
  Quack, 375
  Reed, 378
  Reed Canary, 388
  Ripgut, 378
  Rye. See Ryegrass
  Salt, 381
  Scratch, 387
  Scribner's Panic, 388
  Slender Hair, 381
  Slough, 377
  Smooth Crab, 381
  Stink, 384
  Switch, 388
  Tickle, 376
  Tufted Hair, 381
  Wheat. See Wheatgrass
  Witch, 388
Grass of Parnassus, 316
Gratiola, 323
  neglecta, 323
Grayia, 84
  brandegei, 84
  spinosa, 84

453

Index

Greasebush, 78
Greasewood, 84
Grindelia, 119
  aphanactis, 119
  fastigiata, 119
  laciniata, 119
  squarrosa, 119
Groundsel, 128
Gumweed, 119
Gutierrezia, 119
  microcephala, 119
  sarothrae
    pomariensis, 119
    sarothrae, 119
Gymnocladus, 207
  dioica, 208
Gymnosteris, 249
  parvula, 250
Gypsophila, 74
  elegans, 74
  paniculata, 74

Habenaria, 410
  dilatata, 410, 413
  hyperborea, 410
  obtusata, 410
  saccata, 410
  sparsiflora, 410
  unalascensis, 410
  viridis, 410
Habranthus, 362
  brachyandrus, 362
  robustus, 362
  texanus, 362
Hackelia, 59
  floribunda, 59
  jessicae, 59
  patens, 59
Haemanthus, 362
  albiflos, 362
  coccineus, 362
  katherinae, 362
Halimolobos, 149
  virgata, 150
Halogeton, 84
  glomeratus, 84
Haplopappus, 119
  acaulis, 119
  armerioides, 120
  lanceolatus, 120
  macronema, 120
  nuttallii, 120
  racemosus, 120
  rydbergii, 120
  spinulosus, 120
  watsonii, 120
Haw, Eastern, 299
Hawksbeard, 113
Hawkweed, 122
  Hawthorn, 299
  English, 299
Hazelnut, 44
Healall, 183
Heavenly Blue, 138

Hedeoma, 180
  drummondii, 180
Hedysarum, 208
  boreale, 208
Helianthella, 120
  microcephala, 120
  quinquenervis, 120
  uniflora, 120
Helianthus, 120
  annuus, 121
    annuus, 121
    lenticularis, 121
  anomalus, 121
  canus, 121
  deserticola, 121
  nuttallii, 120
  **peliolaris
    fallax,** 121
Heliotropium, 59
  convolvulaceum, 59
  curassavicum
    obovatum, 59
    oculatum, 59
Hellebore, False, 406
Helleborine, 410
Hemerocallis, 404
  fulva, 404
Hemlock, 21
  Canadian, 22
  Poison, 347
  Water, 347
Henbane, 337
Hens-and-Chickens, 139
Heracleum, 349
  lanatum, 349
Hermidium, 232
  alipes, 232
Heronsbill, 167
Hesperis, 150
  matronalis, 150
Hesperochiron, 170
  Dwarf, 170
  pumilus, 170
Hesperochloa kingii, 385
Heterocodon, 67
  rariflorum, 67
Heterotheca, 121
  grandiflora, 121
  subaxillaris, 121
  villosa, 121
Heuchera, 314
  parviflora, 315
  rubescens, 314
  sanguinea, 315
  versicolor, 315
Hibiscus, 226
  esculentus, 226
  militaris, 226
  syriacus, 226
  trionum, 226
Hickory, 177
  Shagbark, 177

Index

Hieracium, 122
  gracile, 122
  scouleri, 122
Hilaria, 386
  jamesii, 386
Hippurus, 168
  vulgaris, 168, 172
Hoffmanseggia, 208
  repens, 208
Hofmeistera, 122
  pluriseta, 122
Hollyhock, 226
  Wild, 226
Holodiscus, 300
  discolor, 300
  dumosus, 300
  microphyllus, 300
Honeysuckle, 69
  Blue Leaf, 69
  Bush, 69
  European Fly, 70
  Japanese, 69
  Morrow, 70
  Tatarian, 70
  Utah, 69
  Winter, 69
Hop, 231
  American, 231
Hopsage, 84
  Spiny, 84
Hop-Tree, 308
Hordeum, 386
  gussonianum, 386
  jubatum, 386, 389
  murinum, 386
  nodosum, 386
Horehound, 181
  Water, 180
Hornwort, 78
Horsebrush, 132, 133
  Spiny, 133
Horseradish, 153
Horsetail, Meadow, 4
Horseweed, 114
Houndstongue, 57
Houstonia, 306
  rubra, 306
Humulus, 231
  americanus, 231
Hutchinsia, 150
  procumbens, 150, 151
Hyacinth, 404
  Common, 404
  Grape, 406
  Wild, 402
Hyacinthus, 404
  orientalis, 404
Hydrangea, 315
  arborescens, 315
  opuloides, 315
  paniculata, 315
  Pee Gee, 315
  quercifolia, 315
Hydrophyllum, 170

capitatum
  alpinum, 171
  capitatum, 171
  fendleri, 171
  occidentalis, 171, 172
Hymenocallis, 363
  calathina, 363
  littoralis, 363
Hymenoclea, 122
  salsola, 122
Hymenopappus, 122
  filifolius, 122
Hymenoxys, 122
  acaulis, 122
  richardsonii, 122
Hyocyamus, 337
  niger, 337
Hypericum, 176
  formosum, 176
  perforatum, 176
  scouleri, 182
Hyssop, Giant, 179
  Hedge, 323

Iberis, 150
  sempervirens, 150
Iliamna, 226
  rivularis, 227
Impatiens, 42
  balsamina, 42
Indian Paintbrush, 322
Indigo, False, 188
  Wild, 207
Indigo Bush, 207
Ipomoea, 138
  batatas, 138
  purpurea, 138
  tricolor, 138
Iris, 395
  Dutch, 395
  hoogiana, 395
  missouriensis, 395
  pallida, 395
  pumila, 395
  siberica, 395
  Spanish, 395
  spuria, 395
  variegata, 395
  xiphium, 395
Isatis, 150
  tinctoria, 150
Isoëtes, 3
  bolanderi, 3
  echinospora, 3
  howellii, 3, 8
  lacustris, 3
Iva, 122
  axillaris, 122
  xanthifolia, 122
Ivesia, 300
  gordonii, 300
  kingii, 300
  utahensis, 300
Ivy, Poison, 39

455

Index

Jamesia, 315
  americana, 315
Jasmine, Rock, 282
Jimsonweed, 337
Joe Pye Weed, 117
Jonquil, 363
Joshua Tree, 406
Juglans, 177
  cinerea, 177
  nigra, 177
  regia, 177
Juncus, 396
  albescens, 398
  balticus, 396
  bufonius, 397
  confusus, 397
  cooperi, 396
  drummondii, 397
  ensifolius, 398
  filiformis, 396
  gerardii, 397
  hallii, 397
  longistylis, 397
  mertensianus, 398
  nodosus, 398
  orthophyllus, 397
  parryi, 397
  regelii, 397
  saximontanus, 398
  sphaerocarpus, 397
  tenuis, 397
  torreyi, 398
  tracyi, 398
  tweedyi, 398
  uncialis, 397
  xiphoides, 398
Juniper, 15
  Chinese, 15, 16
  Common, 15
  Creeping, 15
  Greek, 16
  Meyer, 15
  Needle, 15
  Pfitzer, 15
  Rocky Mountain, 16
  Tam, 15
  Utah, 16
  Virginia, 15, 16
Juniperus, 15
  chinensis, 15, 16
    pfitzeriana, 15
  communis, 15, 20
  excelsa, 16
  horizontalis, 15
  osteosperma, 16
  rigida, 15
  sabina
    tamariscifolia, 15
  scopulorum, 16
  squamata
    meyeri, 15
  virginiana, 15, 16

Kalanchoe, 139
  pinnata, 139
Kalmia, 160
  Bog, 160
  polifolia, 160
Kelloggia, 306
  galioides, 306
Kerria, 300
  japonica, 300
Kings Crown, 139
Kitten-tails, 322
Knapweed, 108
  Russian, 108
Knautia, 158
  arvensis, 158
Knotweed, 276
Kobresia, 368
  bellardii, 368
Kochia, 84
  americana, 84
  scoparia, 84
Koelaria, 387
  cristata, 387
Kolkwitzia, 69
  amabilis, 69
Krameria, 177
  grayi, 177
  parviflora, 177

Lace-Vine, Silver, 276
Lactuca, 122
  pulchella, 122
  scariola, 123, 129
Ladies' Tresses, 411
Lady's Slipper, 408
  Clustered, 408
  Yellow, 408
Lagenaria, 157
  siceraria, 157
Lamium, 180
  amplexicaule, 180
  purpureum, 180
Lamphamia, 123
  palmeri, 123
  stansburii, 123
Langloisia, 250
  schottii, 250
  setosissima, 250
Lappula, 59
  echinata, 59
  occidentalis
    cupulata, 59
    occidentalis, 59
Larch, 18
  European, 18
  Western, 18
Larix, 18
  decidua, 18
  occidentalis, 18
Larkspur, 289
  Desert, 289
  Low, 289
  Rocket, 289
  Tall, 289
Larrea, 357

Index

tridentata, 357
Lathyrus, 208
  brachycalyx
    brachycalyx, 208
  eucosmus, 209
  zionis, 209
  lanzwertii, 208
  latifolius, 208
  odoratus, 208
  pauciflorus,
    utahensis, 208
  rigidus, 209
  sylvestris, 208
  utahensis, 209
  wrightii, 209
Lavandula, 180
  officinalis, 180
Lavender, 180
Layia, 123
  grandulosa, 123
Leather Flower, Hairy, 289
Lemna, 399
  minor, 399, 405
  trisulca, 399
Leonurus, 180
  cardiaca, 180
Lepidium, 150
  campestre, 150
  densiflorum, 150, 155
  dictyotum, 150
  fremontii, 150
  montanum, 150
  perfoliatum, 150
  virginicum, 150
Leptodactylon, 250
  caespitosum, 250
  nuttallii, 250
  pungens, 250
  watsonii, 250
Lesquerella, 150
  alpina, 152
  fendleri, 150
  garrettii, 152
  intermedia, 152
  ludoviciana, 152
  multiceps, 152
  palmeri, 152
  rectipes, 152
  subumbellata, 152
  utahensis, 152
  wardii, 152
Lettuce, 122
  Blue, 122
  Indian, 279
  Lamb's, 354
  Miners, 279
  Prickly, 123
Leucocrinum, 404
  montanum, 404
Leucojum, 363
  acstivum, 363
  vernum, 363
Lewisia, 279
  brachycalyx, 279
  pygmaea, 279
  rediviva, 279
  triphylla, 279
Libocedrus, 16
  decurrens, 16
Licorice, 207
Ligusticum, 349
  filicinum, 349
  porteri, 349
Ligustrum, 235
  amurense, 236
  obtusifolium, 236
  ovalifolium, 236
  vulgare, 236
Lilac, 236
  Chinese, 237
  Common, 236
  Hungarian, 236
  Late, 236
  Mountain, 293
  Persian, 236
  Tree, 236
Lilium, 404
  bulbiferum, 404
  candidum, 404
  concolor, 404
  dauricum, 404
  longiflorum, 404
  regale, 404
  speciosum, 404
  tigrinum, 404
Lily, 404
  Alp, 404
  Blood, 362
  Candlestick, 404
  Crinum, 362
  Day, 404
  Easter, 404
  Funnel, 402
  Jacobean, 363
  Kaffir, 361
  Leopard, 404
  Madonna, 404
  Orange Day, 404
  Prairie, 362
  Rain, 362
  Regal, 404
  Sego, 403
  Spider, 363
  Star, 404
  Tiger, 404
  Water, 234
  Yellow Pond, 233
  Zephyr, 363
Lily-of-the-Valley, 403
Limosella, 323
  aquatica, 323
Linanthastrum, 250
Linanthus, 250
  bigelovii, 250
  demissus, 250
  harknessii, 250
  liniflorus, 250
  septentrionalis, 250

Index

Linaria, 323
  canadensis, 324
  dalmatica, 324
  texensis, 324
  vulgaris, 324
Linnaea, 69
  borealis, 69, 76
Linum, 221
  aristatum, 221
  kingii, 221
  perenne, 221, 228
  puberulum, 221
  usitatissimum, 221
Liriodendron, 224
  tulipifera, 224
Listera, 411
  borealis, 411
  convallarioides, 411
Lithophragma, 315
  bulbifera, 315
  parviflora, 315
  tenella, 315
Lithospermum, 59
  arvense, 60
  incisum, 60
  multiflorum, 60
  ruderale, 60
Lloydia, 404
  serotina, 405, 406
Lobelia, 67
  erinus, 67
Lobularia, 152
  maritima, 152
Locoweed, 214
Locust, Clammy, 218
  Black, 218
  Honey, 207
  New Mexico, 218
Lolium, 387
  multiflorum, 387, 389
  perenne, 387
Lomatium, 349
  ambiguum, 349
  dissectum, 350
  donnellii, 350
  grayi, 350
  juniperinum, 350
  latilobum, 350
  leptocarpum, 349
  macdougalii, 350
  minimum, 350
  nevadense, 350
  nudicaule, 349
  nuttallii, 350
  parryi, 350
  scabrum, 350
  simplex, 350
  triternatum, 350
Lonicera, 69
  fragrantissima, 69
  involucrata, 69, 76
  japonica, 69
  korolkowii, 69
  morrowii, 70
  tatarica, 70
  utahensis, 69
  xylostema, 70
Loosestrife, 284
Lotus, 209
  corniculatus, 209
Lousewort, 325
Lovage, 349
Lunaria, 152
  annua, 152
Lupine, 209
  Bajada, 209
  Burke's, 211
  King's, 210
  Lodge Pole, 211
  Longspur, 212
  Many Flowered, 213
  Mountain, 213
  Palmer's, 213
  Redstem, 211
  Rusty, 210
  Sand, 211
  Short-stem, 209
  Silky, 213
  Silvery, 212
  Spathulate, 211
  Stemless, 214
  Tailcup, 212
  Velvet, 213
  Wyeth's, 213
  Yelloweye, 210
Lupinus, 209
  alpestris, 213
  ammophilus, 211
  amplus, 211
  arbustus
    calcaratus, 212
  arcticus
    tetonensis, 210
  argenteus
    argenteus, 212
    stenophyllus, 213
    tenellus, 212
  bakeri, 212
  barbiger, 212
  brevicaulis, 209
  burkei, 211
  caespitosus, 214
  caudatus
    argophyllus, 212
    caudatus, 212
  concinnus
    orcuttii, 209
  evermannii, 213
  flavoculatus, 210
  floribundus, 211, 213
  hillii, 213
  holosericeus, 213
  ingratus, 212
  kingii, 210
    argillaceus, 210
  leucophyllus, 213
    tenuispicus, 212
  lyallii

Index

subpandens, 214
maculatus, 211
marianus, 210
palmeri, 213
parviflorus, 211
pusillus
  intermontanus, 210
  pusillus, 210
  rubens, 210
roseolus, 213
rubricaulis, 211, 212
sericeus, 213
spathulatus, 211
uncialis, 209
volutans, 214
wyethii, 213
Luzula, 398
multiflora, 398
spicata, 398
wahlenbergii, 398
Lychnis, 74
alba, 74
apetala, 75
chalcedonica, 74
drummondii, 75
kingii, 75
Lycium, 337
andersonii, 338
halimifolium, 338
pallidum, 338
torreyi, 338
Lycopersicon, 338
esculenta, 338
Lycopus, 180
americanus, 180
lucidus, 180
Lycoris, 363
squamigera, 363
Lygodesmia, 123
grandiflora, 123
juncea, 123
spinosa, 123
Lysimachia, 283
thyrsiflora, 283, 285

Machaeranthera, 123
bigelovii, 124
canescens, 124
commixta, 124
glabriuscula, 125
kingii, 125
linearis, 124
mucronata, 124
parviflora, 124
tagetina, 125
tanacetifolia, 124
tephrodes, 124
tortifolia, 125
venusta, 125
Maclura, 231
pomifera, 231
Madder, 306
Madia, 125
glomerata, 125

Madwort, 143, 323
Magnolia, 224
acuminata, 224
grandiflora, 224
Showy, 224
soulangeana, 224
Star, 224
stellata, 224
virginiana, 224
Mahoberberis, 43
neubertii, 43
Mahonia, 43
aquifolium, 44
bealei, 44
Fremont, 43
fremontii, 43
Holly, 44
repens, 43
Malacothrix, 125
coulteri, 125
glabrata, 125
sonchioides, 126
torreyi, 126
Malcolmia, 152
africana, 152
Mallow, 227
False, 227
Globe, 227
Malus, 300
floribunda, 301
ioensis, 300
sylvestris, 301
Malva, 227
neglecta, 227, 228
Malvastrum, 227
exile, 227
Mamillaria, 64
tetrancistra, 64
Manzanita, 159
Maple, 36
Big Tooth, 37
Devil's, 37
Flowering, 226
Hedge, 36
Japanese, 36
Norway, 37
Red, 37
Rocky Mountain, 36, 37
Silver, 36
Sugar, 37
Sycamore, 37
Vine, 37
Marestail, 168
Marigold, Marsh, 288
Wild, 105
Marijuana, 230
Marrubium, 181
vulgare, 181, 182
Marsilea, 5
vestita, 5
Matricaria, 126
matricarioides, 126
Matricary, 126
Matrimony Vine, 337, 338

459

Index

Matthiola, 152
  bicornis, 153
May Weed, 101
Medic, Black, 214
Medicago, 214
  lupulina, 214
  sativa, 214
Melia, 230
  azedarach, 230
Melica, 387
  bulbosa, 387
Melilotus, 214
  alba, 214
  officinalis, 214, 216
Menodora, 236
  scabra, 236
Mentha, 181
  arvensis, 181, 184
  citrata, 181
  piperita, 181
  spicata, 181
Mentzelia, 221
  albicaulis, 222
  argillosa, 222
  brevicaulis, 222
  dispersa, 222
  humilis, 222
  integra, 222
  longiloba, 222
  multiflora, 222
  nitens, 222
  pterosperma, 222
  pumila, 222
  tricuspis, 222
Mertensia, 60
  arizonica
    arizonica, 60
    leonardii, 61
    subnuda, 60
  bakeri, 61
  brevistyla, 61
  ciliata, 61
  franciscana, 60
  fusiformis, 61
  oblongifolia
    amoena, 62
    nevadensis, 61
    oblongifolia, 62
  viridis, 61
    cana, 61
    dilatata, 61
    viridis, 61
Mesquite, 215
  Vine, 388
Metasequoia, 22
  glyptostroboides, 22
Microseris, 126
  nutans, 126, 129
Microsteris, 250
  gracilis, 250, 252
Mignonette, 293
Milkvetch, 188
Milkweed, 40
Milkwort, 253

Mimulus, 324
  bigelovii, 324
  cardinalis, 324
  eastwoodiae, 324
  floribundus, 325
  glabratus, 324
  guttatus, 324, 336
  lewisii, 324
  parryi, 324
  pilosus, 324
  primuloides, 325
  rubellus, 325
  suksdorfii, 325
  tilingii, 324
Mint, 181
  Bergamot, 181
  Horse, 181
  Pepper, 181
  Spear, 181
Mirabilis, 232
  bigelovii, 232
  jalapa, 232
  multiflora, 233
  oxybaphoides, 233
Mistletoe, 223
  Dwarf, 223
Mitella, 315
  pentandra, 315
  stauropetala, 316
  stenopetala, 316
Mitrewort, 315
Moldavica, 181
  parviflora, 181, 184
Molucella, 181
  laevis, 181
Monarda, 181
  fistulosa, 181
  pectinata, 181
Monardella, 181
  odoratissima, 181
Moneses, 286
  uniflora, 286, 298
Monkeyflower, 324
Monkshood, 287
Monolepis, 84
  nuttallianus, 84, 85
Montia, 279
  chamissoi, 279, 285
  cordifolia, 281
  perfoliata, 281
  sibirica, 281
  spathulata, 281
Moonwort, 5, 6, 152
Morning Glory, 138
Mortonia, 78
  utahensis, 78
Morus, 231
  alba, 231
  nigra, 231
  rubra, 231
Motherwort, 180
Mountain-balm, 170
Mountain Lover, 78
Mountain Mahogany, 297

Index

Alder-leaf, 297
Curl-leaf, 297
Mousetail, 289
Muhlenbergia, 387
  andina, 387
  asperifolia, 387
  curtifolia, 387
  filiformis, 388
  montana, 388
  pungens, 387
  racemosa, 387
  squarrosa, 387
Muhly, Foxtail, 387
  Marsh, 387
  Mat, 387
  Mountain, 388
  Pungent, 387
  Short leaf, 387
  Slender, 388
Mulberry, 231
  Black, 231
  Red, 231
  White, 231
Mules-ears, 136
Mullein, 334
Munroa, 388
  squarrosa, 388
Muscari, 406
  botryoides, 406
Musineon, 351
  divaricatum, 351
  lineare, 351
Muskmelon, 156
Mustard, 145
  African, 152
  Black, 145
  Field, 145
  Haresear, 147
  Leaf, 145
  Stinking, 68
  Tansy, 147
  Treacle, 149
Myosotis, 62
  scorpioides, 62
Myosurus, 289
  minimus, 289
Myriophyllum, 169
  spicatum, 172
    exalbescens, 169
  verticillatum, 169

Najas, 407
  guadalupensis, 407
  marina, 407
Nama, 171
  demissum, 171
  densum
    parviflorum, 171
  hispidum, 171
  retrorsum, 171
Nandina, 44
  domestica, 44
Narcissus, 363
  jonquilla, 363

  poeticus, 363
  Poets', 363
  Polyanthus, 363
  pseudo-narcissus, 363
  tazetta, 363
Nasturtium, Garden, 340
Navarretia, 251
  breweri, 251
  minima, 251
Nemacladus, 67
  glanduliferus, 67
Nemophila, 171
  breviflora
    austinae, 171
    breviflora, 171
Nepeta, 181
  cataria, 181, 184
Nettle, 353
  Dead, 180
  White Horse, 339
Nicotiana, 338
  attenuata, 338
  tabacum, 338
  trigonophylla, 338
Nightshade, 339
  Black, 340
  Enchanters, 238
Ninebark, 301
Notholaena, 10
  fendleri, 10
  jonesii, 10
  limitanea, 10
  parryi, 10
Nuphar, 233
  polysepalum, 233, 244
Nymphaea, 234
  odorata, 234

Oak, 162
  Bur, 162
  Chestnut, 162
  English, 163
  Gambel, 163
  Native Live, 162
  Pin, 163
  Red, 163
  Scarlet, 163
  Swamp White, 163
  Wavy Leaf, 163
  White, 163
Oatgrass, 377, 379
  One-spike, 379
  Tall, 377
  Timber, 379
Oats, 377
  Wild, 377
Ocimum, 181
  basilicum, 181
Oenothera, 239
  albicaulis, 240
  andina, 241
  alyssoides, 242
  boothii, 242
  breviflora, 240

461

Index

caespitosa, 240
chamaenerioides, 242
clavaeformis, 241
contorta, 242
coronopifolia, 240
decorticans, 242
deltoides, 240
flava, 239
heteranthera, 241
hookeri, 240
latifolia, 240
lavandulaefolia, 241
longissima, 239
minor, 242
multijuga, 241
pallida, 240
pallidula, 241
parryi, 241
primiveris, 240
pterosperma, 241
refracta, 242
rydbergii, 240
scapoidea, 241
trichocalyx, 240
Okra, 226
Oleaster, 158
Olive, Russian, 158
Onion, 401
Onobrychis, 214
  viciaefolia, 214
Onopardum, 126
  acanthium, 126
Opuntia, 64
  acanthocarpa, 65
  aurea, 65
  basilaris, 65
  compressa, 65
  covillei, 65
  echinocarpa, 64
  erinacea, 65
    xanthostema, 65
  fragilis, 65
  nicholii, 65
  phaeacantha, 65
  polyacantha, 65
  whipplei, 64
Orange, Common Mock, 316
  Mock, 316
  Osage, 231
Orchid, Bog, 410
Oregon Grape, 43
Oreoxis, 351
  alpina, 351
  bakeri, 351
Ornithogalum, 406
  umbellatum, 406
Orobanche, 243
  cooperi, 243
  corymbosa, 243
  fasciculata, 243, 244
  multiflora, 243
  uniflora, 243
Orogenia, 351
  linearifolia, 351

Orthocarpus, 325
  luteus, 325
  purpureo-albus, 325
  tolmei, 325
Oryzopis, 388
  hymenoides, 388
Osier, Golden, 310
Osmanthus, 236
  fragrans, 236
Osmorhiza, 351
  chilensis, 351, 353
  depauperata, 351
  occidentalis, 351
Oxalis, 243
  corniculata, 243
  stricta, 243
Oxybaphus, 233
  glaber, 233
  linearis, 233
  pumilus, 233
Oxypolis, 351
  fendleri, 351
Oxyria, 275
  digyna, 275, 280
Oxytenia, 126
  acerosa, 126
Oxytheca, 276
  perfoliata, 276
Oxytropis, 214
  campestris, 215
  deflexa, 214
  jonesii, 215
  lambertii, 215
  multiceps, 214
  obnapiformis, 215
  oreophila, 215
  parryi, 215
  sericea, 215
  viscida, 214

Pachystima, 78
  myrsinites, 78
Paeonia, 289
  albiflora, 289
  anomala, 289
  brownii, 289
Pagoda Tree, Japanese, 218
Palafoxia, 126
  linearis, 216
Palm, Sago, 16
Panicum, 388
  capillare, 388
  huachucae, 388
  obtusum, 388
  scribnerianum, 388
  virgatum, 388
Pansy, 356
Papaver, 245
  nudicaule, 245
  orientale, 246
  rhoeas, 246
Paperflower, 127
Parietaria, 353
  obtusa, 353

Index

pennsylvanica, 353
Parnassia, 316
  fimbriata, 316
  palustris, 316
  parviflora, 316
Parrya, 153
  platycarpa, 153
Parryella, 215
  filifolia, 215
Parsley, 352
  Desert, 349, 352
Parsnip, 351
  Cow, 349
  Water, 346, 352
Parthenium, 126
  alpinum
    ligulatum, 126
Pastinaca, 351
  sativa, 351
  sylvestris, 351
Pea, 215
  Golden, 219
  Perennial Sweet, 208
  Rush, 208
  Scurf, 217
  Sweet, 208
  Wild, 208
Peach, 302
  Desert, 302
Pear, 303
  Prickly, 64
Pearlbush, 300
Pearlwort, 75
Pearly Everlasting, 99
Pea Tree, 207
Pecan, 177
Pectis, 126
  papposa, 127
Pectocarya, 62
  heterocarpa, 62
  platycarpa, 62
  recurvata, 62
  setosa, 62
Pedicularis, 325
  centranthera, 325
  crenulata, 325
  grayi, 326
  groenlandica, 325
  parryi, 326
  paysoniana, 326
  racemosa, 325
Pediocactus, 65
  simpsonii, 65
Pelargonium, 168
  hortorum, 168
Pellaea, 10
  breweri, 10
  longimucronata, 10
  suksdorfiana, 10
Pellitory, 353
Pennycress, 156
Pennyroyal, Mock, 180
Penstemon, 326
  abietinus, 331
  acaulis, 331
  ambiguus, 331
  angustifolia, 334
  arenicola, 334
  barbatus, 326
  bridgesii, 326
  caespitosus, 331
  cleburnei, 330
  cobaea, 326
  comarrhenus, 327
  concinnus, 329
  confusus, 332
  crandallii, 331
  cyananthus, 327
  cyanocaulis, 327
  deustus, 329
  dolius, 333
  eatonii, 326
  floridus, 330
  fremontii, 327
  garrettii, 328
  glaber, 327
  grahamii, 330
  grandiflorus, 326, 333
  hallii, 333
  hartwegii x cobaea, 326
  humilis, 332
  jamesii, 330, 333
  kingii, 328
  laevis, 329
  leiophyllus, 332
  lentus, 333
  leonardii, 328
  linarioides, 331
  moffatii, 333
  montanus, 328
  nanus, 331
  newberryi, 328
  osterhoutii, 334
  ovatus, 330
  pachyphyllus, 334
  palmeri, 330
  parvus, 333
  petiolatus, 329
  platyphyllus, 329
  procerus, 332, 333
  radicosus, 332
  rupicola, 328
  rydbergii, 332
  scariosus, 328
  sepalulus, 328
  speciosus, 329
  spectabilis, 330
  strictus, 329
  subglaber, 327
  thompsoniae, 331
  tidestromii, 329
  uintahensis, 327
  utahensis, 332
  virgatus, 332
  wardii, 329
  watsonii, 332, 333
  whippleanus, 330, 332
Peony, 289

Index

Western, 289
Pepper, 337
Peppergrass, 150
Pepperwort, 5
Peraphyllum, 301
   ramosissimum, 301
Perezia, 127
   wrightii, 127
Perfume-Plant, 153
Perideridia, 351
   bolanderi, 352
   gairdneri, 351
   punctatum, 277
Petalostemon, 215
   flavescens, 215
   occidentale, 215
   searlsiae, 215
Peteria, 215
   thompsonae, 215
Petradoria, 127
   pumila, 127, 131
Petroselinium, 352
   crispum, 352
Petunia, 338
   hybrida, 338
Phacelia, 171
   affinis, 173
   alba, 173
   ambigua
      ambigua, 174
   anelsonii, 173
   cephalotes, 175
   coerulea, 173
   constancei, 173
   corrugata, 174
   crenulata, 174
   curvipes, 175
   demissa
      demissa, 175
      heterotricha, 175
   fremontii, 173
   glandulosa
      argillacea, 174
   hastata
      hastata, 175
   heterophylla, 182
      heterophylla, 175
   howelliana, 174
   incana, 176
   indecora, 175
   integrifolia
      integrifolia, 174
   ivesiana, 173
   linearis, 175, 182
   mammalariensis, 174
   palmeri, 173
   parshii, 175
   peirsoniana, 176
   pulchella, 175
   rafaelensis, 173
   rotundifolia, 176
   salina, 173
   saxicola, 176
   scopulina, 173
   serecia, 175
   splendens, 174
   utahensis, 174
   vallis-mortae, 175
Phalaris, 388
   arundinacea, 388, 389
Phaseolus, 215
   vulgaris, 215
Pheasants Eye, 287
Philadelphus, 316
   coronarius, 316
   incanus, 316
   inodorus, 316
   lemoinei, 316
   lewisii, 316
   microphyllus, 316
   virginalis, 316
Phleum, 388
   alpinum, 388, 389
   pratense, 388
Phlox, 251
   austromontana, 251
   bryoides, 251
   caespitosa, 251
   gladiformis, 251
   grahamii, 251
   hoodii, 251
   longifolia, 251
   multiflora, 251
   rigida, 251
   stansburyi, 251
Phoradendron, 223
   californicum, 223
   juniperinum, 224
Phragmites, 390
   communis, 390
Physalis, 338
   crassifolia, 339
   fendleri, 338
   hederaefolia, 339
   heterophylla, 339
   longifolia, 339
Physaria, 153
   australis, 153
   chambersii, 153
   grahamii, 153
   newberryi, 153
Physocarpus, 301
   alternans, 301
   capitatus, 301
   malvaceus, 301
   monogynus, 301
Picea, 18
   abies, 19
   engelmannii, 19
   glauca, 19
   koyamai, 19
   mariana, 19
   omorika, 19
   orientalis, 19
   pungens, 19
   sitchensis, 19
Pickleweed, 80
Pincushion-Flower, 158

Index

Pine, 19
  Austrian, 21
  Bristle-cone, 19
  Chinese, 21
  Himalayan, 19
  Jack, 21
  Japanese Red, 21
  Jeffrey, 21
  Limber, 19
  Lodgepole, 21
  Mugo, 21
  New Caledonian, 14
  Norwalk Island, 14
  Pinyon, 21
  Ponderosa, 21
  Red, 21
  Scots, 21
  Umbrella, 23
  White, 19
Pineapple Weed, 126
Pinedrops, 286
Pink, 74
  Garden, 74
  Rush, 123
  Slender Rush, 123
  Spring Rush, 123
Pinus, 19
  aristata, 19
  banksiana, 21
  contorta, 20, 21
  densiflora, 21
  edulis, 21
  flexilis, 19
  griffithii, 19
  jeffreyi, 21
  monophylla, 19
  mugo, 21
  nigra, 21
  ponderosa, 21
  resinosa, 21
  strobus, 19
  sylvestris, 21
  tabulaeformis, 21
Pinyon, Single-leaf, 19
Pipsissewa, 284
Pisum, 215
  sativum, 215, 216
Pityrogramma, 10
  triangularis, 10
Plagiobothrys, 62
  arizonicus, 63
  jonesii, 62
  leptocladus, 62
  scouleri, 63
  tenellus, 63
Plantago, 246
  elongata, 246
  eriopoda, 246
  insularis, 246
  lanceolata, 246, 252
  major, 246
  purshii, 246
  tweedyi, 246
Plantain, 246

Rattlesnake, 410
  Water, 360
Platycodon, 67
  grandiflorum, 67
Platystemon, 246
  californicus, 246
Plectritis, 353
  macrocera, 354
Pluchea, 127
  sericea, 127
Plum, 303
  Flowering, 302
Plumyew, 14
  Japanese, 14
Poa, 390
  alpina, 391, 394
  ampla, 391
  annua, 390
  bulbosa, 390
  curta, 390
  compressa, 390
  fendleriana, 391
  longiligula, 390
  nervosa, 390
  nevadensis, 391
  palustris, 390
  pratensis, 390
  reflexa, 390
  secunda, 391
Podistera, 352
  eastwoodiae, 352
Poinsettia, 161
Polanisia, 68
  trachysperma, 68
Polemonium, 251
  coeruleum, 252, 253
  delicatum, 253
  foliosissimum, 253
  micranthum, 253
  pulcherrimum, 253
  viscosum, 251
Poliomintha, 183
  incana, 183
Polygala, 253
  acanthoclada, 253
  subspinosa, 253
Polygonum, 276
  amphibium, 277
  aubertii, 276
  aviculare, 276, 280
  bistortoides, 277
  coccineum, 277
  convolvulus, 276
  douglasii, 276
  engelmannii, 276
  hydropiper, 277
  kelloggii, 276
  lapathifolium, 277
  minimum, 276
  persicaria, 277
  ramosissimum, 276
  sawatchense, 276
  viviparum, 277
  watsonii, 276

Index

Polypodium, 10
  hesperium, 10
Polypody, 10
Polypogon, 391
  monspeliensis, 391, 394
Polystichum, 10
  lonchitis, 11
  scopulinum, 11
Pondweed, 411
  Horned, 416
Popcornflower, 62
Poplar, 308
  Bolleana, 309
  Carolina, 308
  Chinese, 309
  Lombardy, 308
  White, 309
Poppy, 245
  California, 245
  Corn, 246
  Desert, 245
  Field, 246
  Oriental, 246
  Prickly, 245
Poppymallow, 226
Populus, 308
  x **acuminata**, 309
  alba, 309
    bolleana, 309
  angustifolia, 309
  candicans, 309
  deltoides, 308
  fremontii, 309
  nigra
    italica, 308
  simonii, 309
  tremuloides, 308
Portulaca, 281
  grandiflora, 281
  oleracea, 281
  retusa, 281
Potamogeton, 411
  alpinus, 412
  filiformis, 411
  foliosus, 412
  gramineus, 412
  illinoensis, 411
  latifolius, 411
  nodosus, 412
  nutans, 412, 413
  pectinatus, 411
  praelongus, 412
  pusillus, 412
  richardsonii, 412
  robbinsii, 411
  strictifolius, 412
  zosteriformis, 412
Potato, 339
  Indian, 351
  Sweet, 138
  Wild, 339
Potentilla, 301
  anserina, 301
  arguta, 302

  biennis, 301
  diversifolia, 301
  fissa, 302
  fruticosa, 301
  glandulosa, 302
  gracilis, 301
  monspeliensis, 301
  ovina, 302
  pectinisecta, 301
  wyomingensis, 302
Poverty Weed, 84
Primrose, 283
  Cave, 284
  Evening, 239
Primula, 283
  incana, 283
  maguirei, 284
  parryi, 283
  specuicola, 283
Prince's Plume, 154
Privet, 235
  California, 236
  Common, 236
Proboscidea, 230
  louisianica, 230
Prosopis, 215
  glandulosa, 217
  pubescens, 215
Prune, 303
Prunella, 183
  vulgaris, 183, 184
Prunus, 302
  americana, 303
  amygdalus, 302
  armeniaca, 302
  avium, 303
  ceracifera, 302
  cerasus, 302
  domestica, 303
  fasciculata, 302
  padus, 303
  pennsylvanica, 302
  persica, 302
  serrulata, 302
  triloba, 302
  virginiana, 307
    melanocarpa, 303
Psathyrotes, 127
  annua, 127
  pilifera, 127
  ramosissima, 127
Pseudocymopterus, 352
  montanus, 352
Pseudotsuga, 21
  menziesii, 21
Psilostrophe, 127
  bakeri, 127
  cooperi, 127
  sparsiflora, 128
Psoralea, 217
  aromatica, 217
  castorea, 217
  epipsila, 217
  juncea, 218

Index

lanceolata
  lanceolata, 218
  purshii, 218
  megalantha, 217
  mephitica, 217
tenuiflora
  bigelovii, 218
  floribunda, 218
Ptelea, 308
  baldwinii, 308
  trifoliata, 308
Pteridium, 11
  aquilinum, 11, 12
Pterospora, 286
  andromedea, 286
Pterostegia, 277
  drymaroides, 277
Pteryxia, 352
  terbinthina, 352
Puccinellia, 391
  airoides, 391
Puccoon, 59
Pumpkin, 157
Puncture Vine, 357
Purshia, 303
  tridentata, 303
Purslane, 281
  Sea, 37
Pusley, 281
Pussy Paws, 281
Pussytoes, 99
Pyracantha, 303
  coccinea, 303
Pyrola, 286
  asarifolia, 286
  chlorantha, 286
  minor, 286
  picta, 286
  secunda, 286
Pyrus, 303
  communis, 303

Quercus, 162
  alba, 163
  bicolor, 163
  borealis, 163
  coccinea, 163
  gambelii, 163
  macrocarpa, 163
  palustris, 163
  robur, 163
  turbinella, 162
  x undulata, 163
  variabilis, 162
Quillwort, 3
Quince, 299
  Japanese, 297

Rabbitbrush, 109
Rabbitfoot, 391
Rafinesquia, 128
  californica, 128
  neomexicana, 128
Ragweed, 98

Ranunculus, 290
  acriformis, 292
  acris, 292
  adoneus, 292
  aquatilis, 290, 298
  arvensis, 291
  cardiophyllus, 292
  circinatus, 290
  cymbalaria, 291
  eschscholtzii, 292
  flabellaris, 291
  glaberrimus, 291
  gmelinii, 291
  inamoenus, 292
  jovis, 292
  juniperinus, 290
  longirostris, 290
  pedatifidus, 292
  ranunculinus, 291
  repens, 292
  testiculatus, 290
Raspberry, Red, 304
  Western Black, 304
  Western Red, 304
Ratany, 177
Rattleweed, 334
Redbud, 207
  Eastern, 207
Redtop, 376
  Ross, 376
  Spike, 376
  Thurber, 376
Redwood, Dawn, 22
Reed, Bur, 414
  Common, 390
Reseda, 293
  lutea, 293
Rhamnus, 294
  betulaefolia, 294
  cathartica, 294
  smithii, 294
Rheum, 277
  rhaponticum, 277
Rhinanthus, 334
  crista-galli, 334
Rhubarb, 277
Rhus, 39
  glabra, 39
  trilobata, 39
Ribes, 316
  aureum, 318, 336
  cereum, 318
  coloradense, 318
  grossularia, 317
  hudsonianum, 317
  inerme, 317
  lacustre, 317
  leptanthum, 317
  mogollonicum, 317
  montigenum, 317
  nigrum, 317
  sativum, 317
  setosum, 317
  velutinum, 317

Index

viscossissmum, 318
Ricinus, 162
  communis, 162
Rigiopappus, 128
  leptocladus, 128
Robinia, 218
  hispida, 218
  neomexicana, 218
  pseudoacacia, 218
  viscosa, 218
Rockbrake, 9
Rockcress, 143
Rocket, 150
  Prairie, 149
  Sand, 147
Roemeria, 246
  refracta, 246
Rorippa, 153
  armoracia, 153
  curvisiliqua, 153
  islandica, 153, 155
  nasturtium-aquaticum, 153
  obtusa, 153
  sinuata, 153
  sphaerocarpa, 153
Rosa, 303
  multiflora, 303
  nutkana, 303
  odorata, 303
  woodsii, 303
Rose, 303
  Cliff, 299
  Tea, 303
Rose Crown, 139
Rosemallow, 226
Rosmarinus, 183
  officinalis, 183
Rosemary, 183
Rose of Sharon, 226
Rubia, 306
  tinctoria, 306
Rubus, 303
  idaeus, 304
  leucodermis, 304
  occidentalis, 304
  parviflorus, 303
  strigosis, 304
Rudbeckia, 128
  hirta, 128
  laciniata, 128
  occidentalis, 128
Rue, Meadow, 292
Rumex, 277
  acetosella, 277, 280
  crispus, 278
  fueginus, 278
  hymenosepalus, 278
  obtusifolius, 278
  occidentalis, 278
  patientia, 278
  paucifolius, 278
  triangulivalvis, 278
  venosus, 278
Ruppia, 412

  maritima, 412, 415
Rush, 396
  Bul, 368
  Spike, 367
  Wood, 398
Rye, 391. See also Wildrye
Ryegrass, 387
  Beardless, 383
  Gray, 383
  Italian, 387
  Pacific, 384
  Perennial, 387
  Sand, 381

Sacaton, Alkali, 392
Safflower, 106
Sage, 183
  Desert, 183
  Garden, 183
  Purple, 183
Sagebrush, 102
  Black, 103
  Big, 103
  Bud, 103
  Old Man, 103
Sagina, 75
  saginoides, 75
Sagittaria, 360
  cuneata, 360, 369
  latifolia, 360
Sainfoin, 214
St. Johnswort, 176
Salicornia, 84
  pacifica, 84, 100
  rubra, 84
Salix, 309
  alba, 310
  amygdaloides, 310
  anglorum
    antiplasta, 312
  babylonica, 309
  bebbiana, 311
  blanda, 309
  caudata, 310
  discolor, 310
  exigua, 310
  fragilis, 310
  geyeriana, 311
  goodingii, 310
  lutea, 311
  matsudana
    tortuosa, 311
    umbraculifera, 311
  melanopsis, 310
  nigra, 310
  nivalis, 312
  pentandra, 310
  pseudocordata, 311
  pseudolapponum, 312
  scouleriana, 311
  subcoerulea, 311, 312
  viminalis, 310
  wolfii, 312
Salpiglossis, 339

Index

sinuata, 339
Salsify, 135
Salsola, 84
  kali, 84
Saltbush, 80
  Fourwing, 80
Saltwort, 283
Salvia, 183
  carnosa, 183
  columbariae, 183
  officinalis, 183
  reflexa, 183
  splendens, 183
Sambucus, 70
  canadensis, 70
  coerulea, 70
  melanocarpa, 70
  nigra, 70
  racemosa, 70, 76
Samphire, 84
Sandbur, 379
Sandpuff, 233
Sandspurry, 77
Sandwort, 72
Sanguisorba, 304
  minor, 304
Saponaria, 75
  officinalis, 75
  vaccaria, 75
Sarcobatus, 84
  vermiculatus, 84
Saxifraga, 318
  adscendens, 319
  arguta, 318
  bronchialis, 319
  caespitosa, 319
  cernua, 319
  chrysantha, 319
  debilis, 319
  flagellaris, 319
  hirculus, 319
  oregana, 318
  rhomboidea, 318
  sarmentosa, 318
  vreelandii, 318
Saxifrage, 318
  Golden, 314
  Strawberry, 318
Scabiosa, 158
  atropurpurea, 158
Scabious, Sweet, 158
Scarlet O'Hara, 138
Schlerochloa, 391
  dura, 391
Sciadopitys, 23
  verticillata, 23
Scirpus, 368
  acutus, 370
  americanus, 370
  caespitosus, 368
  microcarpus, 368
  nevadensis, 370
  olneyi, 370
  pallidus, 368

  paludosus, 368
  validus, 370
Scorpion Weed, 171
Scouring Rush, Common, 4
  Mottled, 4
  Smooth, 4
Scrophularia, 334
  lanceolata, 334
Scutellaria, 183
  angustifolia, 183
  antirrhinoides, 183
  galericulata, 183
Secale, 391
  cereale, 391, 394
Sedge, 365
Sedum, 139
  acre, 139
  debile, 139
  integrifolium, 139
  rhodanthum, 139
  spectabile, 139
  stenopetalum, 139, 146
Seepweed, 84
Selaginella, 3
  densa, 4
  mutica, 4
  underwoodii, 4
  utahensis, 4
  watsonii, 4
Selinocarpus, 233
  diffusus, 233
Sempervivum, 139
  tectorum, 139
Senecio, 128
  ambrosioides, 129, 130
  atratus, 131
  blitoides, 310
  canus, 130
  crocatus, 130
  crassulus, 131
  cymbalarioides, 130
  fremontii, 130
  holmii, 128
  hydrophyllus, 131
  integerrimus, 131
  platylobus, 130
  pseudaureus, 130
  serra, 130
  triangularis, 130
  uintahensis, 128
Senna, Bladder, 207
Sequoiadendron, 23
  giganteum, 23
Sesuvium, 37
  verrucosum, 37
Setaria, 391
  lutescens, 391
  viridis, 391
Shad Bush, 297
Shadscale, 81
Shellflower, 181
Shepherdia, 158
  argentea, 159
  canadensis, 159, 166

Index

rotundifolia, 158
Shepherdspurse, 145
Shinleaf, 285
Shooting Star, 282
Sibbaldia, 304
  procumbens, 304, 307
Sida, 227
  hederacea, 227
Sidalcea, 227
  candida, 227
  neomexicana, 227
  oregana, 227
Silene, 75
  acaulis, 75
  antirrhina, 75, 76
  douglasii, 77
  menziesii, 75
  oregana, 75
  petersonii, 75
  scouleri, 77
  verecunda, 77
Silk-Tree, 188
Sisymbrium, 153
  altissimum, 154, 155
  linifolium, 154
  officinalis, 154
Sisyrinchium, 395
  douglasii, 396
  halophilum, 396
  montanum, 396
  occidentale, 396
  radicatum, 396
  segetum, 396
Sitanion, 391
  hystrix, 392
Sium, 352
  suave, 352
Skullcap, 183
Smelowskia, 154
  calycina, 154, 155
Smilacina, 406
  racemosa, 406
  stellata, 406, 409
Snakeweed, 119
Snapdragon, 321
  Least, 322
Snapweed, 42
Snowball Bush, 71
Snowberry, 70, 71
  Common, 70
  Mountain, 71
  Trailing, 71
Snowdrop, 362
Snowflake, 363
Snow-on-the-Mountain, 161
Soapwort, 75
  Cow, 75
Solanum, 339
  dulcamara, 339
  elaeagnifolium, 339
  jamesii, 339
  nigrum, 340
  rosstratum, 339
  sarachoides, 340

triflorum, 339
tuberosum, 339
Solidago, 131
  canadensis, 131
  decumbens, 131
  parryi, 131
  sparsiflora, 131
Solomonsseal, 406
Sonchus, 131
  asper, 132
Sophora, 218
  japonica, 218
  nuttalliana, 218
  stenophylla, 218
Sorbaria, 304
  sorbifolia, 304
Sorbus, 304
  aucuparia, 304
  hybrida, 304
  scopulina, 304, 307
Sorrel, Mountain, 275
Sparganium, 414
  angustifolium, 414
  emersum, 414, 415
  eurycarpum, 414
  minimum, 414
Spartina, 392
  gracilis, 392
Spectacle Pod, 147
Speedwell, 334
Spergularia, 77
  marina, 77
Sphaeralcea, 227
  ambigua, 229
  caespitosa, 227
  coccinea, 229, 244
  grossulariifolia, 229
  leptophylla, 229
  munroana, 229
  parvifolia, 230
  rusbyi, 229
Spiderwort, 364
Spiraea, 304
  caespitosa, 304
  prunifolia, 304
  vanhoutii, 304
Spiranthes, 411
  romanzoffiana, 413
Spirea, 304
  False, 304
  Rock, 300, 304
Spirodela, 399
  polyrrhiza, 399
Spleenwort, 7
Sporobolus, 392
  airoides, 392
  contractus, 392
  cryptandrus, 392
  flexuosus, 392
Spraguea, 281
  umbellata, 281
Sprekelia, 363
  formosissima, 363
Spring Beauty, 279

Index

Spruce, 18
  Black, 19
  Blue, 19
  Engelmann, 19
  Korean, 19
  Norway, 19
  Oriental, 19
  Serbian, 19
  Sitka, 19
  White, 19
Spurge, 161
Squash, 157
  Fall and Winter, 157
  Winter Crookneck, 157
Squawbush, 39
Squirreltail, 391
Stachys, 183
  palustris, 183, 189
Stanleya, 154
  albescens, 154
  pinata, 154
  viridiflora, 154
Star-of-Bethlehem, 406
Steironema, 284
  ciliatum, 284
Stellaria, 77
  jamesiana, 77, 85
  longifolia, 77
  longipes, 77
  media, 77
  nitens, 77
  obtusa, 78
  umbellata, 77
Stickseed, 59
Stipa, 392
  columbiana, 392
  comata, 392
  lettermannii, 392
  neomexicana, 392
Stocks, 152
Stonecrop, 139
Stoneseed, 59
Storksbill, 168
Stephanomeria, 132
  exigua, 132
  parryi, 132
  pauciflora, 132
  tenuifolia, 132
Streptanthella, 154
  longirostris, 154
Streptanthus, 154
  cordatus, 154
Streptopus, 406
  amplexifolius, 406, 409
Stylocline, 132
  micropoides, 132
Suaeda, 84
  depressa, 84
  fruticosa, 86
  occidentalis, 86
  torreyana, 86
Sullivantia, 319
  purpusii, 319
Sumac, 39

Smooth, 39
Sunflower, 120
  Desert, 118
  Prairie, 121
  Ruderal, 121
Swainsonia, 219
  salsula, 219
Sweetflag, 364
Sweet William, 74
Swertia, 167
  albomarginata, 167
  perennis, 167, 172
  radiata, 167
  utahensis, 167
Symphoricarpos, 70
  albus, 70
  longiflorus, 71
  occidentalis, 70
  orbiculatis, 70
  oreophilus, 71
  parishii, 71
  rotundifolius, 71
  utahensis, 71
  vaccinioides, 71
Synthyris, 334
  pinnatifida, 334
Syntrichopappus, 132
  fremontii, 132
Syringa, 236
  amurensis, 236
  chinensis, 237
  josikaea, 236
  persica, 236
  villosa, 236, 244
  vulgaris, 236

Talinum, 281
  brevifolium, 281
Tamarix, 340
  aphylla, 340
  pentandra, 340
Tanacetum, 132
  diversifolium, 132
  vulgare, 132
Tansy, 132
Taraxacum, 132
  officinale, 132
Tarweed, 125
Taxodium, 23
  distichum, 23
Taxus, 22
  baccata, 22
  cuspidata, 22
  x **media**, 22
Tea, Mormon, 16
  New Jersey, 293
Teasal, 158
Tea Vine, 338
Telesonix, 319
  jamesii, 319
Tetradymia, 132
  glabrata, 133
  nuttallii, 133
  spinosa, 133

Index

Teucrium, 185
  canadense, 185
Thalictrum, 292
  fendleri, 292
Thamnosma, 308
  montana, 308
Thelesperma, 133
  megapotamicum, 133
  subnudum, 133
Thelypodium, 154
  integrifolium, 154
  lilacinum, 156
  satittatum, 154
Thermopsis, 219
  montana, 219
Thistle, Bull, 111
  Canada, 111
  Musk, 108
  New Mexico, 112
  Parry, 112
  Russian, 84
  Rydberg, 111
  Sow, 131
  Utah, 112
Thlaspi, 156
  alpestre, 156
  arvensis, 156, 166
Thorn, Cockspur, 299
  Washington, 299
Threeawn, 377
  Red, 377
Thuja, 16
  occidentalis, 16
  orientalis, 16
  plicata, 20
Thyme, 185
Thymus, 185
  vulgaris, 185
Tidestromia, 38
  lanuginosa, 38
Tidy Tips, 123
Timothy, 388
  Mountain, 388
Toadflax, 323
  Bastard, 312
Tobacco, 338
Tomatilla, 338
Tomato, 388
Townsendia, 133
  annua, 134
  aprica, 133
  arizonica, 135
  exscapa, 134
  florifer, 133, 135
  hookeri, 134
  incana, 134, 135
  jonesii, 134
  leptotes, 134
  mensana, 133, 134
  minima, 134
  montana, 133
  strigosa, 135
Toxicodendron, 39
  radicans, 39

Tradescantia, 364
  occidentalis, 364
  virginiana, 364
Tragopogon, 135
  dubius, 135
  porrifolius, 135
Trautvetteria, 292
  carolinensis, 292
Trefoil, 209
Tribulus, 357
  terrestris, 357
Trichardia, 176
  watsonii, 176
Tridens, 392
  Hairy, 393
  pilosa, 393
  pulchella, 393
Trifolium, 219
  dasyphyllum, 219
  eriocephalum, 220
  fragiferum, 219
  gymnocarpon, 219
  hybridum, 219
  kingii, 219, 220
  longipes, 219, 220
  nanum, 219
  parryi, 219
  pratense, 219, 228
  repens, 219
  variegatum, 219
Triglochin, 398
  maritima, 398
  palustris, 398
Triodanis, 67
  perfoliata, 67
Tripterocalyx, 233
  micranthus, 233
  pedunculatus, 233
Trisetum, 393
  spicatum, 393, 394
Triticum, 393
  aestivum, 393
  durum, 393
Trollius, 292
  laxus, 292
Tropaeolum, 340
  majus, 340
Trumpet-Vine, 45
Tsuga, 21
  canadensis, 22
Tulip, 406
  Common, 406
  Duc Van Thol, 406
Tulipa, 406
  fosteriana, 406
  gesneriana, 406
  suaveolens, 406
Tulip Tree, 224
Twayblade, 411
Twinflower, 69
Twinpod, 153
Twistedstalk, 406
Twistflower, 154
  Little, 154

Index

Typha, 414
  angustifolia, 414
  domingensis, 414
  latifolia, 414, 415
Ulmus, 341
  alata, 341
  americana, 341
  fulva, 341
  glabra, 341
  x hollandica, 341
  parvifolia, 341
  procera, 342
  pumila, 341
  thomasii, 341
Umbrellawort, 232
Urtica, 353
  breweri, 353
  gracilis, 353, 355
Utricularia, 220
  minor, 220
  vulgaris, 220, 228

Vaccinium, 160
  arbuscula, 160
  caespitosum, 160
  globulare, 160
  myrtillus, 160
  occidentale, 160
  scoparium, 160
Valerian, 354
  Common, 354
  Red, 353
Valeriana, 354
  capitata, 354
  edulis, 354
  officinalis, 354
  occidentalis, 354, 355
Valerianella, 354
  locusta, 354
Vanclevea, 135
  stylosa, 135
Velvet-Leaf, 226
Venus Lookingglass, 67
Veratrum, 406
  californicum, 406
Verbascum, 334
  thapsus, 334
Verbena, 354
  bracteata, 354
  hastata, 354
  Sand, 232
Verbesina, 135
  encilioides, 135
Veronica, 334
  americana, 334
  anagallis-aquatica, 334
  arvensis, 335
  peregrina, 335
  persica, 335
  scutellata, 335
  serpyllifolia, 335
  wormskjoldii, 335, 336
Vetch, 220

Milk, 188
Sweet, 208
Viburnum, 71
  alnifolium, 71
  burkwoodii, 71
  opulus, 71
  rytidophyllum, 71
  tomentosum, 71
Vicia, 220
  americana, 220
  exigua, 220
  sativa, 220
  villosa, 220
Viguiera, 135
  annua, 136
  ciliata, 135
  multiflora, 135
  soliceps, 135
Vinca, 40
  minor, 40
Viola, 356
  adunca, 356
  beckwithii, 356
  canadensis, 356
  nephrophylla, 356
  odorata, 356
  palustris, 356
  praemorsa, 356
  purpurea, 356
  tricolor, 356
  vallicola, 356
Violet, 356
  Blue, 356
  Bog, 356
  Dames, 150
  Dogtooth, 403
  Marsh, 356
  Pine, 356
  Sweet, 356
  Western Pansy, 356
  White, 356
  Yellow, 356
Virgin's Bower, 288
  Western, 289

Wallflower, 149
  Western, 149
Walnut, 177
  Black, 177
  English, 177
Wandering Jew, 364
Watercress, 153
Waterleaf, 170
Watermelon, 156
Watermilfoil, 169
Waterstarwort, 66
Waterweed, 393
Weigela, 71
  japonica, 71
Wheat, 393
  Durum, 393
Wheatgrass, 375
  Bearded, 376
  Beardless, 376

473

Index

Bluebunch Wate, 376
Crested, 375
Slender, 376
Thickspike, 375
Whitetop, 145
Whitlowgrass, 147
Wildrye, 381
  Blue, 384
  Canada, 384
  Giant, 381
  Macoun's, 383
  Salina, 383
Willow, 309
  Arctic, 312
  Beak, 311
  Black, 310
  Blue, 311, 312
  Brittle, 310
  Corkscrew, 311
  Desert, 45
  Dixie Black, 310
  Dusky, 310
  Laurel-leaf, 310
  Peach-leaf, 310
  Pussy, 310
  Sandbar, 310
  Sierra, 312
  Weeping, 309
  White, 310
  Wisconsin Weeping, 309
  Yellow, 311
Willowherb, 238
Windflower, 288
Wintercress, 145
Winterfat, 84
Wintergreen, 160
Wisteria, 220
  floribunda, 220
  sinensis, 220
Woad, 150
  Dyer's, 150
Wolfberry, 70
  Anderson, 338
  Pale, 338
  Torrey, 338
Wolfia, 399
  punctata, 399
Woodland Star, 315
Woodnymph, 286
Woodreed, 379
Woodsia, 11
  mexicana, 11
  oregana, 11
  scopulina, 11
Woodsorrel, 243
Wormseed, 149
Wormwood, 102
Wyethia, 136
  amplexicaulis, 136
  arizonica, 136
  scabra, 136
Xanthium, 136
  strumarium, 136

Yarrow, 98
Yellow Wood, 207
Yerba Mansa, 312
Yerba-Santa, 170
Yew, 22
  English, 22
  Hybrid, 22
  Japanese, 22
Yucca, 406
  angustissima, 407
  baccata, 407
  brevifolia, 406
  Datil, 406
  harrimaniae, 407
  smalliana, 406
  standleyi, 407

Zannichella, 416
  palustris, 416
Zauschneria, 242
  garrettii, 242
Zebrina, 364
  pendula, 364
Zelkova, 342
  serata, 342
Zephyranthes, 363
  ajax, 363
  atamasco, 363
  candida, 363
  grandiflora, 363
Zizia, 352
  aptera, 352
Zygadenus, 407
  elegans, 407, 409
  paniculatus, 407
  venenosus, 407